Soil Nutrition and Soil Fertility

Soil Nutrition and Soil Fertility

Edited by Kye Young

SYRAWOOD
PUBLISHING HOUSE

New York

Published by Syrawood Publishing House,
750 Third Avenue, 9th Floor,
New York, NY 10017, USA
www.syrawoodpublishinghouse.com

Soil Nutrition and Soil Fertility
Edited by Kye Young

© 2018 Syrawood Publishing House

International Standard Book Number: 978-1-68286-588-0 (Hardback)

Cataloging-in-Publication Data

Soil nutrition and soil fertility / edited by Kye Young.
    p. cm.
Includes bibliographical references and index.
ISBN 978-1-68286-588-0
1. Soils and nutrition. 2. Soil fertility. I. Young, Kye.
S596.5 .S65 2018
631.41--dc23

TABLE OF CONTENTS

PREFACE

The purpose of the book is to provide a glimpse into the dynamics and to present opinions and studies of some of the scientists engaged in the development of new ideas in the field from very different standpoints. This book will prove useful to students and researchers owing to its high content quality.

Soil is the main source of nutrients for the growth of plants. Some of the nutrients obtained from the soil are nitrogen, potassium, phosphorus, etc. The fertility of soil depends on the amount of nutrients, soil depth and microorganisms present in it. This book explores all the important aspects of soil nutrition and soil fertility in the present day scenario. It strives to provide a fair idea about this discipline and to help develop a better understanding of the latest advances within this field. This book is an essential guide for both academicians and those who wish to pursue this discipline further.

At the end, I would like to appreciate all the efforts made by the authors in completing their chapters professionally. I express my deepest gratitude to all of them for contributing to this book by sharing their valuable works. A special thanks to my family and friends for their constant support in this journey.

Editor

Twenty-Two Years of Warming, Fertilisation and Shading of Subarctic Heath Shrubs Promote Secondary Growth and Plasticity but Not Primary Growth

Matteo Campioli[1]*, Niki Leblans[1], Anders Michelsen[2,3]

1 Department of Biology, University of Antwerp, Wilrijk, Belgium, **2** Department of Biology, University of Copenhagen, Copenhagen, Denmark, **3** Center for Permafrost (CENPERM), University of Copenhagen, Copenhagen, Denmark

Abstract

Most manipulation experiments simulating global change in tundra were short-term or did not measure plant growth directly. Here, we assessed the growth of three shrubs (*Cassiope tetragona*, *Empetrum hermaphroditum* and *Betula nana*) at a subarctic heath in Abisko (Northern Sweden) after 22 years of warming (passive greenhouses), fertilisation (nutrients addition) and shading (hessian fabric), and compare this to observations from the first decade of treatment. We assessed the growth rate of current-year leaves and apical stem (primary growth) and cambial growth (secondary growth), and integrated growth rates with morphological measurements and species coverage. Primary- and total growth of *Cassiope* and *Empetrum* were unaffected by manipulations, whereas growth was substantially reduced under fertilisation and shading (but not warming) for *Betula*. Overall, shrub height and length tended to increase under fertilisation and warming, whereas branching increased mostly in shaded *Cassiope*. Morphological changes were coupled to increased secondary growth under fertilisation. The species coverage showed a remarkable increase in graminoids in fertilised plots. Shrub response to fertilisation was positive in the short-term but changed over time, likely because of an increased competition with graminoids. More erected postures and large, canopies (requiring enhanced secondary growth for stem reinforcement) likely compensated for the increased light competition in *Empetrum* and *Cassiope* but did not avoid growth reduction in the shade intolerant *Betula*. The impact of warming and shading on shrub growth was more conservative. The lack of growth enhancement under warming suggests the absence of long-term acclimation for processes limiting biomass production. The lack of negative effects of shading on *Cassiope* was linked to morphological changes increasing the photosynthetic surface. Overall, tundra shrubs showed developmental plasticity over the longer term. However, such plasticity was associated clearly with growth rate trends only in fertilised plots.

Editor: Bente Jessen Graae, Norwegian University of Science and Technology, Norway

Funding: This work was supported by Methusalem funding (Research Center of Excellence 'Eco', University of Antwerp) and by the Danish Council for Independent Research | Natural Sciences. The funders had no role in study design, data collection and analysis, decision to publish, or preparation of the manuscript.

Competing Interests: The authors have declared that no competing interests exist.

* E-mail: matteo.campioli@ua.ac.be

Introduction

The Arctic is the region which will likely experience the most pronounced alteration in climate and environment due to global change [1]. As arctic ecosystems are very sensitive to changes in environmental conditions and store a significant amount (12%) of the global soil carbon (C), extensive research efforts have been made in the last three decades to understand the future feedback of arctic ecosystems to the greenhouse effect and global climate [1–3]. In particular, manipulation experiments have been set up to mimic the expected changes in arctic climate and their impact on ecosystems [4–6]. Many experiments have focused on the effect of warming during the growing season, of crucial importance for the arctic plant communities adapted to a short and cool summer [7]. Focus has been on plant growth, which (i) can be considered as a surrogate for plant fitness and as such a crucial process for plant subsistence and development, (ii) represents the amount of C taken up annually by the vegetation, and (iii) determines, through the process of C allocation to plant organs with different life-spans and decomposition rates, the C release by the ecosystem in the long-term [8].

The impact of warming on plant growth can be direct or indirect. The direct effect of warming has been mimicked by enhancing air and soil temperature, e.g. with open top chambers [9,10]. Indirect effects of warming are manifold [11,12]. However, the increase in nutrient availability through enhanced net mineralization is thought to be one of the most important indirect effects of warming for arctic plant communities, which are commonly nutrient limited [13,14]. This indirect effect of warming has been mimicked by adding fertilisers during the growing season [15], under ambient or enhanced temperature. A second indirect effect (particularly important in the Subarctic and Low Arctic), is the potential increase in competition due to tree-line advancement and shrub expansion [16–18]. Such impact has been mimicked by shading [9,10]. In the Subarctic and Low Arctic, manipulative experiments have shown that fertilisation has a strong effect on the growth of deciduous species in tussock tundra, and of all vascular species (and particularly graminoids) in heath tundra, whereas warming and shading have small or non significant effects [5,11]. However, these findings rely mainly on studies not longer than a decade.

In this study, we aimed to broaden the current knowledge on the long-term impact of warming, fertilisation and shading in subarctic ecosystems by assessing the long-term responses in growth of the widespread dwarf-shrubs *Cassiope tetragona* (L.) D. Don., *Empetrum hermaphroditum* Hagerup and the low shrub *Betula nana* L., at a tree-line heath in Northern Sweden after 22 years of manipulation. The experiment is unique as we are not aware of similar well-replicated experiments of such duration in the Subarctic and Low Arctic. Furthermore, the experimental site is particularly suited for this analysis as it was intensively investigated in the first decade of manipulation, providing reports on the shorter term responses of shrub growth to manipulations. In the first decade of manipulation, warming yielded a modest positive response in the growth of *Cassiope* and no response in *Empetrum* and *Betula*, fertilisation led to a positive response in *Cassiope* and particularly in *Empetrum* but not in *Betula*, whereas shading gave a negative response, strong for *Betula* and modest for *Cassiope* and *Empetrum* [5,13,19,20,21]. After more than two decades of treatment, we expected the growth responses of arctic shrubs to differ from the short-term responses for three reasons. First, the steady changes in community composition, favouring graminoids, which were observed in fertilized plots in the short-term [22], and the competition for light that graminoids exert on prostrate shrubs [5], are likely to negatively affect the shrub growth over the longer term. Second, the mechanisms that buffered the negative effect of shading in the first years of treatment (e.g. usage of stored resources, short-term acclimations) [19,21], were expected to weaken over the longer term, in particular for the less shade tolerant species such as *Betula* and *Cassiope*. Third, the rate of physiological processes that counterbalanced the positive effect of warming on gross photosynthesis over the shorter term (e.g. respiration, tissue turnover) [9,11] was expected to decrease because of long-term acclimation [9,23].

The stem secondary growth of shrubs (cambial growth or increase in stem diameter) accounts for a significant portion of aboveground net primary production in tundra ecosystems (e.g. up to ~50% at species level (*Salix pulchra* [24]) and ~20% at plant community level (subarctic heath [25])) and it is sensitive to environmental perturbations [26]. Nevertheless, secondary growth is seldom investigated. The physiological function of secondary growth differs from the one of primary growth. In fact, whereas primary growth assures light interception and photosynthetic uptake, secondary growth sustains the C uptake (e.g. by producing new conduits for water and sugar transport) but also provides the essential mechanical support to the canopy [26]. In a recent study on the growth of arctic shrubs, Campioli et al. [27] found that changes in primary and secondary growth between sites with different environmental conditions were not proportional.

In detail, we tested two hypotheses. Hypothesis 1. The growth responses of arctic shrubs to long-term environmental manipulations differ from the short-term growth responses. Over the longer term, warming was expected to have a more positive effect, fertilisation a less positive effect (or even a negative effect) and shading a more negative effect. Hypothesis 2. The responses of primary and secondary growth to environmental manipulations differ according to the concurrent morphological changes. If the manipulations promoted morphological changes implying enhanced mechanical support for the shrub stem (e.g. increase in shrub height or branching), the response of the secondary growth is expected to differ from the response of the primary growth. On the other hand, if the morphology of the shrub was not altered by the manipulations, primary and secondary growth are expected to present similar response patterns.

Results

Growth rates

The long-term environmental manipulations did not affect the total growth of *Cassiope* and *Empetrum* (Fig. 1a,b; Table 1). By contrast, the secondary growth increased under fertilisation for both species (Fig. 1a,b; Table 1). The total growth of *Betula* was significantly reduced (by a factor 2.0–2.4) under shading and fertilisation (Fig. 1c; Table 1). For *Betula*, the primary growth presented significant trends similar as for the total growth, whereas the secondary growth increased under fertilisation (Fig. 1c; Table 1).

Samples from the control treatment showed that the total growth rate was largest for *Betula* (~130% year^{-1}), intermediate for *Empetrum* (~110% year^{-1}) and lowest for *Cassiope* (~70% year^{-1}), and that the primary growth represented 80–85% of the total growth for each species (Fig. 1a,b,c). Leaf production accounted for the large majority of primary growth: 85% for *Cassiope* and *Empetrum* and 98% for *Betula*.

Shrub height and length

Shrub height was positively affected by warming for each species and by fertilisation for *Empetrum* and *Betula* (Fig. 2a,b,c; Table 1). Increase in height was of particular relevance in the warming plus fertilisation treatment (WF), where shrubs were 2–3 times taller than in the control (Fig. 2a,b,c). Shading had no effect on shrub height (Fig. 2a,b,c; Table 1). The impact of the environmental manipulations on shrub length was similar to the impact on shrub height (Fig. 2g,h,i). In contrast to shrub height, shrub length of *Cassiope* was positively affected by shading (Fig. 2g; Table 1) which relates to a significant increase in the number and length of the branches (see below).

Branch number, branch length and apical increment

Shading exerted mainly a positive impact on the number and length of the youngest branches, particularly for *Cassiope* (Fig. 3a,d). Fertilisation had mainly a negative impact on the number and length of old branches of *Empetrum* and, particularly, *Cassiope* and a positive impact on the length of the youngest branches (up to 3 year-old) of all species (Fig. 3). The impact of warming was minor and affecting only *Cassiope* and *Empetrum*, with reduced branch numbers in few old cohorts and increased branch length in few young cohorts (Fig. 3a,b,d,e). The *F* and *p* values of the ANOVA analyses are reported in Table S1.

The apical increment of *Cassiope* was positively affected by shading and warming, whereas the apical increment of *Empetrum* and *Betula* was positively affected by fertilisation (Fig. 2d,e,f; Table 1). Overall, the number of branches was similar for *Cassiope* and *Empetrum* but much lower (a factor of 2–4) for *Betula* (Fig. 3a,b,c). *Cassiope* and *Empetrum* presented branch length larger than *Betula* for young cohorts but similar branch length for older cohorts (Fig. 3d,e,f).

Species coverage

Fertilisation had a positive impact on the coverage of graminoids and *Empetrum* and negative on the coverage of *Cassiope*, whereas warming had a positive effect on *Betula* (Fig. 4; Table 2). Overall, the total vascular cover was positively affected by warming and fertilisation and negatively affected by shading

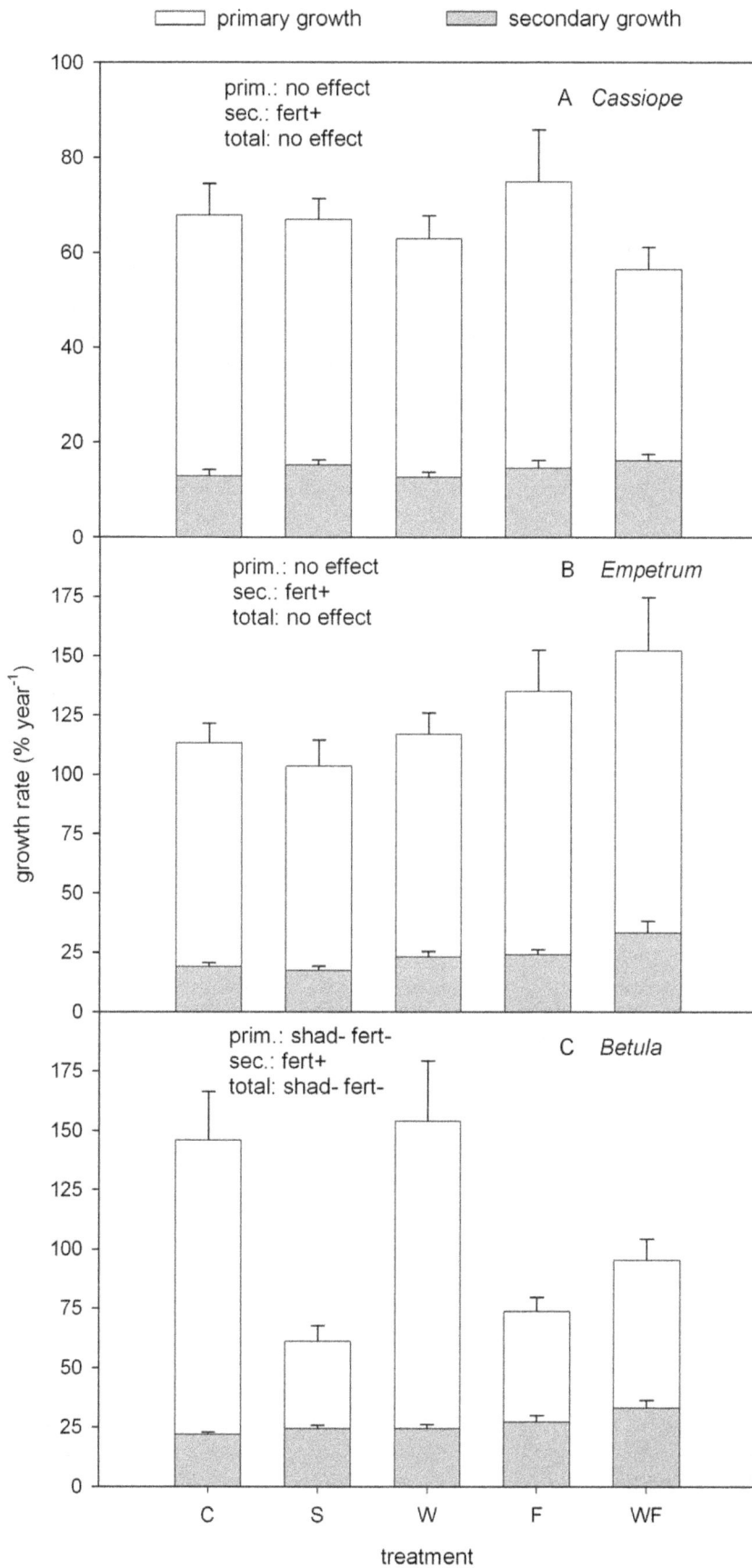

Figure 1. Growth rate of subarctic shrubs in manipulated environment. Growth rate (bars: total aboveground vegetative growth, indicated as 'total'; white stacks: primary growth i.e. leaves plus apical stem, indicated as 'prim.'; grey stacks: secondary growth i.e. stem diameter increment, indicated as 'sec.'; mean+1SE; n = 5–6) of the shrubs *Cassiope tetragona*, *Empetrum hermaphroditum* and *Betula nana* in a subarctic heath in Abisko (Northern Sweden) subjected to 22 years of environmental manipulation: shading (S), warming (W), fertilisation (F), combined warming plus fertilisation (WF). The control is indicated by C. The environmental factors significantly affecting growth are reported on the top left corner of each panel (shad: shading; fert: fertilisation) with the symbols + and − indicating the direction of the response, positive and negative, respectively. Note the different scale between y-axes of panel A and panel B,C.

(Fig. 4; Table 2). Graminoids increased by a factor of 6 in the WF treatment (Fig. 4).

Discussion

Our expectations were only partially confirmed, with differences among treatments and species. Both hypotheses were overall valid under fertilisation but not under warming. For shading, the expectations were confirmed for *Empetrum* and *Betula* only partially and were not confirmed for *Cassiope*. Besides the growth patterns, the study showed that the investigated shrubs developed a significant plasticity over the longer term.

Hypothesis 1: growth responses change over the longer term

Fertilisation. As expected, the effect of fertilisation became less favourable over the longer term for each species. The effect of fertilisation on *Empetrum* and *Cassiope* changed from positive to non-significant (Table 3). This reveals that the positive effect of nutrient addition on the growth of evergreen dwarf-shrubs was transient at our experiment because of the concomitant positive effect of fertilisation on graminoids. A progressive shrub decline concurrent to an increase in graminoids has previously been

observed in other subarctic and low arctic fertilised heaths [28,29]. Furthermore, the growth of evergreen dwarf-shrubs degenerated in tussock tundra after 3 and 9 years of fertilisation because of the progressive increase in competition with *Betula* [9]. However, our study is the first to show that after more than two decades of treatment, aboveground growth of evergreen dwarf-shrubs is not suppressed and ancillary positive growth impacts are still recorded (e.g. increase in *Empetrum* coverage, Fig. 4). This is likely due to the shade tolerance of *Cassiope* and *Empetrum* and to morphological plasticity (see below). *Betula* showed a different dynamics as the effect of fertilisation changed from non-significant to strongly negative (i.e. halving of the growth rate). Despite an important plastic response (see below), *Betula* was suppressed over the longer term because the competition with graminoids was particularly severe for this species characterized by low shade tolerance. However, *Betula* might suffer competition (or other growth limitations) even in control conditions at our site and fertilisation likely exacerbates a natural constrained growth. This is supported by comparing the growth of *Betula* at the experimental site with the growth of *Betula* at other heath sites [27] and by the fact that the growth of fertilised *Betula* was stimulated after 8 years of treatment in a more open tundra heath despite an even larger increase in graminoids abundance [28].

Table 1. Results of ANOVAs on the growth rates (primary, secondary and total), shrub height, apical (current year's stem) increment and shrub length of the shrubs *Cassiope tetragona*, *Empetrum hermaphroditum* and *Betula nana* at a subarctic heath in Abisko (Northern Sweden) after 22 years of environmental manipulation.

effect	primary growth		secondary growth		total growth		shrub height		apical increment		shrub length	
	F	P	F	P	F	P	F	P	F	P	F	P
Cassiope												
shading	0.14[a]	0.71	2.15[a]	0.16	0.01[a]	0.91	<0.01[b]	0.96	4.58[a]	0.045	5.89[a]	0.025
warming	1.96[c]	0.17	0.30[c]	0.58	0.94[c]	0.17	7.05[a]	0.016	6.63[c]	0.014	5.55[c]	0.024
fertilisation	0.46[c]	0.50	4.32[c]	0.044	0.10[c]	0.75	2.77[a]	0.11	0.77[c]	0.39	<0.01[c]	0.95
warm×fert	0.81[c]	0.37	0.54[c]	0.47	0.72[c]	0.40	<0.01[a]	0.99	1.94[c]	0.17	0.35[c]	0.56
Empetrum												
shading	0.35[a]	0.56	0.45[a]	0.51	0.57[a]	0.46	2.08[b]	0.15	0.34[a]	0.57	0.19[a]	0.66
warming	0.04[c]	0.84	2.74[c]	0.11	0.52[c]	0.48	17.75[a]	<0.001	2.73[c]	0.11	5.46[c]	0.025
fertilisation	0.93[c]	0.34	4.70[c]	0.036	1.45[c]	0.24	48.23[a]	<0.001	4.38[c]	0.043	4.10[c]	0.050
warm×fert	0.06[c]	0.81	0.04[c]	0.85	0.26[c]	0.61	4.53[a]	0.046	1.27[c]	0.27	1.46[c]	0.23
Betula												
shading	6.00[d]	0.014	1.42[d]	0.27	12.73[d]	<0.01	0.30[b]	0.58	<0.01[d]	1.00	0.57[d]	0.48
warming	0.41[e]	0.53	2.84[e]	0.11	0.74[e]	0.40	11.27[a]	<0.01	1.96[e]	0.18	1.15[e]	0.30
fertilisation	18.52[e]	<0.001	9.15[e]	<0.01	14.51[e]	<0.01	5.39[a]	0.031	3.30[e]	0.088	3.56[e]	0.078
warm×fert	0.09[e]	0.76	0.56[e]	0.46	0.16[e]	0.69	0.15[a]	0.70	0.59[e]	0.45	0.59[e]	0.45

[a]: degrees of freedom of between group variation or effect (df effect) of 1 and degrees of freedom of within group variation or residuals (df residuals) of 20;
[b]: df effect = 1, df residuals = 10;
[c]: df effect = 1, df residuals = 40;
[d]: df effect = 1, df residuals = 8;
[e]: df effect = 1, df residuals = 16.

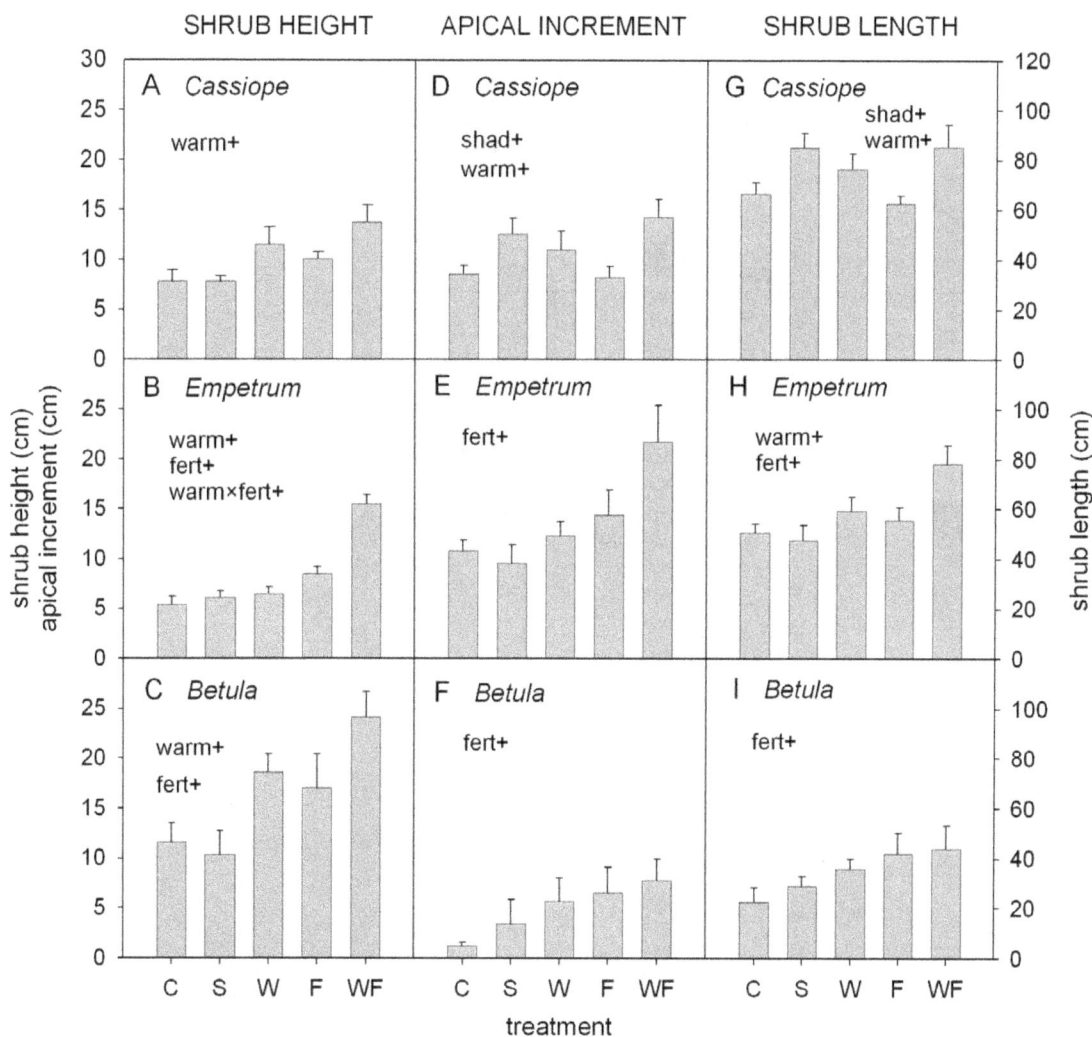

Figure 2. Shrub height, shrub length and apical (current year's stem) increment of subarctic shrubs in manipulated environment. Values (mean+1SE; n = 5–6) of shrub height (left panels), apical increment (central panels) and shrub length (right panel) of three subarctic heath shrubs (*Cassiope tetragona*, *Empetrum hermaphroditum* and *Betula nana*) in Abisko (Northern Sweden) subjected to 22 years of environmental manipulation (shading S, warming W, fertilisation F, combined warming plus fertilisation WF) against the control (C). The environmental factors significantly affecting growth are reported on the upper part of each panel (shad: shading; warm: warming; fert: fertilisation, and warm×fert: warming×fertilisation) with the symbols + and − indicating the direction of the response (positive and negative, respectively).

Shading. Despite some acclimation to shade (e.g. increased leaf nitrogen and chlorophyll [21]), the growth of *Cassiope* and *Empetrum* showed some negative responses in the first 10 years of shading treatment (Table 3). Over the longer term, contrary to our expectations, negative effects disappeared as the response of the growth rate of *Cassiope* and *Empetrum* to shade was non-significant. *Cassiope* was expected to have a limited shade tolerance as it does not grow in shaded habitats [19]. However, *Cassiope* likely compensated the negative effect of shading by substantial changes in allocation pattern. While maintaining the same growth rate (and thus biomass production), shaded *Cassiope* ramets increased greatly in branch numbers and branch length (Fig. 3a,d), hence increasing photosynthesising surface and light interception. *Empetrum* is likely to be shade tolerant (it grows in the understory of taller tundra shrubs and boreal forests [21]) and only very minor changes in morphology were observed over the longer term. On the other hand, the response of *Betula* to shading confirmed our expectations for this species. The reduction of aboveground growth of shaded *Betula* was one of the most

significant responses recorded in our study and it was related to a strong reduction in leaf production. This confirms that the negative response of *Betula* leaves observed earlier at the same plots was not transient (Table 3). On the other hand, the observed continuous growth decline was not associated with alteration in *Betula* cover (Fig. 4). Despite the well known low tolerance of *Betula* to shade [30,31], some compensatory processes might have played a role over the longer term and avoided complete suppression. For instance, shaded *Betula* might have partially benefitted from a reduction in total vascular cover (Fig. 4) or from a reduction in the stem turnover, which is stimulated in shaded *Betula* [9] and is perhaps responsible for the transient reduction in stem biomass recorded earlier in our plots (Table 3).

Warming. Contrary to our expectations, the effect of warming on aboveground growth was non-significant after 22 years of treatment. Despite the consensus on the positive direct effect of warming on the growth of arctic plants, our findings suggest that physiological processes limiting net biomass production (e.g. respiration, stem turnover; [9,11]) do not

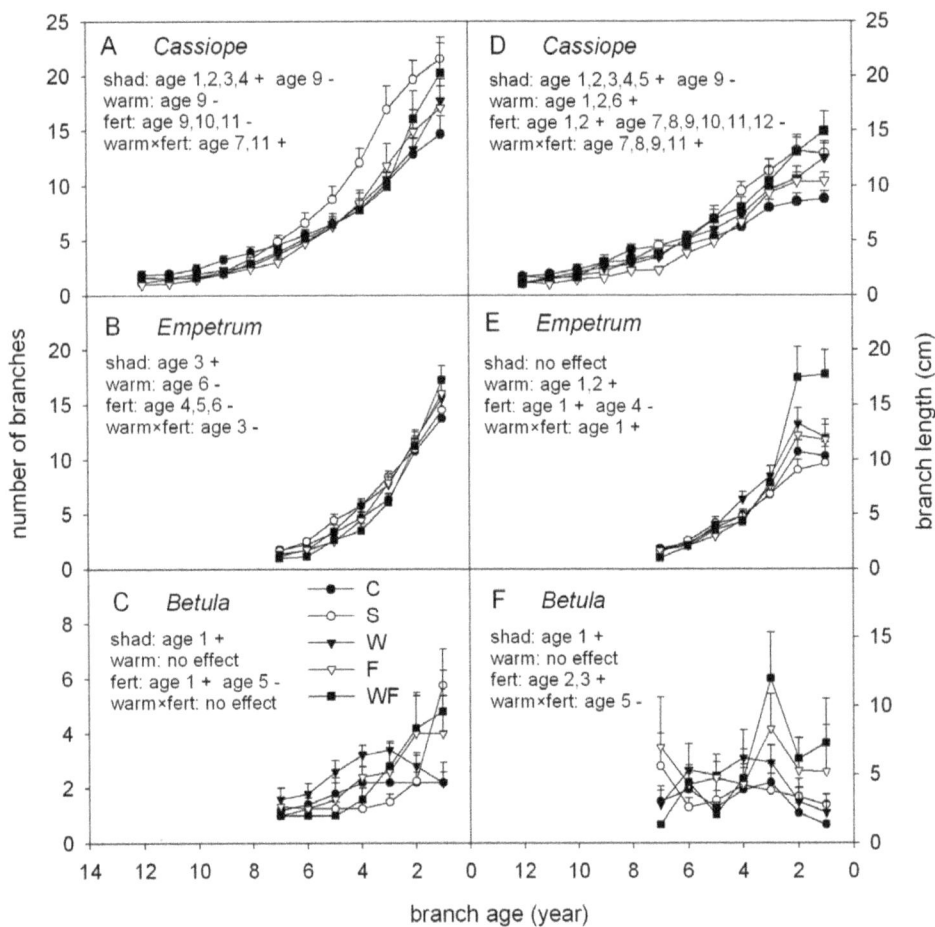

Figure 3. Branch number and length (according to age) of subarctic shrubs in manipulated environment. Number and length of the branches vs. branch age (means+1SE; n = 5–6) of the shrubs *Cassiope tetragona*, *Empetrum hermaphroditum* and *Betula nana* at a subarctic heath in Abisko (Northern Sweden) after 22 years of environmental manipulation. Text on the top left indicates the environmental factors with a significant impact (shad: shading; warm: warming; fert: fertilisation, and warm×fert: warming×fertilisation), their direction (+: positive, −: negative) and the age of the branches affected. Note the different scale between y-axes of panel A,B,D,E and panel C,F.

acclimate over the longer term. Alternatively, other factors might limit growth under long-term warming [11]. For instance, nutrient limitation might be important in our warmed plots due to increased vascular cover and competition (Fig. 4). On the other hand, heat stress might occur on warmed plants during warm summer days [11], as observed for *Ledum palustre* in the Low Arctic [9] and in *Salix arctica* in the Mid Arctic [32]. *Empetrum* is likely to be particularly sensitive to heat stress as it was favoured by warming level of 2.5°C and not by warming level of 4°C at our site after 6 years of treatment [20]. Our results indicate thus that the positive effect of warming on *Cassiope* recorded after 5 years of treatment was transient (Table 3). Transient positive responses to warming have been observed in other short-term tundra warming experiments and associated with temporary increases in mineralization or use of stored resources [11].

Hypothesis 2: responses of primary and secondary growth differ according to morphology

Fertilisation. Fertilised ramets of the three species showed a similar pattern with non-significant variation or decrease in primary growth and increase in secondary growth (Fig. 1). For *Empetrum* and *Betula*, this pattern was accompanied by a significant increase in shrub height, total shrub length and length of the

youngest branches. For these two species our expectations were therefore confirmed. In presence of a lush graminoid canopy, *Empetrum* and *Betula* (procumbent at the site) grew more vertically and explored more lateral space. A more erected and large posture requires more resources to reinforce the stem mechanical strength, implying enhanced secondary growth. This typical morphological plastic reaction prevented these shrubs from being completely confined in the shaded understory. For *Empetrum*, such response likely avoided reduction in the aboveground growth. For *Betula*, such plastic reaction was likely not enough to maintain the same C assimilation as in the control, resulting in fewer resources available for primary growth and in an overall growth reduction. *Cassiope* is likely to have a similar pattern as for *Empetrum* but less marked. In fact, for *Cassiope*, the significant increase in secondary growth was coupled to a non-significant increase in shrub height (p = 0.11) (Table 1).

Shading. Our expectations were confirmed for *Empetrum*, whose growth pattern and morphology were both unaffected by shading. For *Betula*, we did not expect uncoupling in the response of primary and secondary growth as the morphology of *Betula* did not change in shaded plots. However, the prolonged light attenuation significantly decreased *Betula* leaf production, impairing the relationship between primary and secondary

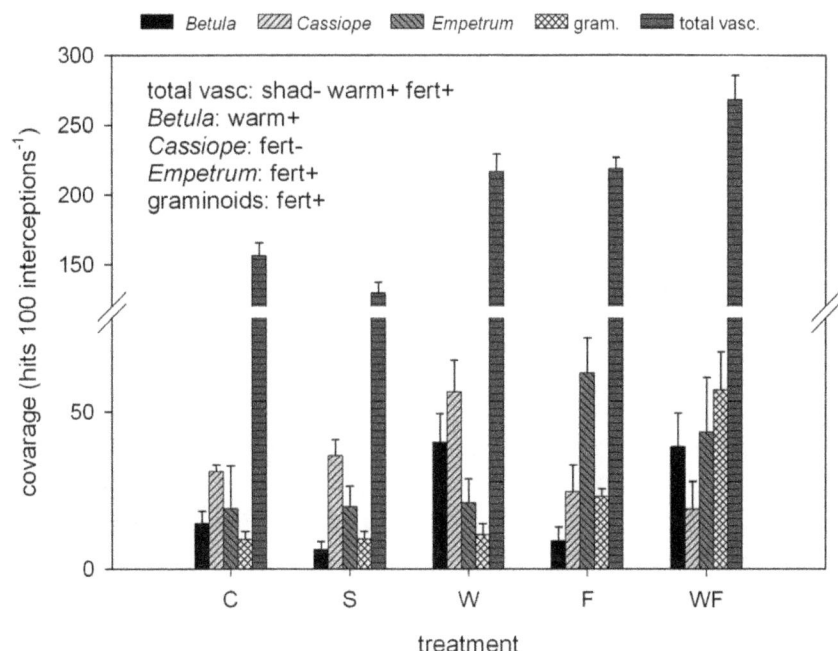

Figure 4. Species coverage in subarctic heath under environmental perturbations. Coverage (mean+1SE; n = 6) of the shrubs *Cassiope tetragona*, *Empetrum hermaphroditum* and *Betula nana*, the graminoids (gram.) and of the total vascular species (total vasc.) at a subarctic heath in Abisko (Northern Sweden) after 22 years of environmental manipulation: shading (S), warming (W), fertilisation (F), combined warming plus fertilisation (WF). Text on the top left indicates the environmental factors with a significant impact on coverage (shad: shading; warm: warming; fert: fertilisation, and warm×fert: warming×fertilisation) and their direction (+: positive, −: negative).

growth. *Cassiope* showed a less clear pattern with no increase in secondary growth and substantial increase in total length and branching. However, the fact that the shaded *Cassiope* ramets had low stature (Fig. 2b) probably resulted in procumbent *Cassiope* branches laid on the moss mat, thus requiring less mechanical support from the stem.

Warming. Warmed ramets showed unaffected primary and secondary growth but increased shrub height and (for *Cassiope* and *Empetrum*) increased shrub length and branch length of the youngest branches. This was unexpected. It is possible that the morphological changes under warming were the result of an overall improved (micro)environment rather than the result of the competition for light as in the fertilised plots. Such supposition is coherent with the species coverage results, which showed no

negative impact of warming, increase in total vascular cover and no impact on graminoids (Fig. 4).

Overall. Previous short term (<10 years) studies on tundra concluded that evergreen dwarf-shrubs have low developmental plasticity, conservative secondary growth and tend to become subcanopy species [9,26,33]. Our study reveals that evergreen dwarf-shrubs can show opposite dynamics over the longer term and posses an overall important plasticity. As acclimation normally occurs through formation of new tissue, it is indeed plausible that slow growing species need long time to acclimate [9,34]. Furthermore, our study showed that apical increment presented a different response than primary growth for each treatment and species. This because length increment does not necessarily correlate to biomass increment in arctic shrubs [25,27]. Such uncoupling calls

Table 2. Results of ANOVAs on the coverage of the dwarf-shrubs *Cassiope tetragona* and *Empetrum hermaphroditum*, the low shrub *Betula nana*, the graminoids and the total vascular species at a subarctic heath in Abisko (Northern Sweden) after 22 years of environmental manipulation.

plant type	shading[a]		warming[b]		fertilisation[b]		warm×fert[b]	
	F	P	F	P	F	P	F	P
Cassiope	0.71	0.42	0.02	0.89	8.88	<0.01	1.35	0.26
Empetrum	0.93	0.34	0.35	0.56	5.35	0.03	0.52	0.48
Betula	2.51	0.14	13.02	<0.01	1.00	0.33	1.00	0.33
graminoids	<0.01	0.97	2.39	0.14	25.36	<0.001	1.99	0.17
total vascular	3.92	0.076	15.95	<0.001	17.14	<0.001	0.19	0.67

[a]: degrees of freedom of between group variation or effect (df effect) of 1 and degrees of freedom of within group variation or residuals (df residuals) of 10;
[b]: df effect = 1 and df residuals = 20.

Table 3. Synthesis of the significant impact of shading (S), warming (W), fertilisation (F) and their interactions on the growth of the tundra shrubs *Cassiope tetragona*, *Empetrum hermaphroditum* and *Betula nana*.

year[a]	aboveground growth response		shrubs species[b]			reference[c]
	variable	meristem (organ)	*Cassiope*	*Empetrum*	*Betula*	
3	mass per shoot	primary (leaf)	F+	n.a.	n.a.	[19]
5	biomass ratio[d]	primary (leaf, stem)	F+	n.a.	n.a.	[21]
5	biomass unit ground	primary plus secondary (leaf, stem)	W+, F+	W×F+	n.a.	[5,21]
6	mass per shoot	primary (leaf)	n.a.	S-	S-	[20]
6	shoot density	primary (stem)	n.a.	F+, W×F+	no effect	[20]
10	biomass unit ground	primary (leaf)	S-	F+	S—	[5]
10	biomass unit ground	primary plus secondary (stem)	no effect	no effect	S—	[5]
10	biomass unit ground	primary plus secondary (leaf, stem)	no effect	F+	S—	[5]
22	growth rate	primary (leaf, stem)	no effect	no effect	S—, F—	this study
22	growth rate	secondary (stem)	F+	F+	F+	this study
22	growth rate	primary plus secondary (leaf, stem)	no effect	no effect	S—, F—	this study

+: positive impact; —: negative impact; no effect: manipulations had no significant impact on the growth variable; W×F— and W×F+: lower and higher effects in the combined treatment, respectively, than expected from the single treatments alone; n.a.: no data available;
[a]: years of treatment at the time of measurement;
[b]: non-significant effects are not listed;
[c]: data refer only to the experimental site investigated in this study (the tree-line heath of Paddustieva, Abisko, Northern Sweden) and are derived from earlier published papers and the current work, and
[d]: green biomass as ratio of grey stem biomass.

for caution when inferring growth responses of tundra shrub to environmental manipulations from apical increment only.

Materials and Methods

Experimental set up

Study site. The study took place at a tree-line heath on sloping terrain (20–30%) at 450 m a.s.l. at Abisko (68°21′N, 18°49′E), in Northern Sweden. The region has a subarctic montane climate, with mean annual temperature and precipitation of −1.0°C and 304 mm, respectively, and the growing season lasting from early-mid June until late August-early September [25,35]. The site is an evergreen dwarf-shrub community dominated by *Cassiope tetragona* and by the co-dominant *Empetrum hermaphroditum*. *Betula nana* is one of the most common deciduous shrubs. Graminoids and forbs are also present, as well as nonvascular plants which form a continuous mat [20]. Bedrock consists of base-rich mica schists [19]. The soil has pH of 7.1 (typical for ecosystems with similar bedrock and topography in the region [36]), an organic layer of about 15 cm and is well drained [25,37,38]. No specific permits were required for the described field studies, as the location is not privately-owned and not protected and the field studies did not involve endangered or protected species.

Environmental manipulations. The experiment started in 1989 in an area of about 400 m². It consisted of eight treatments replicated in six blocks: control, low warming, high warming, shading, fertilisation, fertilisation plus low warming, fertilisation plus high warming and fertilisation plus shading [13]. In this study, we investigated five key treatments: shading (S), high warming (W), fertilisation (F), fertilisation plus high warming (WF) and control (C). Temperature was enhanced by small (1.2×1.2 m, 50 cm high) dome-shaped open top greenhouses of polyethylene film (0.05 mm) supported by PVC tubes. Greenhouses were in place every year from early June (just after snowmelt) until end of

August-early September (leaf fall) enhancing the summer air and soil temperature by 3.9°C and 1.2°–1.8°C, respectively. The greenhouses did not provoke critical side effects on plant growth because: (1) they caused only minor and non-significant reduction in relative soil water content (<6%) as the sloping terrain permitted lateral water movement [19,21]; (2) they did not change significantly the air humidity (<3%); (3) they only led to a 9% reduction in photosynthetically active radiation [19,21]; (4) they did not affect the snow cover as they were not in place in winter and (5) their sheltering effect had a minor impact on shrub morphology as in tundra heath the effect of wind exposure on the shrub structure is much more relevant in winter [7,39]. Furthermore, the greenhouses (resting on the shrub canopy and opened at the top) do not impact the presence of small herbivores (e.g. insects, rodents). Reindeer grazing and moose browsing is generally limited to periods out of the growing season (e.g. reindeers normally do not stay at the tree-line during summer) and are not affected by our manipulative experiment, which is in place only between June and August. The shading was obtained with hessian (jute) fabric, arranged in the same way and in the same period as the polyethylene film of the greenhouses. Hessian fabric reduced the light by 64% without significant effect on air humidity and temperature [19,21]. Fertilisation (10.0 g m^{-2} N, 2.6 g m^{-2} P and 9.0 g m^{-2} K, in the form of NH_4NO_3, KH_2PO_4 and KCl) occurred once per year after snow melt in June (except in 1998, half amount, and in 1993 and 2001, not applied). The dose of N applied was considered similar in magnitude to potential N release from the soil of tundra ecosystems under global change scenario [15].

Sampling and processing of shrub ramets. Sampling took place in mid-late August 2010, after 22 years of experimental manipulation. Individual ramets (i.e. aboveground stem with all lateral branches) of *Cassiope* and *Empetrum* (two ramets) and *Betula* (one ramet) were collected for each block and treatment (n=6). However, for *Betula*, only five replicates were considered in the analysis as most of the ramets of one block were too young or too damaged (i.e. broken stem or branches) to be analyzed. The ramets were sampled in the central part of the experimental plots. This assured that biases due to sampling of ramets supporting biomass outside of the plot or ramets grown into the plots after the treatment began were minor because (i) the colonization rate of these slow growth ericaceous plants is inherently low [19], (ii) the shrub frequency in the area surrounding the plots has been reduced by trampling [40] and (iii) woody functional type as evergreen and multiple-flush deciduous have a high degree of 'shoot autonomy' [41], meaning that relocation of assimilates among ramets is limited.

Stem and branches were divided into segment cohorts of the same age. This was done by counting the apical bud scars for *Empetrum* [24], the stem sections with smaller leaves for *Cassiope* [42] and the annual growth rings in thin stem/branch cross-sections after staining with 0.5% phloroglucinol in 10% HCl for *Betula* [26]. Each stem/branch segment cohort was dried at 70°C for 48 hours and leaves were detached. The number of stem/branch segments of each cohort, their aggregated length and dry weight were recorded [24] as well as the weight of the current-year leaves. Ramets of *Empetrum* and *Betula* were 5–7 year-old. Ramets of *Cassiope* were 10–12 year-old.

Measurement of growth variables

Growth of shrubs was assessed with comparable but refined methods to the ones used to characterize growth responses in the first decade of manipulation, as the latter were too invasive to be repeated (e.g. standing biomass harvest) or had limitations (e.g. lack of assessment of secondary growth and branching pattern). In the current study, shrub growth was determined as primary growth (leaf and apical stem production) and secondary growth,

both expressed as percentage of old stem biomass [24,26,27,43,44]. Data on growth rate were complemented by several measurements of shrub morphology (e.g. shrub height, number and length of branches, total length of stem and branches) and of species coverage. Growth rates were used to evaluate the long-term impact of the environmental manipulations on shrub growth because they provide a direct estimation of plant biomass production. Morphological characteristics were compared to the growth rate estimates to better understand the response pattern of the primary- and secondary growth. The species coverage was used to indirectly infer species abundance. A summary of the growth variables measured, the plant organ and organ age considered and the way the results are expressed (e.g. relative or absolute values, values per plot or per ramet) is reported in Table 4.

Growth rates. We assessed the annual rate of the aboveground vegetative growth as primary-, secondary- and total growth. We determined primary growth rate (% $year^{-1}$) for each ramet as the production of current year's apical biomass (leaves and stem) as a percentage of old standing stem biomass [27]. The data needed to calculate the primary growth were directly available from the harvested material (see above). The stem secondary growth was expressed in the same way but the current-year secondary growth was not directly available from the harvested material. Instead, we derived that following the model of Bret-Harte et al. [26] (see key equations below) who assumed that (i) a stem/branch segment is cylindrical, (ii) the annual increment in stem/branch radius does not vary with the age of the stem/branch segments in the aboveground portion of the ramet, and (iii) the annual increment in stem/branch radius changes under manipulated environmental conditions. For the study species, the performance of the model of Bret-Harte et al. [26] was very good, with average slopes of the regression modelled vs. measured standing stem biomass M of 0.90–1.1 and $r^2 > 0.94$ [45]. The total

Table 4. Summary of the shrub growth variables measured in this study, the plant organ and organ age they refer to and the way the results are expressed.

Growth variables	species	organ[a]	age organ (year)	result type
Primary growth	*Cassiope*	leaves plus stem/branches	0[b]	relative to old stem biomass
	Empetrum, Betula	leaves plus stem/branches	0	relative to old stem biomass
Secondary growth	*Cassiope*	stem/branches	1 to 12	relative to old stem biomass
	Empetrum, Betula	stem/branches	1 to 7	relative to old stem biomass
Total growth	*Cassiope*	leaves plus stem/branches	0 to 12	relative to old stem biomass
	Empetrum, Betula	leaves plus stem/branches	0 to 7	relative to old stem biomass
Shrub height	*Cassiope*	stem/branches	no distinction	absolute per plot
	Empetrum, Betula	stem/branches	no distinction	absolute per plot
Shrub length	*Cassiope*	stem/branches	1 to 12	absolute for each ramet
	Empetrum, Betula	stem/branches	1 to 7	absolute for each ramet
Branch numbers	*Cassiope*	stem/branches	1 to 12	absolute for each cohort
	Empetrum, Betula	stem/branches	1 to 7	absolute for each cohort
Branch length	*Cassiope*	stem/branches	1 to12	absolute for each cohort
	Empetrum, Betula	stem/branches	1 to 7	absolute for each cohort
Apical increment	*Cassiope*	stem/branches	0	absolute for each ramet
	Empetrum, Betula	stem/branches	0	absolute for each ramet
Plant coverage	*Cassiope*	leaves plus stem/branches	no distinction	absolute per plot
	Empetrum, Betula	leaves plus stem/branches	no distinction	absolute per plot

[a]: no distinction was made for stem and branches as they are difficult to differentiate for the clonal species investigated;
[b]: 0 indicates 'current year'.

growth rate was calculated as the sum of the primary and secondary growth rate.

The calculation of the secondary growth of individual ramets at yearly basis is summarized in four steps (for details see [26]). (i) The mass (m) of a stem/branch segment cohort of a given age n (in years) of an individual ramet is estimated as:

$$m = l\left(\alpha^2(n-1)^2 + 2\alpha(n-1)c + c^2\right) \qquad (1)$$

where l is the length of the stem/branch segment cohort, c equals $(m/l)^{1/2}$ of the current-year stem and α the slope of a linear relationship $(m/l)^{1/2}$ vs. n. (ii) The annual mass increment due to secondary growth (Δm) of a stem/branch segment cohort equals:

$$\Delta m = l\left(2\alpha^2(n-1) + \alpha^2 + 2\alpha c\right) \qquad (2)$$

(iii) The mass (M) of an individual ramet is calculated as the sum of m for all the segment cohort age classes and the annual mass increment due to secondary growth (ΔM) as the sum of Δm for all the segment cohort age classes. (iv) The annual stem secondary growth rate equals $\Delta M/M$.

Shrub height and length. The height of nine randomly selected shoots of *Cassiope*, *Empetrum* and *Betula* was measured in two 25×50 cm rectangulars within the central area of each plot in mid-late August 2010 and averaged for each plot (n = 6). The height was measured with a ruler as the perpendicular height from shoot apex to ground. Shrub length refers to the total length of stem and branches of each ramet of *Cassiope*, *Empetrum* and *Betula* sampled for the determination of the growth rate (see above).

Branch number, branch length and apical increment. Branch number and branch length of each ramet of *Cassiope*, *Empetrum* and *Betula* were derived separately for each age class composing the ramet (in years) from the ramets sampled for the determination of the growth rate (see above). Branch length refers to the total aggregated length for a given branch age class. The length of the current year's branches is defined as apical increment. As young stem and young branches are difficult to differentiate for the clonal species investigated, all woody segments of a given age class were considered as branches in these assessments.

Species coverage. Species coverage was measured in mid-late August 2010 with the pin-point method in the same two rectangulars (25×50 cm) per plot investigated for shrub height. A pin was passed vertically at 100 points (5 cm spaced) and all matter touched by the pin was recorded as a hit [46]. Vascular vegetation was recorded at species level, nonvascular vegetation was lumped in bryophytes and lichens, whereas attached or unattached dead tissue was recorded as litter. In this study, we present coverage data for the model shrub species *Cassiope*, *Empetrum* and *Betula*, for the lumped graminoids group and for the total vascular plant cover.

Statistics

The impact of manipulated environmental factors (warming, fertilisation, shading) was assessed with analysis of variance. Due to the incomplete factorial design, the analysis was conducted separately for (i) shading and (ii) warming and fertilisation [13]. The response to shading was tested with a one-way ANOVA, whereas the response to warming and fertilisation with a two-way ANOVA with interaction between warming and fertilisation. The block factor was not considered because preliminary analyses showed that it had no effect on the dependent variables. If prerequisites for analysis of variances (normality, checked with Shapiro test, and homoscedasticity, checked with Bartlett test) were not met, we performed a Kruskal-Wallis test instead of one-way ANOVA or repeated the same analysis after transformation ($\log x$ and in few cases $x^{1/2}$ or x^{-1}) for two-way ANOVA, as no standard non-parametric test fitted our design and transformed data fulfilled the ANOVA prerequisites. Transformation was needed for: (i) growth rates of *Cassiope* and *Empetrum*, (ii) total height of each species, (iii) total length and apical increment of *Cassiope* and *Empetrum*, (iv) coverage of *Cassiope*, *Betula* and graminoids, and (v) datasets on branch number and branch length (see details in Table S1).

Dependent variables tested with the ANOVA analysis were growth rates (primary, secondary and total), shrub height, shrub length, species coverage, apical increment, branch number and length. Primary and secondary growth were tested separately as they proved to be uncorrelated (tested with Pearson's correlation) for any treatment and species. If more within-plot measures of the dependent variable were available per plot (e.g. for growth rates of *Cassiope* and *Empetrum*), a nested level was added to the ANOVAs. Treatment effects on branch number and length were tested separately for each species and age class, except for 7 year-old branches of *Empetrum* and *Betula* which were not analyzed because of the few replicates available. All analyses were performed in R version 2.12.2 (R Development Core Team 2011).

Supporting Information

Table S1 Results of ANOVAs (F and P values) on the number and length of branches of cohort of the same age (up to 12 year-old) of the shrubs *Cassiope tetragona*, *Empetrum hermaphroditum* and *Betula nana* at a subarctic heath in Abisko (Northern Sweden) after 22 years of environmental manipulation.

Acknowledgments

Special thanks are due to the Abisko Scientific Research Station for the logistic support.

Author Contributions

Conceived and designed the experiments: MC AM. Performed the experiments: MC NL AM. Analyzed the data: MC NL. Contributed reagents/materials/analysis tools: MC AM. Wrote the paper: MC NL AM.

References

1. IPCC, Solomon S, Qin D, Manning M, Chen Z, Marquis M, et al. (2007) Climate change 2007: the physical science basis. Contribution of Working Group I to the fourth assessment report of the Intergovernmental Panel on Climate Change. Cambridge and New York: Cambridge University Press.
2. Callaghan TV, Bjorn LO, Chernov Y, Chapin T, Christensen TR, et al. (2004) Effects on the function of arctic ecosystems in the short- and long-term perspectives. Ambio 33: 448–458.
3. Callaghan TV, Bjorn LO, Chernov Y, Chapin T, Christensen TR, et al. (2004) Effects of changes in climate on landscape and regional processes, and feedbacks to the climate system. Ambio 33: 459–468.
4. Arft AM, Walker MD, Gurevitch J, Alatalo JM, Bret-Harte MS, et al. (1999) Responses of tundra plants to experimental warming: Meta-analysis of the international tundra experiment. Ecol Monog 69: 491–511.
5. van Wijk MT, Clemmensen KE, Shaver GR, Williams M, Callaghan TV, et al. (2004) Long-term ecosystem level experiments at Toolik Lake, Alaska, and at Abisko, Northern Sweden: generalizations and differences in ecosystem and plant type responses to global change. Global Change Biol 10: 105–123.
6. Wookey PA (2008) Experimental approaches to predicting the future of tundra plant communities. Plant Ecol Divers 1: 299–307.

7. Sonesson M, Callaghan TV (1991) Strategies of survival in plants of the Fennoscandian tundra. Arctic 44: 95–105.

8. Trumbore S (2006) Carbon respired by terrestrial ecosystems - recent progress and challenges. Global Change Biol 12: 141–153.

9. Chapin FS, Shaver GR (1996) Physiological and growth responses of arctic plants to a field experiment simulating climatic change. Ecology 77: 822–840.

10. Jonasson S, Michelsen A, Schmidt IK (1999) Coupling of nutrient cycling and carbon dynamics in the Arctic, integration of soil microbial and plant processes. Appl Soil Ecol 11: 135–146.

11. Callaghan TV, Bjorn LO, Chernov Y, Chapin T, Christensen TR, et al. (2004) Responses to projected changes in climate and UV-B at the species level. Ambio 33: 418–435.

12. Callaghan TV, Bjorn LO, Chernov Y, Chapin T, Christensen TR, et al. (2004) Effects on the structure of arctic ecosystems in the short- and long-term perspectives. Ambio 33: 436–447.

13. Jonasson S, Michelsen A, Schmidt IK, Nielsen EV (1999) Responses in microbes and plants to changed temperature, nutrient, and light regimes in the arctic. Ecology 80: 1828–1843.

14. Shaver GR, Jonasson S (2001) Productivity of arctic ecosystems. In: Roy J, Saugier B, Mooney HA, eds. Terrestrial global productivity. San Diego: Academic Press. pp 189–209.

15. Mack MC, Schuur EAG, Bret-Harte MS, Shaver GR, Chapin FS (2004) Ecosystem carbon storage in arctic tundra reduced by long-term nutrient fertilisation. Nature 431: 440–443.

16. Tape K, Sturm M, Racine C (2006) The evidence for shrub expansion in Northern Alaska and the Pan-Arctic. Global Change Biol 12: 686–702.

17. Forbes BC, Macias Fauria M, Zetterberg P (2010) Russian Arctic warming and 'greening' are closely tracked by tundra shrub willows. Global Change Biol 16: 1542–1554.

18. Hallinger M, Manthey M, Wilmking M (2010) Establishing a missing link: warm summers and winter snow cover promote shrub expansion into alpine tundra in Scandinavia. New Phytol 186: 890–899.

19. Havström M, Callaghan TV, Jonasson S (1993) Differential growth-responses of Cassiope tetragona, an arctic dwarf shrub, to environmental perturbations among three contrasting High sites and Sub-Arctic sites. Oikos 66: 389–402.

20. Graglia E, Jonasson S, Michelsen A, Schmidt IK (1997) Effects of shading, nutrient application and warming on leaf growth and shoot densities of dwarf shrubs in two arctic-alpine plant communities. Ecoscience 4: 191–198.

21. Michelsen A, Jonasson S, Sleep D, Havström M, Callaghan TV (1996) Shoot biomass, δ¹³C, nitrogen and chlorophyll responses of two arctic dwarf shrubs to in situ shading, nutrient application and warming simulating climatic change. Oecologia 105: 1–12.

22. Graglia E, Jonasson S, Michelsen A, Schmidt IK, Havstrom M, et al. (2001) Effects of environmental perturbations on abundance of subarctic plants after three, seven and ten years of treatments. Ecography 24: 5–12.

23. Atkin OK, Tjoelker MG (2003) Thermal acclimation and the dynamic response of plant respiration to temperature. Trends plant sci 8: 343–351.

24. Shaver GR (1986) Woody stem production in Alaskan tundra shrubs. Ecology 67: 660–669.

25. Campioli M, Michelsen A, Demey A, Vermeulen A, Samson R, et al. (2009) Net primary production and carbon stocks for subarctic mesic-dry tundras with contrasting microtopography, altitude, and dominant species. Ecosystems 12: 760–776.

26. Bret-Harte MS, Shaver GR, Chapin FS (2002) Primary and secondary stem growth in arctic shrubs: implications for community response to environmental change. J Ecol 90: 251–267.

27. Campioli M, Leblans N, Michelsen A (2012) Stem secondary growth of tundra shrubs: impact of environmental factors and relationships with apical growth. Arc Antarc Alp Res 44: 16–25.

28. Gough L, Wookey PA, Shaver GR (2002) Dry heath arctic tundra responses to long-term nutrient and light manipulation. Arc Antarc Alp Res 34: 211–218.

29. Richardson SJ, Press MC, Parsons AN, Hartley SE (2002) How do nutrients and warming impact on plant communities and their insect herbivores? A 9-year study from a sub-arctic heath. J Ecol 90: 544–556.

30. DeGroot WJ, Thomas PA, Wein RW (1997) Betula nana L and Betula glandulosa Michx. J Ecol 85: 241–264.

31. Niinemets U, Portsmuth A, Truus L (2002) Leaf structural and photosynthetic characteristics, and biomass allocation to foliage in relation to foliar nitrogen content and tree size in three Betula species. Ann Bot-London 89: 191–204.

32. Marchand FL, Kockelbergh F, van de Vijver B, Beyens L, Nijs I (2006) Are heat and cold resistance of arctic species affected by successive extreme temperature events? New Phytol 170: 291–300.

33. Parsons AN, Welker JM, Wookey PA, Press MC, Callaghan TV, et al. (1994) Growth responses of four sub-arctic dwarf-shrubs to simulated environmental change. J Ecol 82: 307–318.

34. Berry J, Björkman O (1980) Photosynthetic response and adaptation to temperature in higher plants. Annual Review of Plant Physiology and Plant Molecular Biology 31: 491–543.

35. Campioli M, Michelsen A, Samson R, Lemeur R (2009) Seasonal variability of leaf area index and foliar nitrogen in contrasting dry-mesic tundras. Botany 87: 431–442.

36. Arnesen G, Beck PSA, Engelskjon T (2007) Soil acidity, content of carbonates, and available phosphorus are the soil factors best correlated with alpine vegetation: Evidence from Troms, North Norway. Arc Antarc Alp Res 39: 189–199.

37. Schmidt IK, Jonasson S, Shaver GR, Michelsen A, Nordin A (2002) Mineralization and distribution of nutrients in plants and microbes in four arctic ecosystems: responses to warming. Plant Soil 242: 93–106.

38. Ruess L, Michelsen A, Schmidt IK, Jonasson S (1999) Simulated climate change affecting microorganisms, nematode density and biodiversity in subarctic soils. Plant Soil 212: 63–73.

39. Callaghan TV, Bjorn LO, Chernov Y, Chapin T, Christensen TR, et al. (2004) Biodiversity, distributions and adaptations of arctic species in the context of environmental change. Ambio 33: 404–417.

40. Tybirk K, Nilsson MC, Michelson A, Kristensen HL, Shevtsova A, et al. (2000) Nordic Empetrum dominated ecosystems: Function and susceptibility to environmental changes. Ambio 29: 90–97.

41. Sprugel DG, Hinckley TM, Schaap W (1991) The theory and the practice of branch autonomy. Annual Review of Ecology and Systematics 22: 309–334.

42. Callaghan TV, Carlsson BA, Tyler NJC (1989) Historical records of climate-related growth in Cassiope tetragona from the Arctic. J Ecol 77: 823–837.

43. Shaver GR, Chapin FS (1991) Production - biomass relationships and element cycling in contrasting arctic vegetation types. Ecol Monogr 61: 1–31.

44. Shaver GR, Laundre JA, Giblin AE, Nadelhoffer KJ (1996) Changes in live plant biomass, primary production and species composition along a riverside toposequence in Arctic Alaska, USA. Arc Antarc Alp Res 28: 363–379.

45. Leblans N (2011) Environmental drivers and expected impacts of climate change on the secondary growth of woody shrubs in arctic ecosystems (MS thesis in Dutch). Wilrijk: University of Antwerp.

46. Press MC, Potter JA, Burke MJW, Callaghan TV, Lee JA (1998) Responses of a subarctic dwarf shrub heath community to simulated environmental change. J Ecol 86: 315–327.

2

Response of CH_4 and N_2O Emissions and Wheat Yields to Tillage Method Changes in the North China Plain

Shenzhong Tian[1], Tangyuan Ning[1]*, Hongxiang Zhao[1], Bingwen Wang[1], Na Li[1], Huifang Han[1], Zengjia Li[1], Shuyun Chi[2]*

1 State Key Laboratory of Crop Biology, Shandong Key Laboratory of Crop Biology, Shandong Agricultural University, Taian, Shandong PR, China, **2** College of Mechanical and Electronic Engineering, Shandong Agricultural University, Taian, Shandong PR, China

Abstract

The objective of this study was to quantify soil methane (CH_4) and nitrous oxide (N_2O) emissions when converting from minimum and no-tillage systems to subsoiling (tilled soil to a depth of 40 cm to 45 cm) in the North China Plain. The relationships between CH_4 and N_2O flux and soil temperature, moisture, NH_4^+-N, organic carbon (SOC) and pH were investigated over 18 months using a split-plot design. The soil absorption of CH_4 appeared to increase after conversion from no-tillage (NT) to subsoiling (NTS), from harrow tillage (HT) to subsoiling (HTS) and from rotary tillage (RT) to subsoiling (RTS). N_2O emissions also increased after conversion. Furthermore, after conversion to subsoiling, the combined global warming potential (GWP) of CH_4 and N_2O increased by approximately 0.05 kg CO_2 ha^{-1} for HTS, 0.02 kg CO_2 ha^{-1} for RTS and 0.23 kg CO_2 ha^{-1} for NTS. Soil temperature, moisture, SOC, NH_4^+-N and pH also changed after conversion to subsoiling. These changes were correlated with CH_4 uptake and N_2O emissions. However, there was no significant correlation between N_2O emissions and soil temperature in this study. The grain yields of wheat improved after conversion to subsoiling. Under HTS, RTS and NTS, the average grain yield was elevated by approximately 42.5%, 27.8% and 60.3% respectively. Our findings indicate that RTS and HTS would be ideal rotation tillage systems to balance GWP decreases and grain yield improvements in the North China Plain region.

Editor: Ben Bond-Lamberty, DOE Pacific Northwest National Laboratory, United States of America

Funding: This work was financially supported by the Nature Science Fund of China (30900876 and 31101127), the National Science and Technology Research Projects of China (2012BAD14B17), and Special Research Funding for Public Benefit Industries (Agriculture) of China (201103001). The funders had no role in study design, data collection and analysis, decision to publish, or preparation of the manuscript.

Competing Interests: The authors have declared that no competing interests exist.

* E-mail: ningty@163.com (TN); chishujun1955@163.com (SC)

Introduction

CH_4 and N_2O play a key role in global climate change [1]. The emission of gas from disturbed soils is an especially important contributory factor to global change [2]. N_2O is emitted from disturbed soil, whereas CH_4 is normally oxidized by aerobic soils, making them sinks for atmospheric CH_4 in dry farmland systems [3]. According to estimates of the IPCC [4], CH_4 and N_2O from agricultural sources account for 50% and 60% of total emissions, respectively. Therefore, it is critical to reduce emissions of greenhouse gases (GHG) from agricultural sources. Many studies have reported that soil tillage has significant effects on CH_4 and N_2O emissions from farmland because the production, consumption and transport of CH_4 and N_2O in soil are strongly influenced by tillage methods [5–8].

The North China Plain is one of the most important grain production regions of China. Harrow tillage (HT), rotary tillage (RT) and no-tillage (NT) are frequently used conservation tillage methods in this region because they not only improve crop yield but also enhance the utilization efficiency of soil moisture and nutrients [8–12]. However, successive years of shallow tillage (10–20 cm) exacerbate the risk of subsoil compaction, which not only leads to the hardening of soil tillage layers and an increase in soil bulk density, but also reduced crop root proliferation, limited water and nutrient availability and reduced crop yield [13].

Subsoiling is an effective method that is used to break up the compacted hardpan layer every 2 or 4 years in HT, RT or NT systems [14,15]. Subsoiling significantly increases soil water content and temperature and decreases soil bulk density as well [16,17]. These rotation tillage systems are currently utilized in the North China Plain. Soil moisture and temperature are two factors controlling CH_4 and N_2O emissions [18–22]. In addition, CH_4 and N_2O emissions are normally associated with N application (as fertilizer) under wet conditions [23].

Collectively, reasonable soil tillage methods may reduce GHG emissions and may be important for developing sustainable agricultural practices [24]. However, it is unclear how conversion to subsoiling would affect CH_4 and N_2O emissions and whether subsoiling increases or reduces GHG emissions and the GWP of these agricultural techniques. In addition, there is little information on the soil factors affecting CH_4 and N_2O emissions after conversion to subsoiling in the North China Plain. The aim of this study was to determine whether conversion to subsoiling can reduce CH_4 and N_2O emissions.

Materials and Methods

Ethics Statement

The research station of this study is a department of Shandong Agricultural University. This study was approved by State Key Laboratory of Crop Biology, Shandong Key Laboratory of Crop Biology, Shandong Agricultural University.

Study Site

The study was conducted at Tai'an (Northern China, 36°09′N, 117°09′E), which is characteristic of the North China Plain. The average annual precipitation is 786.3 mm, and the average annual temperature is 13.6°C, with the minimum (-1.5°C) and maximum (27.5°C) monthly temperatures in January and July, respectively. The annual frost-free period is approximately 170–220 days in duration, and the annual sunlight time is 2462.3 hours. The soil is loam with 40% sand, 44% silt and 16% clay. The characteristics of the surface soil (0–20 cm) were measured as follows: pH 6.2; soil bulk density 1.43 g cm^{-3}; soil organic matter 1.36%; soil total nitrogen 0.13%; and soil total phosphorous 0.13%. The meteorological data during the experiment are shown in Figure 1.

Experimental Design

The experiment was designed as HT, RT and NT farming methods that started in 2004. In 2008, each plot was bisected, with one half maintained using the original tillage method as the control and the other half converted to subsoiling, resulting in six treatment plots: HT and HT conversion to subsoiling (HTS); RT and RT conversion to subsoiling (RTS); and NT and NT conversion to subsoiling (NTS) in a split-plot design with three replicates. Each replicate was 35 m long and 4 m wide. After maize was harvested in each plot, straw was returned to the soil by one of the six following tillage operations:

HT - disking with a disc harrow to a depth of 12 cm to 15 cm,
RT - rototiller plowing to a depth of 10 cm to 15 cm,
NT - no tillage,
HTS, RTS, and NTS - plowed using a vibrating sub-soil shovel to a depth of 40 cm to 45 cm,

The experimental site was cropped with a rotation of winter wheat (*Triticum aestivum* Linn.) and maize (*Zea mays L.*). The wheat was sown in mid-October immediately after tilling the soil and was harvested at the beginning of June the following year. The maize was sown directly after the wheat harvest and was harvested in early October. During the wheat growth period, fertilizer was used at a rate of 225 kg N ha^{-1}, 150 kg ha^{-1} P_2O_5 and 105 kg ha^{-1} K_2O, and 100 kg N ha^{-1} was used as topdressing in the jointing stage with 160 mm of irrigation water. During the maize growth period, 120 kg N ha^{-1}, 120 kg ha^{-1} P_2O_5 and 100 kg ha^{-1} K_2O were used as a base fertilizer, and 120 kg N ha^{-1} was used as topdressing in the jointing stage.

CH$_4$ and N$_2$O Sampling and Measurements

CH$_4$ and N$_2$O content was measured using the static chamber-gas chromatography method [25]. The duration of gas sample collection was based on the diurnal variations in this region: the collection of CH$_4$ occurred from 9:00 a.m. to 10:00 a.m., and N$_2$O was collected between 9:00 a.m. and 12:00 p.m. from October 10, 2007, to May 19, 2009 at approximately 1-month intervals [26]. Both CH$_4$ and N$_2$O were sampled at 5 minutes, 20 minutes and 35 minutes after chamber closing. Simultaneously, the atmospheric temperature, the temperature in the static chamber, the land

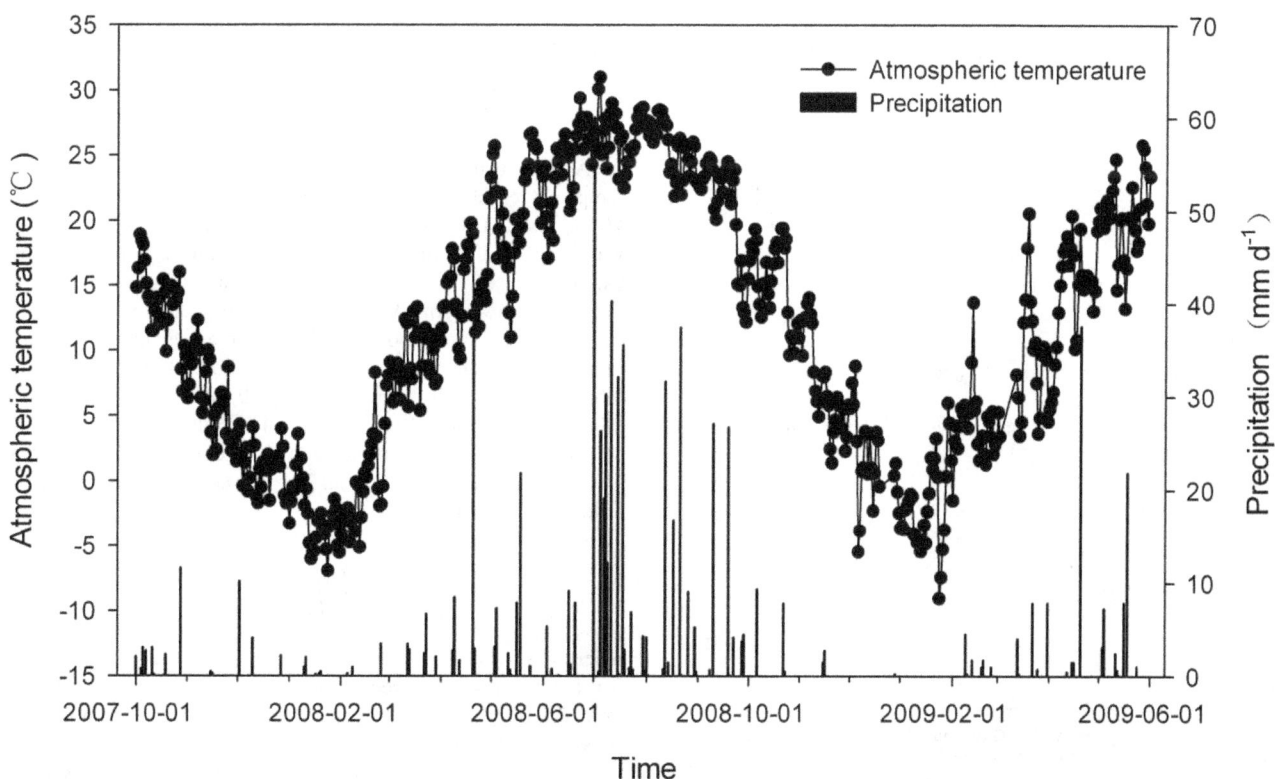

Figure 1. The atmospheric temperature and precipitation at the experiment site. The data were collected by the agricultural meteorological station approximately 500 m from the experiment field.

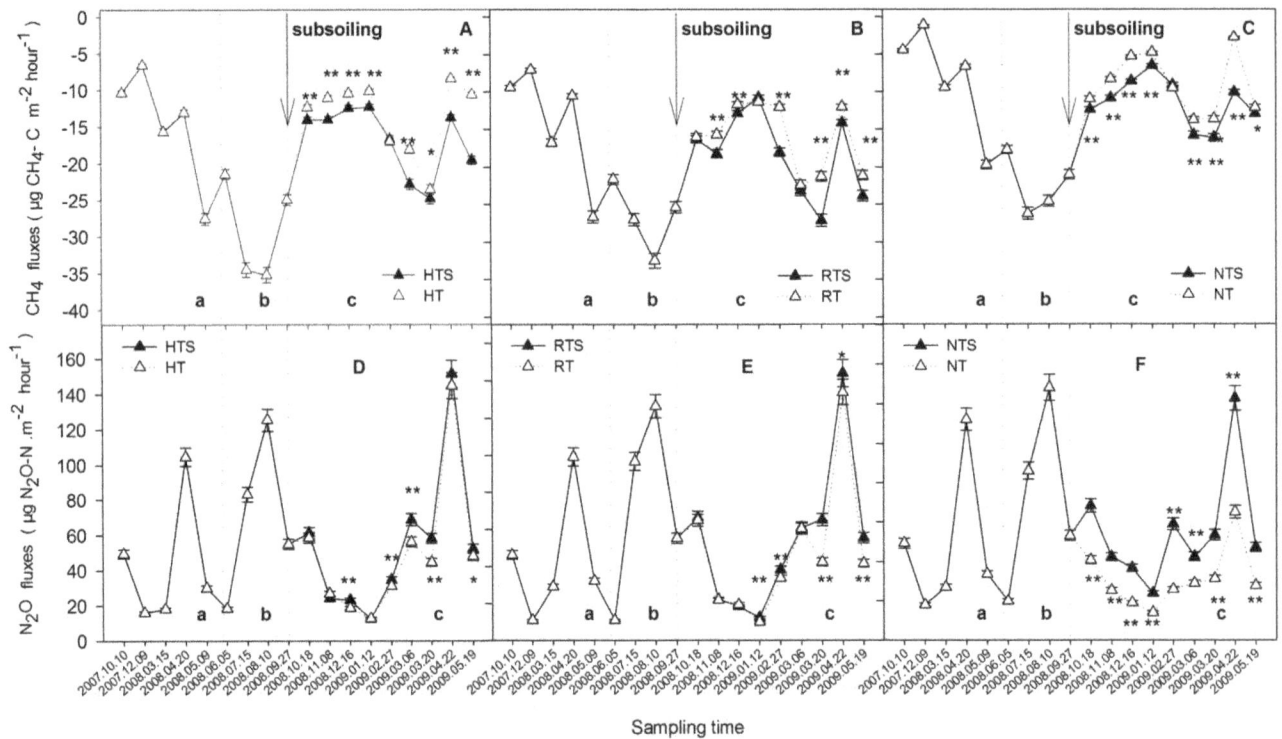

Figure 2. A to C CH$_4$ flux variations of H, R, and N after subsoiling in different periods; D to F N$_2$O flux variations of H, R, and N after subsoiling in different periods. a in Fig. 2 is the wheat growth stage of 2007 to 2008; **b** is the maize growth stage of 2008 to 2009; c is the wheat growth stage of 2008 to 2009. Arrows indicate time of subsoiling. Dotted lines distinguish the growth period of wheat and maize. * indicates $P<0.05$ and **indicates $P<0.01$ between subsoiling and the control.

Table 1. GWP and total changes in CH$_4$ and N$_2$O after subsoiling (2008.10~2009.05).

Treatments	HT	HTS	RT	RTS	NT	NTS
CH$_4$ total emission (kg·ha^{-1})	−0.73	−0.84	−0.64	−0.78	−0.39	−0.52
GWP of CH$_4$ (kgCO$_2$·ha^{-1})	−0.17	−0.19	−0.15	−0.18	−0.09	−0.12
N$_2$O total emission (kg·ha^{-1})	2.14	2.42	2.26	2.46	1.46	2.67
GWP of N$_2$O (kgCO$_2$·ha^{-1})	0.49	0.56	0.52	0.57	0.35	0.61
Total emissions of CH$_4$ and N$_2$O (kg·ha^{-1})	1.41	1.58	1.62	1.68	1.07	2.15
GWP of CH$_4$ and N$_2$O (kgCO$_2$ ha^{-1})	0.32	0.37	0.37	0.39	0.26	0.49
Increased emissions after conversion (kg·ha^{-1})	–	0.17	–	0.06	–	1.08
Increased GWP after conversion (kgCO$_2$·ha^{-1})	–	0.05	–	0.02	–	0.23

Total emissions of CH$_4$ and N$_2$O (kg·ha^{-1}), N$_2$O total emission flux added CH$_4$ total emission flux; **GWP of CH$_4$ and N$_2$O (kgCO$_2$·ha^{-1})**, GWP of N$_2$O added GWP of CH$_4$; **Increased emissions after conversion (kg·ha^{-1})**, difference of total emission of CH$_4$ and N$_2$O before and after conversion; **Increased GWP after conversion (kgCO$_2$·ha^{-1})**, difference of GWP of CH$_4$ and N$_2$O before and after conversion.

surface temperature and the soil temperature at a depth of 5 cm were determined after collecting samples.

The samples were measured using a Shimadzu GC-2010 gas chromatograph. CH$_4$ was measured using a flame ionization detector with a stainless steel chromatography column packed with a 5A molecular sieve (2 m long); the carrier gas was N$_2$. The temperatures of the column, injector and detector were 80°C, 100°C and 200°C, respectively. The total flow of the carrier gas was 30 ml min^{-1}, the H$_2$ flow was 40 ml min^{-1}, and the airflow was 400 ml min^{-1}. N$_2$O was measured using an electron capture detector with a Porapak-Q chromatography column (4 m long); the carrier gas was also N$_2$. The temperatures of the column, injector and detector were 45°C, 100°C and 300°C, respectively. The total flow of the carrier gas was 40 ml min^{-1}, and the tail-blowing flow was 40 ml min^{-1}. The gas fluctuations were calculated by the gas concentration change in time per unit area.

Emission changes in CH$_4$ and N$_2$O were calculated using the following formula [25]:

$$F = \frac{60HMP}{8.314(273+T)}\frac{dc}{dt}$$

where F is the change in gas emission or uptake ($\mu g \cdot m^{-2} \cdot h^{-1}$); 60 is the conversion coefficient of minutes and hours; H is the height (m); M is the molar mass of gas (g·mol^{-1}); P is the atmospheric pressure (Pa); 8.314 is the Ideal Gas Constant (J mol^{-1} K^{-1}); T is the average temperature in the static chamber (°C); and dc/dt is the line slope of the gas concentration change over time.

Table 2. Correlation analysis between changes in CH_4 and N_2O with soil temperature and soil moisture per sampling time.

Sampling time	Soil temperature				Soil moisture			
	CH_4		N_2O		CH_4		N_2O	
	R^2	n	R^2	n	R^2	n	R^2	n
2008.10.18	0.6020*	3	0.3832	3	0.5429*	3	0.1020	3
2008.11.08	0.6180*	3	0.0377	3	0.2945	3	0.1241	3
2008.12.16	0.7314**	3	0.0087	3	0.0085	3	0.5142*	3
2009.01.12	0.6490**	3	0.0723	3	0.2988	3	0.5200*	3
2009.02.27	0.6597**	3	0.3053	3	0.5370*	3	0.0914	3
2009.03.06	0.3824	3	0.1461	3	0.0417	3	0.0005	3
2009.03.20	0.2876	3	0.0257	3	0.4966*	3	0.6132*	3
2009.04.22	0.4476*	3	0.3044	3	0.5154*	3	0.6735**	3
2009.05.19	0.8870**	3	0.0503	3	0.4593*	3	0.5027*	3

*$P<0.05$,
**$P<0.01$.

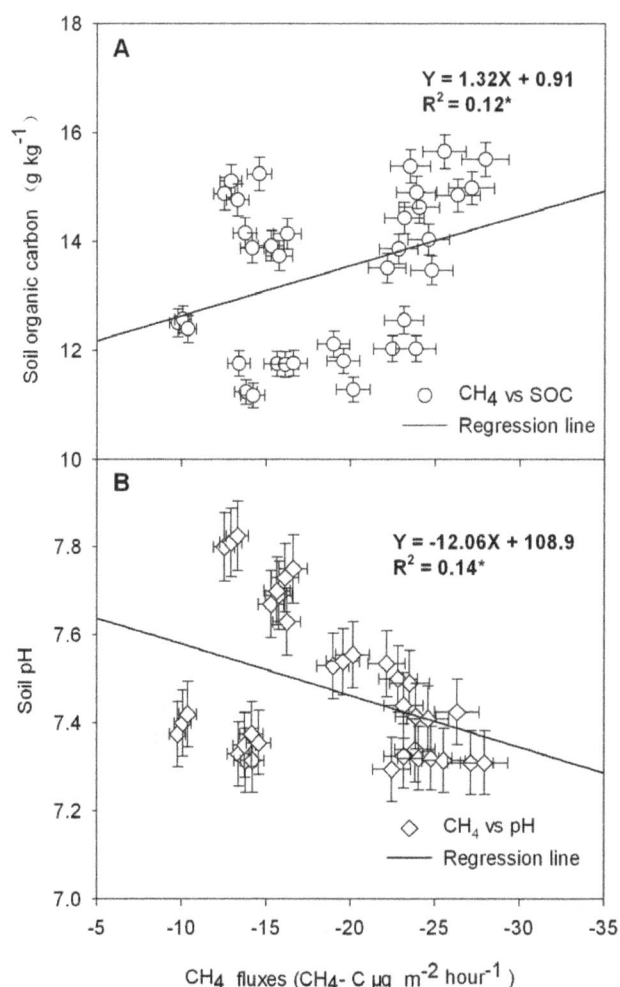

Figure 3. A Linear regression between the CH_4 uptake fluxes and SOC, B Linear regression between the CH_4 uptake fluxes and soil pH. Arrows indicate the regression equation between the CH_4 uptake fluxes and soil organic carbon, soil pH. *indicates $P<0.05$.

GWP of CH_4 and N_2O

The global warming potentials (GWP) were determined by measuring CH_4 and N_2O emissions. The GWP of CH_4 and N_2O are 25 and 298 times higher, respectively, than that of CO_2 (the GWP of CO_2 is 1) [27] and are calculated as follows:

$$GWP(CH_4) = \frac{TF(CH_4) \times 25}{100}$$

$$GWP(N_2O) = \frac{TF(N_2O) \times 298}{100}$$

where $GWP(CH_4)$ is the GWP of CH_4 (kg CO_2 ha^{-1}); $TF(CH_4)$ is the total uptake of CH_4 (kg CO_2 ha^{-1} a^{-1}); 25 is the GWP coefficient of CH_4; 100 is the time scale of climate change (a); $GWP(N_2O)$ is the GWP of N_2O (kg CO_2 ha^{-1}); $TF(N_2O)$ is the total emission of N_2O (kg CO_2 ha^{-1} a^{-1}); and 298 is the GWP coefficient of N_2O.

Soil Factor Measurements

The meteorological data during the experiment were obtained from an agricultural weather station in the experimental area. To evaluate the relation between soil temperature and moisture and CH_4 and N_2O emissions, we measured soil temperature at a depth of 5 cm and the soil moisture in the 0–20 cm soil layers simultaneously using a soil temperature, moisture and electric conductivity instrument (WET brand, made in the UK) as the temperature and moisture data collection tool. The soil samples were collected using a soil sampler with five replicates in each different tillage treatment and were dried and triturated after mixing. This sample was used to determine the SOC, NH_4^+-N and pH using the Potassium Dichromate Heating Method, the UV Colorimetric Method and the Potentiometry Method, respectively [28].

Grain Yield

The grain yield of winter wheat was sampled from the 1.5 m × 6 m portion in the central area of each plot.

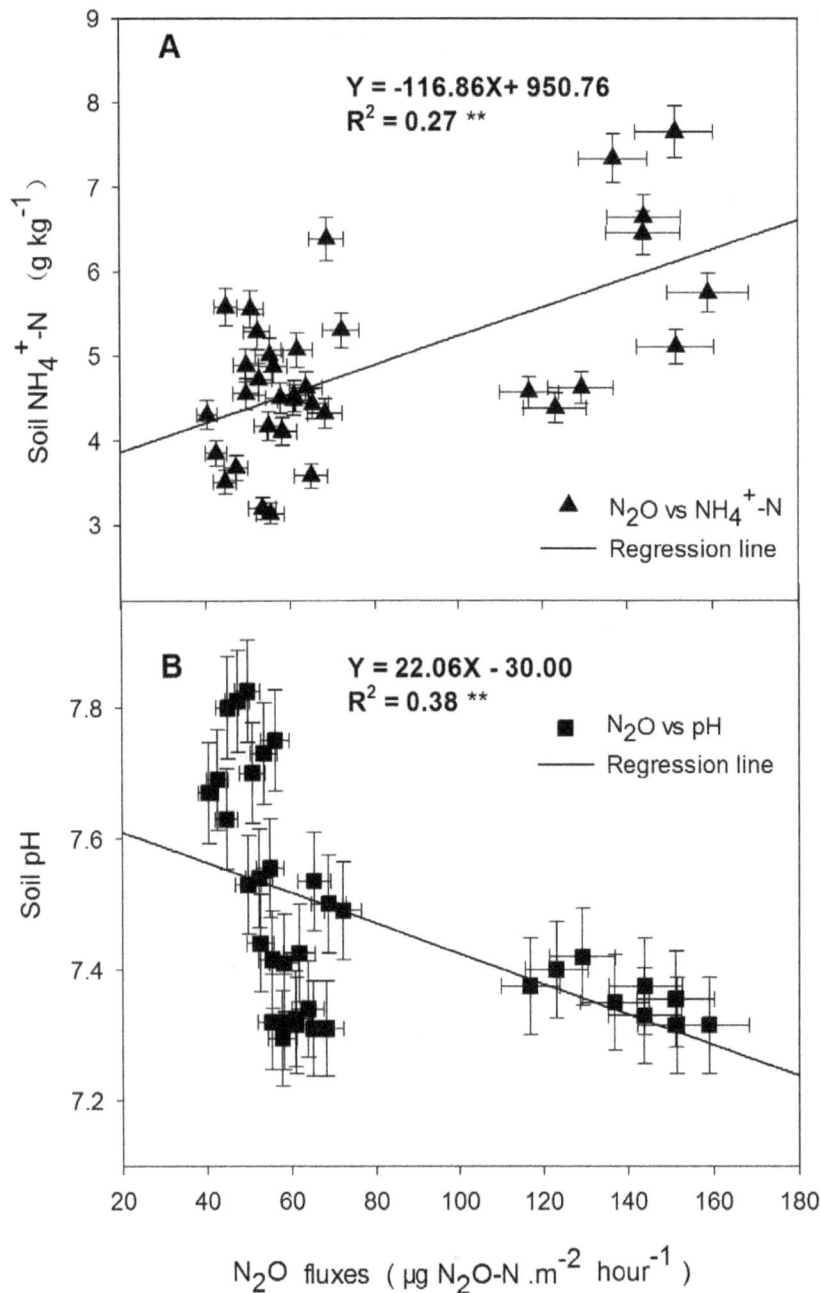

Figure 4. A Linear regression between the N_2O emission fluxes and soil NH_4^+-N, B Linear regression between the N_2O emission fluxes and soil pH. Arrows indicate the regression equation between the N_2O emission fluxes and soil NH_4^+-N, soil pH. **indicates $P<0.01$.

Statistical Analyses

The data were analyzed using analyses of variance and the SPSS 17.0 Statistical Analysis System and were mapped using Sigma Plot 10.0. The mean standard deviation and least significant difference were calculated for comparison of the treatment means.

Results

CH_4 and N_2O

Differences in CH_4 flux were observed when converting from HT to HTS, from RT to RTS and from NT to NTS (Figs. 2 A to

C). The soil absorption of CH_4 increased in different periods after conversion to subsoiling compared with the control. The soil absorption of CH_4 increased from 13.53 $\mu g \cdot m^{-2} \cdot h^{-1}$ under HT to 16.72 $\mu g \cdot m^{-2} \cdot h^{-1}$ under HTS, from 15.59 $\mu g \cdot m^{-2} \cdot h^{-1}$ under RT to 18.20 $\mu g \cdot m^{-2} \cdot h^{-1}$ under RTS and from 9.01 $\mu g \cdot m^{-2} \cdot h^{-1}$ under NT to 11.36 $\mu g \cdot m^{-2} \cdot h^{-1}$ under NTS, respectively. However, N_2O emission also increased after subsoiling (Fig. 2 D to F), which increased from 49.07 $\mu g \cdot m^{-2} \cdot h^{-1}$ under HT to 54.05 $\mu g \cdot m^{-2} \cdot h^{-1}$ under HTS and from 47.49 $\mu g \cdot m^{-2} \cdot h^{-1}$ under RT to 53.60 $\mu g \cdot m^{-2} \cdot h^{-1}$ under RTS. Compared with the above two treatments, however, the N_2O emissions from the

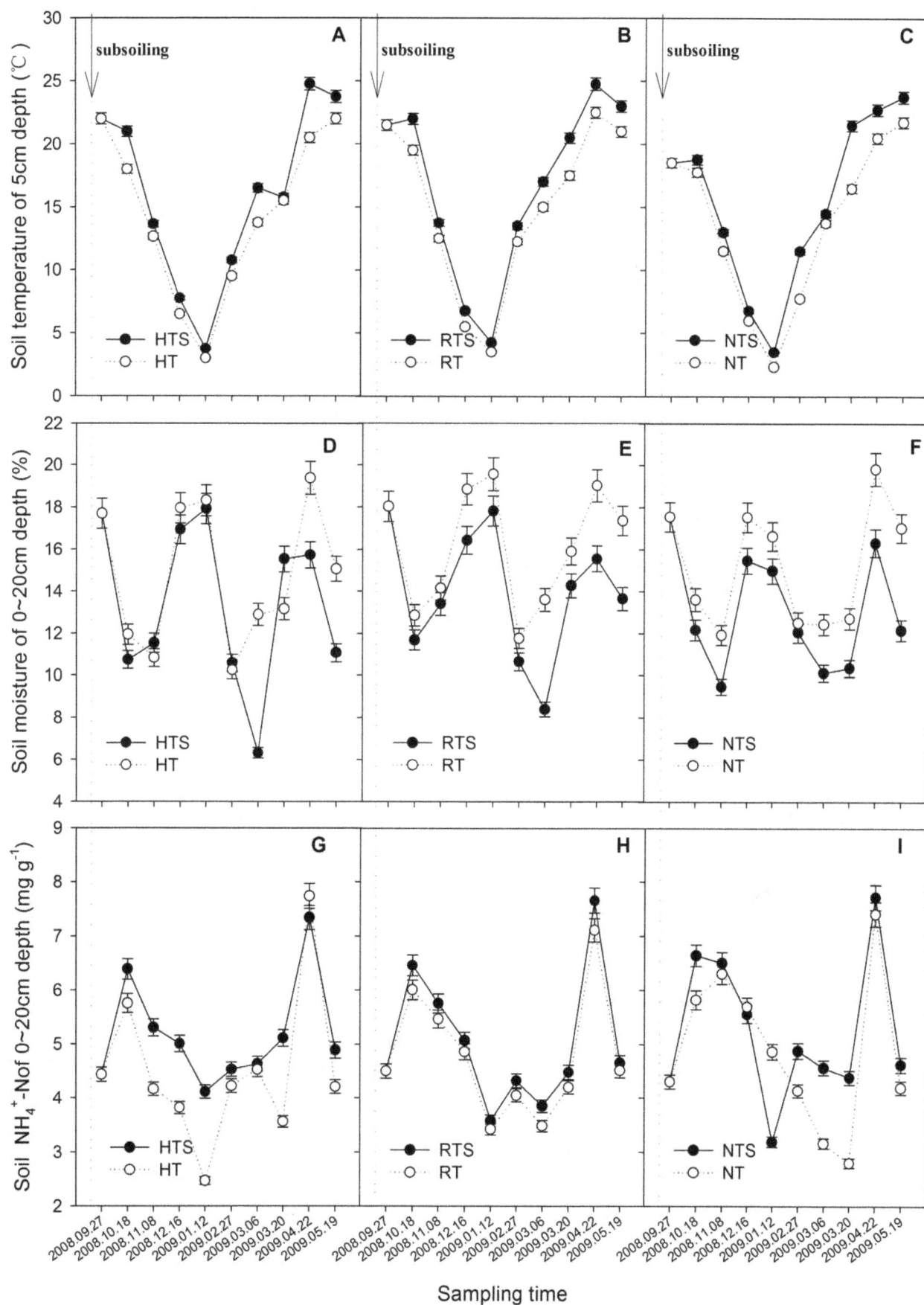

Sampling time

Figure 5. A to C Variation of Soil temperature at a 5 cm depth (°C) after subsoiling; D to F Variation of Soil water content at a 0~20 cm depth (%) after subsoiling; G to I Variation of Soil NH_4^+-N at a 0~20 cm depth (mg·kg^{-1}) after subsoiling. Arrows and the dotted line indicate time of subsoiling.

soil after conversion to NTS increased significantly, from 30.92 $\mu g \cdot m^{-2} \cdot h^{-1}$ under NT to 55.15 $\mu g \cdot m^{-2} \cdot h^{-1}$ under NTS.

GWP of CH_4 and N_2O

CH_4 uptake increased under HTS, RTS and NTS; consequently, the GWP of CH_4 decreased using these tilling methods compared with HT, RT and NT. However, the GWP of N_2O increased under HTS, RTS and NTS (Table 1). Overall, therefore, the GWPs of the CH_4 and N_2O emissions taken together increased from 0.32 kg CO_2 ha^{-1} under HT to 0.37 kg CO_2 ha^{-1} under HTS, from 0.37 kg CO_2 ha^{-1} under RT to 0.39 kg CO_2 ha^{-1} under RTS and from 0.26 kg CO_2 ha^{-1} under NT to 0.49 kg CO_2 ha^{-1} under NTS, respectively.

Correlation Analysis between CH_4 and N_2O and Soil Factors

Soil temperature significantly affected the CH_4 uptake in soils, especially in lower (i.e., December, $R^2 = 0.7314$, $P<0.01$; January, $R^2 = 0.6490$, $P<0.01$; February, $R^2 = 0.6597$, $P<0.01$) or higher (i.e., May, $R^2 = 0.8870$, $P<0.01$) temperatures ($P<0.01$) (Table 2). At other sampling times, however, temperature did not affect on CH_4 uptake, and soil moisture became a main influencing factor on the absorption of CH_4 by the soils, especially in wet soil, such as after rain ($R^2 = 0.5154$, $P<0.05$) and irrigation ($R^2 = 0.5154$, $P<0.05$), when CH_4 absorption was significantly limited ($R^2 = 0.5429$, $P<0.05$). Higher soil moisture generally promoted the emission of N_2O ($R^2 = 0.6735$, $P<0.01$), but there was no obvious correlation between soil temperature and N_2O emissions.

In this study, SOC was also correlated with greater CH_4 uptake ($R^2 = 0.12$, $P<0.05$) (Fig. 3 A), whereas higher soil pH limited its absorption in the soil ($R^2 = 0.14$, $P<0.05$) (Fig. 3 B).

The emission of N_2O was correlated with higher soil NH_4^+-N content ($R^2 = 0.27$, $P<0.01$) (Fig. 4 A), while, similar to CH_4, a higher pH in soil strongly limited the emission of N_2O ($R^2 = 0.38$, $P<0.01$) (Fig. 4 B).

Variation of Soil Factors

The soil factors under HTS, RTS and NTS changed after subsoiling. The soil temperature at a depth of 5 cm rose under HTS, RTS and NTS compared with the temperatures under HT, RT and NT (Fig. 5 A to C). Soil temperature variations followed atmospheric temperature changes, but the average soil temperature during sampling period increased from 13.5°C under HT to 15.3°C under HTS, from 14.4°C under RT to 16.2°C under RTS and from 13.1°C under NT to 15.1°C under NTS, respectively. However, soil moisture decreased in the soil at 0–20 cm when converting to subsoiling that in the order of RTS>HTS>NTS (Fig. 5 D to F). The most obvious decrease, by 15.74%, occurred under the NTS treatment, while HTS and RTS decreased by 10.34% and 14.85%, respectively. The soil NH_4^+-N content increased with subsoiling that was NTS>HTS>RTS. Moreover, two peaks occurring on October 18, 2008, and April 22, 2009 (Fig. 5 G to I), due to the application of nitrogenous base fertilizer and topdressing fertilizer.

The CH_4 uptake and N_2O emission were correlated with the content of soil pH and SOC (Table 3). The pH value decreased after conversions, but with the pH under the NTS treatment being higher than that of the HTS and RTS treatments not only at 0~10 cm but also at 10~20 cm. Conversely, SOC content increased under HTS, RTS and NTS, with the highest values was under RTS, followed by NTS and then HTS. SOC was higher in the soil at 0–10 cm than at 10–20 cm.

Grain Yield

The highest wheat yields under RT were 5937.20 kg ha^{-1} in 2009 and 6164.83 kg ha^{-1} in 2010, which were only 3.8% greater than those under HT and NT (Table 4). However, the wheat yields under HTS, RTS and NTS improved significantly ($P<0.01$) than the control, not only in 2009 but also in 2010. The average yield of the two years increased by approximately 2416.25 kg ha^{-1}, 1695.38 kg ha^{-1}and 2804.33 kg ha^{-1} with subsoiling compared with that under HT, RT and NT, respectively. The increases of average yield were not only related to the number of spikes, which increased by 59×10^4 ha^{-1} after conversions as determined by the average of the three conversion treatments, but were also correlated with the grains per ear and 1000-grain weight, which increased by an average of 6.0 grains and 2.8 g, respectively.

Table 3. Soil pH and SOC variations after conversion to subsoiling.

Treatments		pH						SOC					
		HT	HTS	RT	RTS	NT	NTS	HT	HTS	RT	RTS	NT	NTS
0~10 cm	(i)	7.37^c	7.33^d	7.25^e	7.21^f	7.72^a	7.66^b	8.62^f	9.45^e	9.69^d	11.47^b	11.79^a	10.32^c
	(ii)	7.25^d	7.21^e	7.27^c	7.25^d	7.69^a	7.62^b	10.77^d	12.25^a	9.82^f	10.21^e	11.68^c	11.93^b
	(iii)	7.25^e	7.23^f	7.38^a	7.34^c	7.37^b	7.31^d	11.43^d	12.58^b	12.07^c	13.11^a	10.13^e	9.75^f
	(iv)	7.44^{cd}	7.42^d	7.45^c	7.40^e	7.86^a	7.82^b	9.01^f	9.39^e	10.83^b	12.42^a	10.57^c	10.49^d
10~20 cm	(i)	7.71^c	7.67^d	7.52^e	7.46^f	7.77^a	7.75^b	5.93^f	6.29^e	9.10^b	9.44^a	8.09^d	8.34^c
	(ii)	7.46^c	7.43^d	7.36^e	7.35^f	7.85^a	7.83^b	9.22^f	9.97^d	9.45^e	10.07^c	11.35^b	11.77^a
	(iii)	7.44^c	7.40^d	7.39^e	7.37^f	7.56^a	7.52^b	9.76^f	10.62^c	10.11^e	10.40^d	10.88^b	11.76^a
	(iv)	7.71^c	7.68^d	7.43^e	7.43^e	7.83^a	7.81^b	7.63^f	9.90^a	8.26^d	9.55^b	8.31^c	7.84^e

Different small letter means $P<0.01$; (i), (ii), (iv) and (iii) means time of sample collection in 2008.10.18, 2009.03.17, 2009.04.20 and 2009.05.19 respectively.

Table 4. The wheat yield variations of HT, RT and NT after subsoiling from 2008–2010.

Treatments	Number of spikes ($10^4 \cdot ha^{-1}$)	Grains per ear	1000-grain weight (g)	Grain yield ($kg \cdot ha^{-1}$)	Increased ($kg \cdot ha^{-1}$)
2008–2009					
HT	646.50[bc]	30.05[bc]	33.79[b]	5582.83[b]	
HTS	683.50[a]	34.45[a]	34.31[b]	6866.55[a]	+1283.72
RT	655.00[b]	31.45[b]	33.94[b]	5937.20[b]	
RTS	637.50[c]	35.00[a]	36.83[a]	6985.20[a]	+1048.00
NT	583.00[d]	28.60[c]	32.40[c]	4595.87[c]	
NTS	688.50[a]	34.70[a]	33.96[b]	6895.06[a]	+2299.19
2009–2010					
HT	644.67[e]	30.93[e]	33.73[d]	5716.53[e]	
HTS	741.00[b]	38.59[a]	37.70[a]	9161.94[a]	+3548.77
RT	705.00[c]	31.68[d]	32.47[f]	6164.83[d]	
RTS	754.67[a]	35.78[c]	36.77[b]	8439.35[b]	+2342.76
NT	601.67[f]	28.02[f]	32.70[e]	4685.80[f]	
NTS	682.00[d]	37.72[b]	36.13[c]	7898.86[c]	+3309.46

Different small letter means $P<0.05$.

Discussion

Effect of Conversion to Subsoiling on CH$_4$ Uptake and N$_2$O Emissions

Long periods of shallow or no-tillage have resulted in an increase in soil bulk density and compacted hardpan in this region, especially in the subsoil [29,30], while subsoiling changed the soil structure, allowing increased gas diffusion in the soil. In this study, soils under HT conversion to HTS, RT conversion to RTS and NT conversion to NTS increased CH$_4$ absorption and strengthened the sink capacity of the soils (Fig. 2 A to C); however, these conversions also promoted the emission of N$_2$O (Fig. 2 D to F). This increase may be due to changes in soil conditions as a result of conversion to tillage (Fig. 5). For example, the increase in CH$_4$ absorption after conversion was mainly correlated with soil temperature, soil moisture, soil pH and SOC content according to the correlation analysis (Fig. 3 and Table 2), which is consistent with some previous studies [31–33]. A higher temperature and greater SOC may be advantageous to increasing the amount of CH$_4$ absorbed by the soil (Table 2, Fig. 3A) [34,35]. However, soil moisture and pH were two limiting factors in our study (Table 2, Fig. 3B) that had negative effects on CH$_4$ absorption in the soils [36].

At the same time, subsoiling would reduce subsoil compaction, and some have found improved permeability of soil to increased soil methane sinks [37] and higher bulk density to limit gas diffusion from the soil to the atmosphere, prolonging methane transfer pathways and thereby reducing CH$_4$ and O$_2$ diffusion between the soil and the atmosphere [38]. Sometimes, although increased soil tillage may slightly decrease CH$_4$ uptake [39], this effect is small and can be largely ignored [6,40].

The conditions for the aeration of the soil profile were reduced after irrigation [41,42] that increases emissions of the greenhouse gas N$_2$O through denitrification in farmland [22], the N$_2$O emission peaks also coincided with higher moisture and NH$_4^+$-N content in this study (Fig. 2 D to F, Table 2, Fig. 4A), the emissions of N$_2$O were significantly affected by soil moisture and NH$_4^+$-N content in each treatment. Some studies have indicated that there

is a significant linear relationship between N$_2$O emissions and soil moisture and nitrogenous fertilizer [21,22]. In addition, there was no significant correlation between N$_2$O emission and soil temperature in this study, and similar results were found by Koponen et al. [43]. In contrast, other studies found that at low temperatures, N$_2$O emissions may be hindered by soil N and water content [44,45]. However, in different experimental sites, N$_2$O emission was often related to increased soil temperature [46,47]. These studies demonstrated that when soil moisture and N fertilization were not limiting factors to N$_2$O emission, the rate of N$_2$O emission increased as soil temperature increased [22].

Similarly, soil pH also influenced N$_2$O production in soil (Fig. 4B). N$_2$ was mainly produced through denitrification when the soil pH was neutral, and the N$_2$O/N$_2$ ratio increased when soil pH decreased [48]. In our study, when soil pH values decreased with irrigation, N$_2$O emissions significantly increased, however, there was no relation to N$_2$O emission in periods of without irrigation, so soil pH does not directly cause soil GHG emissions [36] but via affected the action of microbes [49]. On the other hand, the predominant form of nitrogen is NO$_3$-N or NH$_4$-N after sufficient mixed between soil and straw through tillage, which may produced little N$_2$O in soil, particularly near the soil surface, with an important influence on N$_2$O emissions [12].

Therefore, the CH$_4$ uptake and N$_2$O emissions under HTS, RTS and NTS were higher than those under HT, RT and NT, respectively, due to the effect of subsoiling. Moreover, the emission differences of CH$_4$ and N$_2$O between HTS, RTS and NTS were largely due to the original tillage systems, because they had different background value of soil environment factors, these soil factors change extent after conversion highly affected on CH$_4$ and N$_2$O emissions among treatment in this study. Therefore, the variations in CH$_4$ uptake and N$_2$O emissions correlated with subsoiling are mainly due to alterations in soil conditions resulting from subsoiling, including soil temperature, moisture, NH$_4^+$-N, SOC and pH.

GWP of CH_4 and N_2O after Conversion to Subsoiling

Although there was a negative effect on the GWP of N_2O after conversion to subsoiling, the increased CH_4 absorption by soils partially counteracted this negative effect. The total GWP of CH_4 and N_2O increased slightly compare with the original tillage systems, especially under HTS and RTS (Table 1). Some previous studies reported that no-tillage is a better tillage system at mitigating GHG emissions [6,50], and the lowest GWP of CH_4 and N_2O was only measured under NT in this study. However, the GWP of CH_4 and N_2O would increase if NT was converted to NTS.

Yield Variation after Conversion to Subsoiling

In this study, the fields where the HT, RT and NT methods were previously used showed only slight improvements in wheat grain yields between two years (Table 4), possibly due to the subsoil hardpan. However, under HTS, RTS and NTS, the number of spikes, grains per ear and 1000-grain weight significantly increased, which is in agreement with other reports in which subsoiling was found to be an effective method to increase wheat production [51–53].

Conclusions

Significant variations were measured in CH_4 and N_2O emissions after conversion to subsoiling in the North China Plain. While the uptake of CH_4 improved greatly, N_2O emissions also increased after subsoiling. As a result, we demonstrated that the GWP would increase if converted from minimum or no-tillage to subsoiling, especially from no-tillage. Soil temperature, moisture, SOC, NH_4^+-N and pH also varied and were strongly related to CH_4 uptake and N_2O emissions. In addition, the original tillage systems had an important effect on soil factors and GWP variations after conversion to subsoiling. Therefore, the results of our study provide evidence that conversion from rotary tillage to subsoiling (RTS) or harrow tillage to subsoiling (HTS) had a lower GWP for CH_4 and N_2O compared with conversion from no-tillage to subsoiling (NTS), while the grain yields under both RTS and HTS increase. Therefore, we suggest that these two rotation tillage systems be developed in this region.

Author Contributions

Conceived and designed the experiments: ST TN ZL HH SC. Performed the experiments: ST HZ BW NL. Analyzed the data: ST TN. Contributed reagents/materials/analysis tools: ST TN. Wrote the paper: ST TN.

References

1. Forster P, Ramaswamy V, Artaxo P, Berntsen T, Betts R, et al. (2007) Changes in atmospheric constituents and in radiative forcing. In: Solomon S, Qin D, Manning M, Chen Z, Marquis M, et al., eds. Climate Change 2007: The Physical science basis. Contribution of working group I to the fourth assessment report of the intergovernmental panel on climate change. Cambridge University Press, Cambridge, United Kingdom and New York, NY, USA.
2. Bouwman AF (1990) Exchange of greenhouse gases between terrestrial ecosystems and the atmosphere. In: Bouwman AF, eds. Soils and the Greenhouse Effect. Wiley, Chichester, pp. 61–127.
3. Goulding KWT, Hütsch BW, Webster CP, Willison TW, Powlson DS (1995) The effect of agriculture on methane oxidation in the soil. Philip Transaction Royal Society London A 351: 313–325.
4. IPCC (2001) Climate change 2001, The scientific basis–contribution of work group I to the third assessment report of IPCC. Cambridge University Press, Cambridge.
5. Bruce CB, Albert S, John P, Parker (1999) Fields N_2O, CO_2 and CH_4 fluxes in relation to tillage, compaction and soil quality in Scotland. Soil and Tillage Research 53: 29–39.
6. Six J, Ogle SM, Breidt FJ, Conant RT, Mosier AR, et al. (2004) The potential to mitigate global warming with no-tillage management is only realized when practised in the long term. Global Change Biology 10: 155–160.
7. Lee J, Six J, King AP, Van Kessel C, Rolston DE (2006) Tillage and field scale controls on greenhouse gas emissions. Journal of Environment Quality 35: 714–725.
8. Bhatia A, Sasmal S, Jain N, Pathak H, Kumar R, et al. (2010) Mitigating nitrous oxide emission from soil under conventional and no-tillage in wheat using nitrification inhibitors. Agriculture Ecosystem & Environment 136: 247–253.
9. Zhang HL, Gao WS, Chen F, Zhu WS (2005) Prospects and present situation of conservation tillage. Journal of China Agriculture University 10: 16–20.
10. Chatskikh D, Olesen JE (2007) Soil tillage enhanced CO_2 and N_2O emissions from loamy sand soil under spring barley. Soil and Tillage Research 97: 5–18.
11. Elder JW, Lal R (2008) Tillage effects on gaseous emissions from an intensively farmed organic soil in North Central Ohio. Soil and Tillage Research 98: 45–55.
12. Bai XL, Zhang HL, Chen F, Sun GF, Hu Q, et al. (2010) Tillage effects on CH_4 and N_2O emission from double cropping paddy field. Transactions of the CSAE 26: 282–289.
13. Xu YC, Shen QR, Ran W (2002) Effects of no-tillage and application of manure on soil microbial biomass C, N, and P after sixteen years of cropping. Acta Pedologica Sinica 39: 89–96.
14. Bowen HD (1981) Alleviating mechanical impedance. In: Arkin GF, Taylor HM, eds. Modifying the Root Environment to Reduce Crop Stress. Published by the ASAE. St. Joseph, MI, pp. 21–57.
15. Balbuena HR, Aragon A, McDonagh P, Claverie J, Terminiello A (1998) Effect of three different tillage systems on penetration resistance and bulk density. In: Proceedings of the IV CADIR (Argentine Congress on Agricultural Engineering), vol. 1. pp.197–202.
16. Huang M, Li YJ, Wu JZ, Chen MC, Sun JK (2006) Effects of subsoiling and mulch tillage on soil properties and grain yield of winter wheat. Journal of Henan University Science and Technology 27: 74–77.
17. Qin HL, Gao WS, Ma YC, Ma L, Yin CM (2008) Effects of Subsoiling on Soil Moisture under No-tillage 2 Years Later. Science of Agriculture Sinica 41: 78–85.
18. Bradford MA, Ineson P, Wookey PA (2001) Role of CH_4 oxidation, production and transport in forest soil CH_4 flux. Soil Biology & Biochemistry 33: 1625–163.
19. Watanabe T, Kimura M, Asakawa S (2007) Dynamics of methanogenic archaeal communities based on rRNA analysis and their relation to methanogenic activity in Japanese paddy field soils. Soil Biology & Biochemistry 39: 2877–2887.
20. Zheng XH, Wang MX, Wang Y, She R, Shangguan X, et al. (1997) CH_4 and N_2O emissions from rice paddy fields in Southeast China. Scientia atmospherica Sinica 21: 231–237.
21. Merino P, Artetxe A, Castellon A, Menendez S, Aizpurua A, et al. (2012) Warming potential of N_2O emissions from rapeseed crop in Northern Spain. Soil & Tillage Research, 123: 29–34.
22. Gregorich EG, Rochette P, Vandenbygart AJ, Angers DA (2005) Greenhouse gas contributions of agricultural soils and potential mitigation practices in Eastern Canada. Soil and Tillage Research 83: 53–72.
23. Clayton H, Arah JRM, Smith KA (1994) Measurement of nitrous oxide emissions from fertilised grassland using closed chambers. Journal of Geophysical Research 99: 16599–16607.
24. Paustian K, Andren O, Janzen HH, Lal R, Smith P, et al. (1997) Agricultural soil as a C sink to offset CO_2 emissions. Soil Use and Management 13: 230–244.
25. Robertson G (1993) Fluxes of nitrous oxide and other nitrogen trace gases from intensively managed landscapes: a global perspective. In: Harpwr LA, Mosier AR, Duxbury JM, Rolston DE, (eds) Agricultural ecosystem effects on trace gases and global climate change. ASA Special Publication No. 55. ASA, CSSA, SSSA, Madison, wi 95–108.
26. Tian SZ, Ning TY, Chi SY, Wang Y, Wang BW, et al. (2012) Diurnal variations of the greenhouse gases emission and their optimal observation duration under different tillage systems. Acta. Ecol. Sinica. 32, 879–888.
27. IPCC (2007) Climate change 2007: The physical science basis. Contribution of working group I to the fourth assessment report of the intergovernmental panel on climate change. Cambridge University Press, Cambridge, United Kingdom and New York, NY, USA.
28. Bao SD (2000) Soil and Agricultural Chemistry Analysis. China Agriculture Press, Beijing.
29. Han B, Li ZJ, Wang Y, Ning TY, Zheng YH, et al. (2007) Effects of soil tillage and returning straw to soil on wheat growth status and yield. Transactions of the CSAE 23: 48–53.
30. Ahmad S, Li C, Dai G, Zhan M, Wang J, et al. (2009) Greenhouse gas emission from direct seeding paddy field under different rice tillage systems in central China. Soil and Tillage Research 106: 54–61.
31. Qi YC, Dong YS, Zhang S (2002) Methane fluxes of typical agricultural soil in the north china plain. Rural Ecology Environment 18: 56–60.
32. Wu FL, Zhang HL, Li L, Chen F, Huang FQ, et al. (2008) Characteristics of CH_4 Emission and Greenhouse Effects in Double Paddy Soil with Conservation Tillage. Science Agriculture Sinica 419: 2703–2709.
33. Dijkstra FA, Morgan JA, von Fischer JC, Follett RF (2011) Elevated CO_2 and warming effects on CH_4 uptake in a semiarid grassland below optimum soil moisture. Journal of Geophysical Research 116: 1–9.

34. Wang ZP, Han XG, Li LH (2005) Methane emission from small wetlands and implications for semiarid region budgets. Journal of Geophysical Research 110(D13): Art. No. D13304.

35. Bayer CL, Gomes J, Vieira FCB, Zanatta JA, Piccolo MC, et al. (2012) Methane emission from soil under long-term no-till cropping systems. Soil & Tillage Research, 124: 1–7.

36. Ouyang XJ, Zhou GY, Huang ZL, Peng SJ, Liu JX, et al. (2005) The incubation experiment studies on the influence of soil acidification on greenhouse gases emission. China Environment Science 25: 465–470.

37. Dong YH, Ou YZ (2005) Effects of organic manures on CO_2 and CH_4 fluxes of farmland. Chinese Journal of Applied Ecology 16: 1303–1307.

38. Ball BC, Scott A, Parker JP (1999) Field N_2O, CO_2 and CH_4 fluxes in relation to tillage, compaction and soil quality in Scotland. Soil and Tillage Research 53: 29–39.

39. Hütsch BW (1998) Tillage and land use effects on methane oxidation rates and their vertical profiles in soil. Biology and Fertilizer of Soils 27: 284–292.

40. Robertson GP, Paul EA, Harwood RR (2000) Greenhouse gases in intensive agriculture: Contributions of individual gases to the radiative forcing of the atmosphere. Science 289: 1922–1925.

41. Czyz EA (2004) Effects of traffic on soil aeration, bulk density and growth of spring barley. Soil & Tillage Research. 79, 153–166.

42. Berisso FE, Schjønning P, Keller T, Lamande M, Etana A, et al. (2012) Persistent effects of subsoil compaction on pore size distribution and gas transport in a loamy soil. Soil & Tillage Research 122: 41–45.

43. Koponen HT, Flojt L, Martikainen PJ (2004) Nitrous oxide emissions from agricultural soils at low temperatures: a laboratory microcosm study. Soil Biology & Biochemistry 36: 757–766.

44. Conen F, Dobbie KE, Smith KA (2000) Predicting N_2O emissions from agricultural land through related soil parameters. Global Change Biology 6: 417–426.

45. Sehy U, Ruser R, Munch J C (2003) Nitrous oxide fluxes from maize fields: relationship to yield, site-specific fertilization, and soil conditions. Agriculture Ecosystem & Environment 99: 97–111.

46. Groffman PM, Hardy JP, Driscoll CT, Fahey TJ (2006) Snow depth, soil freezing, and fluxes of carbon dioxide, nitrous oxide and methane in a northern hardwood forest. Global Change Biology 12: 1748–1760.

47. Rachhpal S, Jassal T, Andrew B, Real R, Gilbert E (2011) Effect of nitrogen fertilization on soil CH_4 and N_2O fluxes, and soil and bole respiration. Geoderma 162: 182–186.

48. Daum N, Schenk MK (1998) Influence of nutrient solution pH on N_2O and N_2 emissions from a soilless culture system. Plant and Soil 203: 279–287.

49. Robertson LA, Kuenen JG (1991) Physiology of nitrifying and denitrifying bacteria. In: Rogers JE and Whitman WBC, (eds) Microbial production and consumption of greenhouse gases: Methane, Nitrogan oxides and Halo methane. American Society for microbiology Washington D. C., 189–199.

50. Lal R (2004b) Soil carbon sequestration impacts on global climate change and food security. Science 304: 1623–1627.

51. He J, Li HW, Gao HW (2006) Subsoiling effect and economic benefit under conservation tillage mode in Northern China. Transactions of the CSAE 22: 62–67.

52. Gong XJ, Qian CR, Yu Y, Zhao Y, Jiang YB, et al. (2009) Effects of Subsoiling and No-tillage on Soil Physical Characters and Corn Yield. Journal of Maize Science 17: 134–137.

53. Huang M, Wu JZ, Li YJ, Yao YQ, Zhang CJ, et al. (2009) Effects of different tillage management on production and yield of winter wheat in dryland. Transactions of the CSAE 25: 50–54.

3

Nitrogen Addition Regulates Soil Nematode Community Composition through Ammonium Suppression

Cunzheng Wei[1,2,3]*, **Huifen Zheng**[1,3], **Qi Li**[2], **Xiaotao Lü**[2], **Qiang Yu**[2], **Haiyang Zhang**[2], **Quansheng Chen**[1], **Nianpeng He**[4], **Paul Kardol**[5], **Wenju Liang**[2], **Xingguo Han**[1,2]*

1 State Key Laboratory of Vegetation and Environmental Change, Institute of Botany, Chinese Academy of Sciences, Beijing, China, **2** State Key Laboratory of Forest and Soil Ecology, Institute of Applied Ecology, Chinese Academy of Sciences, Shenyang, China, **3** Graduate University of Chinese Academy of Sciences, Beijing, China, **4** Institute of Geographic Sciences and Natural Resources Research, Chinese Academy of Sciences, Beijing, China, **5** Department of Forest Ecology and Management, Swedish University of Agricultural Sciences, Umeå, Sweden

Abstract

Nitrogen (N) enrichment resulting from anthropogenic activities has greatly changed the composition and functioning of soil communities. Nematodes are one of the most abundant and diverse groups of soil organisms, and they occupy key trophic positions in the soil detritus food web. Nematodes have therefore been proposed as useful indicators for shifts in soil ecosystem functioning under N enrichment. Here, we monitored temporal dynamics of the soil nematode community using a multi-level N addition experiment in an Inner Mongolia grassland. Measurements were made three years after the start of the experiment. We used structural equation modeling (SEM) to explore the mechanisms regulating nematode responses to N enrichment. Across the N enrichment gradient, significant reductions in total nematode abundance, diversity (H' and taxonomic richness), maturity index (MI), and the abundance of root herbivores, fungivores and omnivores-predators were found in August. Root herbivores recovered in September, contributing to the temporal variation of total nematode abundance across the N gradient. Bacterivores showed a hump-shaped relationship with N addition rate, both in August and September. Ammonium concentration was negatively correlated with the abundance of total and herbivorous nematodes in August, but not in September. Ammonium suppression explained 61% of the variation in nematode richness and 43% of the variation in nematode trophic group composition. Ammonium toxicity may occur when herbivorous nematodes feed on root fluid, providing a possible explanation for the negative relationship between herbivorous nematodes and ammonium concentration in August. We found a significantly positive relationship between fungivores and fungal phospholipid fatty acids (PLFA), suggesting bottom-up control of fungivores. No such relationship was found between bacterivorous nematodes and bacterial PLFA. Our findings contribute to the understanding of effects of N enrichment in semiarid grassland on soil nematode trophic groups, and the cascading effects in the detrital soil food web.

Editor: Lynn Carta, USDA-ARS BARC-W, United States of America

Funding: This work was supported by the Key Project of National Natural Science Foundation of China (30830026,http://www.nsfc.gov.cn/), the State Key Basic Research Development Program (2007CB106801,http://www.973.gov.cn/Default_3.aspx). The funders had no role in study design, data collection and analysis, decision to publish, or preparation of the manuscript.

Competing Interests: The authors have declared that no competing interests exist.

* E-mail: weicunzheng@ibcas.ac.cn (CZW); xghan@ibcas.ac.cn (XGH)

Introduction

Widespread nitrogen (N) enrichment resulting from anthropogenic activities such as N deposition and fertilization has greatly changed ecosystem processes, structure, and functioning [1,2,3]. Although low-level N addition generally promotes ecosystem functioning, N saturation has been reported to induce forest dieback, soil acidification and inhibit soil biota [4,5,6,7]. It is therefore important to improve our understanding of the dose-response relationship between N enrichment and soil ecosystem functioning.

Soil nematodes are wide-spread, abundant and highly diverse, both taxonomically and functionally [8,9], occupying multiple trophic positions in the soil food web, including root herbivores, bacterivores, fungivores, as well as omnivores and predators [10,11,12]. Changes in nematode community composition are widely acknowledged as indicative of shifts in environmental conditions [13,14]. Moreover, soil nematodes can contribute up to 40% of nutrient mineralization by feeding on microbial populations [15,16]. Thus, understanding the response and underlying mechanisms of soil nematodes to N enrichment will contribute to elucidate potential cascade effects in the soil food web.

Previous studies have documented inhibition of nematodes after N addition [7,17]. Generally, N addition decreased total nematode abundance and diversity, but responses varied among trophic groups. Reduced numbers of root herbivores, fungivores and omnivores -predators, but increased numbers of some opportunistic bacterivores in response to N addition have been shown for forests [18–20], grasslands [17,18,21] and croplands [7,22]. Moreover, responses of soil nematodes to N addition often vary with time after application. For example, in a long-term fertilization experiment in northern China, Liang et al. [7] found that N fertilization initially decreased the abundance of root-feeding nematodes, but increased their abundance the following season.

Soil acidification following N addition has been proposed as one of the important factors inhibiting soil nematode abundance after N addition [21,23]. In other studies, NO_3^--N and NH_4^+-N concentrations were found negatively correlated with root herbivores and fungivores [7,24], suggesting direct effects of N addition on soil nematodes. Importantly, soil nematodes may not only be influenced by changes in physicochemical soil conditions, but also indirectly by shifts in plant community composition [25]. Altered plant community composition and widespread species loss after N addition has been observed across the world [26,27]. As nematodes are heterotrophs they are ultimately dependent on autotrophs, such as higher plants, for their resources (i.e. root exudates and litter input) [28]. Plant species diversity and composition likely affect the composition of soil nematodes through the complementarity in resource quality of the component plant species rather than to an increase in total resource quantity [25]. Understanding how N addition affects taxonomic richness and composition of soil nematodes is challenging because of the complex interactions between the direct effects of N addition on soil biota, and indirect effects mediated by altered plant community composition [29].

We used a 4-year multi-level N addition experiment in a typical Inner Mongolia grassland in Northern China to investigate the soil nematode community, the plant community [27], soil biodiversity [21] and key soil physicochemical factors which have been documented to be significantly affected by N addition. We tested the general hypothesis that N enrichment alters soil abiotic properties and plant community composition and species richness which in turn affects nematode community composition and taxon richness. Nematodes were sampled twice during the growing season to address two primary questions. First, how does N enrichment affect temporal fluctuations in nematode abundance and community composition across an N addition gradient? Second, in effects of N enrichment on soil nematode communities, what is the relative importance of direct effects through changes in ammonium and nitrate concentrations, and indirect effects through changes in soil pH and shifts in plant community composition? The causal relationships between components of the plant-soil-nematode system in response to N addition were evaluated using structural equation modeling.

Materials and Methods

No specific permits were required for the described field studies. No specific permissions were required for these locations/ activities. The location is not privately owned or protected in any way and the field studies did not involve endangered or protected species.

Experimental design

The field experiment was conducted at the Inner Mongolia Grassland Ecosystem Research Station (IMGERS), which is located in the Xilin River Basin, Inner Mongolia Autonomous Region of China (116°42′E, 43°38′N). Topographic relief exhibits little variation, with elevation ranging from 1250 to 1260 m at our experimental site. The region has a semi-arid continental climate with a mean annual temperature of 0.7°C and a mean annual precipitation of 346.1 mm with 60–80% of the precipitation falling during the growing season from May to August. The soil is classified as dark chestnut (Calcic Chernozem according to ISSS Working Group RB, 1998) or loamy sand in terms of texture. Previous experiments have documented that N limitation constrains primary production and regulates plant community composition [27]. The vegetation at the experimental site is

classified as typical steppe grassland [30]. The perennial rhizomatous grass *Leymus chinensis* (Trin.) Tsvelev and the perennial bunchgrass *Stipa grandis* P. Smirn are the dominant plant species. Our N manipulation experiment was conducted in *Leymus chinensis* grassland that had been fenced since 1979 to prevent grazing by large animals.

In 2006, 42 8 m×8 m plots were laid out in a randomized block design. Plots were separated by 1-meter walkways. There were seven treatments with six replicates each, including Control (no nutrient addition), and 6 levels of N addition (0, 0.4, 0.8, 1.6, 2.8, 4.0 mol N m^{-2} yr^{-1}; N was added as urea). Hereafter, treatments will be referred to as: Control, N_0, $N_{0.4}$, $N_{0.8}$, $N_{1.6}$, $N_{2.8}$, and $N_{4.0}$. Each plot, except for Control, also received 0.05 mol P m^{-2} (added as KH_2PO_4) to ensure that N was the only limiting nutrient. The fertilizer was thoroughly mixed with sand and then applied to the plot surfaces as a single dose at the beginning of the growing season every year from 2006 to 2009 [3,27,31].

Plant sampling and analysis

The aboveground vegetation was sampled each year in mid-August by clipping all plants at the soil surface using a 0.5 m×1 m quadrat randomly placed in each plot with the restriction of no spatial overlap of quadrats across years. All living vascular plants were sorted to species, and all plant materials, including litter and standing dead biomass, were oven-dried at 65°C for 48 h and then weighted. The dry mass of all living plants per quadrat averaged over the six replicates for each treatment was used to estimate aboveground biomass. We classified the plants into five functional groups (PFGs) based on life forms, including perennial rhizome grass (PR), perennial bunchgrasses (PB), perennial forbs (PF), shrubs and semi-shrubs (SS), and annuals (AS), as in Bai et al. [30]. Two soil cores with a diameter of 6.5 cm were collected at 0–15 cm depth to determine root biomass from each plot after plant biomass harvest. Root samples were immediately placed in a cooler and transported to the laboratory. In the laboratory, root samples were soaked in deionized water and cleaned from soil residuals using a 0.5-mm sieve. Root biomass was then dried and weighed as described above.

Soil sampling and analysis

In 2009, three years after the start of the experiment, from each plot soil samples were first collected on June 10 (when the rain season started and urea dissolved), and then on August 18 and September 19. From each plot, three soil cores (3 cm diameter, 0–15 cm depth) were collected. Per plot, the three soil cores were bulked comprising one soil sample per plot. Bulked samples were placed in a plastic bag and then immediately transported to the laboratory and stored at 4°C. Soil moisture (SM) was determined as mass loss after drying the soils at 105°C for 24 h. Part of the air-dried and sieved samples was prepared for measurement of pH and total organic carbon. For measurement of soil inorganic N, soil samples were extracted using 2 mol L^{-1} KCl. The concentrations of nitrate N (NO_3^--N) and ammonium N (NH_4^+-N) in the filtrates were determined using a flow injection auto analyzer (FIAstar 5000 Analyzer; Foss Tecator, Hillerød, Denmark). The concentrations of soil inorganic N were calculated based on dry soil weight.

Extraction and identification of soil nematodes

Soil nematodes were extracted from 100 g subsamples collected in August and September 2009 using a modified cotton-wool filter method [32]. After counting the total number of nematodes, the first 100 individuals encountered were identified to genus level using an inverted compound microscope. If the total number of nematodes was less than 100, all nematodes were identified. Nematode populations are expressed as number of nematodes per 100 g dry soil. The nematodes were allocated to the following trophic groups: root herbivores, bacterivores, fungivores, and omnivores–predators [10].

Data analysis

August and September physicochemical soil properties used in the SEM model and linear regression analyses (see below) were calculated as the mean of July and August, and the mean of August and September, respectively.

The following nematode community indices were calculated: 1) Nematode taxon richness (S), i.e., the number of genera identified [25]; 2) The Shannon diversity index $H' = -\sum p_i(\ln p_i)$, where p_i is the proportion of individuals belonging to the ith taxon relative to the total number of individuals in the sample; and 3) The maturity index $MI = \Sigma v(i)f(i)$, where $v(i)$ is the c-p value of taxon i indicating their r and K strategies [33], and $f(i)$ is the frequency of taxon i in a sample. Nematode abundance and ecological indices were $\ln(x+1)$-transformed prior to statistical analysis. The main and interactive

effects of N addition and sampling time (August, September) on soil nematode abundance and ecological indices were determined using two-way analysis of variance (ANOVA). Polynomial and exponential decay regressions were used to test relationships of nematode abundance and diversity indices with N addition rate; results from the best-fitting regression models were presented.

Structural equation modeling (SEM) was used to gain a mechanistic understanding of how soil properties and altered plant community composition mediate effects of N enrichment on soil nematode communities. SEM is based on a simultaneous solution procedure, where the residual effects of predictors are estimated (partial regressions) once common causes from inter-correlations have been statistically controlled for [34–36]. Prior to the SEM procedure we reduced the number of variables for plant and nematode community composition through principal component analyses (PCA) [35]. Plant functional group biomass and abundance of nematode trophic groups were used as raw data for the PCAs. The first principal components (PC) were used in the subsequent SEM analysis to represent nematode community composition (PC1 explained 76.7% of the variation, Fig. S1), and

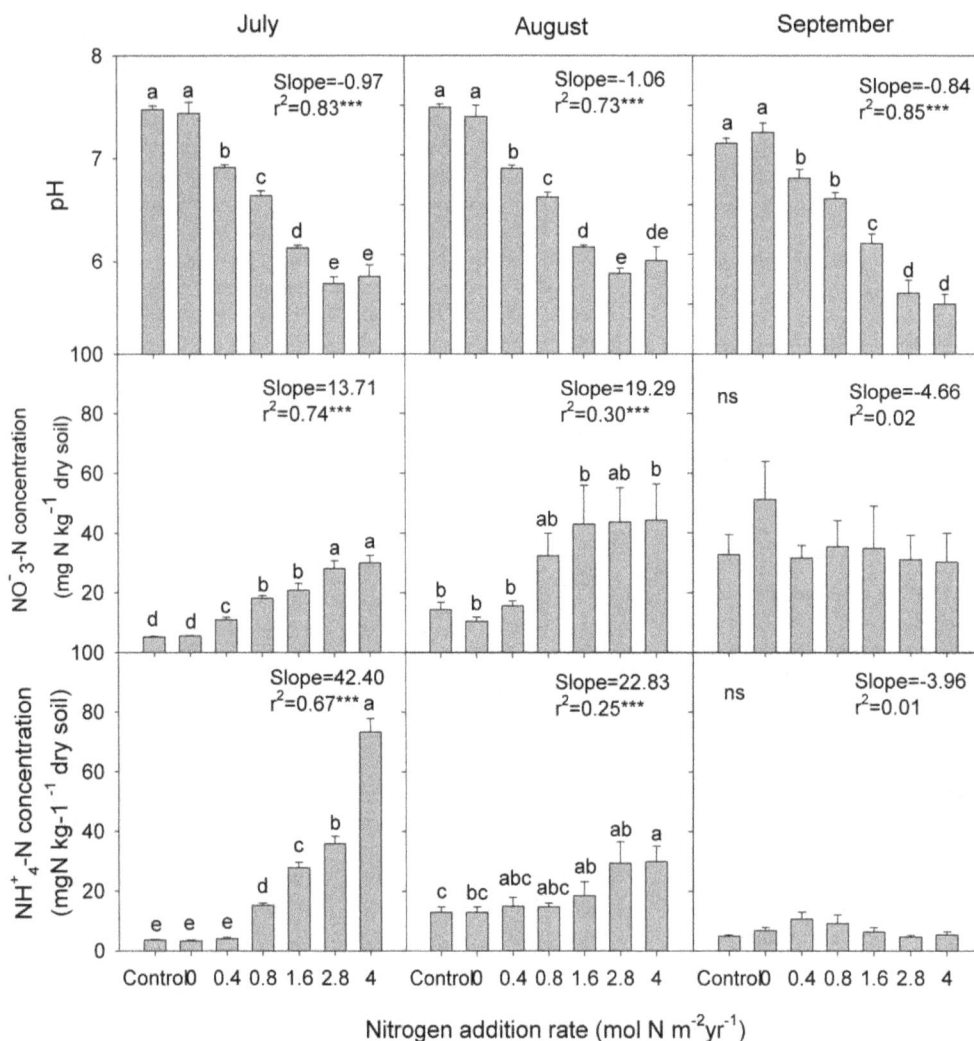

Figure 1. Soil ammonium (NH_4^+), nitrate (NO_3^-) concentration, and pH for Control and N addition treatments in July, August and September 2009. Data are mean ± S.E. (N = 6). Regression parameters were estimated using log-linear models with N treatment as a continuous predictor. Significant regressions are reported as $P<0.05*$, $P<0.01**$, $P<0.001***$.

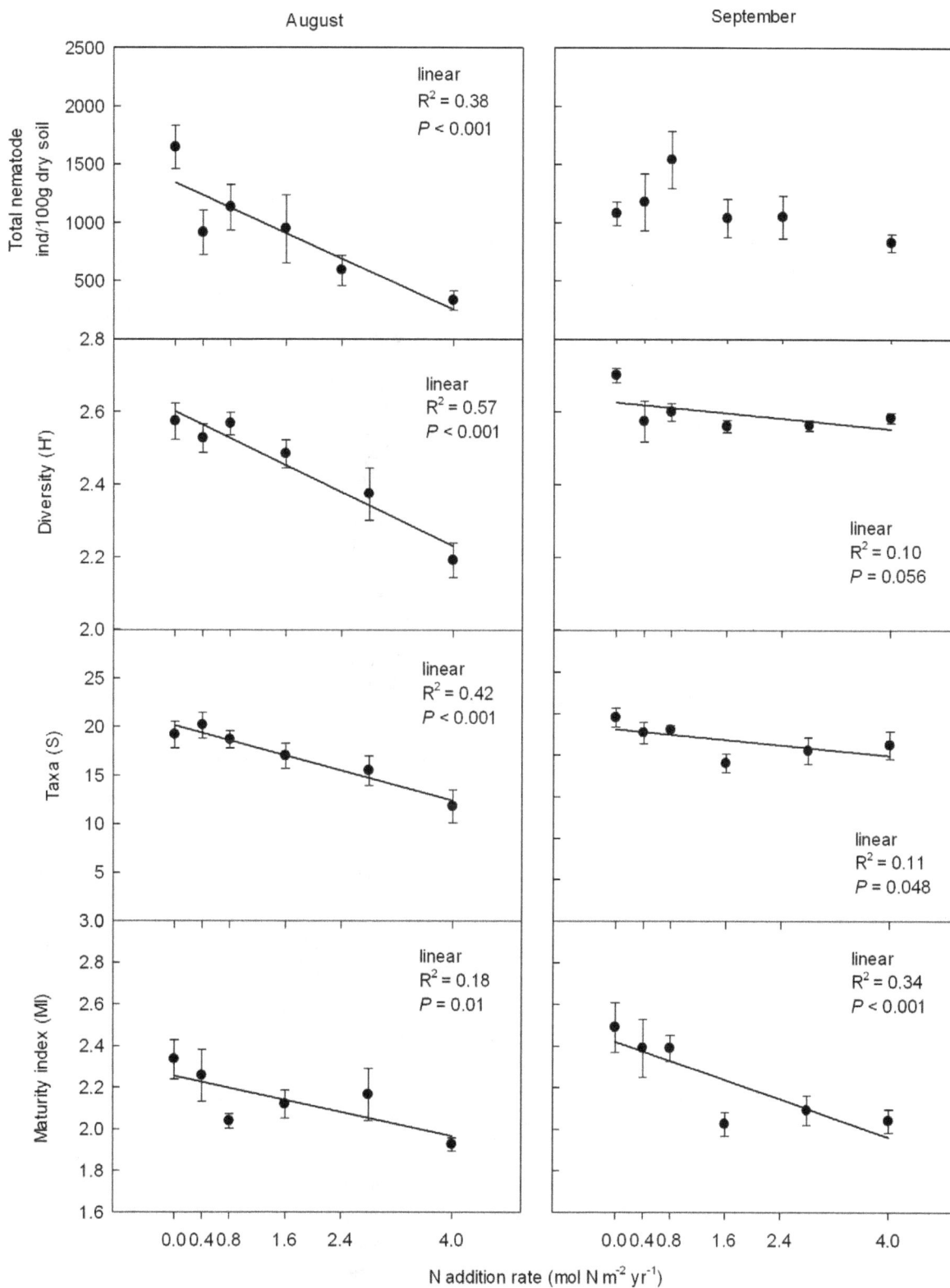

Figure 2. Response of soil nematode abundance, diversity (H'), taxon richness (S), and maturity index (MI) to N addition in August and September 2009. Data are means ± S.E. Linear regressions best fitted the data; regression coefficients are reported if $P<0.05$.

plant community composition (PC1 explained 87.5% of the variation, Fig. S2). We started the SEM procedure with the specification of a conceptual model of hypothetical relationships

based on *a priori* and theoretical knowledge (Fig. S3). In this model, we assumed that grazing alters soil abiotic properties and plant community composition and species richness which in turn affect nematode community composition and taxon richness. In the SEM analysis we compared the model-implied variance-covariance matrix against observed variance-covariance matric. Data were fitted to the models using the maximum likelihood estimation method. The χ^2 goodness-of-fit statistic and its associated P value, Root Mean Square Error of Approximation (RMSEA) and Akaike Information Criterion (AIC) were used to judge the model fit to the data. A large P value and a low RMSEA value indicates that the covariance structure of the data does not differ significantly from what would be expected based on the model.

All univariate analyses were performed using SPSS 17.0 (SPSS, Chicago, IL). PCA was performed using Program CANOCO Version 4.5 (Plant Research International, Wageningen, The Netherlands). SEM analyses were performed using AMOS 7.0 (Amos Development, Spring House, Pennsylvania, USA).

Results

Soil N and plant responses to N addition

Nitrogen addition greatly increased nitrate and ammonium concentrations in July (Fig. 1). However, effects of N addition on nitrate and ammonium concentrations decreased during the growing season. In August, nitrate and ammonium concentrations were only somewhat increased at the highest N addition levels. In September, nitrate and ammonium concentrations did not significantly differ between Control and N addition treatments. Across the growing season, pH was significantly decreased with increasing levels of N addition (Fig. 1).

Nitrogen addition significantly reduced plant species richness from 7 at N_0 to 4 at N_4 plots ($F_{6,35} = 4.95$, $P<0.001$; Table S1). Plant shoot and root biomass were not significantly affected by N addition ($F_{6,35} = 1.34$, $P = 0.27$; $F_{6,35} = 1.57$, $P = 0.19$ respectively; Table S1). Plant functional groups varied in their response to N addition. For example, shoot biomass of perennial bunchgrasses (PB) increased at low levels of N addition, but decreased at high levels of N addition, with a threshold at 0.4 mol N m^{-2} y^{-1} ($F_{6,35} = 2.314$, $P = 0.074$; Table S1). In contrast, shoot biomass of perennial rhizome grasses (PR) showed the exact opposite pattern, although the effects of N additions were not statistically significant ($F_{6,35} = 1.633$, $P = 0.167$; Table S1). No significant responses to N addition were found for perennial forbs (PF) and for (semi-)shrubs (SS) ($F_{6,35} = 0.954$, $P = 0.470$; $F_{6,35} = 1.048$, $P = 0.412$, respectively; Table S1).

Nematode responses to N addition

Thirty nine and fifty nematode genera were identified in August and September respectively (Table S3, Table S4). Effects of N addition on nematode community diversity (H') strongly changed during the growing season, as indicated by the significant treatment×time interaction effect (Table 1). Nematode abundance, diversity (H' and S) and MI all linearly decreased with the level of N addition in August (Fig. 2). In contrast, no effect of N addition on nematode abundance was found in September (Table 1; Fig. 2). Nematode community diversity (H') and taxon richness (S) both decreased with N addition ($P = 0.054$ and 0.048, respectively) in September, but the effects were much weaker than in August (Fig. 2). Both in August and September the MI decreased linearly with N addition, while on average the MI was higher in September than in August (Fig. 2).

Nematode trophic groups showed a diverse response to the N addition treatments. N addition negatively affected the abundance of fungivores; their numbers linearly decreased with N addition. The abundance of fungivores was on average higher in August than in September (Table 1; Fig. 3). N addition also negatively affected the abundance of omnivores-predators, but the effect of N was stronger in August than in September. Across treatments, omnivores-predators were on average more abundant in September than in August (Table 1; Fig. 3). Root herbivores were strongly negatively affected by N addition in August, but not in September, resulting in a significant treatment×time interaction (Table 1). Analysis of variance did not reveal an overall effect of N addition on the abundance of bacterivores (Table 1; Fig. 3). Regression models showed that bacterivorous nematodes had a hump-shaped relationship with the level of N addition, but without a critical threshold (Fig. 3).

Direct and indirect effects of N addition on soil nematode communities

The final SEM model adequately fitted the data describing effects of N addition on the plant-soil-nematode system ($\chi^2_9 = 8.286$, $P = 0.507$, AIC = 62.268, RMSEA<0.001; Standardized path coefficients are given in Fig. 4). Nitrogen addition explained 82%, 32% and 86% of the variation in soil pH, nitrate concentration, and ammonium concentration, respectively. Ammonium concentration (Fig. 4) had a negative relationship with nematode composition and richness ($P<0.01$, Table S2). Ammonium suppression explained 61% of the variation in nematode richness and 43% of the variation in nematode community composition. However, we did not find any significant effects of

Table 1. Results from two-way analysis of variance for total and nematode feeding group abundance and nematode ecological indices using N treatment (N), sampling time (T) and their interaction as fixed factors.

Dependent variable	Error df	Treatment (6df)	Time (1df)	N*T (6df)
Total nematode abundance	84	**4.15 (0.001)**	3.50 (0.066)	1.76 (0.119)
Nematode taxa (S)	84	**6.25 (<0.001)**	**43.30 (<0.001)**	1.61 (0.157)
Diversity (H')	84	**10.06 (<0.001)**	**62.19 (<0.001)**	**5.87 (<0.001)**
Maturity index (MI)	84	**6.96 (<0.001)**	**6.16 (0.015)**	1.64 (0.149)
Root herbivores	84	**2.25 (0.048)**	**4.31 (0.042)**	**2.77 (0.018)**
Fungivores	84	**8.63 (<0.001)**	3.74 (0.057)	1.74 (0.125)
Bacterivores	84	1.42 (0.218)	**54.36(<0.001)**	0.30 (0.936)
Omnivores-predators	84	**4.895 (0.001)**	0.16 (0.687)	1.98 (0.080)

Shown are F-values with significance levels in parentheses. Significant effects (P<0.05) are indicated in bold and marginally significant effects are indicated in italic.

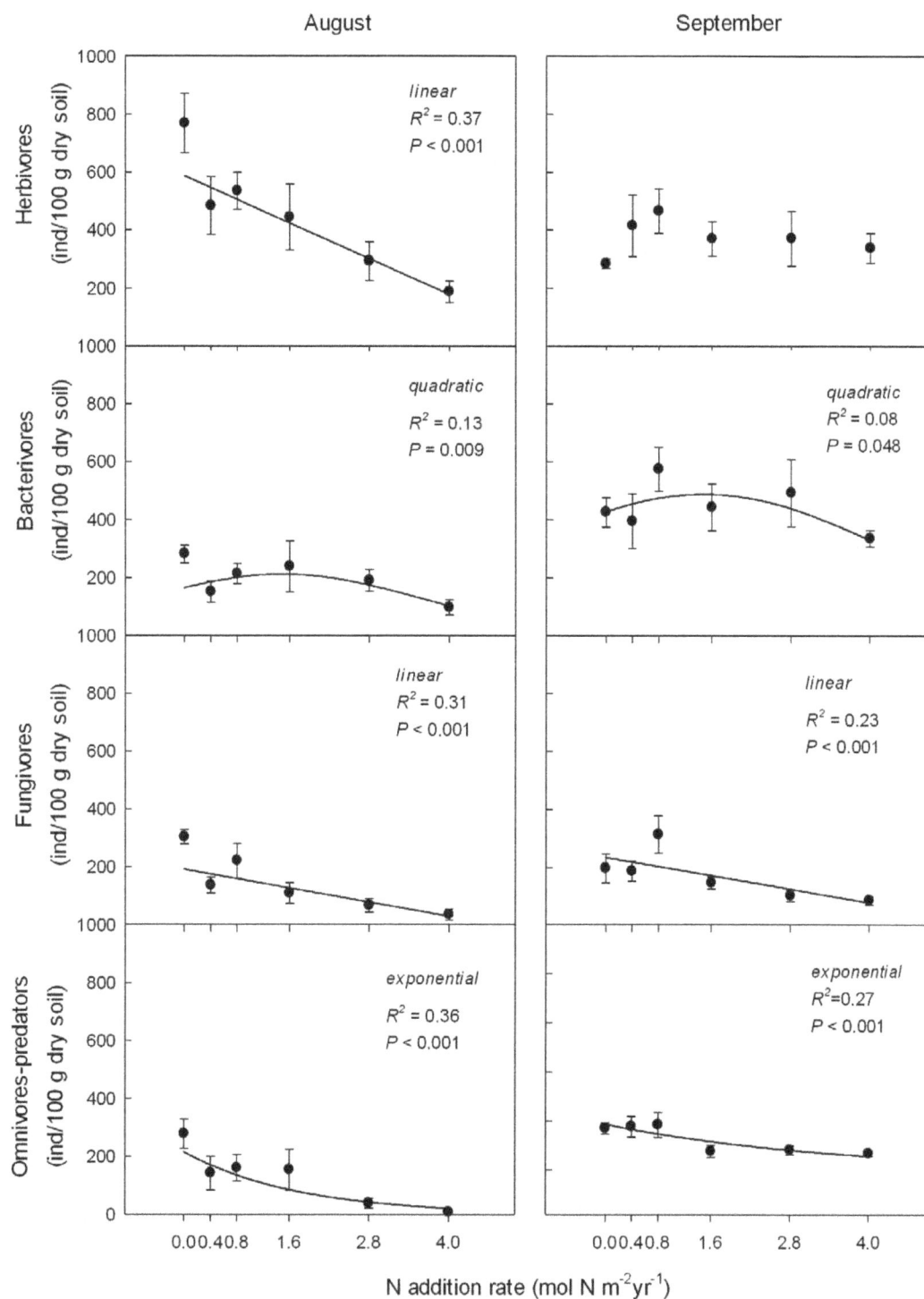

Figure 3. Response of nematode tropic groups to N addition in August and September 2009. Data are means ± S.E.. R^2 values were determined using polynomial and exponential decay regressions. Regression coefficients of the best-fitting models are reported if $P<0.05$.

altered plant community composition on nematode richness and community composition. Also, effects of soil pH and nitrate concentration on nematode richness and community composition were not significant (Fig. 4; Table S2). A further regression model showed that ammonium concentration had a strong negative relationship with abundance of herbivorouos nematodes in August, but not in September (Fig. 5).

Discussion

Ammonium suppression of herbivorous nematodes after N addition

We observed a decrease in total soil nematode abundance and diversity after N addition, which was consistent with previous findings in a nearby NH_4NO_3 addition experiment [21].

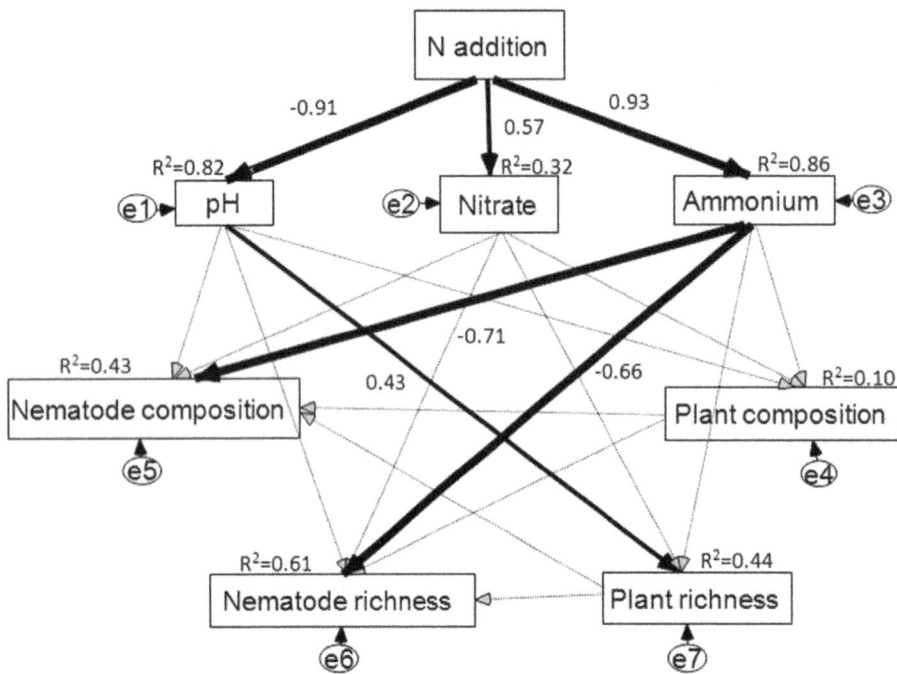

Figure 4. Structural equation model of N addition on soil physicochemical factors, plant community composition and richness, nematode community composition and richness (August data was used in SEM; ammonium, nitrate concentration was calculated as the average of July and August). The model fit the data well: $\chi^2_9 = 8.286$, $P = 0.507$, AIC $= 62.268$, RMSEA < 0.001. Numbers at arrows are standardized path coefficients (equivalent to correlation coefficients). Width of the arrows indicates the strength of the causal influences: black arrows indicate significant standardized path coefficients ($P < 0.05$). Circles indicate error terms (e1–e7). Percentages close to endogenous variables indicate the variance explained by the model (R^2).

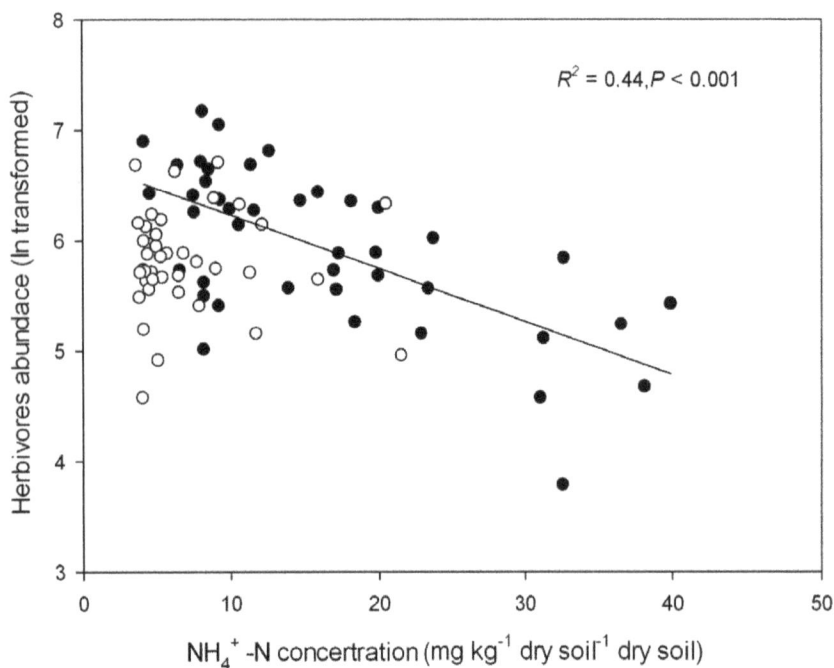

Figure 5. Relationships between soil ammonium (NH_4^+) concentration and abundance of herbivorous nematodes in August and September 2009 (close circles: August, open circles: September; ammonium concentration was calculated as mean value of July and August, August and September respectively). In August, abundance of herbivorous nematodes correlated negatively with NH_4^+-concentration.

Herbivorous nematodes were the primary contributors to the temporal dynamics of nematode community structure across the N addition gradient (Fig. 2, Fig. 3). Persson et al. (1980) found a strong reduction of the obligatory root-feeder *Geocenamus arcticus* in responses to N fertilization in a Scots Pine forest, even though root production increased [37]. In our experiment, plant root biomass did not change across the N gradient (Table S1). So, parasite-host relationships could not explain the decrease of herbivorous nematode abundance following N addition.

Structural equation modeling showed that ammonium concentration alone significantly contributed to the shift in nematode community composition and diversity (Fig. 4). Further analysis showed a negative relationship between ammonium concentration and herbivorous nematodes along the N addition gradient in August (Fig. 5). Ammonium suppression by fertilizers of plant parasitic nematodes has been documented before and ammonium has been used as a nematicide to control root-knot nematodes [24,38]. In a recent paper, Oka et al. (2007) found that ammonium and ammonia toxicity after soil solarization could be used to control root-knot nematodes in organic farming systems [39]. Cumulative ammonium concentrations were observed initially after urea addition (Fig. 1). Ammonium can be taken up by plants and be nitrified by nitrobacteria under certain conditions [40]. Temporal depletion of ammonium was also observed in August and September (Fig. 4). Alleviation of ammonium suppression could explain the temporal recovery of abundance of herbivorous nematodes, which also has been found in long-term N fertilization of maize crops [7].

Responses of non-herbivorous nematodes to N addition

Ammonium has been documented toxic to a wide range of organisms [41]. However, ammonium toxicity could not well explain the responses of microbial-feeding and omnivorous-predatory nematodes to N addition in our experiment. These contrasting responses might be due to their disparate feeding habits. Plants can take up ammonium directly from soil and accumulate it in their roots, stems and leaves [42]. Herbivorous nematodes parasitize plant roots by puncturing plant cells with their stylet, and ingestion of ammonium-rich fluid might cause ammonium toxicity [10]. In contrast, other nematode trophic groups either feed on microorganisms or other microfauna in which case ammonium might already have been transformed to non-toxic compounds.

Bottom up control of fungivores by fungi was evident from the significant negative relationship between fungal PLFA and abundance of fungivores (Fig. S4). Cascading effects are suggested to be a universal mechanism regulating soil food web structure [29,43] and our experiment provides further evidence. However, bacterivores did not show a relationship with bacterial PLFA, which suggest other controlling mechanisms, such as top down controls by their predaceous soil organisms [43–45].

Indirect effects of altered plant community composition on soil nematodes after N addition

In contrast to Bardgett et al. [46] who showed that in a short-term pot experiment abundance and activity of soil organisms were regulated more by plant species traits than by the direct effects of N addition, we did not find evidence for indirect effects of N addition on soil nematode communities through shifts in plant community composition (Fig. 3). Bottom-up effects from the plant community to soil fauna have long been documented, as plants fix carbon, which is often a limiting resource for soil biota [47]. However, nitrogen saturation resulting from high levels of N addition can strongly deteriorate soil properties (e.g., acidification and ammonium toxicity as shown in our experiment) [48], which may outweigh plant effects on soil biota and soil processes. Considering the high levels of N addition in our experiment, soil properties could therefore probably be expected to be more important than shifts in plant community composition in regulating soil nematode communities (Wei et al. unpublished). Similarly, previous microcosm studies have shown that direct effects of N enrichment on litter decomposition and ecosystem functioning can be stronger than indirect effects through changes in the plant community [49,50]. However, it should be noted that N-induced shifts in plant community composition may simply take longer to exert effects on the soil ecosystem than direct effect of N addition; hence, longer-term studies are needed to further evaluate the relative importance of direct and indirect effects of N addition.

Supporting Information

Figure S1 Species-sample bi-plot of principal component analysis (PCA) of soil nematode community composition. Bac = bacterivores abundance, Fun = fungivores abundance, Her = root herbivores abundance, OP = omnivores-predators abundance. Percentages along the axes correspond to the amount of explained variability in functional group composition.

Figure S2 Species-sample bi-plot of principal component analysis (PCA) of plant community composition. TolBio = Total aboveground biomass, PR = perennial rhizome grass, PB = perennial bunchgrasses, PF = perennial forbs, SS = shrubs and semi-shrubs. Percentages along the axes correspond to the amount of explained variability in functional group composition.

Figure S3 Conceptual model of hypothetical interaction pathways in the studied plant-soil-nematode system.

Figure S4 Relationships between bacterial phospholipid fatty acids (PLFA) and abundance of bacterivorous nematodes, fungal PLFA and fungivorous nematodes. Data are ln transformed.

Table S1 Plant richness, and shoot, root, and functional group shoot biomass (g m^{-2}) for Control and N addition treatments. Values shown are mean ± s.d. (N = 6). Significant differences among treatments are indicated by different letter superscripts ($P<0.05$). PR = perennial rhizome grasses, PB = perennial bunch grasses, PF = perennial forbs, SS = shrubs and semi-shrubs. $N_0 = 0$ mol N m^{-2} y^{-1}, $N_{0.4} = 0.4$ mol N m^{-2} y^{-1}, $N_{0.8} = 0.8$ mol N m^{-2} y^{-1}, $N_{1.6} = 1.6$ mol N m^{-2} y^{-1}, $N_{2.8} = 2.8$ mol N m^{-2} y^{-1}, $N_4 = 4.0$ mol N m^{-2} y^{-1}.

Table S2 Results of structural equation modeling of N addition effects on the plant-soil-nematode system as illustrated in Fig. 3 (main text). Given are the unstandardized path coefficients (estimates), standard error of regression weight (S.E.), the critical value for regression weight (C.R.; z = estimate/S.E.) and the level of significance for regression weight (p). For more information on exogenous and endogenous variables as well as model fit, see main text (*** $P<0.001$).

Table S3 Relative abundance of nematode genera (%) for Control and N treatments in August 2009. Data are mean values (N = 6). H = herbivores, Ba = bacterivores, Fu = fungivores, Om = omnivores, Ca = carnivores.

Table S4 Relative abundance of nematode genera (%) for Control and N addition treatments in September 2009. Data are mean values (N = 6). H = herbivores, Ba = bacterivores, Fu = fungivores, Om = omnivores, Ca = carnivores.

Acknowledgments

We are grateful to the Inner Mongolia Grassland Ecosystem Research Station (IMGERS) for providing the experimental sites and elemental analysis. Comments and suggestions from Drs. Lynn Carta, Christian Mulder and two anonymous reviewers greatly improved the quality of this manuscript.

Author Contributions

Conceived and designed the experiments: CZW QY WJL XGH. Performed the experiments: CZW HFZ QY HYZ NPH. Analyzed the data: CZW HFZ QL. Contributed reagents/materials/analysis tools: QSC. Wrote the paper: CZW XTL QL PK.

References

1. Vitousek PM, Mooney HA, Lubchenco J, Melillo JM (1997) Human domination of Earth's ecosystems. Science 277: 494–499.
2. Galloway JN, Dentener FJ, Capone DG, Boyer EW, Howarth RW, et al. (2004) Nitrogen cycles: past, present, and future. Biogeochemistry 70: 153–226.
3. Tilman D (1987) Secondary succession and the pattern of plant dominance along experimental nitrogen gradients. Ecol Monogr 57: 189–214.
4. Aber J, McDowell W, Nadelhoffer K, Magill A, Berntson G, et al. (1998) Nitrogen saturation in temperate forest ecosystems. BioScience 48: 921–934.
5. Guo JH, Liu XJ, Zhang Y, Shen JL, Han WX, et al. (2010) Significant acidification in major Chinese croplands. Science 327: 1008–1010.
6. Treseder KK (2008) Nitrogen additions and microbial biomass: a meta-analysis of ecosystem studies. Ecol Lett 11: 1111–1120.
7. Liang W, Lou Y, Li Q, Zhong S, Zhang X, et al. (2009) Nematode faunal response to long-term application of nitrogen fertilizer and organic manure in Northeast China. Soil Biol Biochem 41: 883–890.
8. Wu T, Ayres E, Bardgett RD, Wall DH, Garey JR (2011) Molecular study of worldwide distribution and diversity of soil animals. Proc Natl Acad Sci U S A 108: 17720–17725.
9. Yeates GW (2003) Nematodes as soil indicators: functional and biodiversity aspects. Biol Fertil Soils 37: 199–210.
10. Yeates GW, Bongers T, De Goede R, Freckman D, Georgieva S (1993) Feeding habits in soil nematode families and genera: an outline for soil ecologists. J Nematol 25: 315–331.
11. Neher DA (2001) Role of nematodes in soil health and their use as indicators. J Nematol 33: 161–168.
12. Mulder C, Den Hollander HA, Hendriks AJ (2008) Aboveground herbivory shapes the biomass distribution and flux of soil invertebrates. PLoS ONE 3(10): e3573.
13. Bongers T, Ferris H (1999) Nematode community structure as a bioindicator in environmental monitoring. Trends Ecol Evol 14: 224–228.
14. Kardol P, Bezemer TM, Van Der Wal A, Van Der Putten WH (2005) Successional trajectories of soil nematode and plant communities in a chronosequence of ex-arable lands. Biol Conserv 126: 317–327.
15. Verhoef H, Brussaard L (1990) Decomposition and nitrogen mineralization in natural and agroecosystems: the contribution of soil animals. Biogeochemistry 11: 175–211.
16. De Ruiter P, Moore J, Zwart K, Bouwman L, Hassink J, et al. (1993) Simulation of nitrogen mineralization in the below-ground food webs of two winter wheat fields. J Appl Ecol 30: 95–106.
17. Sarathchandra SU, Ghani A, Yeates GW, Burch G, Cox NR (2001) Effect of nitrogen and phosphate fertilisers on microbial and nematode diversity in pasture soils. Soil Biol Biochem 33: 953–964.
18. Murray PJ, Cook R, Currie AF, Dawson LA, Gange AC, et al. (2006) Interactions between fertilizer addition, plants and the soil environment: Implications for soil faunal structure and diversity. Appl Soil Ecol 33: 199–207.
19. Sohlenius B, Wasilewska L (1984) Influence of Irrigation and Fertilization on the Nematode Community in a Swedish Pine Forest Soil. J Appl Ecol 21: 327–342.
20. Xu G, Mo J, Fu S, Per G, Zhou G, et al. (2007) Response of soil fauna to simulated nitrogen deposition: A nursery experiment in subtropical China. J Environ Sci 19: 603–609.
21. Qi S, Zhao X, Zheng H, Lin Q (2010) Changes of soil biodiversity in Inner Mongolia steppe after 5 years of N and P fertilizer applications. Acta Ecol Sin 30: 5518–5526.
22. Hu C, Qi Y (2010) Effect of compost and chemical fertilizer on soil nematode community in a Chinese maize field. Eur J Soil Biol 46: 230–236.
23. Li Q, Jiang Y, Liang W, Lou Y, Zhang E, et al. (2010) Long-term effect of fertility management on the soil nematode community in vegetable production under greenhouse conditions. Appl Soil Ecol 46:111–118.
24. Rodriguez-Kabana R (1986) Organic and inorganic nitrogen amendments to soil as nematode suppressants. J Nematol 18: 129–135.
25. De Deyn GB, Raaijmakers CE, Van Ruijven J, Berendse F, Van Der Putten WH (2004) Plant species identity and diversity effects on different trophic levels of nematodes in the soil food web. Oikos 106: 576–586.

26. Xia J, Wan S (2008) Global response patterns of terrestrial plant species to nitrogen addition. New Phytol 179: 428–439.
27. Bai Y, Wu J, Clark CM, Naeem S, Pan Q, et al. (2010) Tradeoffs and thresholds in the effects of nitrogen addition on biodiversity and ecosystem functioning: evidence from inner Mongolia Grasslands. Glob Chang Biol 16: 358–372.
28. Yeates GW (1999) Effects of plants on nematode community structure. Annu Rev Phytopathol 37: 127–49.
29. Wardle DA (2002) Communities and ecosystems: linking the aboveground and belowground components: Princeton University Press.
30. Bai YF, Han XG, Wu JG, Chen ZZ, Li HL (2004) Ecosystem stability and compensatory effects in the Inner Mongolia grassland. Nature 431: 181–184.
31. Yu Q, Chen Q, Elser J, Cease A, He N, et al. (2010) Linking stoichiometric homeostasis with ecosystem structure, functioning, and stability. Ecol Lett 13: 1390–1399.
32. Oostenbrink M (1960) Estimating nematode populations by some selected methods. Nematology 6: 85–102.
33. Bongers T (1990) The maturity index: an ecological measure of environmental disturbance based on nematode species composition. Oecologia 83: 14–19.
34. Clark CM, Cleland EE, Collins SL, Fargione JE, Gough L, et al. (2007) Environmental and plant community determinants of species loss following nitrogen enrichment. Ecol Lett 10: 596–607.
35. Veen GF, Olff H, Duyts H, Van Der Putten WH (2010) Vertebrate herbivores influence soil nematodes by modifying plant communities. Ecology 91: 828–835.
36. Grace JB (2006) Structural equation modeling and natural systems: Cambridge University Press.
37. Persson H (1980) Fine-root dynamics in a Scots pine stand with and without near-optimum nutrient and water regimes. Acta Phytogeogr Suec 68: 101–110.
38. Collange B, Navarrete M, Peyre G, Mateille T, Tchamitchian M (2011) Root-knot nematode (Meloidogyne) management in vegetable crop production: The challenge of an agronomic system analysis. Crop Prot 30: 1251–1256.
39. Oka Y, Shapira N, Fine P (2007) Control of root-knot nematodes in organic farming systems by organic amendments and soil solarization. Crop Prot 26: 1556–1565.
40. Johnson D, Edwards N, Todd D (1980) Nitrogen mineralization, immobilization, and nitrification following urea fertilization of a forest soil under field and laboratory conditions. Soil Sci Soc Am J 44: 610–616.
41. Warren KS (1962) Ammonia toxicity and pH. Nature 195: 47–49.
42. Wall ME, Tiedjens VA (1940) Potassium deficiency in ammonium-and nitrate-fed plants. Science 91: 221–222.
43. Scherber C, Eisenhauer N, Weisser W, Schmid B, Voigt W, et al. (2010) Bottom-up effects of plant diversity on multitrophic interactions in a biodiversity experiment. Nature 468: 553–556.
44. Wardle D, Yeates G (1993) The dual importance of competition and predation as regulatory forces in terrestrial ecosystems: evidence from decomposer food-webs. Oecologia 93: 303–306.
45. Wardle D, Yeates G, Watson R, Nicholson K (1995) The detritus food-web and the diversity of soil fauna as indicators of disturbance regimes in agroecosystems. Plant Soil 170: 35–43.
46. Bardgett R, Mawdsley J, Edwards S, Hobbs P, Rodwell J, et al. (1999) Plant species and nitrogen effects on soil biological properties of temperate upland grasslands. Funct Ecol 13: 650–660.
47. Wardle DA, Bardgett RD, Klironomos JN, Setala H, Van Der Putten WH, et al. (2004) Ecological linkages between aboveground and belowground biota. Science 304: 1629–1633.
48. Aber JD, Nadelhoffer KJ, Steudler P, Melillo JM (1989) Nitrogen saturation in northern forest ecosystems. BioScience 39: 378–286.
49. Manning P, Newington JE, Robson HR, Saunders M, Eggers T, et al. (2006) Decoupling the direct and indirect effects of nitrogen deposition on ecosystem function. Ecol Lett 9: 1015–1024.
50. Manning P, Saunders M, Bardgett RD, Bonkowski M, Bradford MA, et al. (2008) Direct and indirect effects of nitrogen deposition on litter decomposition. Soil Biol Biochem 40: 688–698.

4

Estimating Litter Decomposition Rate in Single-Pool Models Using Nonlinear Beta Regression

Etienne Laliberté[1], E. Carol Adair[2¤*], Sarah E. Hobbie[3]

1 School of Plant Biology, The University of Western Australia, Crawley, Western Australia, Australia, 2 National Center for Ecological Analysis and Synthesis, University of California Santa Barbara, Santa Barbara, California, United States of America, 3 Department of Ecology, Evolution and Behavior, University of Minnesota, Saint Paul, Minnesota, United States of America

Abstract

Litter decomposition rate (k) is typically estimated from proportional litter mass loss data using models that assume constant, normally distributed errors. However, such data often show non-normal errors with reduced variance near bounds (0 or 1), potentially leading to biased k estimates. We compared the performance of nonlinear regression using the beta distribution, which is well-suited to bounded data and this type of heteroscedasticity, to standard nonlinear regression (normal errors) on simulated and real litter decomposition data. Although the beta model often provided better fits to the simulated data (based on the corrected Akaike Information Criterion, AIC_c), standard nonlinear regression was robust to violation of homoscedasticity and gave equally or more accurate k estimates as nonlinear beta regression. Our simulation results also suggest that k estimates will be most accurate when study length captures mid to late stage decomposition (50–80% mass loss) and the number of measurements through time is ≥ 5. Regression method and data transformation choices had the smallest impact on k estimates during mid and late stage decomposition. Estimates of k were more variable among methods and generally less accurate during early and end stage decomposition. With real data, neither model was predominately best; in most cases the models were indistinguishable based on AIC_c, and gave similar k estimates. However, when decomposition rates were high, normal and beta model k estimates often diverged substantially. Therefore, we recommend a pragmatic approach where both models are compared and the best is selected for a given data set. Alternatively, both models may be used via model averaging to develop weighted parameter estimates. We provide code to perform nonlinear beta regression with freely available software.

Editor: Ben Bond-Lamberty, DOE Pacific Northwest National Laboratory, United States of America

Funding: This work was conducted while E.C.A. was a Postdoctoral Associate at the National Center for Ecological Analysis and Synthesis, a Center funded by NSF (Grant #EF-0553768), the University of California, Santa Barbara, and the State of California. E.L. was supported by fellowships from the University of Western Australia and the Australian Research Council (ARC). The funders had no role in study design, data collection and analysis, decision to publish, or preparation of the manuscript.

Competing Interests: The authors have declared that no competing interests exist.

* E-mail: Carol.Adair@uvm.edu

¤ Current address: Rubenstein School of Environment and Natural Resources, University of Vermont, Burlington, Vermont, United States of America

Introduction

Litter decomposition strongly influences carbon and nutrient cycling within ecosystems [1]. Therefore, estimating an accurate decomposition rate is critical to understanding biogeochemical processes. The most widely used model to describe the rate of litter mass loss is the single-pool negative exponential model [2]

$$M(t) = M(0)e^{-kt}, \qquad (1)$$

where $M(t)$ is litter mass at time t, $M(0)$ is initial litter mass, and k is the litter decomposition rate. Because $M(0)$ is generally known, its estimation is unnecessary and can even lead to biased estimates of k, the parameter of interest [3]. Thus, $M(t)$ is best divided by $M(0)$ and the resulting proportional litter mass loss $X(t)$ modeled as [3]

$$X(t) = \frac{M(t)}{M(0)} = e^{-kt}. \qquad (2)$$

In theory, $X(t)$ is bounded such that $0 \leq X(t) < 1$, but in practice values ≥ 1 sometimes result, especially during the early stages of decomposition.

Often, k is estimated by log-transforming $X(t)$ and using a linear regression model with mean μ and normally distributed errors, where k is the slope and σ^2 is the variance

$$\ln[X(t)] \sim N(\mu = -kt, \sigma^2); \qquad (3)$$

this is similar to the use log-log regression for fitting allometric power equations [4] and biological power laws [5]. However, [3] showed that this approach leads to biased k estimates unless errors are log-normally distributed. Instead, they suggested using nonlinear regression on untransformed data, again with normally-distributed errors

$$X(t) \sim N(\mu = e^{-kt}, \sigma^2). \qquad (4)$$

This model was found to give more accurate k estimates in simulations [3], but it assumes that errors are constant and normally distributed – a likely invalid assumption (Figure 1). Indeed, proportional litter mass loss data often shows smaller variance near bounds (0 and 1), which is typical of bounded data [6]. In these cases, fitting a model with constant normal errors may lead to biased k estimates.

One solution could be to model the variance σ^2 as a function of t, but this requires additional parameters. An alternative solution may be to use an error distribution better suited to bounded data, such as the beta distribution [6]. Like the normal distribution, it only has two parameters. Unlike the normal distribution, it is bounded between 0 and 1, and can easily accommodate the type of heteroscedasticity shown in Figure 1 [6]. Its probability density function is a function of two scale parameters, α and β

$$f(x,\alpha,\beta) = \frac{\Gamma(\alpha+\beta)}{\Gamma(\alpha)\Gamma(\beta)} x^{\alpha-1}(1-x)^{\beta-1}, \tag{5}$$

where $\Gamma(n)$ is the gamma function $\Gamma(n)=(n-1)!$ and $0 \leq x \leq 1$. In the context of regression, the beta distribution is re-parameterized [6],[7] to a location parameter μ (the mean) and a precision parameter ϕ (the inverse of dispersion)

$$\mu = \frac{\alpha}{\alpha+\beta} \tag{6}$$

$$\phi = \alpha+\beta. \tag{7}$$

The variance σ^2 depends on μ and ϕ.

$$\sigma^2 = \frac{\mu(1-\mu)}{(1+\phi)}. \tag{8}$$

Consistent with patterns often found in decomposition data (Figure 1), the numerator shows that σ^2 is smaller near the bounds (0 or 1): if $\phi = 1$ and $\mu = 0.01$ or 0.99, $\sigma^2 = 0.005$; if $\mu = 0.5$, $\sigma^2 = 0.125$. The denominator shows that higher precision ϕ reduces σ^2.

In summary, the beta distribution may be better suited than the normal distribution to model proportional litter mass loss data because it is bounded between 0 and 1, its σ^2 is smaller near its bounds, as with decomposition data (Figure 1), and hence it can model this type of heteroscedasticity without additional parameters.

Since the beta distribution is bounded between 0 and 1, proportional litter mass loss data must also be bounded between 0 and 1. However, litter mass loss data often contain values equal to 0 (no mass remaining), or ≥ 1 (no decomposition or sample contamination by soil), so the data, y, must be compressed to the $]0, 1[$ interval (y'') [6]:

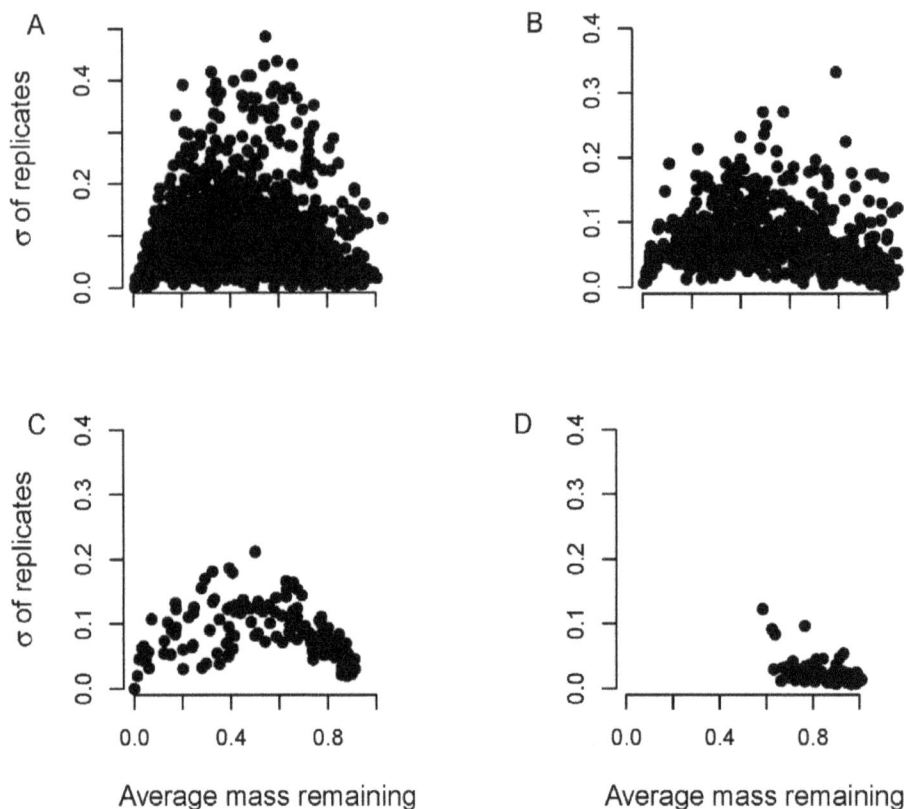

Figure 1. Figure of mean mass remaining versus standard deviation of replicates at each time point for real data. Mean mass remaining versus standard deviation of replicates at each time point for (A) Long-term Intersite Decomposition Experiment Team (LIDET) data, (B) Hobbie data; (C) EL data; and (D) HG data.

$$y' = (y - a)/(b - a) \qquad (9)$$

$$y'' = [y'(N - 1) + 0.5]/N, \qquad (10)$$

where a and b are the y minimum and maximum values, respectively, and N is sample size. Hereafter, we refer to this transformation as Smithson and Verkuilen's [6] (SV) transformation.

The goal of this paper is to compare the normal model (Equation 4) with the beta model

$$X(t) \sim B(\mu = e^{-kt}, \phi). \qquad (11)$$

Specifically, we : (1) compare the performance of the normal vs. beta model in numerical simulations, using different realistic error structures for simulated $X(t)$; (2) investigate the influence of two different transformations to compress $X(t)$ between 0 and 1, namely (i) treating zeros as missing data and setting values ≥ 1 equal to 0.9999, or (ii) Smithson and Verkuilen's transformation [6]; and (3) compare the performance of the normal vs. beta model and evaluate the influence of the transformations mentioned above, using real data from decomposition studies of differing decomposition stage (early, medium, and late based on percent of initial mass remaining: 25, 60, and 72% average mass loss, respectively). Because different decomposition stages encompass different portions of the mean-variance relationships seen in litter decomposition data (Figure 1), we expected that it could influence the fit of beta vs. normal models.

We hypothesized that nonlinear beta regression would provide better fits to proportional mass loss data and give more accurate k estimates than normal nonlinear regression, because of the heteroscedasticity often associated with these data (Figure 1). If so, nonlinear beta regression would provide more reliable k estimates from single-pool models [2].

Materials and Methods

Data Simulation

We simulated $X(t)$ using four values that spanned the range of low to high decomposition rates: 0.0005, 0.002, 0.01, and 0.1 d^{-1}. These k values were chosen by examining the range of k values found in the Adair et al. [3] decomposition review and choosing values that spanned the range from very low to high (Figure S1). The chosen k values resulted in 1% mass remaining at approximately 25, 6, 1.3, and 0.1 years, respectively (using Equation 2; Table 1). We used these k values to simulate $X(t)$ over four different time spans that represented early (80% mass remaining), mid (50% mass remaining), late (20% mass remaining), and end (1% mass remaining) stage decomposition for each k value (Table 1). This strategy allowed us to investigate the ability of each regression type to accurately predict k across a range of k values and decomposition stages (i.e., study lengths or total times).

To investigate whether the number of mass loss measurements taken within a given study would affect a given regression type's ability to accurately estimate k, we generated 2, 5, 7 or 10 "measurements" across each k value and decomposition stage simulation. Because sampling times in decomposition studies are not typically evenly spaced, but are instead weighted towards the beginning of the study (where litter mass loss is most rapid), we used the data gathered during the review completed by Adair et al. [3] to determine sampling times: we (1) recorded total experiment time and all measurement times from each of the 383 references

Table 1. Percent mass remaining at early, mid, late and end stage decomposition for four different decomposition rates (k in d^{-1}).

Stage	Mass remaining	Time (d)			
		$k = 0.0005$	$k = 0.002$	$k = 0.01$	$k = 0.1$
Early	80%	446	112	22	2
Mid	50%	1386	347	69	7
Late	20%	3219	805	161	16
End	1%	9210	2303	461	46
Years to end		25.2	6.3	1.3	0.1

Time is the number of days (d) it takes for mass remaining to reach 80, 50, 20 or 1% for early, mid, late or end stage decomposition, respectively. Time is also provided in years for end stage decomposition.

contained in the review; (2) converted measurement times to proportion of total experiment times; (3) grouped proportional measurement times by the number of times each study made mass loss measurements (i.e., 2, 5, 7 or 10 times); (4) created histograms for each category using bin sizes of 0.1; and (5) selected the most frequent proportional measurement times from each category (2, 5, 7, or 10 measurements; Figure S2). The proportional times used were the averages of the most frequent proportional measurement bins. Thus, for 2 measurements, data was simulated at 0.5 and 1.0 of total time (i.e., at ½ of the total time and at the end of the total time). For 5 measurements, data was simulated at 0.06, 0.14, 0.23, 0.63, 1.0 of total time. For 7 measurements, data was simulated at 0.05, 0.15, 0.24, 0.36, 0.54, 0.65, 1.0 of total time. For 10 measurements, data was simulated at 0.04, 0.11, 0.23, 0.32, 0.43, 0.53, 0.62, 0.84, 0.93, 1.0 of total time.

Finally, we used three different error structures that resembled those found in real data (Figures 1–2). For each error structure we generated data using three different standard deviations (or ϕ's for beta regression) that resulted in low, moderate, and high variation in the simulated $X(t)$s:

1. Normally distributed errors with variable standard deviations (σ). We took random samples from the normal distribution

$$X(t) = N(\mu = e^{-kt}, \sigma^2), \qquad (12)$$

using three different variable σ structures:

$$\text{Var } \sigma_1 = \begin{cases} 0.02 & \text{if } \mu \leq 0.05 \\ 0.04 & \text{if } 0.05 < \mu \leq 0.15 \\ 0.075 & \text{if } 0.15 < \mu < 0.85, \\ 0.05 & \text{if } 0.85 \leq \mu < 0.95 \\ 0.025 & \text{if } \mu \geq 0.95 \end{cases} \qquad (13)$$

$$\text{Var } \sigma_2 = \begin{cases} 0.04 & \text{if } \mu \leq 0.05 \\ 0.08 & \text{if } 0.05 < \mu \leq 0.15 \\ 0.15 & \text{if } 0.15 < \mu < 0.85, \text{ or} \\ 0.10 & \text{if } 0.85 \leq \mu < 0.95 \\ 0.05 & \text{if } \mu \geq 0.95 \end{cases} \qquad (14)$$

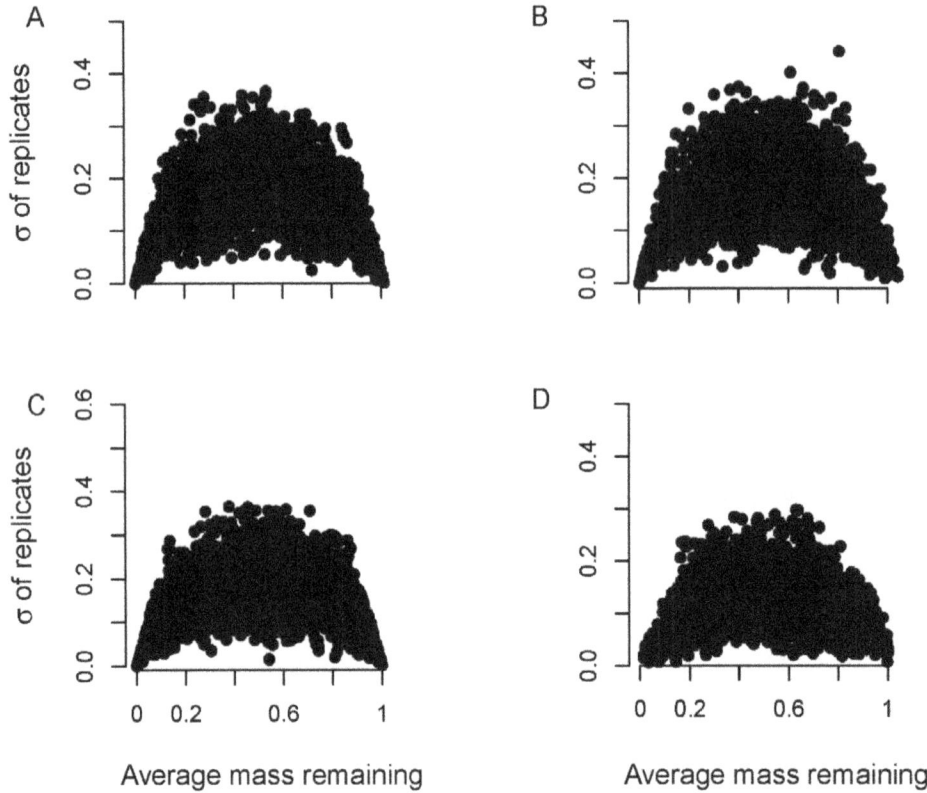

Figure 2. Figure of mean mass remaining versus standard deviation of replicates at each time point for 200 simulations with four different error structures: (A) beta errors + normal errors (option 3a; $\sigma = 0.0125$, $\phi = 5$); (B) beta errors + normal error (option 3b; $\sigma = 0.05$, $\phi = 5$); (C) beta errors (option 2; no 0 or >1 values, $\phi = 5$); and (D) normal error with variable σ (option 1; Var σ_2).

$$\text{Var } \sigma_3 = \begin{cases} 0.06 & \text{if } \mu \leq 0.05 \\ 0.12 & \text{if } 0.05 < \mu \leq 0.15 \\ 0.225 & \text{if } 0.15 < \mu < 0.85 \\ 0.15 & \text{if } 0.85 \leq \mu < 0.95 \\ 0.075 & \text{if } \mu \geq 0.95 \end{cases} \quad (15)$$

where σ increases from Var σ_1 to Var σ_3. Values $X(t) < 0$ were set to 0, whereas values $X(t) > 1.05$ were set equal to 1 (Figure 2d).

2. Beta-distributed errors. We took random samples from the beta distribution, with $\phi = 5$, 8 or 15 (higher values generate less variation in $X(t)$; Figure 2c).

$$X(t) \sim B(\mu = e^{-kt}, \phi). \quad (16)$$

3. Beta-distribution errors with normal errors added. We sampled from the beta distribution ($\phi = 5$, 8 or 15) and added small amounts of normal error (ε; two different σ values) to generate values $X(t) \leq 0$ or ≥ 1, which sometimes occur in real data.

$$X(t) \sim B(\mu = e^{-kt}, \phi) + \varepsilon \quad (17)$$

where ε was either

$$(a)\varepsilon \sim N(\mu = 0, \sigma = 0.0125$$

$$(18)$$

$$(b)\varepsilon \sim N(\mu = 0, \sigma = 0.05$$

Values < 0 were then set equal to 0 (Figure 2a–b).

In total, we ran 768 simulations (four k values; three error options with three variable σ structures for option 1, three ϕ values for option 2 and six $\sigma + \phi$ combinations for option 3; four decomposition stages; four numbers of measurements). Each data set generated within a simulation run had five replicates per measurement time. We generated 12,000 data sets in each simulation run. We estimated parameters via maximum likelihood (ML) estimation with normal and beta distributed errors, using the 'bbmle' package (version 1.0.4.1) [8] and nonlinear least-squares regression (NLS; assumes normal errors), using the 'nls' function in R 2.15.0 [9]. At times, NLS and beta ML regression failed to converge. Thus, to compare regression methods, we used the first 10,000 simulations where all regression types successfully estimated k. NLS only failed in cases where simulated data sets contained many missing values (see REP transformation below). However, beta ML regression often failed to converge during early decomposition, regardless of the number of measurements that were used (2, 5, 7 or 10) to estimate k. This was especially true in simulations that used only beta-distributed errors (option 2). In these cases, we used <10,000 simulated data sets to compare regression methods (Table 2).

Additionally, when using ML estimation with beta errors to estimate the low k value (0.0005 d^{-1}), optimization algorithms often failed to converge. We therefore estimated the low rate as a yearly rate (this solved the convergence problems) and converted it back to a daily rate for analyses, figures and tables.

For simulation runs that generated data sets with values $X(t) = 0$ or ≥ 1 (options 1, 3a, 3b), we compared two data transformations:

Table 2. Simulations for which ML estimation with beta errors (option 2) failed to converge for 10,000 out of 12,000 generated data sets.

Error	Stage	# meas	k = 0.1			k = 0.01			k = 0.002			k = 0.0005		
			$\phi=5$	$\phi=8$	$\phi=15$	$\phi=5$	$\phi=8$	$\phi=15$	$\phi=5$	$\phi=8$	$\phi=15$	$\phi=5$	$\phi=8$	$\phi=15$
Beta only														
	Early	5	5690	9784		6906			7123			6942		
	Early	7	4325	8519		5168	9351		5367	9517		5281	9409	
	Early	10	2581	6638		3274	7454		3390	7802		3335	7641	

The number of generated data sets for which ML estimation with beta errors converged is shown.

the Smithson and Verkuilen (SV) [6] transformation (Equations 9 and 10) or, following [3], converting all values ≥ 1 to 0.9999 and treating zeros as missing data (the 'replacement' or REP transformation). For simulations with values $0 < X(t) < 1$ (option 2), no transformations were necessary. This resulted in 576 additional simulations, for a total of 1344 simulations.

Because the generated data sets had small sample sizes (i.e. $N/p < 40$, where p is the number of parameters), which is typical for litter decomposition studies, we used the corrected Akaike Information Criterion (AIC_c) to compare models fitted via ML

$$AIC_c = AIC + \frac{2p(p+1)}{N-p-1}. \qquad (19)$$

To determine how well the different approaches estimated the litter decomposition rate, k_e, relative to the true k_t (here, 0.002 d^{-1}) we calculated the average percent (%) bias

$$\% \text{ Bias} = 100 \frac{\sum_{i=1}^{s}(k_{ei}-k_t)/k_t}{s}, \qquad (20)$$

where s is the number of simulations (here, $s = 10,000$ or as in Table 2), and the percent relative error (%RE)

$$\% \text{ RE} = 100 \frac{\sum_{i=1}^{s}|k_{ei}-k_t|/k_t}{s}. \qquad (21)$$

Bias measured whether a particular approach over- or underestimated k_t, whereas %RE measured the magnitude of the difference between k_t and all k_e, regardless of direction.

Because results (% bias, % RE, and average k estimates, and AIC_c results) were very similar among k values (e.g., Figures S3,S4,S5,S6,S7,S8,), we present results from one k value ($k = 0.0002$).

Analysis of Real Decomposition Data

We used three real data sets that reflected the range of time frames used in the data simulation: early, mid, and late stage decomposition data, based on the proportion of initial litter mass still present at the end of each study (Table 1).

For the early stage decomposition data set, we used the Hobbie and Gough [10] litter bag decomposition data set. The average percent of initial mass remaining at the end of this experiment was 75.4% (standard error, SE = 3.1%), indicative of early decompo-

sition. The Hobbie and Gough [10] experiment was conducted at two arctic tundra sites near Toolik Lake, Alaska (68 38'N, 149 43'W). Mean annual temperature (MAT) at Toolik Lake is $-7°C$ with low annual precipitation (200–400 mm) [11]. In this experiment, nine litter types were decomposed over 1082 days. Five bags of each litter type were collected from each site on days 308, 361, 717, and 1082. Experiment details are presented in [10].

Mid stage decomposition data were provided by Laliberté and Tylianakis' [12] 560-day litter bag decomposition experiment conducted on the AgResearch Mount John trial site, in the Mackenzie Basin of New Zealand's South Island (43°59'S, 170°27'E). The climate is semi-continental with a MAT of 8.7°C and mean annual precipitation (MAP) of 601 mm. Litterbags of mixed senesced "community litter" were decomposed within a larger fertilization and grazing experiment (described in detail by [12]). The experiment is a split-plot design where fertilizer treatment is the whole-plot treatment and sheep grazing intensity were the sub-plot treatments. Four replicates were collected from each sub-plot after 1, 3, 6, 12, and 18 months. Litterbags were also collected from adjacent unfertilized and ungrazed control sites. Average percent of initial mass remaining at the end of the experiment was 40% (SE = 0.002). This experiment is described in detail in [13].

We used the Hobbie [14] data set for late stage or long-term decomposition. Average mass remaining at the final collection was 27.6% (SE = 0.60%). These data consisted of the data within [14] plus Hobbie's unpublished filter paper mass loss data from the same experiment (hereafter, the Hobbie data set). Briefly, the Hobbie [14] experiment was established at Cedar Creek Ecosystem Science Reserve in central Minnesota, USA (45.40° N, 93.20° W; MAT = 6.7°C, MAP = 800 mm). Eight litters were decomposed for five years (1763 days) at eight sites (two old fields, a hardwood forest, two oak stands, two pine stands, and an aspen stand), with a nitrogen addition treatment at each site (6 replicates per treatment/time point). Details are presented in [14],[15].

Because NLS and ML estimation using normal errors produced nearly identical results in the data simulations for early to late stage decomposition (Figures 3,4,5,6,7,8), we only compared k estimates obtained using ML estimation with normal and beta errors (k estimates/decomposition models were compared using AICc; see below). As in the simulations, when using beta errors to estimate low k values ($k < 0.0015$ d^{-1}), optimization algorithms often failed to converge. In these cases, estimating k in years solved the problem. Thus, while all other k values were estimated as daily rates, k values in the Hobbie and Gough [10] data set were estimated as yearly rates and converted to daily rates for figures and tables.

We fit single pool models (Equation 2) to all litter mass loss curves within each data set using ML estimation with normal or

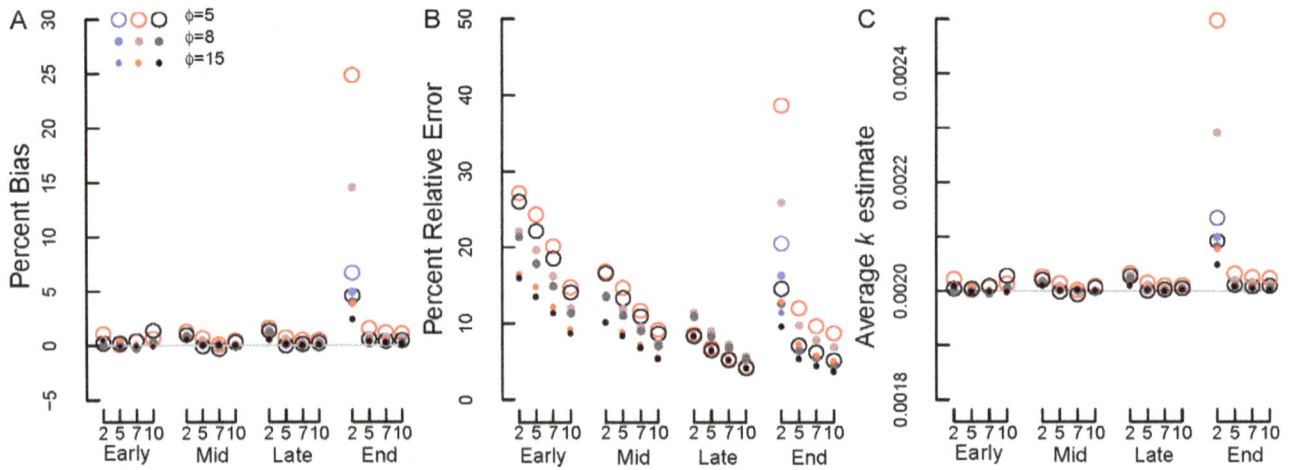

Figure 3. Simulation results for beta-distributed errors (option 2), *k* = 0.002. (A) Percent bias, (B) percent relative error, and (C) average *k* estimate. Early, mid, late and end are early, mid, late and end stage decomposition simulations. The numbers 2, 5, 7 and 10 are the numbers of measurements used in each simulation. Blue circles = NLS, Red circles = Normal ML, gray/black circles = Beta ML. In most cases, nls = Normal ML so that the red circles cover the blue circles. In panel (A), the gray line shows 0% bias. In panel (C), the gray line shows the true *k* value, 0.002 d^{-1}.

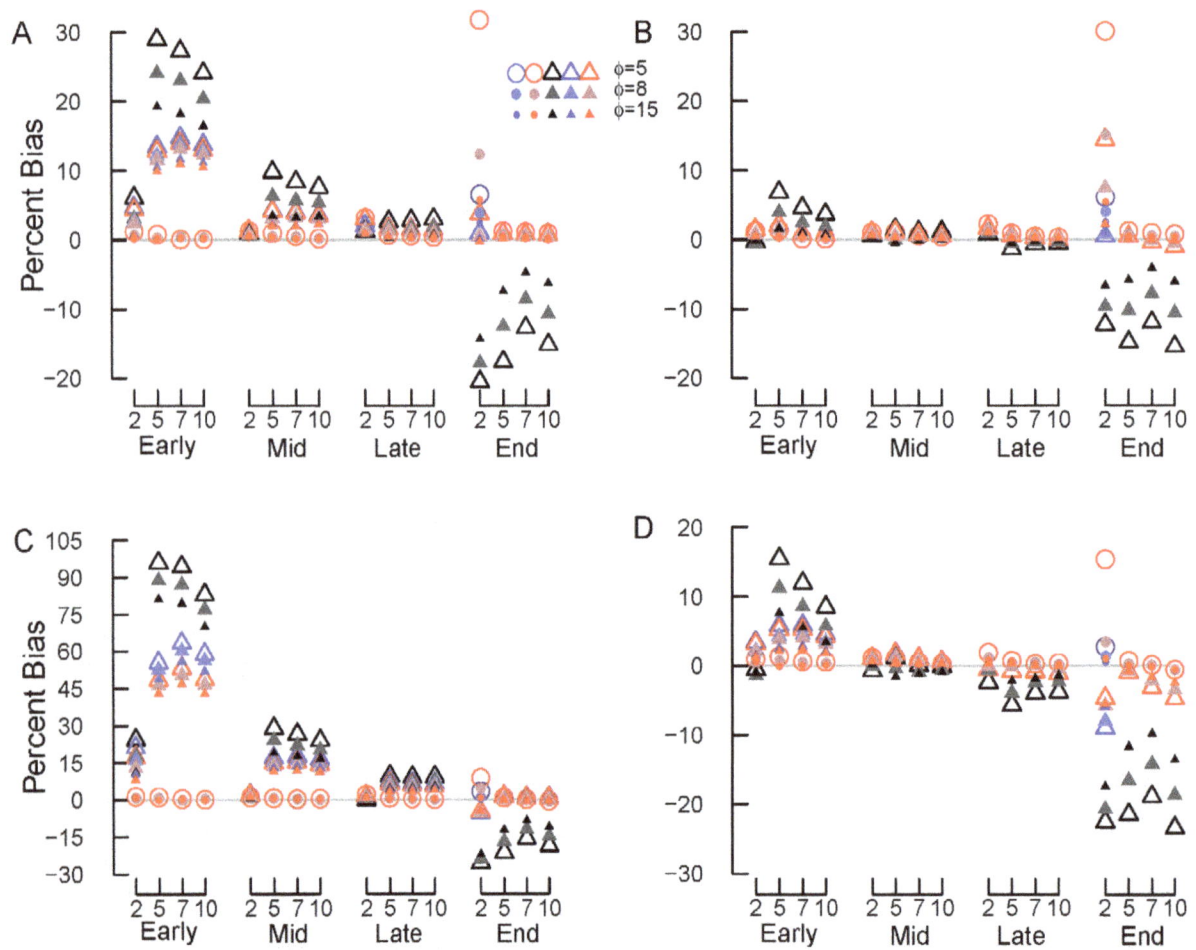

Figure 4. Percent bias for beta-distributed errors plus normal errors. (A) standard deviation (σ) = 0.0125 (option 3a) and SV transformation, (B) σ = 0.0125 (option 3a) and REP transformation, (C) σ = 0.05 (option 3b) and SV transformation, (D) σ = 0.05 (option 3b) and REP transformation. Early, mid, late and end are early, mid, late and end stage decomposition simulations. The numbers 2, 5, 7 and 10 are the numbers of measurements used in each simulation. Blue circles = NLS, Red circles = Normal ML, gray/black circles = Beta ML. In most cases, nls = Normal ML so that the red circles cover the blue circles. Gray lines show 0% bias.

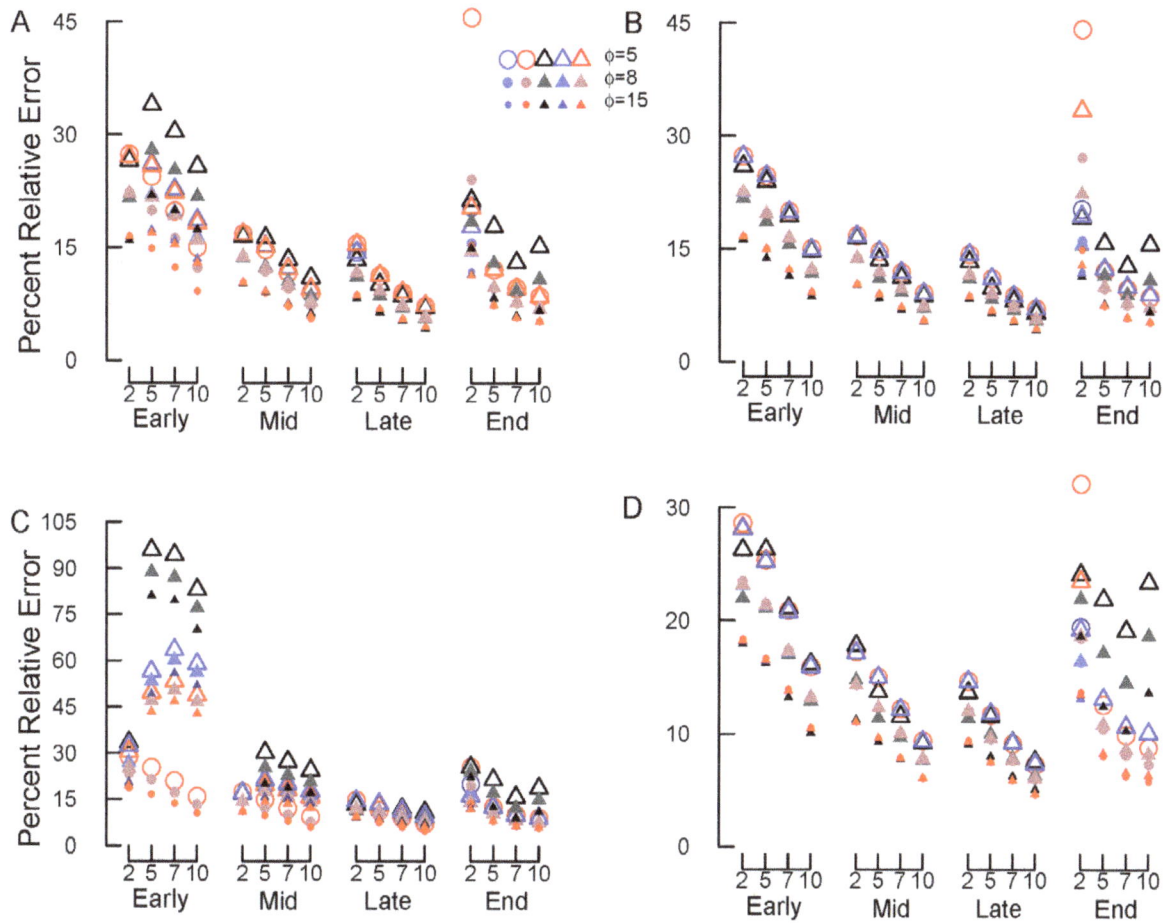

Figure 5. Percent relative error for beta-distributed errors plus normal errors with different σ and transformations. (A) σ = 0.0125 (option 3a) and SV transformation, (B) σ = 0.0125 (option 3a) and REP transformation, (C) σ = 0.05 (option 3b) and SV transformation, (D) σ = 0.05 (option 3b) and REP transformation. Early, mid, late and end are early, mid, late and end stage decomposition simulations. The numbers 2, 5, 7 and 10 are the numbers of measurements used in each simulation. Blue circles = NLS, Red circles = Normal ML, gray/black circles = Beta ML. In most cases, nls = Normal ML so that the red circles cover the blue circles.

beta errors ('bbmle' package version 1.0.4.1) [8]. The early, mid and late stage decomposition data sets contained 18, 64, and 128 litter mass loss curves, respectively. Whenever possible (i.e. all $X(t)$ >0 and <1), we used untransformed data. When transformation was required, we used both the SV and REP data transformations. Within transformed or untransformed data sets, we compared model fit using AIC_c. We considered models with AIC_c between 4 and 7 apart ($4< \Delta AIC_c <7$) as clearly distinguishable and models with $\Delta AIC_c >10$ as definitely different, following previous recommendations [16].

We also examined model fit to the untransformed data using fractional bias (FB)

$$FB = \frac{\overline{M(t)}_{pred} - \overline{M(t)}_{obs}}{0.5[\overline{M(t)}_{pred} + \overline{M(t)}_{obs}]}, \quad (22)$$

and relative bias (RB)

$$RB = \frac{\overline{M(t)}_{pred} - \overline{M(t)}_{obs}}{\sqrt{var[\overline{M(t)}_{obs}]}}, \quad (23)$$

where $\overline{M(t)}_{pred}$ is the mean of predicted values, $\overline{M(t)}_{obs}$ is the mean of all observations, and var[M(t)obs] is the sample variance of all observations [17]. These metrics express the average amount of bias in the model predictions (compared to the observations) and thus describe the 'model-data' discrepancy [17].

Results

Simulations

Beta-distributed errors. For beta-distributed errors (option 2; no data transformations needed), the accuracy of k estimates generally increased with the number of measurements (from two to ten) and declining error (from $\phi = 5$ to 15; Figure 3). Bias and RE decreased with increasing number of measurements and with decreasing error (ϕ). In general, k estimates also improved (reduced bias and RE) from early to late decomposition (Figure 3). However, when study lengths were the longest (end stage decomposition), using only two measurements often resulted in inaccurate k estimates, particularly when using ML estimation with normal errors.

Across all simulations, using ML regression with beta errors resulted in very similar or more accurate k estimates than NLS or ML with normal errors (Figure 3). This was particularly true for

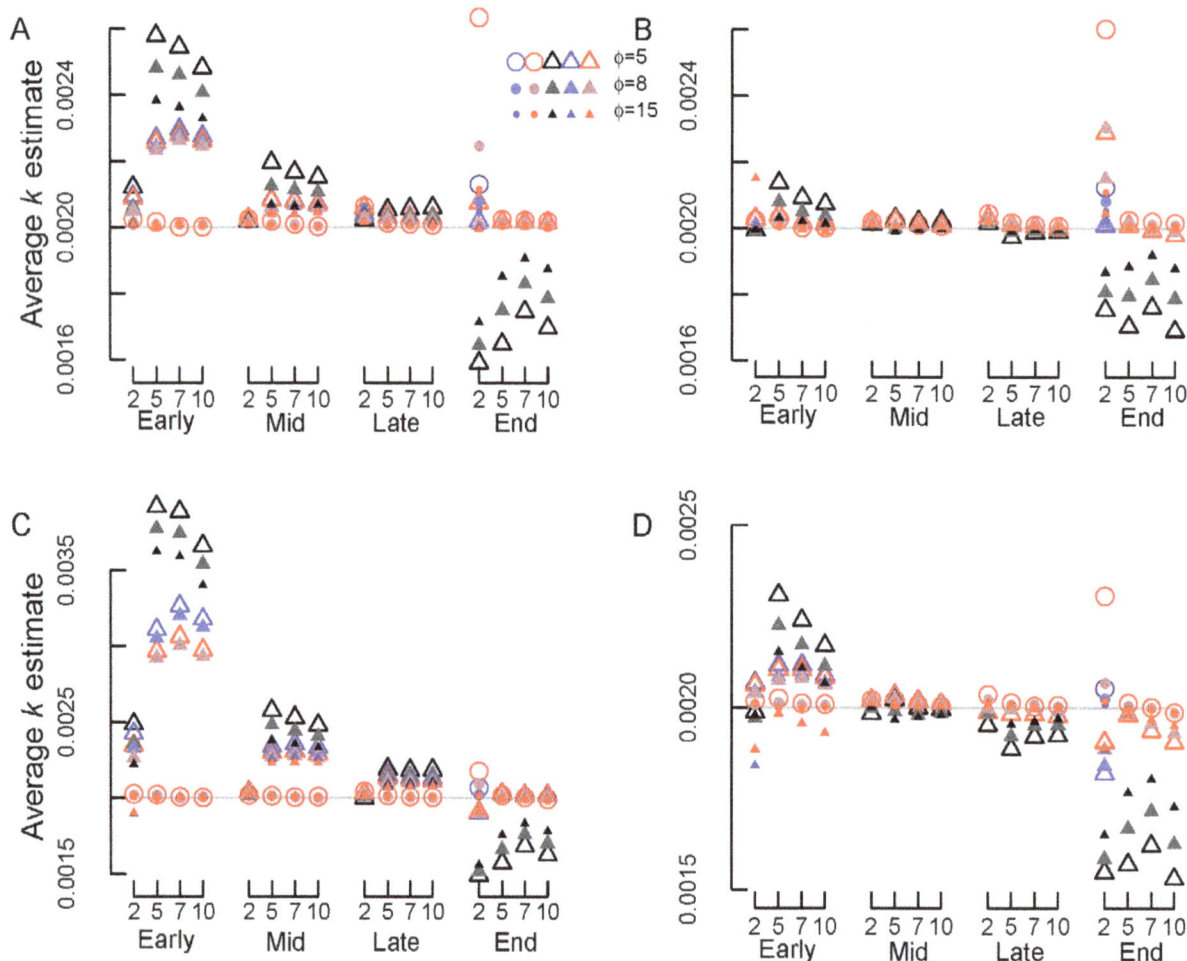

Figure 6. Average k estimates for beta-distributed errors plus normal errors with different σ and transformations. (A) σ = 0.0125 (option 3a) and SV transformation, (B) σ = 0.0125 (option 3a) and REP transformation, (C) σ = 0.05 (option 3b) and SV transformation, (D) σ = 0.05 (option 3b) and REP transformation. Early, mid, late and end are early, mid, late and end stage decomposition simulations. The numbers 2, 5, 7 and 10 are the numbers of measurements used in each simulation. Blue circles = NLS, Red circles = Normal ML, gray/black circles = Beta ML. In most cases, nls = Normal ML so that the red circles cover the blue circles. Gray lines show the true k value of 0.002 d^{-1}.

end stage decomposition with two measurements, where the beta model provided more accurate k estimates than the normal models (i.e., lower bias and RE, average k closer to true k of 0.002; Figure 3). However, beta ML regression did not successfully converge for all the data sets produced by the simulations (Table 2). Beta ML regression most often failed to converge during early decomposition, when ϕ <15, and the number of measurements was >2 (Table 2). In contrast, both NLS and normal ML estimation consistently successfully estimated k. In general, NLS and normal ML estimation produced nearly identical results. The exception was end stage decomposition with two measurements – in this case NLS produced slightly more accurate k estimates than did ML estimation with normal errors.

In most cases, AIC$_c$ identified ML estimation with beta errors as the best model (Table 3). In the majority of simulations, ML estimation with beta errors was identified as the best model in 90–100% of cases (Table 3). In the remaining simulations, AIC$_c$ generally showed either no difference between ML estimation with beta and normal errors (13.3–99.2% of cases) or found ML estimation with beta errors to be the best model (0–86.5% of cases; Table 3). Across all simulations, ML estimation with normal errors was only identified as the best model in 0–3% of cases.

Beta-distribution errors with normal errors added. For simulations with beta-distributed plus normal errors (option 3), the accuracy of k estimates again tended to increase (i.e., bias and RE declined) with the number of measurements and declining error (from ϕ = 5 to 15 and normal error σ = 0.05 to 0.0125; Figures 4,5,6). Estimates of k also improved from early to late stage decomposition (Figures 4,5,6). However, during end stage decomposition, k estimates became more variable, particularly when k was estimated using only two measurements (Figures 4,5,6).

With few exceptions, estimating k using NLS and ML estimation with normal errors on untransformed data produced similar (to one another) and more accurate k estimates (lower bias and RE) than did transforming the data and using NLS or ML estimation with normal or beta errors (Figures 4,5,6). The only exception was for the end stage decomposition simulation with only two measurements, where using ML estimation with normal errors produced high bias and RE (Figures 4–5). In general, beta regression on transformed data resulted in high bias and RE (Figures 4,5,6). These differences were most apparent in the early and end stage decomposition simulations; the smallest amount of bias and RE among estimation and transformation techniques (and thus k estimates) occurred during mid and late stage decomposition.

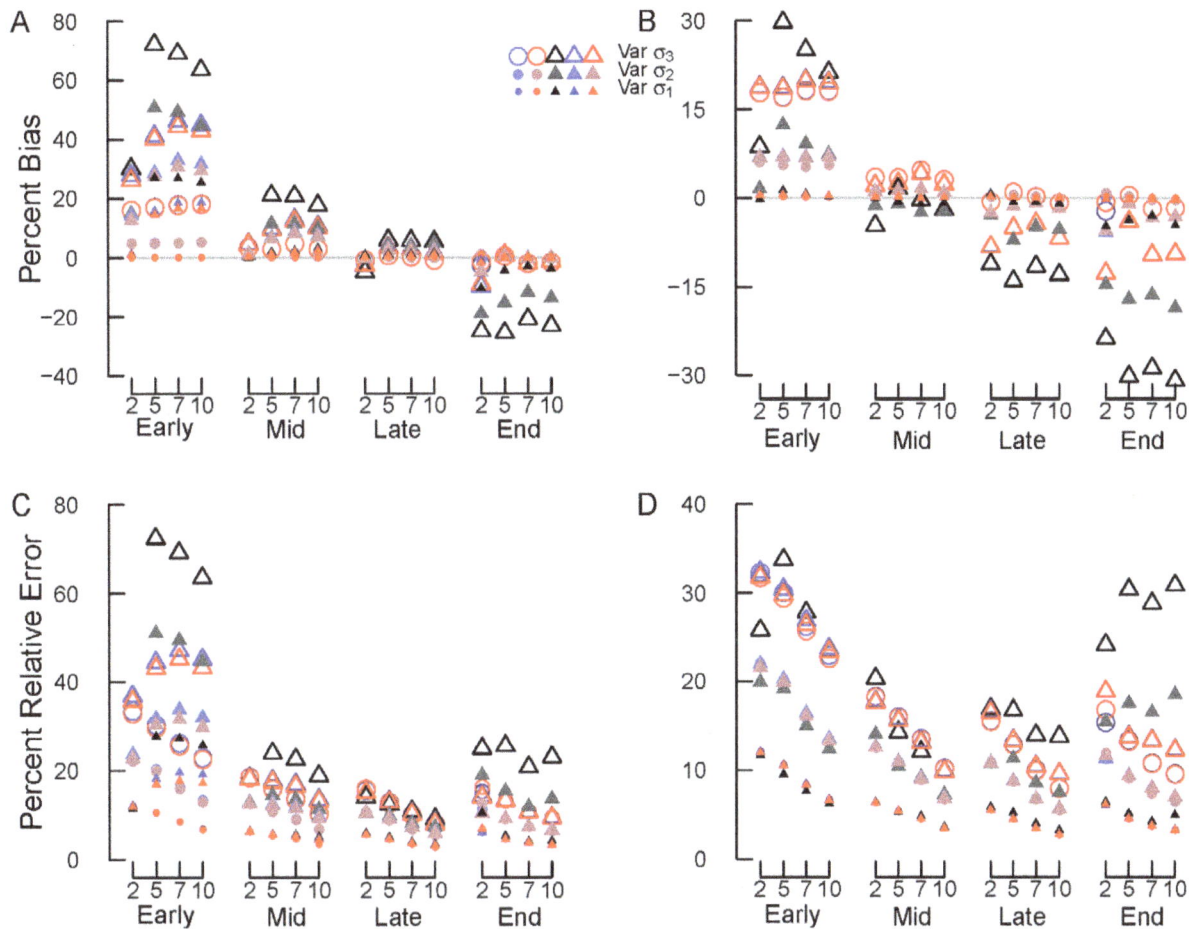

Figure 7. Results for simulations with variable normal errors (option 1). Percent bias using (A) SV and (B) REP transformations and relative error using (C) SV and (D) REP transformations. Early, mid, late and end are early, mid, late and end stage decomposition simulations. The numbers 2, 5, 7 and 10 are the numbers of measurements used in each simulation. Blue circles = NLS, Red circles = Normal ML, gray/black circles = Beta ML. In most cases, nls = Normal ML so that the red circles cover the blue circles. Gray lines in panels (A) and (B) show 0% bias.

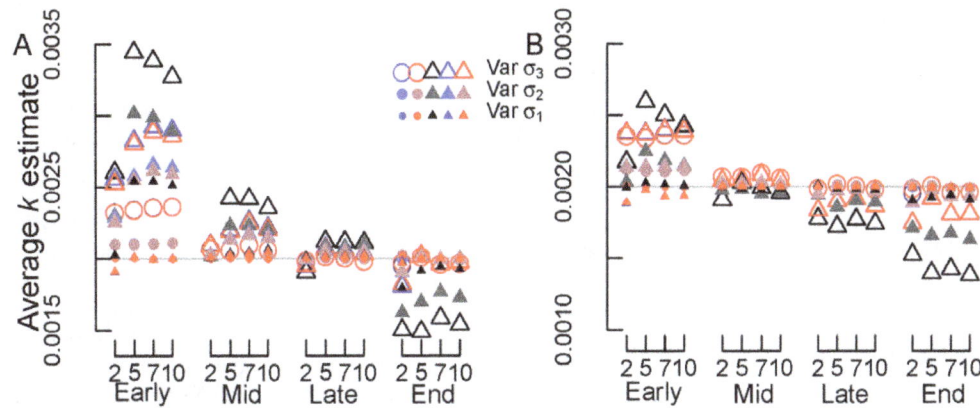

Figure 8. Average k estimates for simulations with variable normal errors (option 1). (A) SV and (B) REP transformations. Early, mid, late and end are early, mid, late and end stage decomposition simulations. The numbers 2, 5, 7 and 10 are the numbers of measurements used in each simulation. Blue circles = NLS, Red circles = Normal ML, gray/black circles = Beta ML. In most cases, nls = Normal ML so that the red circles cover the blue circles. Gray lines in panels (A) and (B) show the true k value of 0.002 d^{-1}.

Table 3. Percent of simulations using beta errors (option 2) for which AICc selected maximum likelihood (ML) estimation with beta or normal errors best or found no difference between the two models (Same) from each simulation (k = 0.0002).

Stage	# meas²	$\phi=5$ Same	Beta ML	Norm ML	$\phi=8$ Same	Beta ML	Norm ML	$\phi=15$ Same	Beta ML	Norm ML
Early	2	35.3	64.3	0.4	66.5	32.5	1.0	91.7	6.0	2.3
	5	0.0	100.0	0.0	0.0	100.0	0.0	0.1	99.9	0.0
	7	0.0	100.0	0.0	0.0	100.0	0.0	0.0	100.0	0.0
	10	0.0	100.0	0.0	0.0	100.0	0.0	0.0	100.0	0.0
Mid	2	98.3	1.2	0.6	99.2	0.0	0.8	99.2	0.0	0.8
	5	0.2	99.8	0.0	1.3	98.7	0.0	8.4	91.3	0.3
	7	0.1	99.9	0.0	0.6	99.4	0.0	4.2	95.6	0.2
	10	0.0	100.0	0.0	0.1	99.9	0.0	1.1	98.9	0.1
Late	2	87.6	11.6	0.8	97.0	1.6	1.4	97.8	0.0	2.2
	5	6.3	93.6	0.1	24.3	75.3	0.4	62.5	35.8	1.8
	7	3.4	96.6	0.1	13.3	86.5	0.2	40.7	58.2	1.2
	10	0.4	99.6	0.0	3.8	96.2	0.1	18.6	80.9	0.5
End	2	0.0	100.0	0.0	0.2	99.8	0.0	1.0	98.9	0.1
	5	0.0	100.0	0.0	0.0	100.0	0.0	0.1	99.9	0.0
	7	0.0	100.0	0.0	0.0	100.0	0.0	0.1	99.9	0.0
	10	0.0	100.0	0.0	0.0	100.0	0.0	0.0	100.0	0.0

Results were similar across all k values.
[1] Norm = normal.
[2] meas = measurements.

Overall, the REP transformation resulted in less bias and RE than did the SV transformation. This was especially apparent in early, mid and late stage decomposition. The amount of bias and RE generated by the REP and SV transformations was similar during end stage decomposition.

Despite the fact that ML estimation using beta errors tended to generate less accurate k estimates than ML estimation using normal errors, AIC_c generally showed either no difference between ML estimation using beta and normal errors or found ML estimation with beta errors to be the best model. For SV and REP transformed data with low normal error ($\sigma = 0.0125$), AIC_c either selected ML estimation with beta errors as the best model or found no difference between ML selection with beta and normal errors. Only during end stage decomposition with two measurements was ML estimation using normal errors selected as the best model more than 3% of the time (Table 4).

In SV transformed data with high normal error ($\sigma = 0.05$), AIC_c more frequently selected ML estimation with normal errors as the best model, particularly in early decomposition simulations with more than two measurements and end stage decomposition simulations with only two measurements (Table 5). In REP transformed data with high normal error ($\sigma = 0.05$), AIC_c again found either no difference between models or ML estimation with beta errors as the best model in the majority of cases across all simulations. Only during end stage decomposition was ML estimation with normal errors selected as the best model more than 8% of the time.

Variable σ Normal Error. Again, percent bias and RE declined from early to late stage decomposition and RE declined with number of measurements (Figures 7–8). Estimates of k also improved with declining error (from Var σ_1 to Var σ_3; Figures 7–8). However, increasing the number of measurements within decomposition stage failed to reduce percent bias and did not

typically improve average k estimates (Figures 7–8). Again, bias and RE increased during end stage decomposition (Figures 7–8), relative to mid and late stage decomposition simulations.

Using NLS or ML estimation with normal errors on untransformed data yielded the most consistently accurate k values with low bias and relative error across all decomposition stages and numbers of measurements (Figures 7–8). For transformed data, using the SV or REP transformation combined with NLS or ML estimation with normal errors frequently resulted in less bias and relative error than using ML estimation with beta errors (Figure 7). In certain cases, using beta regression on transformed data resulted in k values that were just as or more accurate than other methods: most frequently this occurred during mid and late stage decomposition.

In general, using the REP transformation resulted in less bias and relative error than did using the SV transformation (Figure 7). This was especially true during early to late stage decomposition. During end stage decomposition, both transformations generated similar levels of bias and relative error (Figure 7).

When the data were SV transformed, across all decomposition stages, numbers of measurements, and amounts of error used to create the simulated data, AICc generally identified ML estimation with beta errors as the best model or found no difference between ML estimation with beta or normal errors (Table 6). However, ML estimation with normal errors was identified as the best model more frequently than when other error structures were used to generate the data (i.e., beta or beta plus normal errors). In particular, AIC_c identified ML estimation with normal errors as the best model more frequently during early decomposition with more than two measurements, in end stage decomposition with only two measurements, and in mid and late decomposition when error was low (Var σ_1 and Var σ_2).

Table 4. Percent of beta error simulations with normal error ($\sigma = 0.0125$) added, for which AICc selected maximum likelihood (ML) estimation with beta or normal errors best or found no difference between the models (Same) from each simulation ($k = 0.0002$).

Tr[1]	Stage	# meas	$\phi = 5$			$\phi = 8$			$\phi = 15$		
			Same	Beta ML	Norm ML	Same	Beta ML	Norm ML	Same	Beta ML	Norm ML
SV[2]	Early	2	50.6	49.0	0.4	74.6	24.4	0.9	91.6	5.7	2.7
		5	4.6	94.6	0.8	8.7	90.3	1.0	16.4	81.5	2.0
		7	1.1	98.9	0.1	3.3	96.6	0.2	10.4	88.5	1.1
		10	0.3	99.7	0.0	1.5	98.4	0.1	6.1	93.0	0.9
	Mid	2	98.0	1.2	0.8	98.9	0.0	1.1	99.1	0.0	0.9
		5	2.3	97.6	0.1	6.2	93.7	0.1	16.8	82.7	0.5
		7	1.0	99.0	0.0	3.5	96.4	0.1	10.9	88.6	0.5
		10	0.5	99.5	0.0	1.6	98.4	0.1	5.5	94.2	0.3
	Late	2	89.2	10.1	0.7	96.9	1.5	1.6	97.8	0.0	2.2
		5	9.5	90.4	0.1	28.8	70.8	0.5	62.7	35.4	1.9
		7	5.4	94.5	0.1	17.1	82.5	0.4	43.9	55.0	1.1
		10	1.0	99.0	0.0	5.9	94.0	0.1	21.4	77.7	0.9
	End	2	34.3	43.6	22.1	41.8	35.5	22.6	49.0	25.6	25.5
		5	1.9	97.9	0.2	3.1	96.6	0.3	6.3	93.3	0.4
		7	0.1	99.9	0.0	0.6	99.4	0.0	2.1	97.8	0.1
		10	0.0	100.0	0.0	0.0	100.0	0.0	0.0	100.0	0.0
REP[3]	Early	2	32.8	66.9	0.4	62.2	36.5	1.3	89.2	7.4	3.5
		5	0.0	100.0	0.0	0.0	100.0	0.0	0.4	99.5	0.1
		7	0.0	100.0	0.0	0.0	100.0	0.0	0.2	99.8	0.0
		10	0.0	100.0	0.0	0.0	100.0	0.0	0.1	99.9	0.0
	Mid	2	97.7	1.6	0.8	99.2	0.1	0.8	99.1	0.0	0.9
		5	0.4	99.6	0.0	1.7	98.2	0.1	9.4	89.8	0.8
		7	0.2	99.8	0.0	1.2	98.8	0.1	5.5	94.0	0.5
		10	0.0	100.0	0.0	0.5	99.5	0.0	1.9	97.9	0.2
	Late	2	90.6	8.7	0.8	97.1	1.5	1.4	97.6	0.0	2.4
		5	6.9	93.0	0.2	24.0	75.3	0.7	62.3	34.8	2.9
		7	3.9	96.0	0.1	13.6	85.9	0.5	43.4	54.9	1.7
		10	0.7	99.3	0.0	4.4	95.5	0.2	19.5	79.2	1.4
	End	2	39.1	53.3	7.7	44.9	46.3	8.9	53.1	36.3	10.6
		5	8.2	91.1	0.8	11.0	88.2	0.8	14.5	84.4	1.1
		7	1.2	98.8	0.0	3.1	96.7	0.1	7.2	92.4	0.4
		10	0.1	100.0	0.0	0.2	99.8	0.0	1.1	98.8	0.2

[1]Tr = transformation.
[2]SV = Smithson and Verkuilen [6] transformation.
[3]REP = transformed by replacing values ≥ 1 with 0.9999 and treating zeros as missing data.

When using the REP transformation, AIC_c usually selected ML estimation with beta errors as the best model or found no difference between the models, especially when error was high or moderate (Var σ_2 and Var σ_3) and the number of measurements was more than two (Table 6). When error was low (Var σ_1), AIC_c more frequently showed ML estimation with normal errors to be the best model.

Real Data

Early stage decomposition data (hobbie and gough [10]). Overall, normal and beta errors produced similar k estimates within the transformed and untransformed data sets (Figure 9, Table 7). Fractional and relative bias for all transformation and error combinations were relatively small, but the SV transformation resulted in either similar or slightly more bias than the REP transformation or no transformation (Table 7). Within the untransformed and REP transformed data sets, using beta errors produced less bias than using normal errors.

In 13 of 18 cases, the beta distribution could be used on untransformed data (all values >0 and <1). In these cases the beta model was best ($\Delta \mathrm{AIC}_c \geq 4$) in four cases. In the nine remaining cases, the models were indistinguishable based on AIC_c. When the data were SV transformed, the beta distribution produced the best model in five cases, but in the remaining 13 cases the models were indistinguishable. For REP transformed data, the beta model was best in nine cases; for the remaining nine cases the models were indistinguishable.

Table 5. Percent of beta error simulations with normal error ($\sigma = 0.05$) added, for which AICc selected maximum likelihood (ML) estimation with beta or normal errors best or found no difference between the models (Same) from each simulation ($k = 0.0002$).

Tr[1]	Stage	# meas	$\phi = 5$			$\phi = 8$			$\phi = 15$		
			Same	Beta ML	Norm ML	Same	Beta ML	Norm ML	Same	Beta ML	Norm ML
SV[2]	Early	2	73.7	25.9	0.4	85.1	13.9	1.0	92.8	4.3	2.9
		5	33.7	19.3	46.9	31.7	13.2	55.1	25.5	6.9	67.6
		7	36.3	27.8	35.8	33.4	17.6	49.0	26.9	8.6	64.5
		10	31.4	25.9	42.7	26.9	14.6	58.4	17.3	6.2	76.6
	Mid	2	97.8	1.3	0.9	98.4	0.2	1.5	98.2	0.0	1.9
		5	40.1	55.1	4.9	46.9	44.9	8.2	53.1	31.8	15.2
		7	34.2	60.9	4.9	43.1	48.6	8.4	50.0	33.8	16.2
		10	33.9	58.7	7.3	40.2	46.8	13.0	43.6	33.9	22.5
	Late	2	90.5	8.7	0.9	96.7	1.9	1.4	96.3	0.2	3.5
		5	26.4	73.2	0.4	46.8	51.1	2.1	71.4	22.4	6.2
		7	20.3	79.0	0.8	38.8	58.8	2.4	60.4	32.9	6.7
		10	9.1	90.5	0.4	25.1	72.5	2.4	47.8	43.8	8.4
	End	2	41.5	20.8	37.8	45.6	13.7	40.7	44.4	6.2	49.4
		5	12.2	85.4	2.4	18.0	78.6	3.4	28.7	66.0	5.3
		7	1.9	97.9	0.3	6.1	93.2	0.7	15.5	81.8	2.7
		10	0.1	99.9	0.0	0.5	99.4	0.1	2.8	96.7	0.5
REP[3]	Early	2	27.1	72.1	0.8	47.2	50.7	2.1	72.5	21.8	5.7
		5	0.1	99.8	0.0	0.4	99.5	0.1	1.2	98.4	0.4
		7	0.0	100.0	0.0	0.1	99.9	0.0	0.9	98.9	0.3
		10	0.0	100.0	0.0	0.0	100.0	0.0	0.4	99.5	0.1
	Mid	2	93.3	5.7	1.0	97.2	0.8	2.0	97.6	0.0	2.4
		5	0.8	99.1	0.1	3.2	96.5	0.3	10.6	87.3	2.1
		7	0.6	99.4	0.1	2.7	96.8	0.5	9.5	87.5	3.0
		10	0.2	99.7	0.1	1.8	97.8	0.5	5.9	91.9	2.3
	Late	2	93.1	5.8	1.0	97.1	1.4	1.6	96.4	0.1	3.5
		5	9.1	90.4	0.5	24.1	73.7	2.2	54.3	38.7	7.1
		7	6.6	92.7	0.8	15.8	82.1	2.2	39.4	54.8	5.8
		10	2.0	97.8	0.2	7.0	91.9	1.1	23.4	71.7	4.9
	End	2	61.9	17.0	21.1	63.1	12.9	24.0	61.3	7.5	31.2
		5	36.1	56.0	8.0	46.3	41.2	12.5	53.5	28.3	18.2
		7	16.2	81.7	2.1	30.2	64.7	5.1	42.7	44.6	12.7
		10	6.0	93.1	0.9	14.1	82.6	3.3	28.6	59.3	12.1

[1]Tr = transformation.
[2]SV = Smithson and Verkuilen [6] transformation.
[3]REP = transformed by replacing values ≥ 1 with 0.9999 and treating zeros as missing data.

Mid stage decomposition data (laliberté and tylianakis [12]). Using normal and beta errors generally produced very similar k estimates (Figure 9). Notable exceptions were when the data were transformed and k was greater than ~0.01 d^{-1} (Figure 9b,d), in which case the normal model gave larger k estimates than the beta model (Figure 9b,d). This was particularly evident when using the SV transformation (Figure 9b). In these cases, the beta model produced more biased predictions than the normal model (Table S1). For the SV transformed data, FB and RB were 14–60% larger for the beta than normal model. For the REP transformed data, FB and RB were 1.4 to 18 times larger for the beta than normal model (Table S1). Despite the larger bias associated with the beta model for the SV data, the beta model was identified as best in three cases ($\Delta AIC_c \geq 4$; Table S1). In the remaining cases the models were indistinguishable ($\Delta AIC_c < 4$; Table S1). For the REP transformed data, the models were indistinguishable in all cases (Table S1).

In general, using the SV data transformation resulted in similar or slightly more bias than the REP or no transformation (Table 7). Within the untransformed and REP transformed data, using beta errors produced predictions with similar or less bias than did using normal errors.

In 40 of 64 cases, the data did not need to be transformed to use beta errors. Based on AIC$_c$, the beta model was best ($\Delta AIC_c \geq 4$) in only three of these cases. In 18 cases the normal model was best. For the remaining cases, the models were indistinguishable. With

Table 6. Percent of variable normal σ simulations for which AICc selected maximum likelihood (ML) estimation with beta or normal errors best or found no difference between the models (Same) from each simulation ($k = 0.0002$).

Tr[1]	Stage	# meas	Var σ_3			Var σ_2			Var σ_1		
			Same	Beta ML	Norm ML	Same	Beta ML	Norm ML	Same	Beta ML	Norm ML
SV[2]	Early	2	75.1	24.1	0.8	90.1	7.1	2.8	90.7	0.1	9.2
		5	39.3	32.5	28.1	39.3	21.7	39.0	39.7	22.6	37.8
		7	37.3	44.2	18.5	39.3	21.1	39.6	35.5	16.7	47.8
		10	31.8	51.3	16.9	33.1	20.7	46.2	26.8	14.9	58.3
	Mid	2	95.8	3.0	1.3	96.5	0.1	3.5	98.3	0.0	1.7
		5	36.7	57.3	6.0	45.7	47.7	6.6	48.1	41.4	10.5
		7	29.4	66.6	3.9	44.6	43.6	11.8	43.7	38.9	17.4
		10	30.7	60.1	9.3	39.9	41.7	18.5	37.1	47.9	15.0
	Late	2	91.5	7.6	0.9	95.8	0.9	3.3	89.5	0.0	10.5
		5	30.5	67.4	2.1	63.6	27.8	8.6	74.7	1.3	24.0
		7	28.1	68.8	3.0	56.5	31.4	12.2	68.4	3.5	28.1
		10	14.0	84.2	1.8	46.1	39.1	14.8	55.0	6.1	38.9
	End	2	54.5	8.3	37.2	49.1	5.2	45.8	56.7	5.8	37.5
		5	21.4	73.0	5.6	24.2	70.8	5.0	23.0	74.8	2.1
		7	6.2	92.9	0.9	14.6	82.7	2.7	20.0	75.5	4.5
		10	0.6	99.3	0.1	1.9	97.7	0.4	2.3	97.3	0.4
REP[3]	Early	2	23.2	74.8	2.0	53.1	39.6	7.3	85.6	0.5	14.0
		5	0.6	99.2	0.2	1.9	97.4	0.7	10.3	86.3	3.4
		7	0.1	99.8	0.0	1.2	98.3	0.6	13.2	79.6	7.2
		10	0.0	100.0	0.0	0.4	99.4	0.1	8.8	85.0	6.2
	Mid	2	77.9	19.3	2.8	92.7	1.1	6.3	98.4	0.0	1.7
		5	3.9	95.1	1.0	13.9	82.7	3.5	42.4	37.7	19.9
		7	1.5	98.1	0.4	10.6	84.8	4.5	35.2	42.3	22.6
		10	1.3	98.2	0.5	8.6	86.4	5.0	26.2	55.3	18.6
	Late	2	95.6	2.9	1.5	95.9	0.3	3.8	90.2	0.0	9.9
		5	11.7	86.0	2.4	39.4	49.4	11.2	66.2	1.3	32.5
		7	10.3	86.4	3.3	32.3	56.1	11.7	55.0	3.5	41.6
		10	6.0	91.8	2.2	21.3	69.5	9.3	42.4	4.9	52.8
	End	2	70.4	6.6	23.0	65.4	7.1	27.6	67.7	12.1	20.2
		5	37.3	48.7	14.0	48.7	35.4	16.0	35.5	60.3	4.3
		7	26.2	68.1	5.8	39.0	49.5	11.6	30.9	60.8	8.4
		10	9.2	88.4	2.4	16.5	78.3	5.2	7.4	86.7	5.9

[1]Tr = transformation.
[2]SV = Smithson and Verkuilen [6] transformation.
[3]REP = transformed by replacing values ≥ 1 with 0.9999 and treating zeros as missing data.

the SV transformed data, the beta model was best in 14 cases, while the normal model was best in 18 cases. Using the REP transformation, the beta model was best in four cases, while the normal model was best in 18 cases. The models were indistinguishable in all remaining cases.

Late stage decomposition data (hobbie [14]). Again, normal and beta distributed errors produced largely similar k estimates within the same data set (Figure 9). However, at high k values (>0.0015 d^{-1}), beta models produced slightly lower k estimates than normal models (Fig 9b,c). Unlike medium stage data, this was also true for untransformed data, and there were no consistent patterns in bias for these points (Table S1). Of the 21 cases where k estimated using normal errors was ≥ 0.0015 d^{-1}, for both the SV and REP transformations, the beta model was best in

11 cases, the normal model was best in nine cases, and there was no difference between the models in one case (Table S1). For the untransformed data, when k could be estimated using beta errors (9 of 21 cases), using the normal model resulted in less bias and was the best model (data not shown).

Using the SV transformation resulted in predictions with similar or more bias than using no transformation or the REP transformation (Table 7). The only exception was for the beta model, where less bias was generated using the SV than REP transformation. In the untransformed data set, normal and beta errors produced similar bias; in the REP transformed data, using beta errors produced slightly less bias than using normal errors (Table 7).

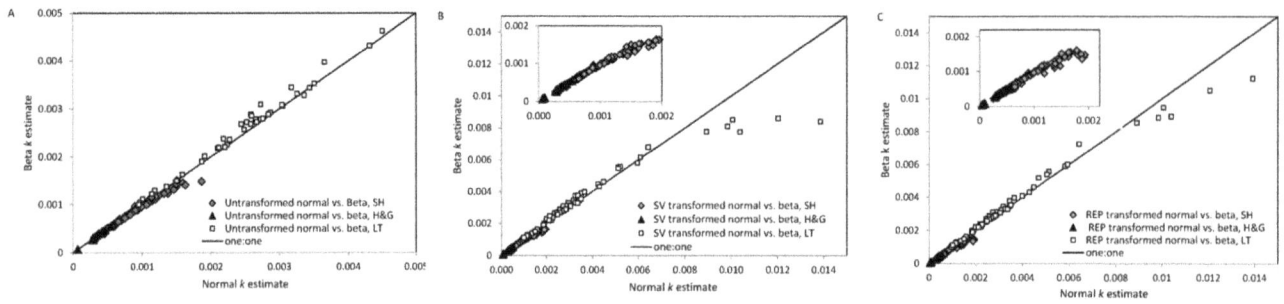

Figure 9. Daily decomposition rate (*k*) estimates for the Hobbie (SH) [14], Laliberté and Tylianakis (LT) [12] and Hobbie and Gough (H&G) [10] data compared by error distribution (beta or normal) used to estimate *k* (A) untransformed (B) Smithson and Verkuilen (SV) [6] transformed and (C) replacement (zeros = missing data; values ≥1 = 0.9999) transformed data sets. Insets in (b) and (c) show only the SH and H&G data.

Of the 79 cases where the beta model could be used on untransformed data, it was best in 33, whereas the normal model was best in 22 cases. The models were indistinguishable in 24 cases. Using SV transformation, in 25 out of 128 cases there was no substantial difference between the models. In the majority of cases (69) the beta model was best. The normal model was best in 34 cases. Using the REP transformation, the beta model was best in 75 cases, the normal model was best in 29 cases, and the models were indistinguishable in 24 cases.

Discussion

Proportional litter mass loss data generally show reduced variance near its bounds (i.e. 0 and 1), but researchers generally use single pool decomposition models that ignore such heteroscedasticity, potentially leading to biased *k* estimates [3]. For example, the most recent recommendation to use standard nonlinear regression on untransformed proportional mass loss data still assumes constant, normally-distributed errors (a problem acknowledged by these authors [3]). We therefore evaluated the

Table 7. Mean *k* (decomposition rate), fractional bias (FB) and relative bias (RB) produced by each data transformation and error structure using the Hobbie [14], Laliberté and Tylianakis [12] and Hobbie and Gough [10] data sets.

Data	Transformation	Error	Mean *k*(d^{-1})	Mean FB	σ FB	Mean RB	σ RB
Hobbie & Gough							
	None	Beta	0.00055	0.0001	0.0020	0.0047	0.0190
		Normal	0.00054	0.0026	0.0088	0.0308	0.1163
	SV[1]	Beta	0.00055	−0.0042	0.0083	−0.1394	0.2908
		Normal	0.00054	−0.0006	0.0121	−0.0758	0.2923
	REP[2]	Beta	0.00055	−0.0002	0.0018	−0.0049	0.0276
		Normal	0.00054	0.0024	0.0089	0.0263	0.1216
Laliberté & Tylianakis							
	None	Beta	0.00258	0.1090	0.0481	0.2778	0.0985
		Normal	0.00363	0.1682	0.1082	0.3399	0.1219
	SV[1]	Beta	0.00349	0.1646	0.1438	0.3256	0.1979
		Normal	0.00361	0.1693	0.1098	0.3416	0.1244
	REP[2]	Beta	0.00357	0.0192	0.0449	0.0279	0.0647
		Normal	0.00356	0.0286	0.0303	0.0561	0.0495
Hobbie							
	None	Beta	0.00088	0.0123	0.0293	0.0249	0.0660
		Normal	0.00091	−0.0142	0.0227	−0.0236	0.0574
	SV[1]	Beta	0.00090	−0.0097	0.0604	−0.0291	0.1247
		Normal	0.00095	−0.0343	0.0562	−0.0636	0.1192
	REP[2]	Beta	0.00084	0.0220	0.0453	0.0330	0.0675
		Normal	0.00091	−0.0148	0.0232	−0.0251	0.0588

[1]SV = Smithson and Verkuilen [6] transformation.
[2]REP = data transformed by replacing values ≥1 with 0.9999 and treating zeros as missing data.

potential of beta regression, which is well suited to bounded data and its associated heteroscedasticity [6],[7]. We hypothesized that nonlinear beta regression would provide a better fit to proportional litter mass loss data, and more accurate k estimates, than standard nonlinear regression in simulated and real decomposition data sets.

Contrary to our hypothesis, we found that standard nonlinear regression with constant, normal errors proved very robust to violations of homoscedasticity. In our simulations, k estimates obtained via the normal model (NLS or ML estimation) on untransformed data were equally or more accurate as those obtained with the beta model, regardless of error structure and data transformation. On transformed and untransformed (beta errors only) data, ML estimation using beta errors tended to generate less accurate k estimates than ML estimation using normal errors. This occurred despite the beta model being clearly equal or superior in nearly all cases to the normal model, as determined by AIC_c. Thus, our concern that standard nonlinear regression may lead to biased k estimates in the presence of heteroscedasticity appears to be unjustified by our simulation results. However, we do not imply that researchers should use standard nonlinear regression even when its assumptions are violated, simply because these results did not show systematic biases in k estimates. Still, it is important to note that k values previously estimated in the presence of heteroscedasticity using standard nonlinear regression should not be strongly biased.

Our simulations also provided information for the design of decomposition experiments, suggesting that the accuracy of k estimates increases with the number of measurements and with the length of the study, from early to late decomposition. Estimates of k from end stage decomposition were less accurate (or at least more variable between estimation methods), perhaps due to an increasing number of zero measurements or missing data (REP transformation), which may bias estimates [3]. In general, mid and late stage decomposition had the least amount of between method and data transformation variation in k estimates, suggesting that studies in these ranges will be less impacted by regression method choice.

Obviously, with real proportional litter mass loss data we cannot evaluate how "biased" k estimates are, because we do not know the "true" k value (which is why it must be estimated from data). Yet, we must make an informed decision on which model provides the best estimate of k. Tools at our disposal include various measures of model fit such as AIC [18], and visual inspection of model predictions and residuals to evaluate model assumptions [19]. In untransformed or REP transformed real data, the beta model produced slightly less bias than did the normal model. Using AIC_c, we found that the models were indistinguishable from each other in the majority of cases. Therefore, we recommend a pragmatic approach where both models are compared and the best one is selected for a given data set (particularly when k estimates are high and normal and beta model k estimations diverge). Alternatively, one may use model averaging to calculate the weighted average of k using both the beta and normal models [18]. This technique has been successfully used to estimate accurate parameters for biological power functions, where similar error structure issues are encountered (normal vs. lognormal models/errors) [5].

While we have focused on the beta distribution because it suits bounded data especially well [6],[7], several other distributions could be used to suit particular situations [16]. Yet, the beta distribution will be especially useful to estimate decomposition rates in single pool models because it easily accommodates the type of heteroscedasticity encountered in proportional mass loss data. In practice, a particular statistical model is often favored by researchers not just because it fits the data better, but for other pragmatic reasons such as computational simplicity [16],[19]. Unlike standard nonlinear regression, nonlinear beta regression is not widely implemented in mainstream statistical packages. This does not mean, however, that nonlinear beta regression is more complex than standard nonlinear regression with normal errors. Like the normal distribution, the beta distribution contains only two parameters and can be easily parameterized with location and precision (the inverse of dispersion) parameters [6,7] (see Introduction). To facilitate the use of nonlinear beta regression in single pool decomposition models, we provide code to implement this approach in the freely available R environment [9] (Appendix S1). Because the beta distribution does not allow values ≤ 0 or ≥ 1, which often occur in proportional litter mass loss data, transformations to constrain the data in the $]0, 1[$ interval may be required. We evaluated two such transformations: the SV [6] and REP transformations. The SV transformation simultaneously standardizes all values, so the transformed data stay perfectly correlated with the untransformed data. In contrast, the REP transformation removes data points, treating zeros as missing values, and converts values ≥ 1 to 0.999. However, using the SV transformation resulted in slightly more error (simulations) or bias (real data) than did using the REP or no transformation.

The potential negative impacts of rapid increases in atmospheric CO_2 require a better understanding of the critical role of litter decomposition in the global carbon cycle. This, in turn, requires accurate estimates of litter decomposition rates. Our results show that nonlinear beta regression is a useful method for estimating these rates. However, with the data explored to date, it did not often produce dramatically different results from standard nonlinear regression. Yet, given the type of heteroscedasticity found in most decomposition data, we suggest that the two methods should be considered alongside one another. Furthermore, our results suggest that regression method choice will have the smallest impacts during mid and late stage decomposition.

Supporting Information

Figure S1 Minimum and maximum decomposition rates (k) versus total experiment time from the Adair et al. [3] single pool decomposition review.

Figure S2 Histograms of sampling times as a proportion of total time from the Adair et al. [3] single pool decomposition review.

Figure S3 Percent bias for simulations using beta error only with k estimated by each regression technique.

Figure S4 Percent bias for simulations using beta error only with k estimated by each regression technique.

Figure S5 Average k value for simulations using beta error only with k estimated by each regression technique.

Figure S6 Percent bias for simulations using beta error with normal error ($\sigma = 0.05$) added with k estimated by each regression technique (SV transformation).

Figure S7 Percent relative error for simulations using beta error with normal error ($\sigma = 0.05$) added with k estimated by each regression technique (SV transformation).

Figure S8 Average k value for simulations using beta error with normal error ($\sigma = 0.05$) added with k estimated by each regression technique (SV transformation).

Table S1 Bias and ΔAIC_c values for Hobbie [14] and Laliberté and Tylianakis [12] data where normal k estimates were greater than beta k estimates.

Appendix S1 R code to perform nonlinear beta regression.

Acknowledgments

We thank B.M. Bolker for contributing the beta regression R code. We appreciate the thorough and thoughtful comments provided by Ben Bond-Lamberty and anonymous reviewers.

Author Contributions

Conceived and designed the experiments: EL ECA. Performed the experiments: EL ECA. Analyzed the data: EL ECA. Contributed reagents/materials/analysis tools: EL ECA SEH. Wrote the paper: EL ECA SEH. Contributed data: EL SEH.

References

1. Swift MJ, Heal OW, Anderson JM (1979) Decomposition in Terrestrial Ecosystems. Oxford, UK: Blackwell.
2. Olson JS (1963) Energy storage and balance of producers and decomposers in ecological systems. Ecology 44: 322-&.
3. Adair EC, Hobbie SE, Hobbie RK (2010) Single-pool exponential decomposition models: potential pitfalls in their use in ecological studies. Ecology 91: 1225–1236.
4. Hayes JP, Shonkwiler JS (2006) Allometry, antilog transformations, and the perils of prediction on the original scale. Physiological and Biochemical Zoology 79: 665–674.
5. Xiao X, White EP, Hooten MB, Durham SL (2011) On the use of log-transformation vs. nonlinear regression for analyzing biological power laws. Ecology 92: 1887–1894.
6. Smithson M, Verkuilen J (2006) A better lemon squeezer? Maximum-likelihood regression with beta-distributed dependent variables. Psychological Methods 11: 54–71.
7. Paolino P (2001) Maximum liklelihood estimation of models with beta-distributed dependent variables. Political Analysis 9: 325–346.
8. Bolker BM (2010) bbmle: Tools for general maximum likelihood estimation. The Comprehensive R Archive Network (CRAN). Vienna, Austria.
9. R Development Core Team (2011) R: A language and environment for statistical computing. R Foundation for Statistical Computing, Vienna, Austria.
10. Hobbie SE, Gough L (2004) Litter decomposition in moist acidic and non-acidic tundra with different glacial histories. Oecologia 140: 113–124.
11. Shaver GR (2001) NPP Tundra: Toolik Lake, Alaska, 1982. Data set.: Oak Ridge National Laboratory Distributed Active Archive Center, Oak Ridge, Tennessee, USA.
12. Laliberté E, Tylianakis JM (2012) Cascading effects of long-term land-use changes on plant traits and ecosystem functioning. Ecology 93: 145–155.
13. Scott D (1999) Sustainability of New Zealand high-country pastures under contrasting development inputs. 1. Site, and shoot nutrients. New Zealand Journal of Agricultural Research 42: 365–383.
14. Hobbie SE (2008) Nitrogen effects on decomposition: A five-year experiment in eight temperate sites. Ecology 89: 2633–2644.
15. Hobbie SE (2005) Contrasting effects of substrate and fertilizer nitrogen on the early stages of litter decomposition. Ecosystems 8: 644–656.
16. Bolker BM (2008) Ecological Models and Data in R. Princeton, New Jersey, USA: Princeton University Press.
17. Janssen PHM, Heuberger PSC (1995) Calibration of process-oriented models. Ecological Modelling 83: 55–66.
18. Burnham KP, Anderson DR (2004) Multimodel inference - understanding AIC and BIC in model selection. Sociological Methods & Research 33: 261–304.
19. Zuur AF, Ieno EN, Walker NJ, Saveliev AA, Smith GM (2009) Mixed Effects Models and Extensions in Ecology with R. New York, USA: Springer.

5

Costs of Defense and a Test of the Carbon-Nutrient Balance and Growth-Differentiation Balance Hypotheses for Two Co-Occurring Classes of Plant Defense

Tara Joy Massad[1]*[¤], Lee A. Dyer[2], Gerardo Vega C.[2]

1 Department of Ecology and Evolutionary Biology, Tulane University, New Orleans, Louisiana, United States of America, **2** Department of Biology, University of Nevada, Reno, Nevada, United States of America

Abstract

One of the goals of chemical ecology is to assess costs of plant defenses. Intraspecific trade-offs between growth and defense are traditionally viewed in the context of the carbon-nutrient balance hypothesis (CNBH) and the growth-differentiation balance hypothesis (GDBH). Broadly, these hypotheses suggest that growth is limited by deficiencies in carbon or nitrogen while rates of photosynthesis remain unchanged, and the subsequent reduced growth results in the more abundant resource being invested in increased defense (mass-balance based allocation). The GDBH further predicts trade-offs in growth and defense should only be observed when resources are abundant. Most support for these hypotheses comes from work with phenolics. We examined trade-offs related to production of two classes of defenses, saponins (triterpenoids) and flavans (phenolics), in *Pentaclethra macroloba* (Fabaceae), an abundant tree in Costa Rican wet forests. We quantified physiological costs of plant defenses by measuring photosynthetic parameters (which are often assumed to be stable) in addition to biomass. *Pentaclethra macroloba* were grown in full sunlight or shade under three levels of nitrogen alone or with conspecific neighbors that could potentially alter nutrient availability via competition or facilitation. Biomass and photosynthesis were not affected by nitrogen or competition for seedlings in full sunlight, but they responded positively to nitrogen in shade-grown plants. The trade-off predicted by the GDBH between growth and metabolite production was only present between flavans and biomass in sun-grown plants (abundant resource conditions). Support was also only partial for the CNBH as flavans declined with nitrogen but saponins increased. This suggests saponin production should be considered in terms of detailed biosynthetic pathway models while phenolic production fits mass-balance based allocation models (such as the CNBH). Contrary to expectations based on the two defense hypotheses, trade-offs were found between defenses and photosynthesis, indicating that studies of plant defenses should include direct measures of physiological responses.

Editor: Martin Heil, Centro de Investigación y de Estudios Avanzados, Mexico

Funding: This work was funded by an Environmental Protection Agency STAR Fellowship; support was also provided by the National Science Foundation grant, CHE 0849369, and the Organization for Tropical Studies. The funders had no role in study design, data collection and analysis, decision to publish, or preparation of the manuscript

Competing Interests: The authors have declared that no competing interests exist.

* E-mail: tmassad77@gmail.com

¤ Current address: Program on the Global Environment, University of Chicago, Chicago, Illinois, United States of America

Introduction

Herbivory and neighboring plant competition for resources are two of the most important biotic forces affecting plant distributions and fitness [1]. Competition, resource availability, and herbivory can affect levels of defensive compounds in plants, since chemical defense is a plastic response. Production of secondary metabolites is often associated with reduced fitness in terms of lower growth and reproduction [2–10]. This trade-off between investment in plant defense versus growth and reproduction is termed an allocation cost [10,11]. However, comparisons between defense and growth or reproduction may be insufficient to quantify the costs of defense because natural selection may strongly favor reductions in trade-offs between such important activities as growth, reproduction, and defense. Physiological parameters can be more useful than growth rates for quantifying the cost of plant defenses [12–16,8,10] (but see [17]). Physiological costs, such as

reductions in photosynthetic enzymes or the biosynthesis of other proteins required for primary metabolism are said to arise from 'metabolic competition' between defense production and primary metabolic functions [18]. Further examination of physiological costs is important for determining the mechanisms underlying allocation costs and for understanding interactions between pathways leading to primary and secondary metabolites. In addition, despite the notable contributions of induced defense literature to understanding costs of chemical defense, it may be particularly interesting to study costs in constitutive defenses to understand the baseline value plants place on tissue retention.

In terms of physiological costs, photosynthesis is among the most important variables to quantify as it forms the foundation of a plant's carbon budget. Studies combining measures of plant defense and photosynthesis can also help clarify two prominent mass-balance based hypotheses of secondary metabolite production. The carbon-nutrient balance hypothesis (CNBH) [19] and the growth-differentiation balance hypothesis (GDBH) [11] were formulated to address differences in defense concentrations among individuals within a species; both hypotheses stem from the assumption that an imbalance in nutrients and carbon will allow plants to invest excess resources in defense as growth becomes limited before photosynthesis. Plants that produce nitrogen-containing defensive compounds (N-based defenses) are expected to increase their production of defenses when available nitrogen is more abundant than carbon; likewise, plants capable of synthesizing carbon-based secondary metabolites (C-based defenses) should increase production when fixed carbon exceeds requirements for growth [11,19]. Nitrogen-rich enzymes and nitrogen-containing precursors are involved in the production of what are termed C-based defenses [20–23], however, so this classification of defenses as C- or N-based may be an oversimplification and confound interpretation of responses to resources in the framework of the CNBH or GDBH. There has, in fact, been much debate as to the utility of the CNBH [24,25], and it has also been erroneously applied [26]. Nonetheless, the empirical support for this hypothesis shows predicted patterns of phenotypic changes in defenses for temperate woody [27,28], herbaceous [29], and tropical [30–33] species.

The GDBH is more detailed than the CNBH and predicts a negative correlation between growth and defense under conditions of moderate to high resource availability [11]. The GDBH is difficult to test because: 1) a broad range of resource availability must be included in studies, 2) most variables assessed are merely correlates of the plastic physiological processes that are part of the hypothesis (e.g., biomass is often a proxy for resource allocation to growth, but it can include tissues and compounds important in defense and storage as well), and 3) it is difficult to ensure the maintenance of experimental resource conditions throughout a plant's growth [34]. Despite these challenges, valuable insights on trade-offs and priorities in plant resource allocation can be gained from studies addressing aspects of the GDBH [35–37].

A key postulate of the CNBH and the GDBH is that defenses will increase under conditions of limited growth when photosynthesis continues to function at normal levels. This mechanistic aspect of the hypotheses is difficult to test, yet some studies have measured photosynthesis, growth, and defense simultaneously. Results from these studies show a variety of patterns. Light can increase photosynthesis and N-based defenses but decrease C-based defenses [38]; available nitrogen can increase photosynthesis and monoterpene production (except during the leaf expansion stage) [39], and high nitrogen can have inverse effects on photosynthesis (positive) and phenolic defenses (negative) [40,41]. In addition, the down-regulation of genes important to photosynthesis has been shown to accompany herbivore induced up-regulation of defenses in *Nicotiana attenuata* (Solanaceae) [42,43], although resource conditions mediate changes in transcription such that they do not always correspond to equivalent changes in the products encoded for [43]. Nevertheless, the paradigm persists that growth is more sensitive to a plant's resource environment than is photosynthesis, and decreased growth with concomitant increases in defenses has been documented many times [11,33,44–47]. The sensitivity of photosynthesis to environmental conditions and the connection between photosynthesis and growth and defense production merit more empirical study.

Here we present experimental results quantifying saponin (terpenoid) and flavan (phenolic) production in a neotropical tree, *Pentaclethra macroloba* Kuntze (Fabaceae: Mimosoideae), a shade-tolerant species with nitrogen-fixing root nodules [48] that produces high levels of saponins which function as an antiherbi-

vore defense [49,50] as well as flavonoids. Saponins are a class of glycosylated triterpenoid, steroid, or steroidal alkaloid C-based compounds produced primarily via the mevalonic acid pathway [51], and flavans are flavonoids known to serve as plant defenses in a related genus, *Inga* [52,53]. Most studies addressing the CNBH and GDBH have focused on phenolics [54], making studies of other classes of defense important. Terpenoids are especially interesting because they are produced by the mevalonic acid pathway, and defenses from this pathway do not fit predictions of the CNBH and GDBH as well as the phenolics produced via the shikimic acid pathway [20,54–56].

We tested the hypothesis that saponin production in *P. macroloba* seedlings incurs both physiological (photosynthetic) and allocation (biomass) costs. We measured saponin and flavan production under different light regimes in response to changes in nutrients and plant density to test ecological predictions made by the CNBH and GDBH. Tests of the CNBH and GDBH have been criticized for not measuring complete costs of secondary defenses [57]; by quantifying relationships between two separate classes of defense, as well as photosynthesis and growth, we do not escape this criticism, but attempt to provide a more complete measure of these trade-offs.

Materials and Methods

We collected seeds of *Pentaclethra macroloba* from multiple individuals distributed throughout the forest of La Tirimbina Rainforest Center, Sarapiqui, Heredia, Costa Rica (10° 23 N, 84°8 W) in January 2008. *Pentaclethra macroloba* was selected for this study because of its dominance in tropical forests where it is found [58], its diverse defensive chemistry, and the ease with which seeds can be found and propagated. La Tirimbina contains 345 ha of tropical wet forest (*sensu* [59]) with an average of 4000 mm annual precipitation, 26°C mean annual temperature, and an average day length of 12 hours.

Planting Design

One hundred eighty seeds were planted in 60 6-liter pots with 1.5 kg of sterile peat moss (Berger BM4: Sphagnum peat moss (coarse), dolomitic and calcitic lime, initial fertilizer charge, wetting agent; pH 5.4–6). A general fertilizer containing 25% phosphate, 41% potassium, 0.02% boron, 8.27% sulfur, 0.1% iron, 0.05% copper, 0.05% magnesium, 0.05% zinc, 0.001% molybdenum, and 25.459% inert ingredients (Miller Chemical and Fertilizer Corporation) was added to each pot at a concentration of 0.35 g/kg soil. Three levels of nitrogen fertilizer (urea: $(NH_2)_2CO$) were also applied: low = 0.002% N, intermediate = 0.004% N, and high = 0.008% N (20 pots per treatment). Seeds were planted alone (30 pots) and in competition pots (30 pots–5 seeds per pot). Half of the pots were then placed in full sunlight (~1175 PAR (μmol/m^2/s)) and half at 24% full sunlight (~282 PAR) in a shadehouse at La Tirimbina (Figure 1). The plants in full sun were exposed to natural rain, and those in the shadehouse were watered regularly to ensure they received adequate moisture. The seedlings never appeared water-stressed. Only one shadehouse was available at the research station, so seedlings exposed to low light were grown together. Therefore, the two levels of the light treatment were analyzed as separate experiments to avoid pseudoreplication. Within each light regime, each combination of the nitrogen and competition treatments was replicated five times, with an individual pot as a replicate. Fertilizer was applied a second time in May 2008. Seedlings were routinely examined, and aboveground herbivore damage was not

	Low N		Intermediate N		High N	
	High density	Low density	High density	Low density	High density	Low density
100% light						
20% light						

A

Sun-grown   Shade-grown

B

Figure 1. Schematic of experimental design. Seeds were planted individually or in competition with four conspecific neighbors and growth at low, intermediate, or high nitrogen levels. There were ten replicate pots per nitrogen x competition combination, five of which were grown in a shadehouse, and five of which were grown in full sunlight. The two light levels were analyzed as separate experiments (a). The photograph shows a sun-grown (left) and shade-grown plant (right) side by side (b).

detected, so it is very unlikely that induction of defenses affected the data. Results indicate the competition treatment increased available nitrogen rather than decreasing it (because *P. macroloba* has N-fixing root nodules), and other work shows legumes can enhance the performance of neighboring plants [60].

Seedling Measurements

Seedling height (cm), leaf area (cm^2), the light saturated rate of photosynthesis (A_{max}; µmol CO_2/m^2/s), and dark respiration (µmol CO_2/m^2/s) were measured for each replicate after six months of growth. Leaf samples were also collected at this time for chemistry analyses. The area of all the leaves on each seedling was measured as the length and width of the leaves multiplied together (cm^2); the leaves are bipinnately compound, so this measurement was used to compare leaf sizes but not to determine actual leaf area. For pots with competition, the average height and leaf area of individuals in the same pot were used in analyses. Plant biomass was determined using regression equations from field collected seedlings (sun n = 10; shade n = 14). PAR at the seedlings was measured between 11:00 and 13:00, and the shade collected plants had an average PAR of 20% while the sun collected plants had an

average PAR of 84%. The height and leaf area of the collected seedlings was measured, and the stems and leaves of the seedlings were then oven dried at 40 degrees Celsius for 72 hours and weighed. Regressions of aboveground biomass by stem height plus leaf area were then created (sun plants R^2 = 0.76, P = 0.001; shade plants R^2 = 0.55, P = 0.002). The resulting regression formulas were used to calculate aboveground biomass for the experimental seedlings.

A_{max} and dark respiration were measured with a LI-COR 6400 gas exchange system (LI-COR, Nebraska, USA), and only one individual was measured in pots with competition. The third leaf from the apical meristem was measured for consistency in leaf age. Measurements were made between 7:00 and 13:00 hours. Leaves were clamped into an airtight cuvette with a red-blue LED light source. Incoming CO_2 was set to 380 µmol/mol from a CO_2 cartridge. Light response curves were made from darkness to 10, 25, 50, 100, 150, 200 µmol/s and continued in increments of 200 µmol/s until an asymptote was reached. Leaves were given 120 seconds to adjust to each light level, and the CO_2 differential was recorded when flow rate, CO_2 and humidity were constant. The flow rate was set to 550 µmol/s, and humidity was between

Table 1. MANOVA and profile analysis results for the response of *Pentaclethra macroloba* photosynthesis, biomass, and carbon-based metabolites (sugars, flavans, and saponins) to light, fertilizer, and competition.

Photosynthesis and biomass

Factor	df	MANOVA F	P	Profile F	P
Sun plants					
Fertilizer	2	0.2	0.8	2.4	0.1
Competition	1	0.05	0.8	2.8	0.1
Error	23				
Shade plants					
Fertilizer	2	0.97	0.4	0.8	0.5
Competition	1	8.0	0.01	0.4	0.5
Error	23				

Metabolites

Factor	df	MANOVA F	P	Profile F	P
Sun plants					
Fertilizer	2	0.2	0.8	0.2	0.8
Competition	1	19.5	0.0002	19.6	0.0002
Error	21				
Shade plants					
Fertilizer	2	0.8	0.5	1.1	0.4
Competition	1	0.2	0.6	0.03	0.9
Fert. * comp.	2	5.7	0.01	0.7	0.5
Error	17				

65 and 75%. All necessary permits and permissions were obtained for the described field studies.

Chemical Analyses

We collected and air-dried leaves for saponin and flavan quantification. In preparation for chemical extraction, leaf samples were dried overnight in an oven at low temperature and ground to a coarse powder. We utilized a new isolation and quantification procedure for saponin content [61]. One hundred milligrams of dry leaf powder were measured into a centrifuge tube and compounds were extracted from the leaf material in 30 ml of 80% ethanol with stirring. The samples were then centrifuged and the extracted compounds plus solvent were separated from the leaf material and dried. The process was repeated to extract any remaining compounds from the plant material. The dried samples were then dissolved in 15 ml methanol and defatted by shaking the solution with hexanes. The hexane layer was pipetted-off and the process was repeated. The defatted methanol layer was dried, and the samples were dissolved in 20 ml water. This solution was centrifuged to separate any remaining leaf material from the dissolved sample.

C-18 SepPak cartridges (Waters Corp., Massachusetts, USA) were then preconditioned with 15 ml acetone followed by 15 ml

water. The water with dissolved sample was passed through the cartridge, and the elution was dried. The cartridge was then sequentially eluted with 20 ml each of 35%, 60%, and 100% methanol. The 35% and 60% methanol elutions were also dried. The 100% methanol elution was transferred directly to a pre-weighed scintillation vial and dried. The dried samples were redissolved in methanol, transferred to pre-weighed scintillation vials, and dried a final time. Samples were stored in the freezer. The water elution contains sugars and organic acids. The 35% elution contains flavans, the 60% layer is comprised of flavones, and the 100% layer contains saponins [61] (Lokvam, pers. comm.). Subsequent work demonstrated that the 60% elution contains mostly sapongenins (saponins without the attached polysaccharides), so it was not used in further flavonoid analyses. Samples were completely dried overnight in an oven at low temperature. Vials with samples were then weighed to determine the mass of each class of compound contained in the leaf material. The weights of samples from the elutions were used in analyses.

HPLC was utilized to examine five samples from the 100% elutions to confirm the presence of saponins and purity of the samples. The HPLC system consisted of a Hitachi LaChrom Elite HPLC System (Hitachi High Technologies America, California, USA) with a diode array detector (DAD; Hitachi High Technologies America) and an evaporative light scattering detector (ELSD; SEDEX 55 Evaporative Light-Scattering Detector; S.E.D.E.R.E., Alfortville, France). The column was packed with C8 coated beads (2×50 mm; 3 µm particle size; 100 Å pore size; Advanced Chromatography Systems, South Carolina, USA). Dry samples from the 100% methanol elution were dissolved in 250 µl 100% methanol and 10 µl were run on the HPLC at a gradient of $MeOH:H_2O$ (1:1) to 100% methanol over 40 minutes with a 0.25 ml/min flow rate. The DAD absorption was set between 225–400 nm. The ELSD detector was maintained at 40°C and the pressure was at 2.3 bar. Nitrogen was the nebulizing gas.

Statistical Analyses

Shade-grown plants and sun-grown plants were treated as two experiments and analyzed separately to avoid problems with pseudoreplication. Biomass and photosynthesis variables (A_{max} and dark respiration) were analyzed together using multivariate analysis of variance (MANOVA) followed by profile analysis with fertilizer, competition, and their interaction as independent variables; when interactions were not significant, main effects models were run. Primary (sugars) and secondary (flavans and saponins) metabolite production were likewise analyzed together with MANOVA and profile analysis. MANOVA analyzes the response of multiple dependent variables to experimental treatments, and profile analysis tests for differences in the magnitude or direction of response of different dependent variables. Three outliers were removed from dark respiration for the shade-grown plants for normality. A_{max} was square-root transformed for normality in the sun-grown plant dataset, and biomass was log-transformed in the sun dataset as well. Sugars and flavans were square-root transformed in both the shade and sun datasets.

We used structural equation models [62] to test hypothesized relationships between photosynthesis (A_{max}), biomass, flavans, saponins, and planting treatments. Models were sequentially run testing *a priori* hypotheses of treatment effects and relationships between response variables, and the best fitting model is presented. The SAS Calis (Covariance Analysis of Linear Structural Equations) procedure was utilized to determine the fit of the models. The Calis procedure uses normal theory maximum likelihood procedures to estimate fit, and parameter vectors are estimated iteratively with a nonlinear optimization algorithm to

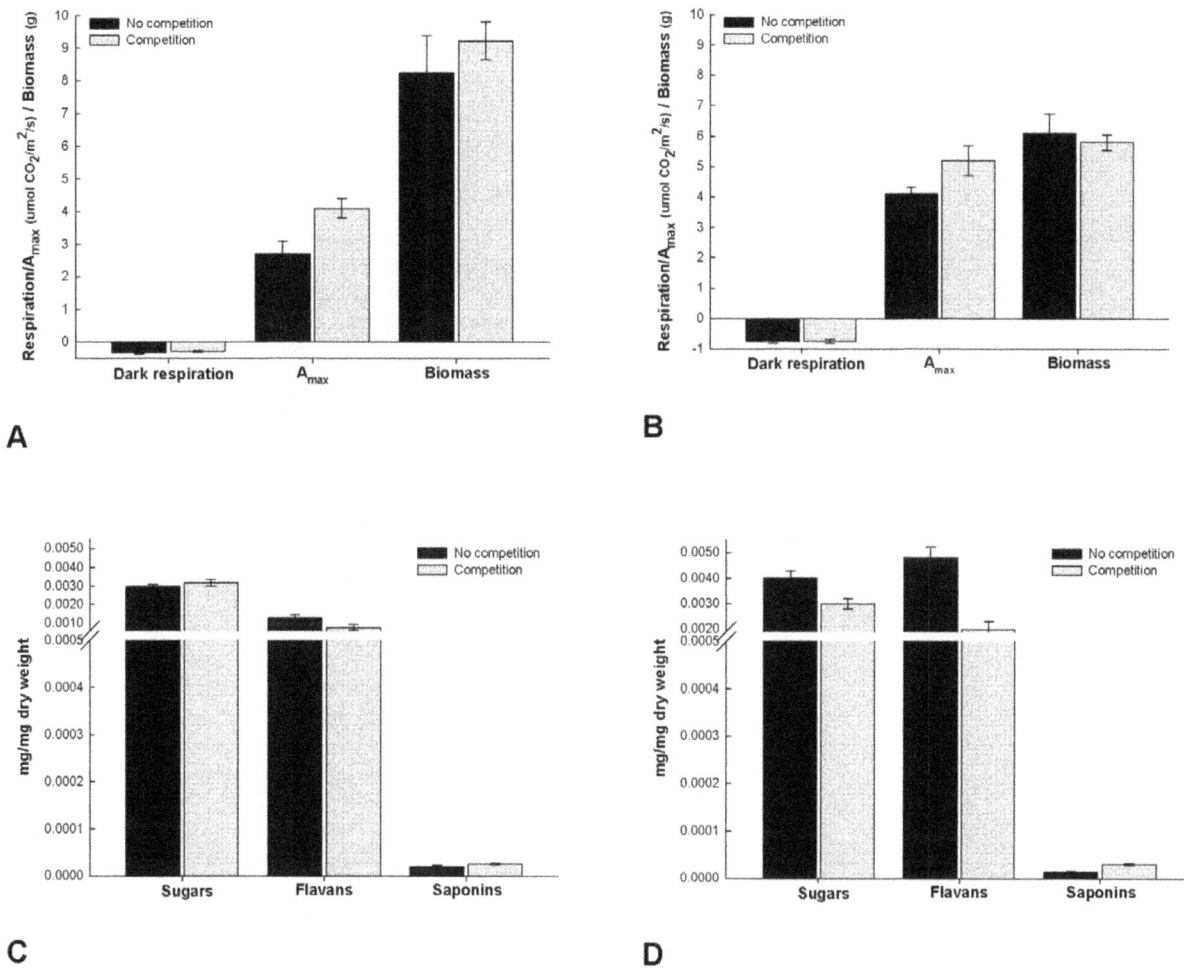

Figure 2. Means (SE) of photosynthesis, dark respiration, biomass, and carbon-based metabolites. Values are from *Pentaclethra macroloba* seedlings grown in shade (a, c) or full sunlight (b, d) with and without competition.

optimize a goodness of fit function. Chi-square values are calculated for the maximum likelihood goodness of fit to determine the fit of the models. P-values greater than 0.05 indicate a good fit of the data to the model. We accepted the model with the highest P-value as the best description of the relationships between variables. All analyses were done with SAS 9.1 (SAS Institute Inc. 2003).

Results

Photosynthesis and biomass of the shade-grown plants were highest with competition, but dark respiration was slightly higher without competition (Table 1; Figure 2a). The fertilizer treatment did not have an effect on the response variables. Neither photosynthesis, respiration, nor biomass of plants grown in the sun changed with the competition or fertilizer treatments (Table 1; Figure 2b).

The interaction between fertilizer and competition was significant for shade-plant metabolite production (Table 1). Sugars were higher in plants with competition and low or intermediate levels of fertilizer. They were lowest also with competition but with high levels of fertilizer. The two groups of secondary metabolites responded in the opposite direction to increased nitrogen. Flavans were highest in low nitrogen conditions (no competition, low fertilizer) and lowest in conditions of high nitrogen (competition

and high fertilizer levels). In contrast, saponins were highest with competition and high fertilizer levels and lowest without competition and with low fertilizer levels (Figure 2c).

Metabolites of sun-grown plants were affected by competition such that sugars and flavans were all higher without competition (low nitrogen), and saponins were higher with competition (Table 1; Figure 2d).

The best-fitting structural equation models differed for plants grown in the sun or the shade (Figure 3). In both datasets, competition increased A_{max} and saponins, but decreased levels of flavans. Higher levels of nitrogen in the competition treatment likely allowed for increased photosynthesis and saponin production by providing nitrogen necessary for enzymatic processes. This increase in N may have instigated a diversion of C from flavan production to processes or pools that were limited under lower N conditions (negative effect of competition on flavans). Competition did not have a direct effect on biomass, however. The effect of competition in the sun was not quite significant for A_{max} (t-value of relationship = 1.55; a significant relationship is described by a t-value ≥ 1.96), but the best-fitting model included this pathway. The best-fitting model for shade-grown plants also included a slightly non-significant causal pathway indicating fertilizer increased saponin levels (t-value = 1.92). Both sun and shade plants showed evidence of a trade-off between photosynthesis and

A

B

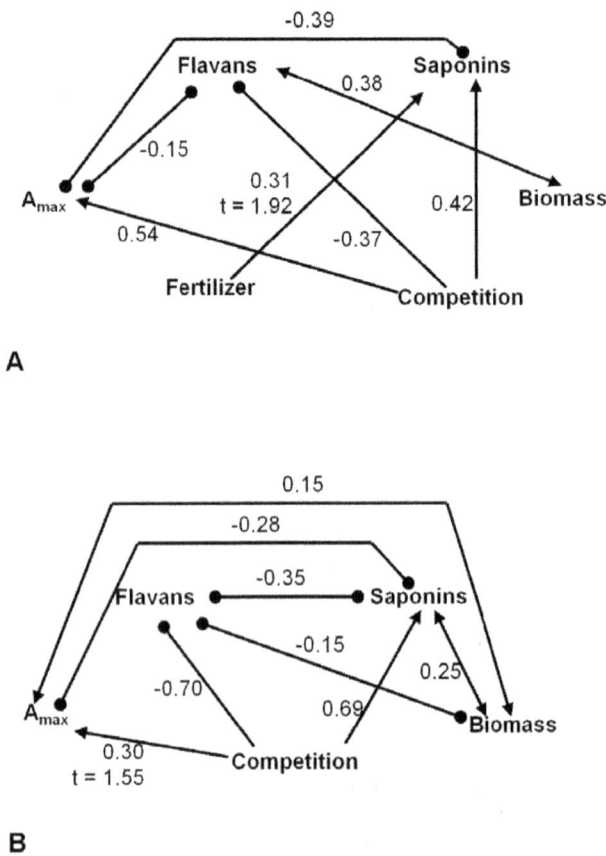

Figure 3. Interactions between experimental treatments and trade-offs in photosynthesis, growth, and defense production. Path diagrams showing causal relationships (single headed arrows) and correlations (double headed arrows) between competition, fertilizer, photosynthesis (A_{max}), flavans, saponins, and biomass in *Pentaclethra macroloba* seedlings grown in the shade (A; $\chi^2 = 0.8$, df = 5, $P = 0.98$) or the sun (B; $\chi^2 = 0.02$, df = 1, $P = 0.89$). Bullets indicate negative relationships and arrows indicate positive relationships. Numbers are the standardized parameter estimates for relationships between variables. All relationships were significant with t-values >1.96 except where smaller t-values are indicated.

saponin production, and a negative relationship was also present between A_{max} and flavans in shade-grown plants. Flavans in shade-grown plants were also positively correlated with biomass, contrary to expectations of a growth-defense trade-off. Plants grown in the sun showed the opposite pattern, and flavans and biomass were negatively correlated. Perhaps full sunlight promoted increased growth (positive relationship between photosynthesis and biomass), creating demands on pathways of plant allocation that limited production of flavans. When the correlation between A_{max} and biomass was included in the model for shade-grown plants, the relationship was negative, although weak (PE = −0.02), and the model fit less well. The shade plant model including the relationship between saponins and biomass also fit less well, and the correlation between saponins and biomass was negative (PE = −0.07).

In summary, in sun-grown plant photosynthesis and biomass were positively correlated as were saponins and biomass. In the shade, both these relationships were negative. In contrast, flavans and biomass were negatively related in the sun, and their relationship was positive in the shade. Light therefore seems to be the limiting factor which, when abundant, allows for positive

relationships between saponins and biomass and A_{max} and biomass or, when restricted, leads to a situation in which trade-offs between these processes and pools become apparent. Flavans, however, show a trade-off with growth only under full sun conditions.

Discussion

Trade-offs between growth and defense differed with the light conditions seedlings were grown under, and the GDBH [11] and CNHB [18] were supported only by comparisons between flavans and biomass. The GDBH predicts growth and defenses will be positively correlated when resources are limited and negatively correlated when resources are abundant. As expected, we found that growth and defense were positively correlated in the shade, while the predicted trade-off between flavans and biomass became apparent in the sun. Both the CNBH and the GDBH assume that growth is limited before photosynthesis, allowing excess resources to accumulate and serve in defense production. By measuring photosynthesis, growth, and two classes of defense, however, we uncovered trade-offs between photosynthesis and defense when biomass and defenses were positively correlated. A similar trade-off between defense and photosynthesis rather than defense and growth was found for an imide (a N-based defense) in *Piper cenocladum* [63]. This is consistent with the hypothesis that costs of defense are not only manifested in growth and reproduction but exist at a physiological level. One important caveat is that we do not have data on root biomass. Overall results may change with the inclusion of information on allocation to below-ground growth; however, because *P. macroloba* have N-fixing root nodules and were grown in pots, differences in below-ground biomass were probably minimal.

The correlations between secondary metabolites and biomass suggest that flavans or saponins are not costly to a plant, except under conditions of full sunlight (contrary to expectations, costs should be most evident when resources are limited) [47]. However, including physiological data showed relationships between defenses and photosynthesis were negative under both shade and full sun. The trade-off between photosynthesis and defense production may occur because defense production, regardless of a compound's classification as C- or N-based, requires nitrogen for enzymes involved in the metabolic pathways. The majority of nitrogen in a plant is contained in Rubisco, the primary enzyme in photosynthesis, which accounts for roughly 25% of leaf nitrogen in C_3 plants. Rubisco content increases with leaf nitrogen and is sometimes, but not always, produced in excess of photosynthetic requirements as a means of nitrogen storage [64]. It is therefore possible that leaf nitrogen in *P. macroloba* is sufficiently limited, such that trade-offs between different cellular demands for nitrogen exist. In addition, flavan production decreased with presence of neighboring plants (higher nitrogen), following predictions of the CNBH and the GDBH that investments in C-based defenses decline as nitrogen availability increases.

In the shade experiment, a positive correlation between growth and defense was present for flavans. This relationship changed in the full sun, and flavans were negatively correlated with biomass. Initially, it would seem the negative relationship is due to increased growth at high light. Biomass was greater, however, in shade grown plants while dark respiration and photosynthesis were higher in the sun. *Pentaclethra macroloba* is a shade-adapted plant, so the increase in respiration may result from metabolic processes necessary to avoid photoinhibition, and the trade-off with biomass may be related to an underlying relationship with respiration. Including dark respiration in the structural equation model for sun-grown plants yielded a significant model ($\chi^2 = 6.0$, df = 3, $P = 0.1$), and flavans and respiration were negatively related while biomass and respiration were positively correlated. Flavan levels were at their highest in plants grown with full sunlight and no competition; this increased production could result from a greater need for the defensive role of flavonoids as UV-B protectants (e.g., [65]), and flavonoid production increases in full sunlight in other species as well [66].

Unlike flavans, triterpenoid saponin levels did not fit predictions of the two defense hypotheses, increasing with nitrogen and having a positive relationship with biomass. Phenolics are the class of secondary metabolites most often found to fit predictions of the CNBH [17,54,67–69], and it has been suggested that the CNBH and GDBH are more relevant to phenolics because they are produced via the shikimic acid pathway which competes directly with protein synthesis (growth) for nitrogen via metabolism of phenylalanine [54,70], while terpenoids are produced by different biosynthetic pathways. Biosynthesis of saponins is initiated via the mevalonic acid and methylerythritol phosphate pathways [51,70], which do not experience a direct trade-off with growth based on available nitrogen [71,72]. Our data suggest saponins and photosynthesis compete for nitrogen before carbon is divided between growth and 'excess' carbohydrates (as per [54]). This may explain why fewer data from terpenoid studies fit predictions of the CNBH and GDBH.

Gershenzon speculated that the CNBH would apply to terpenoids only when they are substrate limited [20], but our data suggest saponin production was more limited by nitrogen resources required for synthesis rather than carbon required as a substrate, and this was also true in the shade for flavans. Overall, we found restricted support for the GDBH and the CNBH but have demonstrated that investigations of costs of defense should focus on the physiological level where many trade-offs appear to take place. In spite of context dependent support of the GDBH and CNBH based on terpenoids and phenolics, the appropriate application of these hypotheses should continue to guide experiments that enhance a clear understanding of plant defensive investments. Basic and applied ecology will benefit from advances in studies that document costs of defense against parasites, and further investigations of interactions between resource availability and physiological trade-offs will demonstrate the strength of both ecological and evolutionary influences on investments in defense– issues of particular contemporary importance due to rapid changes in carbon and nitrogen availability in the environment.

Acknowledgments

Massad and Dyer would like to dedicate this work to their co-author, Gerardo Vega, who sadly passed away before publication. His extensive knowledge of tropical forests helped many researchers over the years. Special thanks to John Lokvam for sharing his chemical analysis methods and the Coley/Kursar laboratory for sharing their laboratory facilities. We would also like to thank Ryan Massad and several EarthWatch volunteers for their assistance in measuring the plants. Jeffrey Chambers and Karen Holl provided valuable comments on this manuscript. La Tirimbina Rainforest Center generously provided facilities for the experiment, and the Max Planck Institute for Biogeochemistry enabled collection of biomass data. Finally, we would like to thank the Organization for Tropical Studies for supporting the work.

Author Contributions

Conceived and designed the experiments: TJM LAD. Performed the experiments: TJM GVC. Analyzed the data: TJM. Contributed reagents/ materials/analysis tools: LAD TJM. Wrote the paper: TJM LAD.

References

1. Hambäck PA, Beckerman AP (2003) Herbivory and plant rescource competition: A review of two interacting interactions. Oikos 101: 26–37. doi: 10.1034/j.1600-0706.2003.12568.x.

2. Cates R, Orians G (1975) Successional status and palatability of plants to generalized herbivores. Ecology 56: 410–418. doi:10.2307/1934971.

3. Levin D (1976) Chemical defenses of plants to pathogens and herbivores. Annu Rev Ecol Syst 7: 121–159. doi:10.1146/annurev.es.07.110176.001005.

4. Fox L (1981) Defense and dynamics in plant-herbivores systems. Am Zool 21: 853–864.

5. Gould F (1988) Genetics of plant-herbivore systems: Interactions between applied and basic study. In: Spencer K, editor. Variable Plants and Herbivores in Natural and Managed Systems. New York: Academic Press. 13–55.

6. Gershenzon J (1994) The cost of plant chemical defense against herbivory: A biochemical perspective. In: Bernays E, editor. Insect-plant Interactions. Boca Raton: CRC Press. 105–173.

7. Sagers C, Coley P (1995) Benefits and costs of defense in a neotropical shrub. Ecology 76: 1835–1843. doi:10.2307/1940715.

8. Bergelson J, Purrington C (1996) Surveying patterns in the cost of resistance in plants. Am Nat 148: 536–558. doi:10.1086/285938.

9. Koricheva J (2002) Meta-analysis of sources of variation in fitness costs of plant antiherbivore defenses RID G-6754–2011. Ecology 83: 176–190. doi:10.2307/2680130.

10. Strauss S, Rudgers J, Lau J, Irwin R (2002) Direct and ecological costs of resistance to herbivory. Trends Ecol Evol 17: 278–285. doi:10.1016/S0169-5347(02)02483-7.

11. Herms D, Mattson W (1992) The dilemma of plants – to grow or defend. Q Rev Biol 67: 283–335. doi:10.1086/417659.

12. Smedegaard-Petersen V, Stolen O (1981) Effect of energy-requiring defense reactions on yield and grain quality in a powdery mildew-resistant barley cultivar. Phytopathology 71: 396–399.

13. Brown D (1988) The cost of plant defense: An experimental analysis with inducible proteinase inhibitors in tomato. Oecologia 76: 467–470.

14. Baldwin I, Sims C, Kean S (1990) The reproductive consequences associated with inducible alkaloidal responses in wild tobacco. Ecology 71: 252–262. doi:10.2307/1940264.

15. Zangerl AR, Arntz AM, Berenbaum MR (1997) Physiological price of an induced chemical defense: Photosynthesis, respiration, biosynthesis, and growth. Oecologia 109: 433–441.

16. Delaney KJ, Haile FJ, Peterson RKD, Higley LG (2009) Seasonal patterns of leaf photosynthesis after insect herbivory on common milkweed, *Asclepias syriaca*: Reflection of a physiological cost of reproduction, not defense? Am Midl Nat 162: 224–238. doi:10.1674/0003-0031-162.2.224.

17. Hemming J, Lindroth R (1999) Effects of light and nutrient availability on aspen: Growth, phytochemistry, and insect performance. J Chem Ecol 25: 1687–1714. doi:10.1023/A:1020805420160.

18. Heil M, Baldwin IT (2002) Fitness costs of induced resistance: Emerging experimental support for a slippery concept. Trends Plant Sci. 7: 61–67. doi:10.1016/S1360-1385(01)02186-0.

19. Bryant J, Chapin F, Klein D (1983) Carbon/nutrient balance of boreal plants in relation to vertebrate herbivory. Oikos 40: 357–368. doi:10.2307/3544308.

20. Gershenzon J (1994) Metabolic costs of terpenoid accumulation in higher plants. J Chem Ecol 20: 1281–1328. doi:10.1007/BF02059810.

21. Jones C, Hartley S (1999) A protein competition model of phenolic allocation. Oikos 86: 27–44. doi:10.2307/3546567.

22. Phillips DR, Rasbery JM, Bartel B, Matsuda SP (2006) Biosynthetic diversity in plant triterpene cyclization. Curr Opin Plant Biol 9: 305–314. doi:10.1016/j.pbi.2006.03.004.

23. Degenhardt J, Koellner TG, Gershenzon J (2009) Monoterpene and sesquiterpene synthases and the origin of terpene skeletal diversity in plants. Phytochemistry 70: 1621–1637. doi:10.1016/j.phytochem.2009.07.030.

24. Hamilton J, Zangerl A, DeLucia E, Berenbaum M (2001) The carbon-nutrient balance hypothesis: Its rise and fall. Ecol Lett 4: 86–95. doi:10.1046/j.1461-0248.2001.00192.x.

25. Lerdau M, Coley P (2002) Benefits of the carbon-nutrient balance hypothesis. Oikos 98: 534–536. doi:10.1034/j.1600-0706.2002.980318.x.

26. Endara M-J, Coley PD (2011) The resource availability hypothesis revisited: a meta-analysis. Funct Ecol 25: 389–398. doi:10.1111/j.1365-2435.2010.01803.x.

27. McDonald E, Agrell J, Lindroth R (1999) CO$_2$ and light effects on deciduous trees: growth, foliar chemistry, and insect performance. Oecologia 119: 389–399. doi:10.1007/PL00008822.

28. Agrell J, McDonald E, Lindroth R (2000) Effects of CO$_2$ and light on tree phytochemistry and insect performance. Oikos 88: 259–272. doi:10.1034/j.1600-0706.2000.880204.x.

29. Agrell J, Anderson P, Oleszek W, Stochmal A, Agrell C (2004) Combined effects of elevated CO$_2$ and herbivore damage on alfalfa and cotton. J Chem Ecol 30: 2309–2324. doi:10.1023/B:JOEC.0000048791.74017.93.

30. Mole S, Ross J, Waterman P (1988) Light-induced variation in phenolic levels in foliage of rain-forest plants. I. Chemical Changes. J Chem Ecol 14: 1–21. doi:10.1007/BF01022527.

31. Coley P, Massa M, Lovelock C, Winter K (2002) Effects of elevated CO_2 on foliar chemistry of saplings of nine species of tropical tree. Oecologia 133: 62–69. doi:10.1007/s00442-002-1005-6.

32. Kurokawa H, Kitahashi Y, Koike T, Lai J, Nakashizuka T (2004) Allocation to defense or growth in dipterocarp forest seedlings in Borneo. - Oecologia 140: 261–270.

33. Massey F, Press M, Hartley S (2005) Long- and short-term induction of defences in seedlings of *Shorea leprosula* (Dipterocarpaceae): Support for the carbon:nutrient balance hypothesis. J Trop Ecol 21: 195–201. doi:10.1017/S026667404002111.

34. Stamp N (2004) Can the growth-differentiation balance hypothesis be tested rigorously? Oikos 107: 439–448.

35. Barto EK, Cipollini D (2005) Testing the optimal defense theory and the growth-differentiation balance hypothesis in *Arabidopsis thaliana*. Oecologia 146: 169–178.

36. Hale BK, Herms DA, Hansen RC, Clausen TP, Arnold D (2005) Effects of drought stress and nutrient availability on dry matter allocation, phenolic glycosides, and rapid induced resistance of poplar to two lymantriid defoliators. J Chem Ecol 31: 2601–2620.

37. Massad TJ, Fincher RM, Smilanich AM, Dyer L (2011) A quantitative evaluation of major plant defense hypotheses, nature versus nurture, and chemistry versus ants. Arthropod-Plant Interactions 5: 125–139.

38. Burns A, Gleadow R, Woodrow I (2002) Light alters the allocation of nitrogen to cyanogenic glycosides in *Eucalyptus cladocalyx*. Oecologia 133: 288–294. doi:10.1007/s00442-002-1055-9.

39. Lerdau M, Matson P, Fall R, Monson R (1995) Ecological controls over monoterpene emissions from douglas-fir (*Pseudotsuga menziesii*). Ecology 76: 2640–2647. doi:10.2307/2265834.

40. Glynn C, Herms D, Egawa M, Hansen R, Mattson W (2003) Effects of nutrient availability on biomass allocation as well as constitutive and rapid induced herbivore resistance in poplar. Oikos 101: 385–397. doi:10.1034/j.1600-0706.2003.12089.x.

41. Donaldson J, Kruger E, Lindroth R (2006) Competition- and resource-mediated tradeoffs between growth and defensive chemistry in trembling aspen (*Populus tremuloides*). New Phytol 169: 561–570. doi:10.1111/j.1469-8137.2005.01613.x.

42. Heidel AJ, Baldwin IT (2004) Microarray analysis of salicylic acid- and jasmonic acid-signalling in responses of *Nicotiana attenuata* to attack by insects from multiple feeding guilds. Plant, Cell, and Env 27: 1362–1373.

43. Lou Y, Baldwin IT (2004) Nitrogen supply influences herbivore-induced direct and indirect defenses and transcriptional responses in *Nicotiana attenuata*. Plant Phys 135: 496–506.

44. Iason G, Hester A (1993) The response of heather (*Calluna vulgaris*) to shade and nutrients-predictions of the carbon-nutrient balance hypothesis. J Ecol 81: 75–80. doi:10.2307/2261225.

45. Cornelissen T, Fernandes G (2001) Defence, growth and nutrient allocation in the tropical shrub *Bauhinia brevipes* (Leguminosae). Austral Ecol 26: 246–253. doi:10.1046/j.1442-9993.2001.01109.x.

46. Ruuhola T, Julkunen-Tiitto R (2003) Trade-off between synthesis of salicylates and growth of micropropagated *Salix pentandra*. J Chem Ecol 29: 1565–1588. doi:10.1023/A:1024266612585.

47. Osier TL, Lindroth RL (2006) Genotype and environment determine allocation to and costs of resistance in quaking aspen. Oecologia 148: 293–303. doi:10.1007/s00442-006-0373-8.

48. Walter C, Bien A (1989) Aerial root nodules in the tropical legume, *Pentaclethra macroloba*. Oecologia 80: 27–31. doi:10.1007/BF00789927.

49. Folgarait P, Dyer L, Marquis R, Braker H (1996) Leaf-cutting ant preferences for five native tropical plantation tree species growing under different light conditions. Entomol Exp Appl 80: 521–530.

50. Massad T (2009) The efficacy and environmental controls of plant defenses and their application to tropical reforestation. PhD Dissertation. Tulane University.

51. Augustin JM, Kuzina V, Andersen SV, Bak S (2011) Molecular activities, biosynthesis and evolution of triterpenoid saponins. Phytochemistry 72: 435–457.

52. Coley P, Lokvam J, Rudolph K, Bromberg K, Sackett T, et al. (2005) Divergent defensive strategies of young leaves in two species of *Inga*. Ecology 86: 2633–2643. doi:10.1890/04-1283.

53. Lokvam J, Kursar T (2005) Divergence in structure and activity of phenolic defenses in young leaves of two co-occurring *Inga* species. J Chem Ecol 31: 2563–2580. doi:10.1007/s10886-005-7614-x.

54. Koricheva J, Larsson S, Haukioja E, Keinanen M (1998) Regulation of woody plant secondary metabolism by resource availability: hypothesis testing by means of meta-analysis. Oikos 83: 212–226. doi:10.2307/3546833.

55. Roth S, Lindroth R (1994) Effects of CO_2-mediated changes in paper birch and white pine chemistry on gypsy moth performance. Oecologia 98: 133–138. doi:10.1007/BF00341464.

56. Honkanen T, Haukioja E, Kitunen V (1999) Responses of *Pinus sylvestris* branches to simulated herbivory are modified by tree sink/source dynamics and by external resources. Funct Ecol 13: 126–140. doi:10.1046/j.1365-2435.1999.00296.x.

57. Stamp N (2004) Can the growth-differentiation balance hypothesis be tested rigorously? Oikos 107: 439–448. doi:10.1111/j.0030-1299.2004.12039.x.

58. Clark D, Clark D (2000) Landscape-scale variation in forest structure and biomass in a tropical rain forest. For Ecol Manage 137: 185–198. doi:10.1016/S0378-1127(99)00327-8.

59. Holdridge L (1947) Determination of world plant formations from simple climatic data. Science 105: 367–368. doi:10.1126/science.105.2727.367.

60. Schmidtke A, Rottstock T, Gaedke U, Fischer M (2010) Plant community diversity and composition affect individual plant performance. Oecologia 164: 665–677. doi:10.1007/s00442-010-1688-z.

61. Kursar TA, Dexter KG, Lokvam J, Pennington RT, Richardson JE, et al. (2009) The evolution of antiherbivore defenses and their contribution to species coexistence in the tropical tree genus *Inga*. Proc Natl Acad Sci U S A 106: 18073–18078. doi:10.1073/pnas.0904786106.

62. Shipley B (2000) Cause and correlation in biology. A user's guide to path analysis, structural equations and causal inference. Cambridge: Cambridge University Press.

63. Fincher R (2007) Patterns of plant defense in the genus *Piper*. PhD Dissertation. Tulane University.

64. Sage RF, Pearcy RW, Seemann JR (1987) The nitrogen use efficiency of C_3 and C_4 plants III. Leaf nitrogen effects on the activity of carboxylating enzymes in *Chenopodium album* (L.) and *Amaranthus retroflexus* (L.). Plant Physiol 85: 355–359.

65. Landry L, Chapple C, Last R (1995) *Arabidopsis* mutants lacking phenolic sunscreens exhibit enhanced ultraviolet-B injury and oxidative damage. Plant Physiol 109: 1159–1166.

66. Jaakola L, Maatta-Riihinen K, Karenlampi S, Hohtola A (2004) Activation of flavonoid biosynthesis by solar radiation in bilberry (*Vaccinium myrtillus* L.) leaves. Planta 218: 721–728. doi:10.1007/s00425-003-1161-x.

67. Nichols-Orians C (1991) Environmentally induced differences in plant traits: consequences for susceptibility to a leaf-cutter ant. Ecology 72: 1609–1623. doi:10.2307/1940961.

68. Lindroth R, Kinney K, Platz C (1993) Responses of deciduous trees to elevated atmospheric CO_2: Productivity, phytochemistry, and insect performance. Ecology 74: 763–777. doi:10.2307/1940804.

69. Boege K, Dirzo R (2004) Intraspecific variation in growth, defense and herbivory in *Dialium guianense* (Caesalpiniaceae) mediated by edaphic heterogeneity. Plant Ecol 175: 59–69. doi:10.1023/B:VEGE.0000048092.82296.9a.

70. Liang Y, Zhao S (2008) Progress in understanding of ginsenoside biosynthesis. Plant Biol 10: 415–421. doi:10.1111/j.1438-8677.2008.00064.x.

71. Haukioja E, Ossipov V, Koricheva J, Honkanen T, Larsson S, et al. (1998) Biosynthetic origin of carbon-based secondary compounds: cause of variable responses of woody plants to fertilization? Chemoecology 8: 133–139. doi:10.1007/s000490050018.

72. Muzika R, Pregitzer K (1992) Effect of nitrogen fertilization on leaf phenolic production of grand fir seedlings. Trees-Struct Funct 6: 241–244.

6

Temporal and Spatial Profiling of Root Growth Revealed Novel Response of Maize Roots under Various Nitrogen Supplies in the Field

Yunfeng Peng, Xuexian Li, Chunjian Li*

Key Laboratory of Plant-Soil Interactions, Ministry of Education, Department of Plant Nutrition, China Agricultural University, Beijing, China

Abstract

A challenge for Chinese agriculture is to limit the overapplication of nitrogen (N) without reducing grain yield. Roots take up N and participate in N assimilation, facilitating dry matter accumulation in grains. However, little is known about how the root system in soil profile responds to various N supplies. In the present study, N uptake, temporal and spatial distributions of maize roots, and soil mineral N (N_{min}) were thoroughly studied under field conditions in three consecutive years. The results showed that in spite of transient stimulation of growth of early initiated nodal roots, N deficiency completely suppressed growth of the later-initiated nodal roots and accelerated root death, causing an early decrease in the total root length at the rapid vegetative growth stage of maize plants. Early N excess, deficiency, or delayed N topdressing reduced plant N content, resulting in a significant decrease in dry matter accumulation and grain yield. Notably, N overapplication led to N leaching that stimulated root growth in the 40–50 cm soil layer. It was concluded that the temporal and spatial growth patterns of maize roots were controlled by shoot growth and local soil N_{min}, respectively. Improving N management involves not only controlling the total amount of chemical N fertilizer applied, but also synchronizing crop N demand and soil N supply by split N applications.

Editor: Carl J. Bernacchi, University of Illinois, United States of America

Funding: The authors thank the State Key Basic Research and Development Plan of China (No. 2007CB109302), the National Natural Science Foundation of China (No: 30671237) and the Innovative Group Grant of National Natural Science Foundation of China (No. 31121062) for financial support. The funders had no role in study design, data collection and analysis, decision to publish, or preparation of the manuscript.

Competing Interests: The authors have declared that no competing interests exist.

* E-mail: lichj@cau.edu.cn

Introduction

Doubling of the world food production over the past four decades is associated with a seven-fold increase in consumption of synthetic nitrogen (N) fertilizer in agricultural systems [1]. In China, a 71% increase in total annual grain production from 283 to 484 MT (million tons) from 1977 to 2005 was achieved at the cost of 271% increase in synthetic N fertilizer application (from 7.07 to 26.21 MT) over the same period [2]. Maize is one of three major cereal crops in China. Its average grain yield per hectare increased rapidly from 962 kg in 1949 to 5,166 kg in 2007 [3]. The consumption of synthetic N fertilizer in China increased rapidly during the same period, exceeding 32 MT in 2007, accounting for 30% of global N fertilizer production [4]. However, the average maize grain yield per hectare of 5166 kg was much lower than that in Western countries such as the USA, where it was 9359 kg in 2006 [5]. Although the high yield records are more than 15 Mg ha^{-1} in some experimental plots [6,7], and even reached 21 Mg ha^{-1} in Shangdong Province in 2005 [8], this was obtained in small experimental plots and with high input costs. The amount of topdressing N fertilizer applied in the high-yield experimental plots varied from 450 to 720 kg N ha^{-1} [6,8]. The continuous increase in fertilizer supply promotes yield increase on the one hand, and brings serious environmental problems on the other hand. Excessive N fertilization in intensive Chinese agricultural systems does not make significant contributions to

grain yield but decreases nitrogen use efficiency (NUE), and increases the risk of N leaching to ground water and soil acidification [2,9–11].

In addition to overapplication, N is often applied incorrectly in China. A survey of chemical N fertilizer application in five major maize-producing provinces in North China during 2001 and 2003 revealed that 31.2–78.3% of the farmers used only a single N application as base fertilizer before sowing [12]. A study of the effects of single N application as base fertilizer on spring maize yield in Jilin province with 110 field experiments in 2004 and 2005 indicated that single N application significantly reduced maize yield compared with optimized N management based on soil mineral N (N_{min}) [13]. In maize 45–65% of the grain N is from pre-existing N in the stover before silking. The remaining 35–55% of the grain N originates from post-silking N uptake [1]. Nitrogen stress at a critical stage may lead to irreversible yield loss. In a greenhouse experiment, Subedi and Ma [14] found that restriction of N supply from seeding to 8-leaf stage could cause an irreparable reduction in maize ear size and kernel yield; however, there was no yield reduction when N was restricted from silking, or 3 weeks after silking, to physiological maturity. Newly developed maize hybrids often show reduced rates of visible leaf senescence, which allows a longer duration of photosynthesis and has a positive effect on N uptake during the grain-filling period [15–18]. Whether N fertilizer application after silking is needed in order to meet the increased N demand of plants in the reproductive growth stage, and how split application of chemical N fertilizer influences root growth and N uptake by plants as well as N movement in the soil, are questions that require addressing.

Chemical N fertilizer applied in the soil is taken up by roots and then assimilated and used by plants. Better root growth and synchronized N supply throughout the crop growing season are beneficial for maximizing fertilizer uptake, optimizing grain yield, and reducing N losses. Many scientists are starting to see roots as central to their efforts to produce crops with a better yield, efforts that go beyond the Green Revolution [19]. However, less attention has been paid to the temporal and spatial dynamics of root growth in the soil profile and how root growth responds to various N supplies [1,20], partially because roots are tangled underground and difficult to study [19]. Few studies have reported root growth plasticity of cereals under different N regimes [1,20]. In a short-term experiment under controlled conditions, N deficiency stimulates root growth, while N oversupply inhibits root growth [21]. Localized nitrate application stimulates lateral root growth (the localized stimulatory effect) [21,22]. Unfortunately, these unsystematic experiments under controlled conditions may not represent real situations in the field. It is interesting to know how roots perform in the soil profile with heterogeneous N distribution in time and space, and whether the responses of root growth to the above-mentioned N applications are repeated in long-term field experiments.

Successful N management requires better understanding of N uptake by roots and synchronized N supply throughout the crop growing season. We hypothesized that N deficiency stimulated early root growth but reduced grain yield. By contrast, N over-application inhibited early root growth and increased potential risk of N leaching without yield increase. Improving N management involved not only controlling the amount of applied N fertilizer, but also synchronizing plant demand and N applications for better root growth and high grain yield. To test this hypothesis and further dissect response strategies of maize roots to various N supplies in the field, comprehensive field studies in three consecutive years (2007–2009) were conducted in the present work to examine temporal and spatial distribution patterns of maize roots, plant N uptake, and N_{min} in the soil profile during the whole growth period, under different chemical N regimes, especially by split application of N fertilizer.

Materials and Methods

Experimental design

The field experiments were conducted in three consecutive years (2007–2009) in three adjacent experimental sites at the Shangzhuang Experimental Station of the China Agricultural University, Beijing. The soil type at the study site is a calcareous alluvial soil with a silt loam texture (FAO classification) typical of the region. The soil N_{min} and related chemical properties of the experimental soils are shown in Table S1. Maize hybrid DH 3719 ('stay-green' cultivar), a popular hybrid in North China, was used in the experiments and sown on 28 April 2007, 27 April 2008, and 27 April 2009, and harvested on 23 September 2007, 19 September 2008, and 21 September 2009. Flooding irrigation before sowing was used to keep the available soil water content above 75%. The amount of rainfall during the maize growing season in the three years was 428 mm, 608 mm, and 216 mm, respectively. In addition, 50 mm and 43 mm of irrigation were applied on 17 June 2007 and 2 July 2009, respectively. The monthly rainfall during the study period is shown in Table S2. Maize was overseeded (three seeds) with hand planters and the

plots were thinned at the seedling stage to a stand of 100,000 plants ha^{-1}. The seeds were sown in alternating 20-cm- and 50-cm-wide rows. The distance between plants was 28 cm in each row. A randomized complete block design with four replicates in each treatment in each year was used. The plot sizes were 56 m^2 (5.6×10 m), 40 m^2 (5×8 m), and 56 m^2 (5.6×10 m) in 2007, 2008, and 2009, respectively.

Fertilization and treatments

There were four (2007 and 2008) or three (2009) N treatments: 1) 0 N as control; no chemical N fertilizer was applied. 2) N topdressing at and after tasseling (TDAT), and 3) N topdressing before tasseling (TDBT). In order to determine the importance of timing of N topdressing, a treatment with delayed N topdressing at and after tasseling was set. In 2007, 175 kg N ha^{-1} as base fertilizer was applied in the TDAT and TDBT treatments, and total amount of N fertilization was 230 and 395 kg N ha^{-1} in TDAT and TDBT, respectively. According to the results of N accumulation in plants and soil N_{min} after the last harvest in 2007, 250 kg N ha^{-1} was set in TDAT and TDBT in 2008 and 2009, in which 60 kg N ha^{-1} was applied as base fertilizer. The remaining N was applied before tasseling (TDBT) at V8 (the eighth leaf emerged with ligule visible) and V12 (the twelfth leaf emerged), or at and after tasseling (TDAT) at VT (tasseling stage) and R2 (grain blister stage), respectively. 4) Traditional N practice (450 N); according to numerous high-yield studies in China, the application rate in the traditional N practice was set at 450 kg N ha^{-1}, in which 175 kg N ha^{-1} was applied as base fertilizer, 50, 170, and 55 kg N ha^{-1} in 2007 and 2008, and 120, 70, 85 kg N ha^{-1} in 2009 were applied in wide interrows by hand as topdressings at the V8, V12 and VT, respectively. Detailed rates and times of N application are shown in Table S3.

The rate and timing of phosphorus and potassium fertilization in each year were the same. In addition, zinc (Zn) was applied in each year as base fertilizer because of the slight Zn deficiency in the experimental region. A total of 135 kg ha^{-1} of P_2O_5 as triple superphosphate [$Ca(H_2PO_4)_2 \cdot H_2O$], 120 kg ha^{-1} of K_2O as potassium sulfate [K_2SO_4], and 30 kg ha^{-1} of $ZnSO_4 \cdot 7H_2O$ were applied. Before sowing, 90 kg ha^{-1} P_2O_5, 80 kg ha^{-1} of K_2O and 30 kg ha^{-1} of $ZnSO_4 \cdot 7H_2O$ were broadcasted and incorporated into the upper 0–15 cm of the soil by rotary tillage. Another 45 kg ha^{-1} of P_2O_5 at V12 and 40 kg ha^{-1} of K_2O at VT were applied in wide interrows by hand as topdressings. Each topdressing (NPK) was applied after harvest.

Harvest 2007. Plants were harvested at 38 (the eighth leaf emerged with ligule visible, V8), 57 (the twelfth leaf emerged, V12), and 74 (tasseling, VT) days after sowing (DAS) before fertilization and at 105 (grain blister stage, R2) and 147 (physiological maturity, R6, when 50% of the plants showed black layer formation in the grains from the mid-portion of the ears) DAS. At harvest, six consecutive plants (three plants each from two narrow rows) were cut at the stem base, chopped to a fine consistency, dried to a constant weight at 60°C and ground into powder to determine aboveground dry weight and N content. To estimate grain yield, ears in the central part of 21 m^2 (2007 and 2009) or 14 m^2 (2008) in each plot were hand-harvested at physiological maturity. Kernels from six randomly selected ears were harvested individually by hand, weighed and calculated to 15.5% moisture content. N content in each plant sample was analyzed by using a modified Kjeldahl digestion method [23]. Briefly, 0.3–0.4 g oven-dried plant tissue was digested with H_2SO_4 (98%)+H_2O_2 at 380°C for 3–4 h in a digestion tube. The digested solution was cooled to room temperature and added deionized water to 100 ml. An aliquot of 5 ml uniform solution was distilled and titrated with standardized 0.01 N sulphuric acid. The total N content was calculated from the concentration of standardized sulphuric acid. After shoot excision at each harvest, three whole root systems were excavated from each plot and washed free of soil with tap water. Two root systems were dried immediately after harvest and used to assess dry weight and N content, and the other root system was stored at $-20°C$ for measuring root length, including embryonic and different whorls of shoot-borne roots [24]. At root harvest, each root system was excavated with a soil volume of 28 cm (14 cm on each side of the plant base in intrarow direction)×35 cm (10 cm in narrow interrow and 25 cm in wide interrow) and a depth of 40 cm. The area of 28 cm×35 cm was the soil surface occupied by each plant at the plant density of 100,000 plants ha^{-1}. In addition, at each harvest, five 2-cm-diameter soil cores per plot were collected and mixed to measure soil N$_{min}$ (auger method, [25,26]). Samples were collected from the 0–90 cm soil layers (in 30 cm increments) in the interrow area. All fresh samples were crushed, sieved through a 3 mm sieve in the field, and extracted immediately after transfer to the laboratory with 0.01 mol L^{-1} CaCl$_2$ solution and analyzed for soil N$_{min}$ (NH$_4^+$-N+NO$_3^-$-N) by continuous flow analysis (TRAACS 2000, Bran and Luebbe, Norderstedt, Germany) [9,10].

Harvest 2008. Plants were harvested on 53 (V8), 71 (V12), 86 (VT) and 111 (R2) DAS before fertilization and on 130 and 145 (R6) DAS. At each sampling date, shoot harvest and determination of dry weight and N content as well as final dry grain yield were performed as in 2007. In addition, two whole root systems were sampled from each plot at each harvest as in 2007 to determine root dry weight and N content. In order to study the temporal and spatial distribution of maize roots and soil N$_{min}$ during the whole growth period, a different method from that in 2007 was used to obtain root and soil samples at each harvest after shoot excision. Soil samples of 28 cm (width)×35 cm (length)×50 cm (depth) with 10 cm increments in each plot under different treatments were collected. There were five soil blocks of 28 cm×35 cm×10 cm in each plot. All visible roots in each soil block were separated in the field by hand and placed in individual marked plastic bags. These roots were washed free of soil after transfer to the laboratory and then frozen at $-20°C$ until root length analyses were performed [24]. After root harvest, the soil in each block was crushed by hand and sieved through a 3 mm sieve in the field. A representative sample of the mixed soil was placed in a marked plastic bag for N$_{min}$ extraction and analysis as performed in 2007.

Harvest 2009. Plants were harvested on 33, 45 (V8), 61 (V12), 80 (VT), 110 (R2) and 147 (R6) DAS. The methods for shoot and root harvest, dry weight and N content determination, root and soil sampling, and soil N$_{min}$ measurement were identical to those used in 2008. The only difference was that the soil was excavated to a depth of 60 cm.

Statistical analysis

Data were analyzed using analysis of variance with the SAS package (SAS Institute, 1996). Differences between data in all tables were tested with PROC ANOVA. N treatments were treated as fixed effects and means of different N treatments were compared based on least significant difference (LSD) at the significance level of 0.05.

Results

Temporal and spatial distribution patterns of maize roots

The total root length of maize plants increased dramatically after the V8 stage, peaked at the VT stage, and then declined rapidly until the R6 stage. The dynamic pattern of total root length over the entire growth period was consistent in all three years, irrespective of N regimes (Fig. 1). However, the total root length in the early growth stage was differentially regulated by base N treatments. N Deficiency (0 N) stimulated root growth in the early growth stage (V8 stage), and the total root length peaked before the VT stage, followed by an early decline compared to other treatments with base N fertilizer and N topdressing before tasseling in all three years. Similarly, the treatment with base N fertilizer and delayed N topdressing in 2008 also caused an early decrease in the total root length. In contrast, 175 kg ha^{-1} base N fertilizer (450 N treatment) inhibited root growth in the early growth period. The total root length in the following growth stages under 450 N treatment was comparable with that under TDBT treatment (with 60 kg base N fertilizer) in 2008 and 2009 (Fig. 1).

To further analyze dynamic changes in root structure under different N treatments, the total length of the embryonic roots and each whorl of nodal roots was monitored in 2007 by whole root excavation with a soil volume of 35 cm (length)×28 cm (width)×40 cm (depth) at each growth stage (Fig. 2). The embryonic root began to shorten after the first harvest; therefore, the total root length was largely determined by nodal roots that initiated with root development before VT (except for the 7th whorl of nodal roots of N-deficient plants). The total length of the embryonic root and most nodal roots peaked at the VT stage and then decreased simultaneously until maturity. Although N deficiency (0 N) enhanced embryonic root growth before V12 (57 DAS), it negatively regulated nodal root growth and mortality. In particular, initiation and growth of the 7th whorl of nodal roots of N-deficient plants after the VT stage was almost completely suppressed (Fig. 2).

Monitoring of temporal and spatial root distribution in 2008 and 2009 (Figs. 3, 4) with a different method also confirmed that N deficiency stimulated, while overapplication of base N fertilizer (450 N) inhibited, root growth in the early growth stage. More roots were distributed in upper soil layers than in deep soil layers in all growth stages. Moderate N input promoted root proliferation in the nutrient-rich soil profile during the fast root growth period in 2009 (V8-tasseling; Fig. 4). Maize roots had already grown into the 50–60 cm soil layer one month after sowing in 2009, regardless of N treatment. Interestingly, deep root growth was also stimulated at the reproductive growth stage in 2008, when N was overapplied (Fig. 3).

Dry matter and N accumulation in the shoot and grain yield

The changes in shoot biomass and N accumulation were represented by a typical 'S curve'. However, the increase in shoot biomass and N content was not synchronized. The rapid increase in shoot biomass began at the V8 stage and peaked during the V12–R2 stages, while the N content increased during sowing-V8 and peaked during the V8–VT stages (Tables 1, 2). The shoot biomass and N accumulation patterns were the same in all three years regardless of N regimes. The study site is located on the North China Plain. In this region, total environmental N inputs (atmospheric and irrigation contributions) reach about 104 kg ha^{-1} year^{-1}. Ammonia volatilization and nitrate leaching are the main N loss pathways [2]. Based on N accumulation (Table 2) and soil N$_{min}$ (see below) after the final harvest in 2007, total environmental N inputs in the growing season, and amount of N loss in the region, 250 kg N ha^{-1} fertilizer was applied in delayed N topdressing (TDAT) and moderate N (TDBT) treatments in 2008 and 2009. N deficiency (0 N) significantly reduced shoot biomass, N content, and grain yield; while N

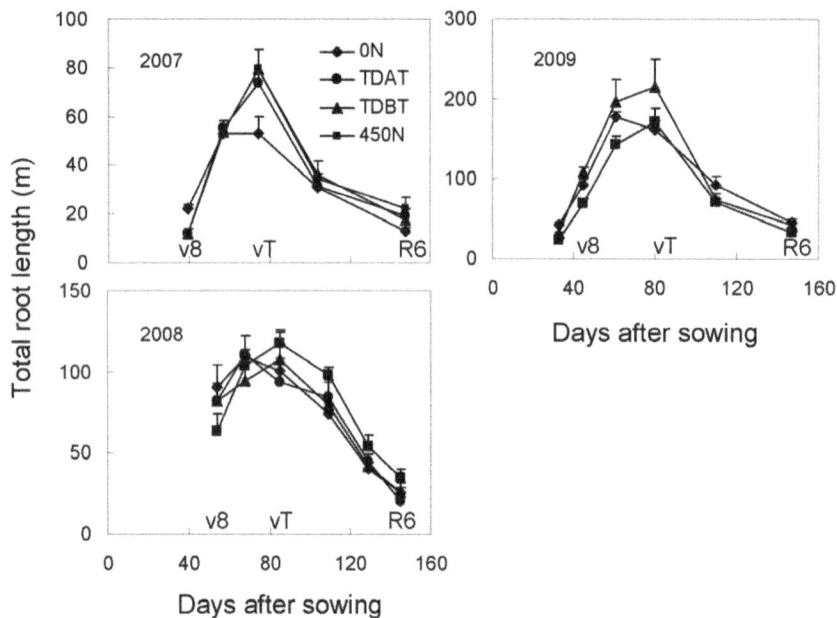

Figure 1. Total root length of maize plants during the whole growth period in response to N fertilization in three consecutive years.
In 2007, the whole root system was excavated with a soil volume of 28 cm×35 cm and a depth of 40 cm. In 2008 and 2009, root systems were excavated within a soil volume of 28 cm×35 cm and a depth of 50 cm (2008) or 60 cm (2009) with 10 cm increments. The bars represent the standard error of the mean, $n = 4$. TDAT means N top dressing after tasseling. The total amount of N applied in TDAT treatment was 230 and 250 kg ha^{-1} in 2007 and 2008, respectively. TDBT means N top dressing before tasseling. The total amount of N applied in TDBT treatment was 395, 250 and 250 kg ha^{-1} in 2007, 2008 and 2009, respectively, and is the same in the following figures.

overapplication (450 N) failed to increase N uptake, shoot biomass, as well as grain yield, compared with TDBT in all three years. The timing of N topdressing affected shoot growth and N accumulation. Delayed N topdressing decreased N uptake and biomass accumulation during the V8–V12 period, and caused a decrease in final dry matter and grain yield, compared with the TDBT treatment in 2008.

Temporal and spatial dynamics of soil N$_{min}$

In 2007, 0–90 cm soil samples were collected in 30 cm increments in the interrow area at each harvest following the auger method. In 2008 and 2009, soil samples were obtained by digging soil blocks within a soil volume of 35 cm (length)×28 cm (width)×50 (or 60) cm (depth), with 10 cm increments at each time after shoot harvest. Irrespective of methods used for soil sample collection, our three-year experiment consistently demonstrated that the soil N$_{min}$ in each soil layer at each growth stage was positively correlated with the amount of N application (Figs. 5–7). In the 0 N treatment, soil N$_{min}$ was very low and remained relatively steady at each harvest in all three years, in spite of the continuous increase in plant N content with time (Table 2). In the 450 N treatment in all three years, however, high N accumulation in soil profiles was observed, even after the last harvest, indicative of N overapplication. This was also true for the TDBT treatment in 2007 when 395 kg N ha^{-1} was applied. N overapplication led to obvious temporal and spatial fluctuations in soil N$_{min}$ and strong N leaching to the deeper soil layer occurred upon heavy rainfall in 2007 and 2008 growth seasons (Figs. 5, 6). Additionally, late N topdressing also caused N leaching to deep soil layers in 2008, although 250 kg N in total was applied (TDAT, Fig. 6). In comparison, moderate N application (TDBT) ensured plant demand without N leaching.

Discussion

Influence on maize root growth

The root system plays predominant roles in nutrient uptake for plant growth and yield formation [24]. Previous studies indicated that optimized N application is beneficial for maize shoot growth and grain yield [2,10,27,28]. However, little is known about how root development responds to different N supply in the soil profile. The present work showed that N deficiency stimulated early root growth as indicated by increased length of early-initiated nodal roots. This stimulatory effect lasted only for a short time and the total root length began to decrease when maize plants were still in the rapid vegetative growth stage (Fig. 1) with high N uptake activity (Table 2). The early decline in total root length under N deficiency was due to early mortality of the early-initiated nodal roots and growth suppression of the later-initiated nodal roots (Fig. 2). Root growth is closely associated with assimilate supply from the shoot [29]. The stimulated root growth in the early growth stage under N deficiency was achieved at the cost of reduced shoot growth [30], which led to insufficient carbon supply for continuous growth of early-initiated nodal roots and rapid elongation of later-initiated nodal roots. As a result, total plant N content was significantly reduced (Table 2). In contrast, the total root length of maize plants supplied with sufficient N didn't decrease until tasseling (Fig. 1), which favored robust nutrient uptake early in the growing season, and nutrient translocation from roots to reproductive organs later in the season [24,31–34].

Overapplication of base N fertilizer (175 kg ha^{-1} in the 450 N treatment) inhibited early root growth of maize plants, compared with other treatments in all three years (Figs. 1–4). The results suggested the 'systemic inhibitory effect' of high external N concentration on root growth [21] applied in the field. It is envisaged that a single application of total N as base fertilizer in the five major maize producing provinces in North China [12]

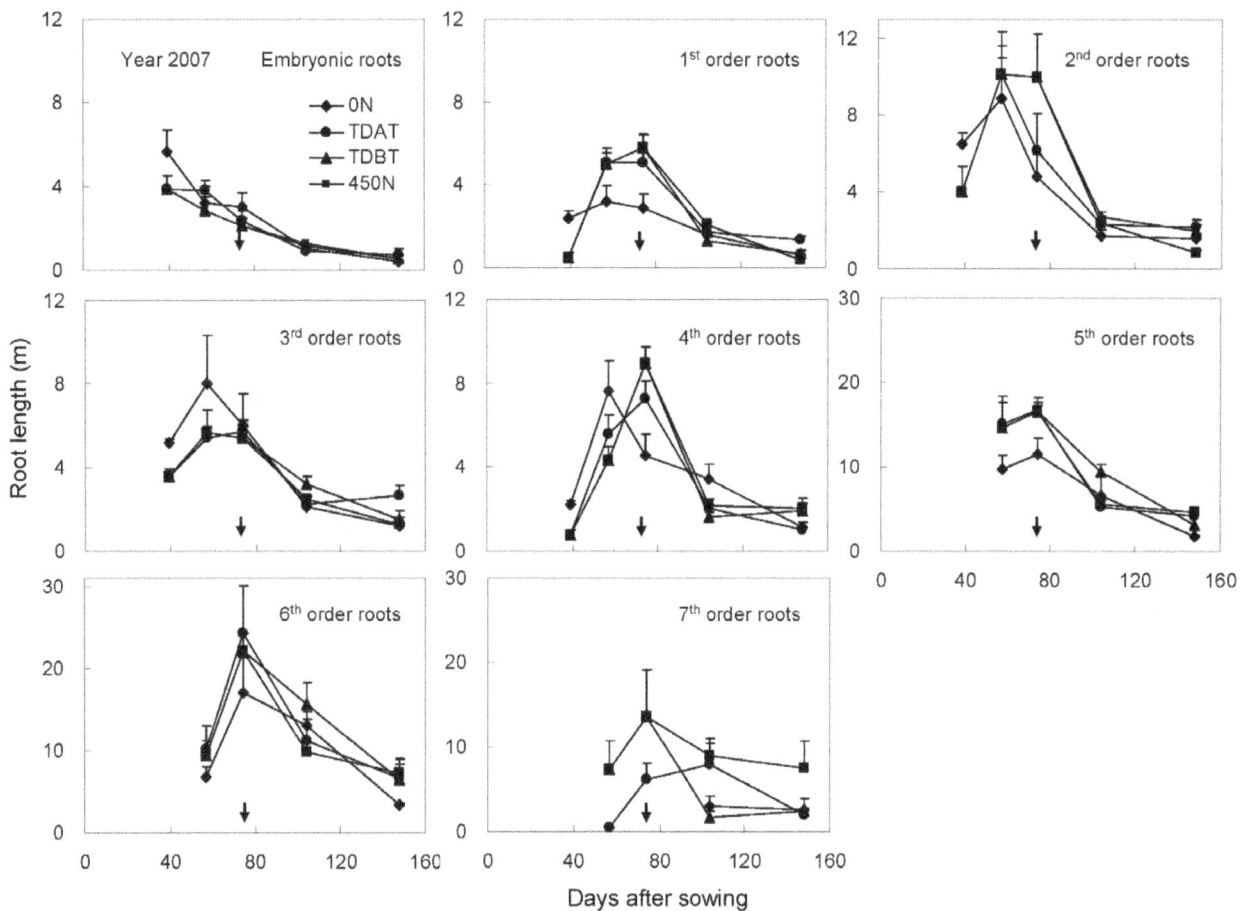

Figure 2. Length of embryonic roots and different whorls (1st to 7th orders) of nodal roots of maize plants in response to N fertilization in 2007. Arrows indicate the time of tasseling. Whole root systems were excavated with a soil volume of 28 cm×35 cm and a depth of 40 cm, and then separated into embryonic roots and different whorls of nodal roots. The bars represent the standard error of the mean, $n = 4$.

would more dramatically inhibit early root growth. A single application of total N before sowing reduced maize yield significantly compared with split applications of chemical N fertilizer based on the soil N_{min} test. However, this reduction of grain yield was not only because of the inhibited root growth in the early growth stage owing to N toxicity, but also because of N deficiency in the reproductive stage owing to N losses by different ways [2,35].

Although maize rooting depth at anthesis varies from around 0.7 m to close to 1 m, approximately 90% of roots grow in the upper 0.3 m soil [36]. Consistently, the present results in 2008 and 2009 (Figs. 3, 4) showed that most of the root length was distributed in the upper 30 cm soil, regardless of N treatments. The decrease in the total root length after tasseling indicates rapid root death that is mainly attributed to lateral root mortality [24], especially in the upper 30 cm soil layer. The root length in deep soil layers (40–60 cm) was quite constant during the whole growth period. Notably, maize roots could sense the changes in soil N_{min}. A localized stimulatory effect of nitrate-N patches on root growth has been reported when the whole root system suffered from N deficiency [21,22]. In this process, nitrate serves as a signal, and the nitrate transporter CHL1 functions as a nitrate sensor [21,37,38]. The majority of the total root length of N-treated plants was distributed in the upper 30 cm soil layer before tasseling in 2008 and during the whole growth period in 2009, because of high soil N_{min} in this soil layer (Figs. 3, 4, 6, 7). However, there was

an obvious increase in root length of maize plants supplied with 450 N in the 40–50 cm soil layer from 109 d after sowing to the last harvest in 2008 (Fig. 3). During the same period increased soil N_{min} owing to N leaching in the same soil profile was also observed (Fig. 6). Together, our study provides direct evidence that N-sufficient maize plants could respond to temporally and spatially heterogeneous soil N_{min} via enhanced root proliferation in the soil profile with higher N_{min}.

Water stress significantly reduces maize N uptake, accelerates leaf senescence, and thus reduces grain yield, compared with the well-watered plants [39,40]. This is supported by the results in the present study that shoot dry weight, N content and grain yield of all treated plants in 2009 were the lowest among the three years (Tables 1, 2), because of the limited precipitation in this year (Table S2). By contrast, the total root length of all treated maize plants in 2009 was extremely high (Fig. 1). Roots are less sensitive to water deficits than leaf, stem or silks growth [41,42]. The results in the present study indicated that maize root growth could be stimulated by low water potential. On the other hand, soil water content influences nutrient availability [30]. Low soil water availability causes low N flux to the root surface. Enhanced root growth (Fig. 1) is beneficial for plants to capture more N under dry soil conditions.

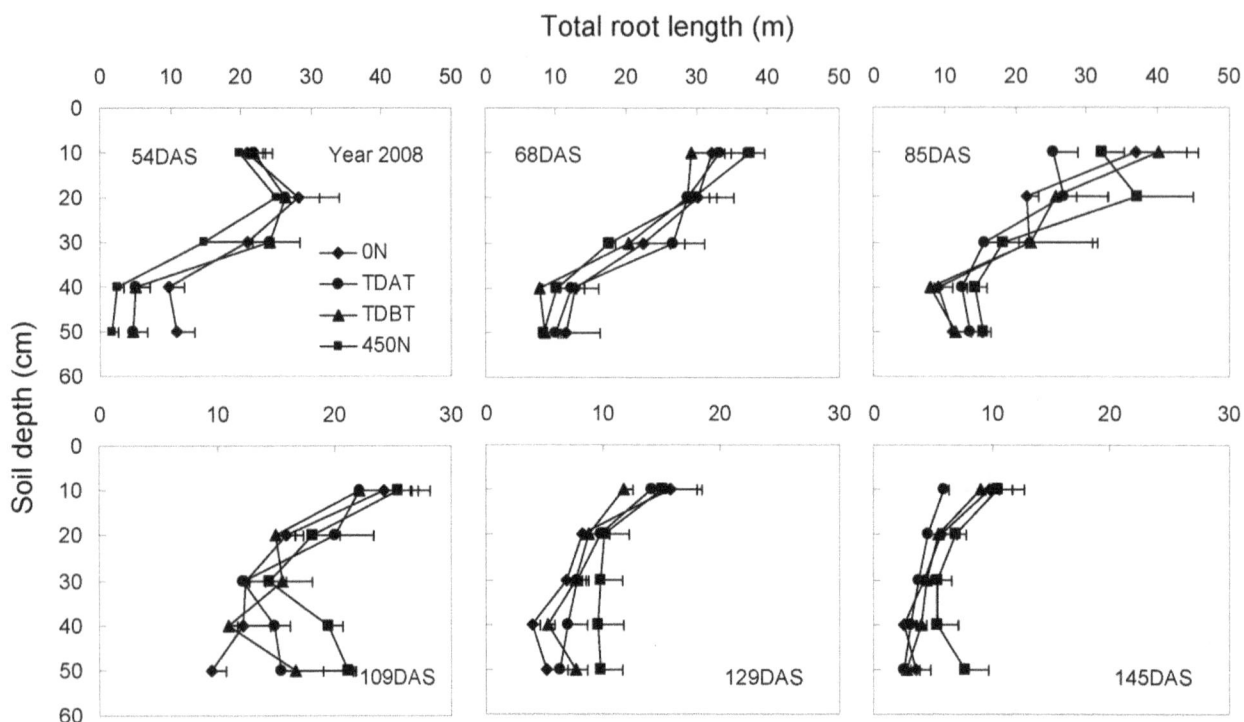

Figure 3. Total root length of maize plants in each soil layer at different growth stages in response to N fertilization in 2008. Root systems were excavated within a soil volume of 28 cm×35 cm and a depth of 50 cm with 10 cm increments. The bars represent the standard error of the mean, $n = 4$.

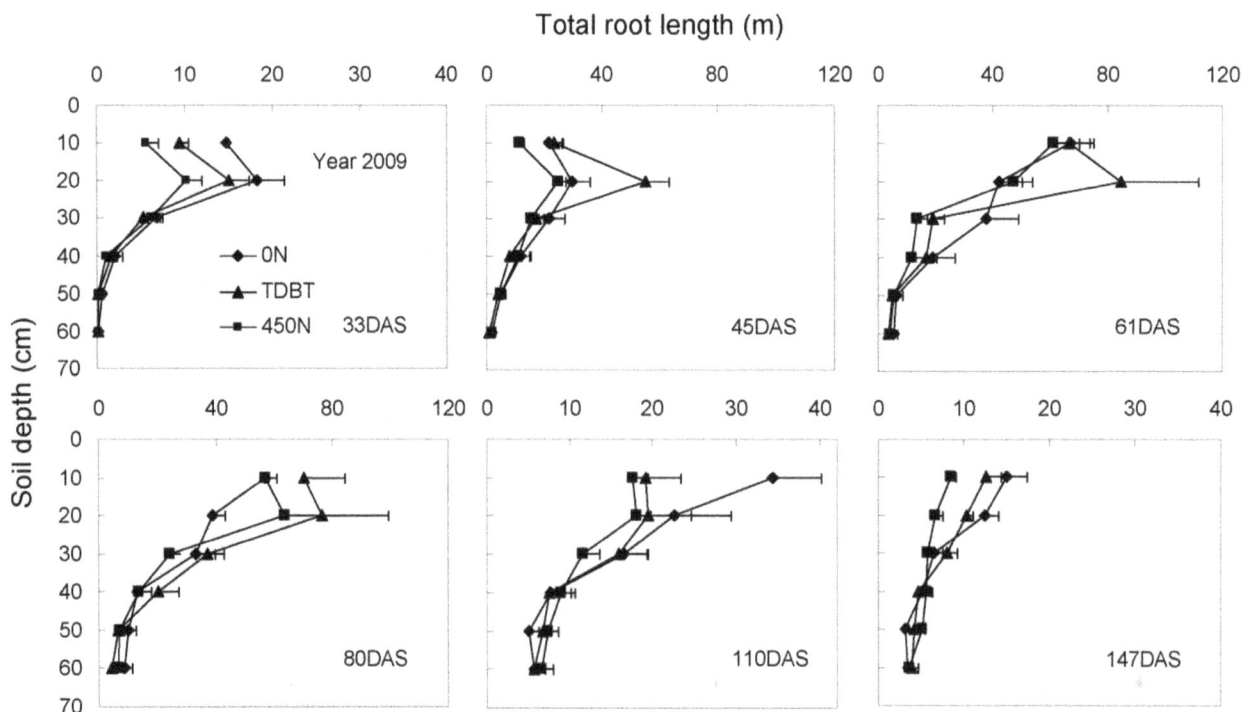

Figure 4. Total root length of maize plants in each soil layer at different growth stages in response to N fertilization in 2009. Roots were excavated within a soil volume of 28 cm×35 cm and a depth of 60 cm with 10 cm increments. The bars represent the standard error of the mean, $n = 4$.

Table 1. Shoot dry matter accumulation (t/ha) in different growth periods, final shoot dry weight (DW) and grain yield (t/ha) of maize plants supplied with different N rates in three years.

Year	Treatments	Growth period					Total DW	Grain yield
		Sowing-V8	V8-V12	V12-VT	VT-R2	R2-R6		
2007	0 N	0.9a	3.8c	3.9b	10.1b	−0.1a	18.6b	9.7b
	TDAT*	0.7a	4.8b	5.0b	13.9a	3.4a	27.8a	12.8a
	TDBT**	0.7a	5.5a	7.3a	8.5b	6.7a	28.7a	12.4a
	450 N	0.7a	5.5a	7.3a	9.2b	6.4a	29.0a	13.3a
2008	0 N	1.2b	2.9b	4.8a	6.8b	4.9b	20.6b	11.0c
	TDAT	1.6a	2.6b	5.2a	8.4ab	5.3ab	23.2ab	12.1b
	TDBT	1.6a	3.8a	4.3a	10.5a	7.0ab	27.2a	13.8a
	450 N	1.4ab	3.5a	4.2a	7.9ab	8.6a	25.6a	13.1a
2009	0 N	1.0a	2.2b	3.6b	4.7b	3.4a	14.9b	6.3b
	TDBT	1.1a	2.9a	5.4a	9.2a	2.0a	20.6a	10.7a
	450 N	1.1a	3.5a	5.4a	9.7a	1.5a	21.2a	11.0a

Values in columns in each year followed by a different letter represent a significant difference between N treatments ($P<0.05$). Values are means ± SE ($n=4$).
*TDAT, N topdressing after tasseling. The total amount of N applied in the TDAT treatment was 230 and 250 kg/ha in 2007 and 2008, respectively, and is the same in the following tables.
**TDBT, N topdressing before tasseling. The total amount of N applied in the TDBT treatment was 395, 250 and 250 kg/ha in 2007, 2008 and 2009, respectively, and is the same in the following tables.
V8, the eighth leaf emerged with ligule visible; V12, the twelfth leaf emerged; VT, tasseling; R2, grain blister stage; R6, physiological maturity, and they are the same in the following table.

N application, uptake and grain yield

Under N deficiency, grain yield is negatively correlated with early root growth due to competition for N resources [1,43]. In order to obtain high grain yield, fertilizer overapplication in Chinese intensive agricultural systems is very common, since farmers believe that additional fertilizers further improve crop yield [2]. In fact, N overapplication not only inhibited early root growth as discussed above, but also failed to increase shoot dry weight and grain yield of maize plants (Table 1; [44]). N overapplication did not increase total plant N content either, compared with the moderate N treatment (TDBT) in 2008 and 2009 (Table 2). The shoot N concentration of maize plants was the same under 450 N and TDBT treatments at each sampling time. Therefore, excessive N could not be taken up by plants and used to increase grain yield, but instead would increase the risk of N leaching and potential environmental pollution.

Besides quantity control, timing of fertilization is also important. The results in the present study indicated that the increases in shoot biomass and N content were not synchronized. Approximately 60–86% of the total N in maize plants (except the 0 N treatment in 2009) was taken up before tasseling, whereas 53–64% of the dry matter was accumulated after tasseling in all three years (Table 3). Therefore, N topdressing after the V8 stage (TDBT) was necessary to ensure adequate soil N supply for rapid plant growth.

Table 2. Shoot N accumulation (kg/ha) in different growth periods and the final shoot N content of maize plants supplied with different N rates in three years.

Year	Treatments	Growth period					Total N content
		Sowing-V8	V8-V12	V12-VT	VT-R2	R2-R6	
2007	0 N	22a	25c	53b	32ab	−8b	124b
	TDAT	22a	123a	35b	86a	5ab	271a
	TDBT	22a	111b	97a	11b	27ab	267a
	450 N	22a	111b	97a	25b	27a	282a
2008	0 N	36b	35b	37a	41a	27a	176b
	TDAT	52a	33b	53a	52a	26a	218ab
	TDBT	52a	64a	49a	61a	46a	273a
	450 N	44ab	73a	47a	35a	55a	254a
2009	0 N	19a	23c	18b	25a	24a	109b
	TDBT	25a	46b	67a	53a	11a	202a
	450 N	30a	58a	72a	46a	5a	210a

Values in columns in each year followed by a different letter represent a significant difference between N treatments ($P<0.05$). Values are means ± SE ($n=4$).

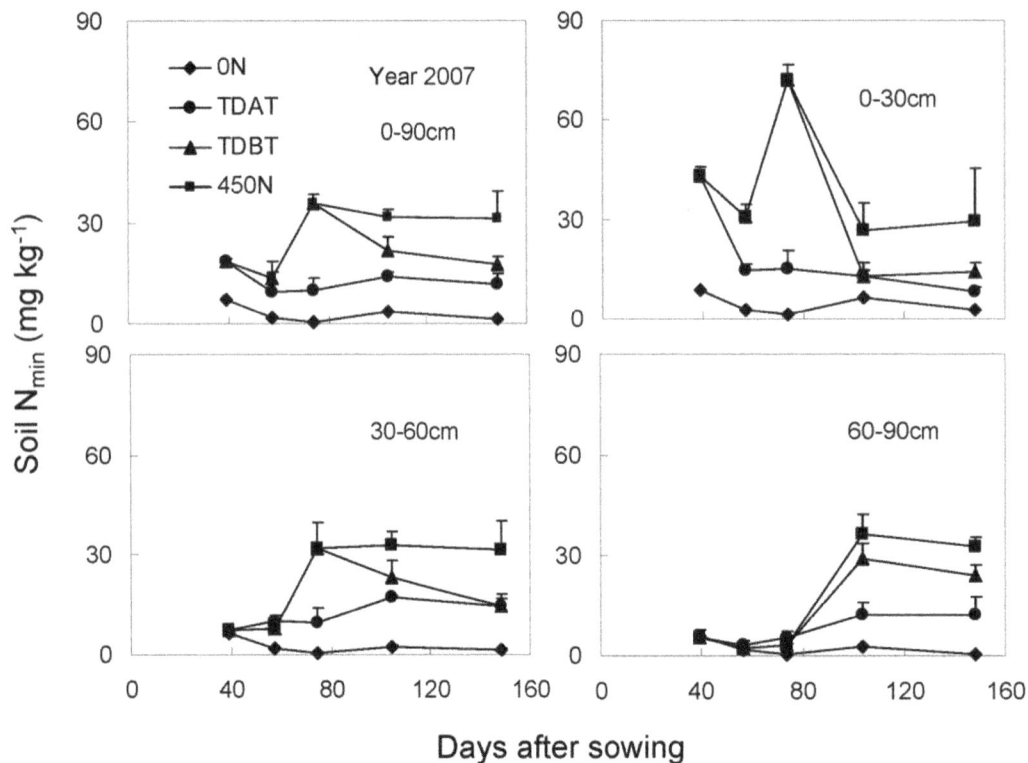

Figure 5. Soil mineral nitrogen (N$_{min}$; NH$_4^+$-N+NO$_3^-$-N) in the 0–90 cm soil profile in response to N fertilization in 2007. The soil samples were obtained using the soil auger method at the same time in each plant harvest. The bars represent the standard error of the mean, $n = 4$.

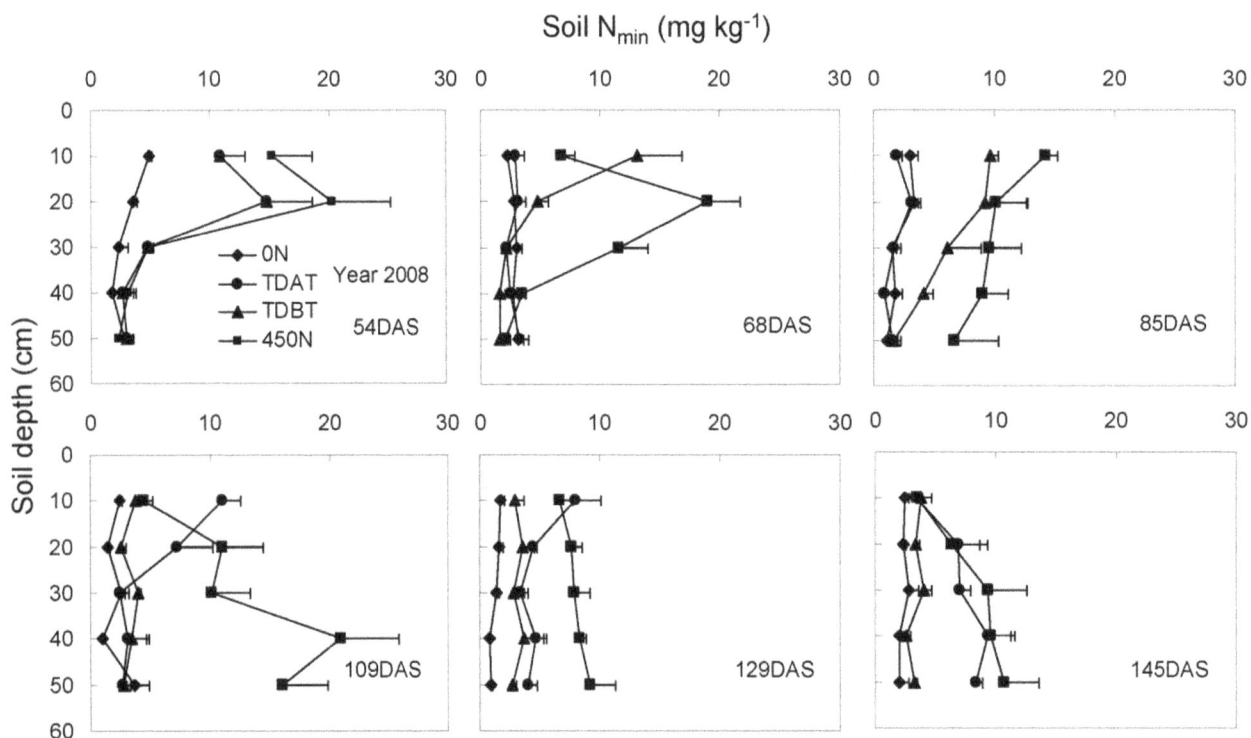

Figure 6. Temporal and spatial distribution of soil mineral nitrogen (N$_{min}$; NH$_4^+$-N+NO$_3^-$-N) in the 0–50 cm soil profile in response to N fertilization in 2008. The soil samples were obtained by excavating soil layers within a soil volume of 28 cm×35 cm and a total depth of 50 cm with 10 cm increments at each time after shoot harvest. The bars represent the standard error of the mean, $n = 4$.

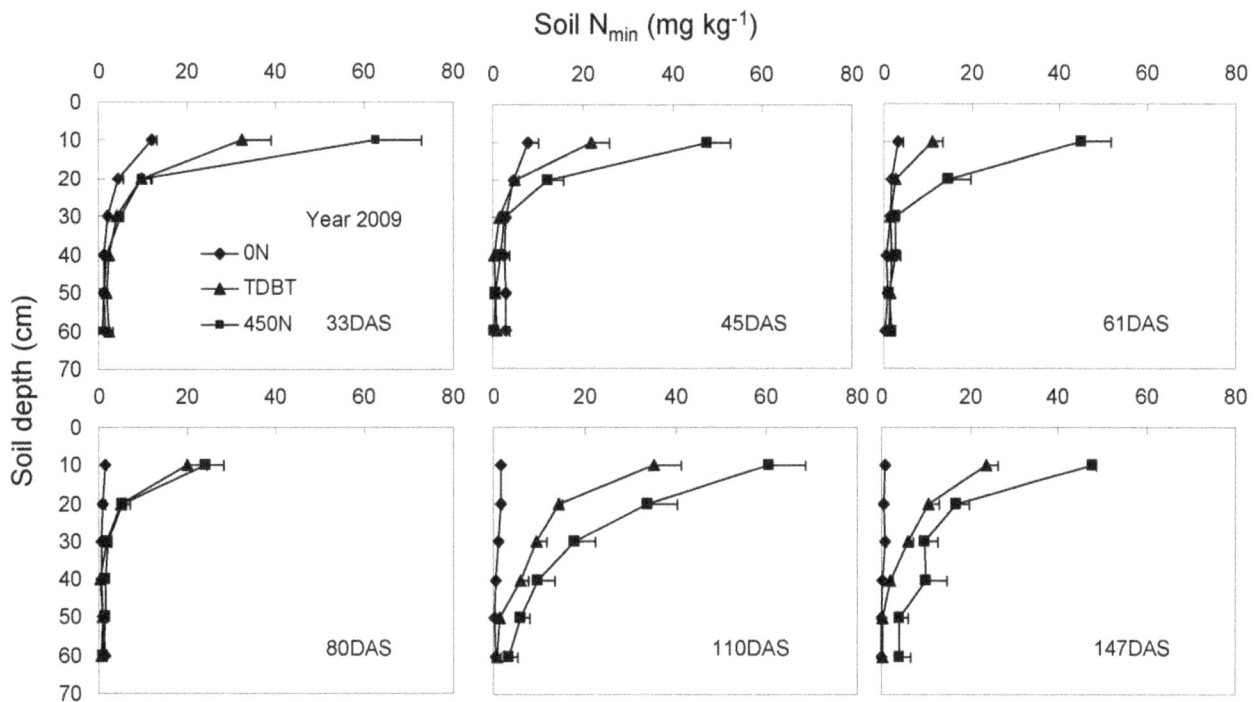

Figure 7. Temporal and spatial distribution of soil mineral nitrogen (N_{min}; NH_4^+-N+NO_3^--N) in the 0–60 cm soil profile in response to N fertilization in 2009. The soil samples were obtained by excavating soil layers within a soil volume of 28 cm×35 cm and a total depth of 60 cm with 10 cm increments at each time after shoot harvest. The bars represent the standard error of the mean, $n=4$.

Delayed N topdressing (TDAT) decreased not only plant N content and dry matter accumulation during the V8–V12 period, but also caused a decrease in final dry matter and grain yield, compared with the TDBT treatment in 2008 (Tables 1, 2). It is reported that N application rate influences production of maize spikelets and the size of the developing ears before flowering [45]. The period from 1–2 weeks before to 2 weeks after silking is critical for establishment of a large grain sink of maize plants [46]. TDAT could not compensate for the reduction in plant N content and shoot growth caused by insufficient N supply in the vegetative growth stage, although total N supply was sufficient. Moreover, because of the rapid mortality of the maize root system (Fig. 1) and the slower plant N content increase (Table 3) after tasseling, remaining N fertilizer and mineralized N from the soil were sufficient to meet the plant demand during the reproductive growth stage. Additional unnecessary N application during this stage would increase the risk of N losses. The results indicate that even in intensive agricultural systems with 'stay-green' maize cultivars, N topdressing should be applied before tasseling to maximally synchronize with crop N demand.

Table 3. Ratio of DW and N uptake presilking and total accumulation in maize plants with different N treatments in 2007, 2008 and 2009.

Year	Treatments	Ratio DW pre silking/total accumulation	Ratio N uptake pre silking/total accumulation
2007	0 N	0.46a	0.81a
	TDAT	0.38a	0.66a
	TDBT	0.47a	0.86a
	450 N	0.47a	0.82a
2008	0 N	0.43a	0.61a
	TDAT	0.41ab	0.63a
	TDBT	0.36b	0.60a
	450 N	0.36b	0.65a
2009	0 N	0.46a	0.55b
	TDAT	0.46a	0.68ab
	450 N	0.47a	0.76a

Values in columns represent the significant differences between N treatments, ($P<0.05$). Means ± SE, $n=4$.

Influence on soil N_{min}

In the 0 N treatment in all three years, soil N_{min} values were very low and remained almost constant in the studied soil profile during the whole maize growth period, except in the 0–10 cm soil layer at the elongation stage (Figs. 5–7). The continuous increase in shoot N content of the 0 N-treated maize plants indicated rapid soil N mineralization in the soil surrounding roots. Studies using the annual grass *Avena barbata* show that the rate of gross N mineralization in rhizosphere soil is about 10-fold higher than that in bulk soil [47]. By contrast, overapplication of chemical N fertilizer in all three years caused obvious N fixation (adsorption in soil lattice and transformation into organic N by soil microorganism) and accumulation in the soil. The more base N fertilizer was applied, the more N was fixed in the soil (Figs. 5–7). High N accumulation in soil profiles not only decreases NUE, but also causes environmental pollution in intensive agricultural systems [48,49]. N leaching is closely correlated to precipitation and soil moisture. With heavy summer rainfall in 2007 and 2008, soil N_{min} moved downward, resulting in high nitrate accumulation in deep soil profiles (Figs. 5, 6; [9]). Even delayed N topdressing in 2008 (TDAT, moderated N input) caused N leaching to deep soil layers (Fig. 6). In comparison with overapplication of chemical N fertilizer, soil N_{min} in the root zone remained at a relatively low level under the TDBT treatment, while maintaining grain yield (Table 1).

Conclusions

With sufficient or very frequently excessive N supply, the highest maize yield reached in North China in recent years was 21 Mg ha^{-1} [8]. The root system has played essential roles in N uptake and dry matter transformation in improving maize grain yield. However, it remained unclear how the root system responded to current fertilization regimes. The present study showed that vast majority of roots grow in the upper 30 cm soil layer. The root length in deep soil layers (40–60 cm) was quite constant during the whole growth period. The decrease in the total root length after tasseling was mainly due to rapid lateral root death in the top 30 cm soil layer. N deficiency stimulated early initiated nodal root growth only for a short time; however, it completely suppressed initiation and growth of the 7th whorl of nodal roots and accelerated older nodal root death, causing an early decrease in the total root length when maize plants were still

in the rapid vegetative growth stage with high N uptake activity. In contrast, 175 kg ha^{-1} base N fertilizer (450 N treatment) inhibited root growth in the early growth period. Importantly, root length increased in the 40–50 cm soil layer in response to N leaching, even under sufficient N supply (450 N) in 2008, indicating that N-sufficient maize plants responded to local N resources via enhanced root proliferation in the soil profile with higher N_{min}. Further, N uptake and accumulation primarily occurred before tasseling, prior to substantial dry matter accumulation and grain formation. Therefore, appropriate N topdressing before tasseling (after the V8 stage) was necessary to ensure adequate N supply for robust plant growth and development. Early N excess, deficiency or delayed N topdressing reduced total N uptake, resulting in a significant decrease in dry matter accumulation and grain yield. Lastly, the soil in the field has great buffering capacity to maintain relatively constant N_{min} in most soil profiles while supporting maize growth under the 0 N treatment in all three years. N overapplication or delayed N topdressing caused N accumulation in the soil and leaching towards deep soil profiles and groundwater upon heavy rainfall in the growth season. TDBT treatment appears to be a good N application strategy because it maintains superior root growth relative to other N application strategies during the maize growing period, and significantly reduces N loss without sacrificing grain yield.

Supporting Information

Table S1 Total soil mineral nitrogen and selected soil chemical properties before maize planting in 2007, 2008 and 2009.

Table S2 Monthly rainfall during the maize growing period in 2007, 2008 and 2009.

Table S3 Rates and times of chemical N application in the field experiments in 2007, 2008 and 2009.

Author Contributions

Conceived and designed the experiments: CL. Performed the experiments: YP. Analyzed the data: YP CL. Contributed reagents/materials/analysis tools: CL. Wrote the paper: YP XL CL.

References

1. Hirel B, Gouis JL, Ney B, Gallais A (2007) The challenge of improving nitrogen use efficiency in crop plants: towards a more central role for genetic variability and quantitative genetics within integrated approaches. J Exp Bot 58: 2369–2387.

2. Ju XT, Xing GX, Chen XP, Zhang SL, Zhang LJ, et al. (2009) Reducing environmental risk by improving N management in intensive Chinese agricultural systems. Proc Nat Acad Sci USA 106: 3041–3046.

3. Li SK, Wang CT (2009) Evolution and development of maize production techniques in China. (In Chinese). Sci Agric Sin 42: 1941–1951.

4. FAO FAOSTAT–Agriculture Database. Available: http://faostat.fao.org/ site/ 291/default.aspx. Accessed 2 October 2010.

5. Liu ZQ, Li WL, Lu LP, Shen HP, Zhou GL, et al. (2007) Revelation of American national maize yield contest in 2006. (In Chinese). J Maize Sci 15: 144–145.

6. Chen GP, Wang HR, Zhao JR (2009) Analysis on yield structural model and key factors of maize high-yield plots. (In Chinese). J Maize Sci 17: 89–93.

7. Chen GP, Yang GH, Zhao M, Wang LC, Wang YD, et al. (2008) Studies on maize small area super–high yield trails and cultivation technique. (In Chinese). J Maize Sci 16: 1–4.

8. Wang YJ (2008) Study on population quality and individual physiology function of super high-yielding maize (Zea mays L.). Shandong Agricultural University. Ph.D. Dissertation.

9. Zhao RF, Chen XP, Zhang FS, Zhang HL, Schroder J, et al. (2006) Fertilization and nitrogen balance in a wheat–maize rotation system in north China. Agron J 98: 938–945.

10. Cui ZL, Zhang FS, Mi GH, Chen FJ, Li F, et al. (2009) Interaction between genotypic difference and nitrogen management strategy in determining nitrogen use efficiency of summer maize. Plant Soil 317: 267–276.

11. Guo JH, Liu XJ, Zhang Y, Shen JL, Han WX, et al. (2010) Significant acidification in major Chinese croplands. Science 327: 1008–1010.

12. Li SK, Wang CT Report of survey on demand for science and technology by farmers in maize production. (In Chinese). Available at http://chinamaize.con. cn/tishengxd/2006ku/2005-77-14/htm.

13. Gao Q, Li DZ, Wang JJ, Bai BY, Huang LH (2007) Studies on the effects of single fertilization on growth and yield of spring maize. (In Chinese). J Maize Sci 15: 125–12.

14. Subedi KD, Ma BL (2005) Nitrogen uptake and partitioning in stay-green and leafy maize hybrids. Crop Sci 45: 740–747.

15. Borrell AK, Hammer GL, Van Oosterom E (2001) Stay-green: a consequence of the balance between supply and demand for nitrogen during grain filling. Ann Appl Biol 138: 91–95.

16. Echarte L, Rothstein S, Tollenaar M (2008) The response of leaf photosynthesis and dry matter accumulation to nitrogen supply in an older and a newer maize hybrid. Crop Sci 48: 656–665.

17. Ma BL, Dwyer ML (1998) Nitrogen uptake and use in two contrasting maize hybrids differing in leaf senescence. Plant Soil 199: 283–291.

18. Rajcan I, Tollenaar M (1999) Source: sink ratio and leaf senescence in maize. II. Nitrogen metabolism during grain filling. Field Crops Res 60: 255–265.

19. Gewin V (2010) An underground revolution. Science 466: 552–553.

20. Amos B, Walters DT (2006) Maize root biomass and net rhizodeposited carbon: an analysis of the literature. SSSAJ 70: 1489–1503.

21. Zhang HM, Jennings A, Barlow PW, Forde BG (1999) Dual pathways for regulation of root branching by nitrate. Proc Nat Acad Sci USA 96: 6529–6534.

22. Drew MC (1975) Comparison of the effects of a localized supply of phosphate, nitrate, ammonium and potassium on the growth of the seminal root system, and the shoot, in barley. New Phytol 75: 479–490.

23. Nelson D W, Somers L E (1973) Determination of total nitrogen in plant material. Agron J 65: 109–112.

24. Peng YF, Niu JF, Peng ZP, Zhang FS, Li CJ (2010) Shoot growth potential drives N uptake in maize plants and correlates with root growth in the soil. Field Crops Res 115: 85–93.

25. Böhm W (1979) Methods of Studying Root Systems. Berlin, New York: Springer-Verlag.

26. Wiesler F, Horst WJ (1993) Differences among maize cultivars in the utilization of soil nitrate and the related losses of nitrate through leaching. Plant Soil 151: 193–203.

27. Chen XP, Zhang FS, Römheld V, Horlacher D, Schulz R, et al. (2006) Synchronizing N supply from soil and fertilizer and N demand of winter wheat by an improved N_{min} method. Nutr Cycl Agroecosys 74: 91–98.

28. Cui ZL, Zhang FS, Chen XP, Miao YX, Li JL, et al. (2008) On-farm evaluation of an in-season nitrogen management strategy based on soil N_{min} test. Field Crops Res 105: 48–55.

29. Ogawa A, Kawashima C, Yamauchi A (2005) Sugar accumulation along the seminar root axis as affected by osmotic stress in maize: A possible physiological basis for plastic lateral root development. Plant Prod Sci 8: 173–180.

30. Marschner P (2011) Mineral Nutrition of Higher Plants, Ed 3. London, UK: Academic Press.

31. Liedgens M, Richner W (2001) Relation between maize (Zea mays L.) leaf area and root density observes with minirhizotrons. Eur J Agron 15: 131–141.

32. Liedgens M, Soldati A, Stamp P, Richner W (2000) Root development of maize (Zea mays L.) as observed with minithizotrons in lysimeters. Crop Sci 40: 1665–1672.

33. Wells CE, Eissenstat DM (2003) Beyond the roots of young seedlings: the influence of age and order on fine root physiology. J Plant Growth Regul 21: 324–334.

34. Niu JF, Peng YF, Li CJ, Zhang FS (2010) Changes in root length at the reproductive stage of maize plants grown in the field and quartz sand. J Plant Nutr Soil Sci 173: 306–314.

35. Gao Q, Li DZ, Wang JJ, Bai BY, Huang LH (2007) Studies on the effects of single fertilization on growth and yield of spring maize. (In Chinese). J Maize Sci 15: 125–12.

36. Dwyer LM, Ma BL, Stewart DW, Hayhoe HN, Balchin D, et al. (1996) Root mass distribution under conventional and conservation tillage. Can J Soil Sci 76: 23–28.

37. Remans T, Nacry P, Pervent M, Filleur S, Diatloff E, et al. (2006) The Arabidopsis NRT1.1 transporter participates in the signaling pathway triggering root colonization of nitrate-rich patches. Proc Nat Acad Sci USA 103: 19206–19211.

38. Ho CH, Lin SH, Hu HC, Tsay YF (2009) CHL1 functions as a nitrate sensor in plants. Cell 138: 1184–1194.

39. Wolfe DW, Henderson DW, Hsiao TC, Alvino A (1988a) Interactive Water and Nitrogen Effects on Senescence of Maize. I. Leaf Area Duration, Nitrogen Distribution, and Yield. Agron J 80: 859–864.

40. Wolfe DW, Henderson DW, Hsiao TC, Alvino A (1988b) Interactive Water and Nitrogen Effects on Senescence of Maize. II. Photosynthetic Decline and Longevity of Individual Leaves. Agron J 80: 865–870.

41. Sharp RE, Davies WJ (1979) Solute regulation and growth by roots and shoots of water-stressed maize plants. Planta 147: 43–49.

42. Sharp RE, Silk WK, Hsiao TC (1988) Growth of the maize primary root at low water potentials. I. Spatial distribution of expansive growth. Plant Physiol 87: 50–57.

43. Gallais A, Coque M (2005) Genetic variation for nitrogen use efficiency in maize: a synthesis. Maydica 50: 531–547.

44. Boomsma CR, Santini JB, Tollenaar M, Vyn TJ (2009) Maize per-plant and canopy-level morpho-physiological responses to the simultaneous stresses of intense crowding and low nitrogen availability. Agron J 101: 1426–1452.

45. Jacobs BC, Pearson CJ (1992) Pre-flowering growth and development of the inflorescences of maize. I. Primordia production and apical dome volume. J Exp Bot 43: 557–563.

46. Cantarero MG, Cirilo AG, Andrade FH (1999) Night temperature at silking affects kernel set in maize. Crop Sci 39: 703–710.

47. Herman DJ, Johnson KK, Jaeger CH, Schwartz E, Firestone MK (2006) Root influence on nitrogen mineralization and nitrification in Avena barbata rhizosphere soil. SSSAJ 70: 1504–1511.

48. Halvorson AD, Follett RF, Bartolo ME, Reule CA (2005) Corn response to nitrogen fertilizer in a soil with high residual nitrogen. Agron J 97: 1222–1229.

49. Hong N, Scharf PC, Davis JG, Kitchen NR, Sudduth KA (2007) Economically optimal nitrogen rate reduces soil residual nitrate. J Environ Qual 36: 354–362.

7

Emissions of CH$_4$ and N$_2$O under Different Tillage Systems from Double-Cropped Paddy Fields in Southern China

Hai-Lin Zhang[1]*, Xiao-Lin Bai[1,2], Jian-Fu Xue[1], Zhong-Du Chen[1], Hai-Ming Tang[3], Fu Chen[1]*

1 College of Agronomy and Biotechnology, China Agricultural University, Key Laboratory of Farming System, Ministry of Agriculture, Beijing, China, **2** Patent Examination Cooperation Center of the Patent Office, SIPO, Beijing, China, **3** Soil and Fertilizer Institute of Hunan Province, Changsha, China

Abstract

Understanding greenhouse gases (GHG) emissions is becoming increasingly important with the climate change. Most previous studies have focused on the assessment of soil organic carbon (SOC) sequestration potential and GHG emissions from agriculture. However, specific experiments assessing tillage impacts on GHG emission from double-cropped paddy fields in Southern China are relatively scarce. Therefore, the objective of this study was to assess the effects of tillage systems on methane (CH$_4$) and nitrous oxide (N$_2$O) emission in a double rice (*Oryza sativa* L.) cropping system. The experiment was established in 2005 in Hunan Province, China. Three tillage treatments were laid out in a randomized complete block design: conventional tillage (CT), rotary tillage (RT) and no-till (NT). Fluxes of CH$_4$ from different tillage treatments followed a similar trend during the two years, with a single peak emission for the early rice season and a double peak emission for the late rice season. Compared with other treatments, NT significantly reduced CH$_4$ emission among the rice growing seasons ($P<0.05$). However, much higher variations in N$_2$O emission were observed across the rice growing seasons due to the vulnerability of N$_2$O to external influences. The amount of CH$_4$ emission in paddy fields was much higher relative to N$_2$O emission. Conversion of CT to NT significantly reduced the cumulative CH$_4$ emission for both rice seasons compared with other treatments ($P<0.05$). The mean value of global warming potentials (GWPs) of CH$_4$ and N$_2$O emissions over 100 years was in the order of NT<RT<CT, which indicated NT was significantly lower than both CT and RT ($P<0.05$). This suggests that adoption of NT would be beneficial for GHG mitigation and could be a good option for carbon-smart agriculture in double rice cropped regions.

Editor: Ben Bond-Lamberty, DOE Pacific Northwest National Laboratory, United States of America

Funding: This research was supported by Special Fund for Agro-scientific Research in the Public Interest Grant (200903003), Ministry of Agriculture of China. The funders had no role in study design, data collection and analysis, decision to publish, or preparation of the manuscript.

Competing Interests: The authors have declared that no competing interests exist.

* E-mail: hailin@cau.edu.cn (HLZ); chenfu@cau.edu.cn (FC)

Introduction

With the current rise in global temperatures, numerous studies have focused on greenhouse gases (GHG) emissions [1–3]. Agriculture production is an important source of GHG [4]. In addition to carbon dioxide (CO$_2$), methane (CH$_4$) and nitrous oxide (N$_2$O) also play an important role in global warming. The global warming potentials (GWPs) of CH$_4$ and N$_2$O are 25 and 298 times that of CO$_2$ in a time horizon of 100 years, respectively [5]. In addition to industrial emissions, farmland is another important source of atmospheric GHG [6–9]. Numerous results indicate rice (*Oryza sativa* L.) paddy field is a significant source of CH$_4$ [9,10]. The anaerobic conditions in wetland rice field are favorable for fostering CH$_4$ emission [11].

A considerable number of studies have shown that some farm operations can influence CH$_4$ and N$_2$O emission. For example, water/nitrogen (N) management, organic matter application and tillage can regulate CH$_4$ and N$_2$O emission [12–14]. Tillage and crop residues retention have a great influence on CH$_4$ and N$_2$O emission through the changes of soil properties (e.g., soil porosity, soil temperature and soil moisture, etc.) [15,16]. In some experiments, conversion of conventional tillage (CT) to no-till

(NT) can significantly reduce CH$_4$ and N$_2$O emission [17,18]. However, tillage effects on CH$_4$ and N$_2$O emission are not always consistent among different studies. Dendooven et al. reported that CH$_4$ emission were not significantly affected by tillage [19]. In addition, some studies show that crop residues retention can increase CH$_4$ and N$_2$O emission from paddy fields [20–22].

Most previous studies of CH$_4$ and N$_2$O emissions in paddy field have focused on the effects of water and N management on GHG emission [23–26]. However, tillage can result in changes to GHG emission through the alteration of soil properties and biochemical processes. Although CT is widely adopted around the world, it strongly disturbs the soil, consumes more energy, and even leads to disaster (i.e., the 1930s Dust Bowl in the U.S.). Conservation tillage is increasingly being adopted in the world because of the numerous benefits (e.g., saving time/energy/fuel, controlling soil erosion and increasing water use efficiency). Presently, more and more countries in Asia are facing the problem of labor shortages and high labor cost in planting rice. Conservation tillage in paddy fields (e.g., NT, direct seeding) has increasingly been adopted in Asia, especially in Southern China. Currently, the labor shortage in agriculture has been a major constraint confronting rural

China. Because of energy and labor savings, NT has been widely adopted as a principal conservation technology in China. Furthermore, it is estimated that about 2.18×10^8 Mg yr^{-1} of rice crop residues are generated in China, accounting for 27.51% of the gross crop residue production [27]. Xiao et al. [28] reported that only 9.81% of crop residue was returned to croplands as fertilizer, but >20% of crop residue was burned directly in the field or thrown away, thus increasing environmental pollution and threatening public safety. Therefore, rational use of tillage and crop residues is of great importance for GHG emission mitigation in China.

Until now, most studies on GHG emissions in paddy fields have been based on single rice (one rice cropping in one year) or rice–wheat (*Triticum aestivum* L.) cropped fields and very few studies have involved tillage impacts on emissions of CH_4 and N_2O in double rice (two rice crops in one year, early rice and late rice) cropped fields [4,12,29]. The lower Yangtze region is a typical double rice cropped area in China, accounting for 40–60% of total arable land in this region [30]. Due to the important role of rice paddies in global agriculture, adopting reasonable agricultural management is of great importance in the mitigation of global GHG emissions. Therefore, it is valuable to examine GHG emissions in paddy fields under different tillage systems and to improve reasonable practices for mitigation of GHG emissions. The objective of this paper was to assess tillage effects on emissions of CH_4 and N_2O, and to identify the influencing factors controlling CH_4 and N_2O emission under different tillage methods.

Materials and Methods

Ethics Statement

This experiment was established in a long-term experiment site (Ningxiang, 112°18′E, 28°07′N, Hunan province, China), which belongs to Soil and Fertilizer Institute of Hunan Province. This research was performed in cooperation with China Agricultural University and Soil and Fertilizer Institute of Hunan Province. The farm operations of this experiment were similar to rural farmers' operations and did not involve endangered or protected species. The experiment was approved by the Key Laboratory of Farming System, China Agricultural University and Soil and Fertilizer Institute of Hunan Province.

Site Description

The experimental area has a subtropical monsoonal humid climate, with an annual average precipitation of 1358.3 mm and annual average temperature of 16.8°C. The typical cropping system in this area is double rice cropping in a year (i.e., early rice and late rice). Normally, rotary tillage is conducted one or two days before rice seedling transplanting. Principal properties of the surface soil (0–20 cm) are presented in Table 1. The experimental site had been cultivated with rice under rotary tillage (RT) without crop residue retention for ~30 years before the initiation of the experiment. Generally, early rice is transplanted in early April and harvested in early July and late rice is immediately transplanted

after the early rice harvest and is subsequently harvested in middle October.

Experimental Design and Treatments

The field experiment was established in 2005 with three tillage treatments: conventional tillage (CT), rotary tillage (RT) and no-till (NT). The treatments were laid out in a randomized complete block design with three replications and the area of each plot was 66.7 m^2. For all treatments, rice residue was retained on the soil surface after rice harvest until tillage operations were conducted. No-till operation was conducted in NT and the rice residue was retained on the soil surface throughout the entire study period. The CT plots were plowed once to a depth of ~15 cm using a moldboard plow and rotavated twice to a depth of ~8 cm on the day of rice seedling transplanting. The RT plots were rotavated four times to a depth of ~8 cm on the day of rice seedling transplanting.

Early rice (*Zhongjiazao 32#*) was transplanted on April 7, 2007 and April 10, 2008. Late rice (*Xiangwanshan 13#*) was transplanted on July 10 both in 2007 and 2008. All plots received 375 kg ha^{-1} compound fertilizer(N:P$_2$O$_5$:K$_2$O = 20:12:14)as basal fertilizer at seedling transplanting. One week after seedling transplanting, the plots were top-dressed with urea (46% of N), 150 kg ha^{-1} for the early rice and 75 kg ha^{-1} for the late rice. Selective herbicides (34% Quinclorac, 4% Bensulfuron-methyl) were applied prior to rice transplanting in all treatments. The planting density was ~803 640 strains ha^{-1} and ~12 500 kg ha^{-1} yr^{-1} of rice residue was retained to the soil during the experimental years.

Data Collection

Soil temperature was measured by thermometers (DF-201A, Beijing Dongfang Mingguang Electronic Science And Technology Co., Ltd) with a measuring range of −30°C to +100°C. The thermometers were inserted into the 5 cm and 10 cm soil depth and data were recorded at 10-day intervals after rice seedling transplanting. Soil bulk densities (ρ_b) at 0–5 cm, 5–10 cm and 10–20 cm depth were determined by the core method.

Soil porosity (SP, m^3 m^{-3}) was calculated by using the formula below:

$$SP = 1 - \rho_b/\rho_s \qquad (1)$$

Where, ρ_s is soil particle density, Mg m^{-3}.

Soil samples were collected from each treatment plot prior to rice seedling transplanting and at the rice harvest.

Fluxes of CH_4 and N_2O were measured with the closed chamber method [31]. For each plot, three chamber bases were inserted into the soil (5 cm depth) after tillage operations. To avoid soil disturbance, every chamber base was placed at a fixed position until rice harvest. A removable wooden bridge (2 m long and 0.5 m wide) was placed near the chamber base for convenience of sampling. The chamber base had a 5 cm deep groove for installation. A chamber made with polymethyl methacrylate was placed at the chamber base. The cross-sectional area of each

Table 1. Principal soil properties of the test soil.

Soil layer (cm)	Bulk density (g cm^{-3})	Soil organic matter (g kg^{-1})	Available N (mg kg^{-1})	Available P (mg kg^{-1})	Available K (mg kg^{-1})	pH (H$_2$O)
0–20	1.21	34.90	224.10	4.38	97.10	6.26

Table 2. Mean monthly precipitation and air temperature from April to October between 2005 and 2008 at the experimental site.

Month	Precipitation (mm)				Air temperature (°C)			
	2005	2006	2007	2008	2005	2006	2007	2008
April	92.2	235.0	38.0	26.3	20.6	19.9	25.8	18.7
May	400.8	125.0	119.0	27.3	22.6	23.6	26.6	24.5
June	272.1	201.0	119.0	25.6	27.2	27.0	26.6	26.6
July	66.7	133.0	44.0	30.9	30.2	30.1	30.8	30.0
August	80.4	154.0	126.0	58.1	27.0	29.5	29.6	28.7
September	47.5	18.0	121.0	43.2	24.6	24.0	23.5	25.6
October	64.4	40.0	3.0	18.2	18.2	21.3	19.4	20.2
Mean	146.3	129.4	81.4	32.8	24.3	25.1	26.0	24.9

Source: China Meteorological Data Sharing Service System. These data represent the mean monthly precipitation and temperature. The early and late rice growing period was April to October.

chamber was 0.36 m^2 ($0.6 \text{ m} \times 0.6 \text{ m}$) and the height was 0.8 m. Chambers were closed by filling the groove of the base with water during gas sampling, and the chamber was equipped with a small fan to mix air inside the chamber. Gas samples were collected with vacuum vials. In order to minimize the underestimation of gas fluxes with the closed chamber method, the time-course of each gas sampling was kept within 10 min [32]. Measurements were conducted every 4 hours on each sampling day. Gas samples were collected at least three times per month. During the tillage period (~1 week) and the field drainage period (~10 days), gas collection was conducted daily. The gas samples were analyzed for CH_4 and N_2O using a gas chromatography with FID and ECD (model 6890N, Agilent Technologies, CA).

The fluxes of CH_4 and N_2O emissions were calculated by using the formula below [33]:

$$F = \frac{Mw}{Mv} \times \frac{Tst}{Tst + T} \times \frac{dc}{dt} \times h \qquad (2)$$

Where F is the emission fluxes (mg m^{-2} min^{-1}); M_w is the molar mass of trace gas (g mol^{-1}); Mv is the molar volume of trace gas (L mol^{-1}); T_{st} is the absolute temperature (273.2 K); T is the air temperature at sampling (°C); dc/dt is the change in the rate of CO_2 or CH_4 concentrations (ppbv min^{-1}); and h is the height of the chamber (m).

The cumulative emissions within one year were calculated assuming the existence of linear changes in gas fluxes between two successive sampling dates. Meteorological data were obtained from China National Meteorological Bureau.

GWPs is defined as the cumulative radiative forcing both direct and indirect effects integrated over a period of time from the emission of a unit mass of gas relative to some reference gas [34]. Carbon dioxide was chosen as this reference gas. The GWPs conversion parameters of CH_4 and N_2O (over 100 years) were adopted with 25 and 298 kg ha^{-1} CO_2-equivalent [35].

Statistical Analyses

Statistical analyses were performed with SPSS 11.0 analytical software package (SPSS Inc., Chicago, IL, US). Statistical analysis was performed with ANOVA to analyze the effects of tillage on ρ_b, SP, CH_4 and N_2O flux among the treatments. The Tukey-HSD was calculated for comparison of the treatment means. With regard to CH_4 and N_2O fluxes, data for each sampling day were analyzed separately. Differences among treatments were declared to be significant at $P < 0.05$.

Results

Air Temperature and Precipitation

In general, air temperature during May and September ranges from 22 to 30°C in this region. April and October are the coldest months during the rice growing period, with mean air temperature ~20°C. The mean air temperature in 2007 was higher than that of other years, but the air temperatures were slightly lower than the average of other years in September and October of 2007 (Table 2). Mean precipitation changed dramatically compared with the two years, 81.4 mm in 2007 and 32.8 mm in 2008. The precipitation is mainly distributed between May and August, especially during May and June in this region. However, the precipitation in August and September of 2007 was more than the average and these months had the highest precipitation in 2007. Precipitation in 2008 was much less compared to that of other years (Table 2).

Soil Bulk Density

Regardless of tillage practice, ρ_b increased with soil depth, but ρ_b increased more in NT than the other tillage treatments. Among the tillage treatments, ρ_b varied in the order of RT>CT>NT at 0–5 cm depth (Fig. 1), but varied in the order of NT>CT>RT at 5–10 cm depth for both the early and the late growing season. Compared with NT, ρ_b was lower at 5–10 cm and 10–20 cm depth under RT and CT. Figure 1 indicated that ρ_b under RT changed dramatically during the rice growing season, especially at 0–10 cm depth. At 0–5 cm and 5–10 cm depth, ρ_b under RT were higher in the early rice season than in the late rice season (0.23 vs. 0.13 g cm^{-3}). In both the early and the late rice growing season, ρ_b under RT was significantly different from that under NT (Tukey HSD. early rice season: 0–5 cm, df=8 F=31.907 $P < 0.05$; 5–10 cm, df=8 F=20.100 $P < 0.05$; 10–20 cm, df=8 F=10.323 $P < 0.05$. Late rice season: 0–5 cm, df=8 F=35.083 $P < 0.05$; 5–10 cm, df=8 F=43.017; $P < 0.05$; 10–20 cm df=8 F=8.089 $P < 0.05$). Because of minimal soil disturbance, ρ_b under NT increased greatly in the deeper soil layers (Fig. 1). The significant change of ρ_b in RT may be due to soil disturbance and crop residue incorporation, whereas NT had the crop residue remaining on the soil surface.

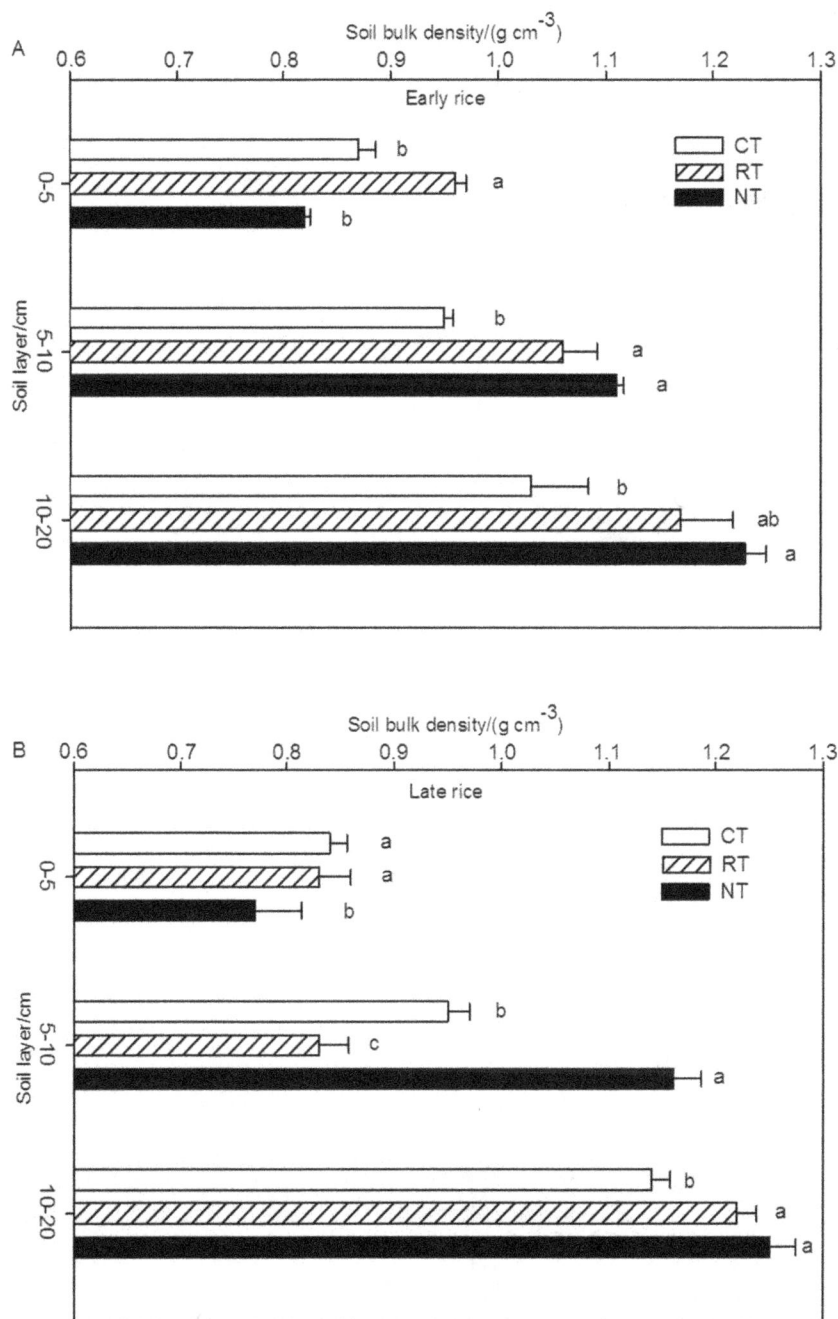

Figure 1. Soil bulk density of different tillage treatments in 2008 (A for the early rice season and B for the late rice season). Data are means of three replications; means followed by different letters are significantly different at $P<0.05$. Sampling was done during the harvest of the early and late rice in 2008.

Soil Porosity

Soil porosity decreased with soil depth among all the treatments (Fig. 2). For the early rice season, SP at $0-5$ cm depth was 68.48%, 63.18% and 61.03% for NT, CT and RT, respectively. Tukey HSD statistical test showed that SP for NT and CT significantly differed with that of RT ($0-5$ cm, df = 8 F = 69.651 $P<0.05$; $5-10$ cm, df = 8 F = 18.589 $P<0.05$; $10-20$ cm, df = 8 F = 10.393 $P<0.05$). The order of SP at depths of $5-10$ cm and $10-20$ cm varied with CT>RT> NT; and SP for CT and RT were 11.5% and 8.9% higher than that of NT, respectively. The

trend of SP in the late rice season varied similarly with that of the early rice season ($0-5$ cm, df = 8 F = 30.167 $P<0.05$; $5-10$ cm, df = 8 F = 195.166 $P<0.05$; $10-20$ cm df = 8 F = 6.957 $P<0.05$). Conversion of traditional tillage to NT, SP at $5-10$ cm depth was higher 1.83% and 7.27% than that for CT and RT, respectively. Compared with NT, SP for CT significantly increased at $10-20$ cm depth in the early rice season. During the early rice growing season, SP at $5-10$ cm depth varied in the order of CT>RT>NT ($P<0.05$). However, during the late rice season, SP at $5-10$ cm depth followed in the order of NT>RT>CT

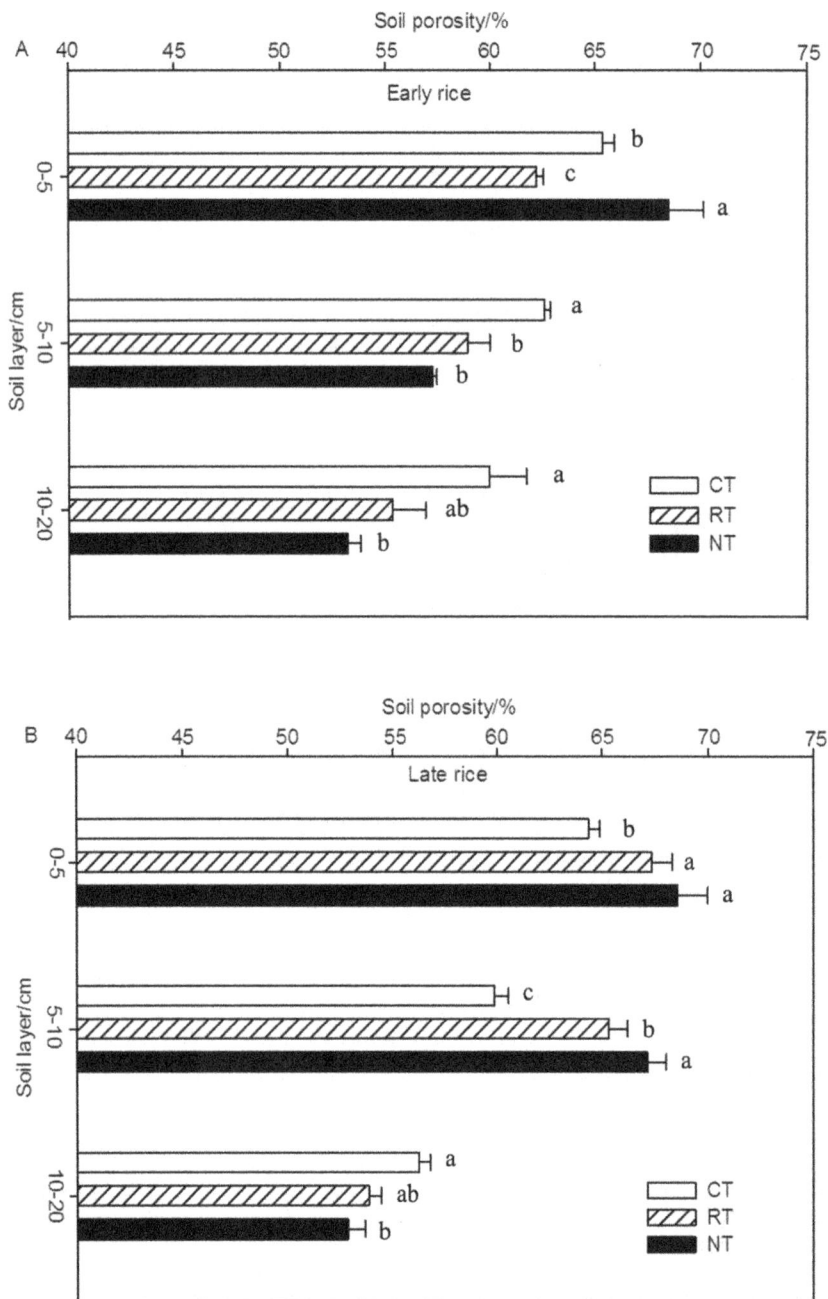

Figure 2. Soil porosity of different tillage treatments in 2008 (A for the early rice season and B for the late rice season). Data are means of three replications; means followed by different letters are significantly different at $P<0.05$.

($P<0.05$) and 9.84% and 6.35% higher for NT and RT than for CT, respectively.

CH$_4$ Emission

For the early rice season, paddy soil was the atmospheric source of CH$_4$ under all treatments in both years. The flux of CH$_4$ showed a single peak pattern characterized by three stages (Fig. 3−a, b). The first stage was the increasing stage of CH$_4$ emission. The flux of CH$_4$ showed a continuous increase under all the treatments and attained the highest fluxes during the aeration stage. The CH$_4$ emissions from both CT and RT displayed similar trends and were higher than that from NT (Fig. 3−a, b). The second stage was the decreasing stage of CH$_4$ emission. The flux of CH$_4$ decreased rapidly from the aeration stage to the flooding stage during the early rice season. The emission fluxes in 2007 and 2008 were in the same order of RT>CT>NT and significant differences among the treatments were observed in 2008 ($P<0.05$). The third stage was characterized by stable CH$_4$ emission. The flux of CH$_4$ remained at a low level and tended to be stable from the flooding stage to the harvest stage. In 2008, the cumulative emissions were 228.3, 276.3 and 188.1 kg ha^{-1} for CT, RT and NT, respectively and were 17.9%, −1.7% and 16.2% lower in 2007, respectively. The difference between 2007 and 2008 was possibly due to weather differences.

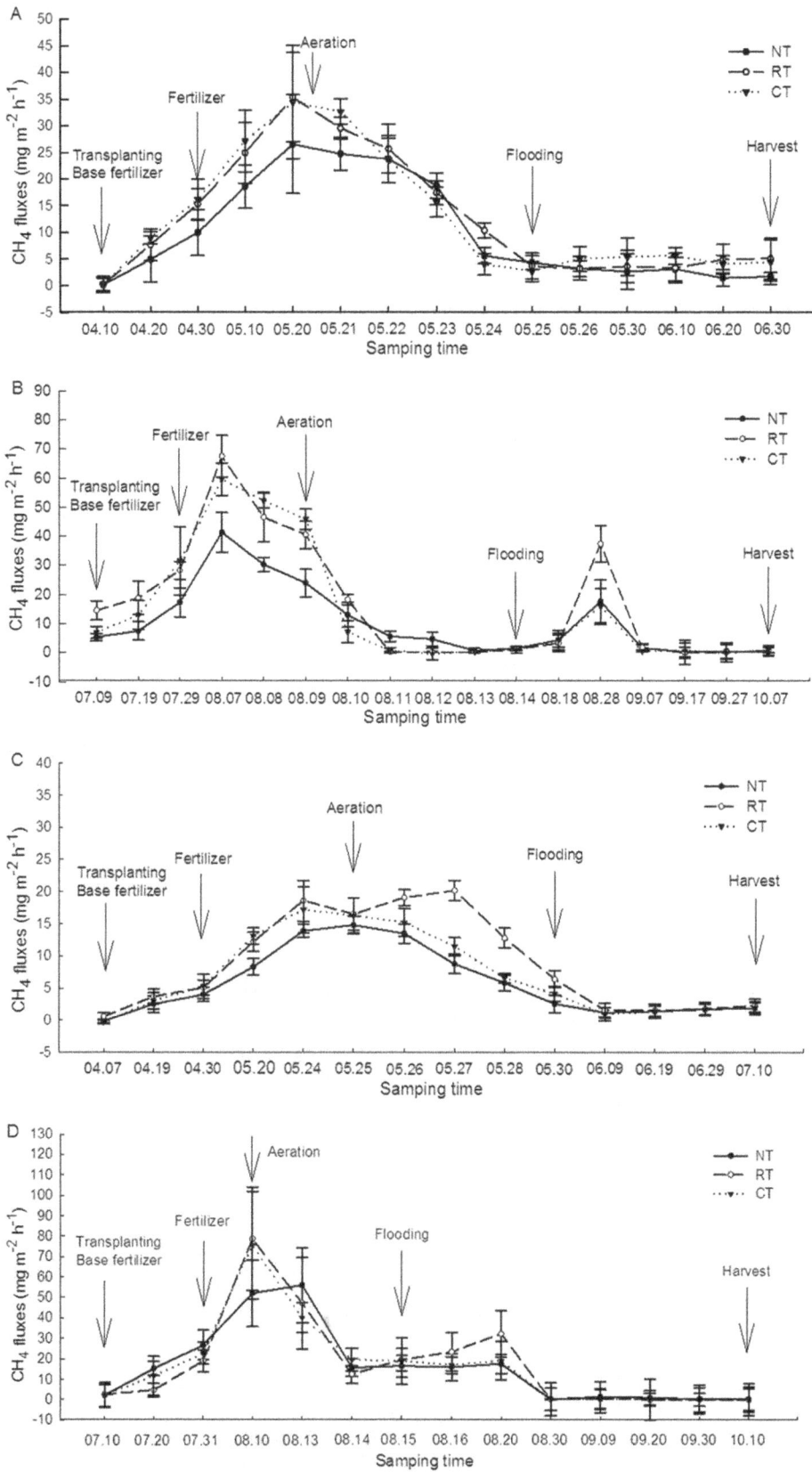

Figure 3. CH₄ flux under different tillage during the rice growing seasons (A, B for the early rice season and the late rice season in 2007; C, D for the early rice season and the late rice season in 2008, respectively). Vertical bars represent standard errors of the mean (n = 3).The arrows in the figures indicate the time of field operation.

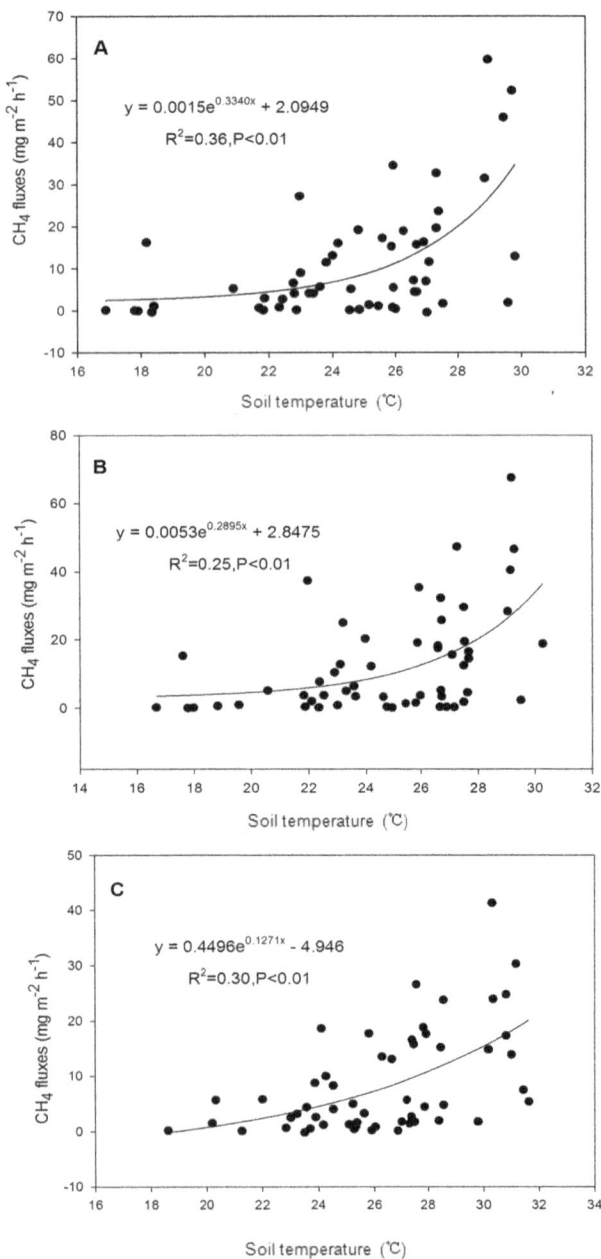

$y = 0.0015e^{0.3340x} + 2.0949$

$R^2 = 0.36, P < 0.01$

$y = 0.0053e^{0.2895x} + 2.8475$

$R^2 = 0.25, P < 0.01$

$y = 0.4496e^{0.1271x} - 4.946$

$R^2 = 0.30, P < 0.01$

Figure 4. Relationship between soil temperature and CH$_4$ emission from paddy fields (A for CT at 5 cm depth soil, B for RT at 5 cm depth soil, and C for NT at surface soil). R^2: coefficient of determination.

The flux of CH$_4$ for the late rice season (Fig. 3-a, b) showed a double emission peak. Before flooding, the CH$_4$ emission flux exhibited similar trends to that of the early rice season. However, there was another small peak emission after the flooding stage which was lower than the first peak emission. For both years, CT had higher CH$_4$ emission in the second peak fluxes than that of RT and NT. The cumulative emissions of CH$_4$ for the late rice season in 2008 were 526.2, 565.5 and 506.2 kg ha^{-1} for CT, RT and NT, respectively; and 68.5%, 39.3% and 140.8% higher than in 2007, respectively.

The cumulative CH$_4$ emission under NT was lower than that under CT and RT (Fig. 3-a, b), and the difference was significant

at the peak emission ($P < 0.05$). In contrast, CT emitted more CH$_4$ during the early and the late rice growing seasons, with a higher peak emission than that of NT and RT (Fig. 3-a, b).

The emission of CH$_4$ was greatly correlated with soil temperature (Fig. 4). There were significant correlations between CH$_4$ emission and soil temperature among the treatments. There was a significant correlation between CH$_4$ emission and soil temperature at 5 cm depth for CT and RT, while significant correlation for NT was at the soil surface.

Compared with the GWPs of CH$_4$ emission (over 100 years), the mean value of 2007 and 2008 for NT was significantly lower than for CT and RT ($P < 0.05$) with 16814, 18988 and 14112 kg ha^{-1} CO$_2$-equivalent for NT, RT and CT, respectively.

N$_2$O Emission

The N$_2$O emission exhibited an impulse type for both the early and the late rice season in 2007 and 2008 (Fig. 5-a, b). Regardless of tillage methods, the N$_2$O emission exhibited an emission peak after tillage, aeration and flooding. The first peak of N$_2$O emission appeared ~10 days after tillage. The emission varied in the order of RT>CT>NT in 2008, and RT was significantly higher than NT ($P < 0.05$). The emission order was NT>CT>RT in 2007, but no significant differences among treatments ($P < 0.05$) were observed. The N$_2$O emission fluxes decreased after fertilizer application, but aeration and flooding triggered emission peaks.

All the three tillage treatments were weak sources of N$_2$O (Table 3). In 2008, the cumulative N$_2$O emission was 0.01, 0.30 and 0.30 kg ha^{-1} for CT, RT and NT, respectively. However, the emissions in 2008 were nearly 60% lower than those in 2007 for all the treatments. The annual difference of the cumulative emission was possibly due to influences from meteorological factors (i.e., temperature, precipitation). Regardless of the year, the N$_2$O emission fluxes for NT was more stable than that for CT and RT, ranging from 13.1–33.0 μg m^{-2} h^{-1} in the late rice season. On the other hand, the emission fluxes for RT and CT changed greatly from day to day. However, aeration strongly influenced N$_2$O emissions for all the treatments. In general, about 68%–81% of the cumulative N$_2$O emissions occurred from aeration to harvest in 2007. Compared with CT and RT, NT significantly increased the N$_2$O emission from aeration to harvest in both 2007 and 2008 ($P < 0.05$).

Compared with the GWPs of N$_2$O emission (over 100 years), the mean value of 2007 and 2008 for CT was significantly lower than that for RT and NT ($P < 0.05$). The values were 126.7, 166.9 and 152.0 kg ha^{-1} CO$_2$-equivalent for NT, RT and CT, respectively.

Discussion

CH$_4$ Emission

Large variations in CH$_4$ emission were observed during the rice growing seasons, which may be attributed to differences in meteorological conditions. However, soil tillage had significant effects on CH$_4$ emission across the entire rice growing seasons. In this study, NT had a lower CH$_4$ emission compared with other treatments ($P < 0.05$), which is consistent with the results of Zhang et al. [36]. Gregorich et al. attributed the differences in gas fluxes between NT and CT to differences in the physical environment [37]. Wang et al. indicated that the major differences in CH$_4$ production zone resulted from the disturbed depth by the different tillage methods [38]. Therefore, the CH$_4$ production zone may vary according to the adopted tillage method. Wang et al. also reported that the main oxidation zone of CH$_4$ was the root surface and the interface between soil and water [38]. The rice residues

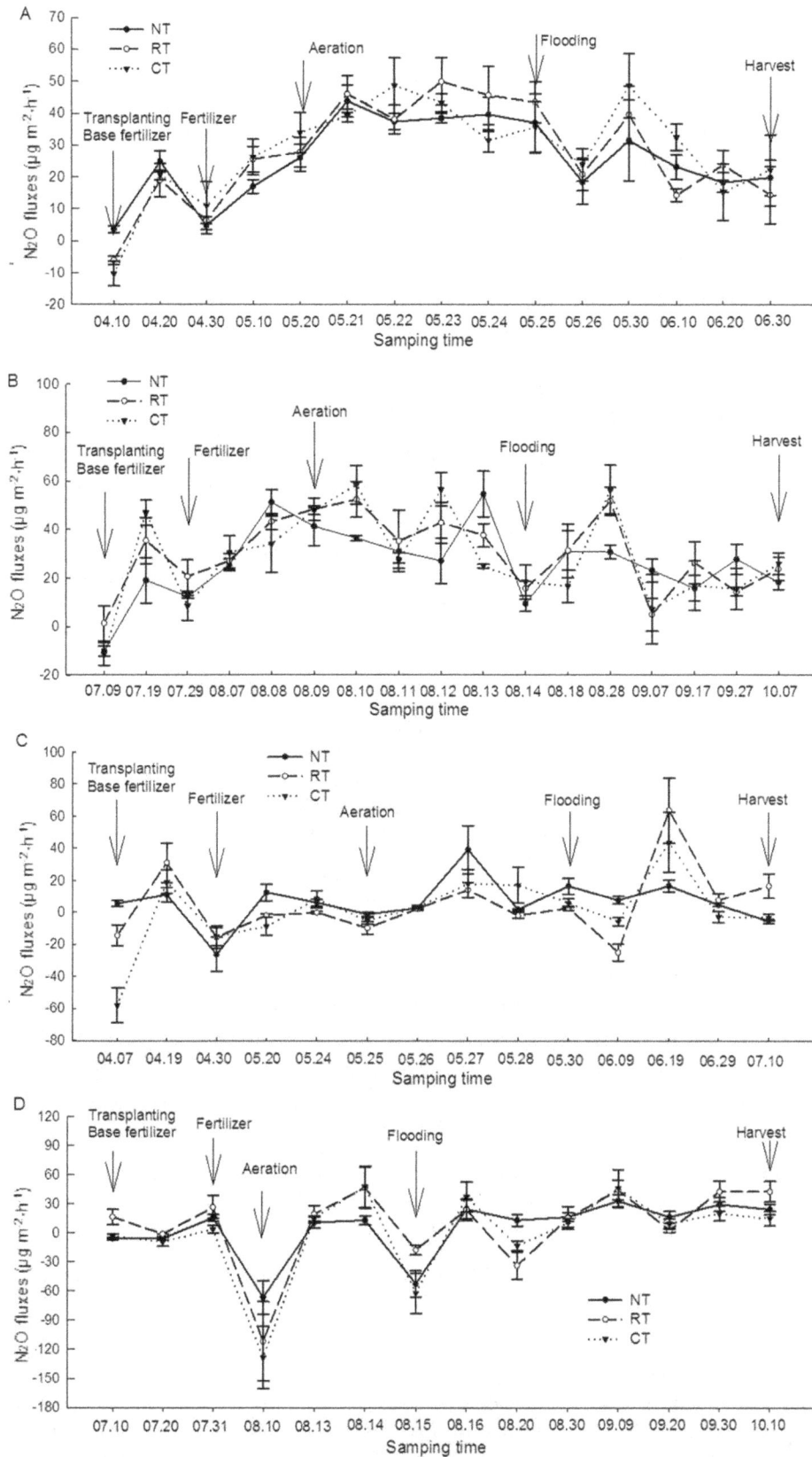

Figure 5. N$_2$O flux under different tillage during the rice growing seasons (A, B for the early rice season and the late rice season in 2007; C, D for the early rice season and the late rice season in 2008, respectively). Vertical bars represent standard errors of the mean (n = 3). The arrows in the figures indicate the time of field operations.

Table 3. Cumulative N$_2$O emissions of each farm operation phase during the rice growing period.

Year			Treatments		
			CT (kg ha^{-1})	RT (kg ha^{-1})	NT (kg ha^{-1})
2007	Early rice	Before aeration	0.09b	0.08c	0.10a
		During aeration	0.13a	0.12a	0.10b
		After aeration	0.24a	0.19b	0.19b
	Late rice	Before aeration	0.12b	0.13a	0.06c
		During aeration	0.10b	0.11a	0.10b
		After aeration	0.16c	0.18b	0.18a
Total emission			0.84a	0.82b	0.72c
2008	Early rice	Before aeration	−0.11c	0a	−0.03b
		During aeration	0b	0b	0.02a
		After aeration	0.09b	0.13a	0.09b
	Late rice	Before aeration	−0.16b	−0.07a	−0.05a
		During aeration	−0.04c	−0.02a	−0.03b
		After aeration	0.22c	0.26b	0.29a
Total emission			0.01c	0.30a	0.30b

Values are means of three replications for each treatment; means followed by different letters are significantly different at $P<0.05$.

retention may have increased the soil oxide layer. In this study, NT significantly increased the SP at 0−5 cm depth (Fig. 2), and thus had a larger oxide layer than other treatments, which may be beneficial to the oxidization of CH$_4$. Regina et al. indicated that CH$_4$ oxidation rate was higher when there were more macro-pores or fewer micro-pores in the soil [39]. In addition, CH$_4$ emission was influenced by soil temperature and soil redox potential (Eh). Yu et al. [40] reported that CH$_4$ emission showed an exponential decrease by an Eh increase.

In this study, the crop residues were distributed on the soil surface under NT. Furthermore, the decomposition of residues consumed limited soil dissolved oxygen. All these factors discussed above resulted in Eh decrease and consequently a reduction of CH$_4$ emission under NT. Khalil et al. [41] observed an increase in CH$_4$ emissions from paddy fields with increasing soil temperature. In this study, temperature was another major factor affecting CH$_4$ emission (Fig. 4). In general, NT decreased soil temperature especially during the hotter days. Therefore, low temperatures also reduced the CH$_4$ emission when compared with other treatments.

In this study, CH$_4$ emission from the late rice season was 65% higher than that from the early rice season, which indicates that the late rice paddy is the principal CH$_4$ source in double paddy fields. Temperature was the major reason for the differences in the CH$_4$ emission pattern between the early and the late rice season. The soil temperature had a predictive functional relationship with CH$_4$ emission. Zhu et al. [42] and Bossio et al. [43] reported a strong correlation between CH$_4$ emission and soil temperature. Furthermore, Whalen and Reeburgh [44] reported that temperature had important influence on CH$_4$ emission from soils and the combination of high soil moisture and low temperature was favorable to decrease CH$_4$ emission. In this study, an exponential model was used for fitting CH$_4$ emission and soil temperature. Our results showed that there was a significant correlation between CH$_4$ emission and soil temperature. But the coefficient of determination was not high, and this may be due to the fluctuation of soil temperature influenced by the alternation of wetting and drying in paddy. In this experimental area, the late rice season was the hottest time of the summer. Therefore, high temperatures enhanced the decomposition rate of crop residues in the moist environment. During the decomposition process of crop residues, a large number of organic compounds are produced and oxygen is consumed, thus decreasing the soil Eh, leading to an increase in the possibility of CH$_4$ emission. In contrast to the warm temperatures of the late rice season, the air temperatures of the early rice season were lower, which resulted in slower crop residue decomposition and therefore little CH$_4$-substrate. Hence, these differences in weather factors (e.g., temperature) resulted in the different characteristics of CH$_4$ between the early and the late rice seasons.

N$_2$O Emission

In our study, the fluxes of N$_2$O emission show a great fluctuation during the rice growth seasons, but it remained at a low level. Indeed, the N$_2$O emission was strongly influenced by external factors and many emission peaks occurred during the rice growing season. The emission of N$_2$O was dramatically different between the two years. This difference is possibly due to the variations in weather. Some studies show that extreme precipitation and drying could increase N$_2$O emission [45,46]. Hao et al. [47] reported that aeration and water flooding led to outbreaks of emissions. The precipitation in 2007 was much higher than the precipitation in 2008. This precipitation difference may explain the fluctuations of N$_2$O emissions between the two years.

The N$_2$O emission differences among the treatments were possibly due to farm operations (e.g., tillage, drainage). Some results indicated that N$_2$O production and emission was greatly influenced by tillage because of the breaking of the soil uniformity [48]. Nitrogen (mainly as NO$_3^-$-N or NH$_4^+$-N) can remain stable in homogeneous soil and thus may decrease N$_2$O production. Tillage practices change the soil nutrients and crop residue distribution. The distribution of soil nutrients was relatively even under CT and RT by cutting, mixing, overturning the soil and crop residues. However, the crop residues were well-distributed only in the 0–8 cm soil layer under RT because of the shallow tilled depth. High stratification ratio of soil nutrients (e.g., N, SOC) across different depths is observed in NT systems [48,49], which means that the soil nutrient distributions are not even among different depths. Therefore, the different distribution of crop residues and soil nutrients among the treatments influences the N$_2$O production and emission. In addition, similar to CH$_4$, N$_2$O emission is also influenced by soil Eh. Weier et al. reported that the rate of N$_2$O emission decreased with increasing soil reducibility [49]. Generally, crop residues in CT are mainly distributed within the plow layer (0–20 cm) and had a strong redox potential due to decomposition of crop residues. Therefore, N$_2$O produced from CT soils tended to be further deoxidized to N$_2$, which consequently decreased N$_2$O emission. Similar results were also reported by Steinbach and Alvarez [50] who observed NT increased N$_2$O emission.

Conclusion

Paddy fields with rice residues retention were a source of atmospheric CH$_4$, regardless of the tillage practice. Compared with other treatments, NT reduced CH$_4$ emission among the rice growing seasons. The GWPs (based on CH$_4$ emission) under NT was significantly *(P<0.05)* lower than that of CT and RT. The N$_2$O emission was vulnerable to external influences and varied greatly during the rice growing seasons. Although the cumulative emission under NT was more than other treatments, GWPs of

N$_2$O was relative low compared to that of CH$_4$. Therefore, N$_2$O emission was a weak source of GHG in paddy fields. The GWPs (based on CH$_4$ and N$_2$O) of NT is lower than that of CT and RT. Thus, adoption of NT is beneficial in GHG mitigation and could be a good practice of carbon-smart agriculture in double rice cropped regions.

Acknowledgments

We would like to express our sincere thanks to Mr. Shadrack Dikgwatlhe and Mr. Jay Lytle for language assistance.

Author Contributions

Conceived and designed the experiments: HLZ FC. Performed the experiments: XLB HMT. Analyzed the data: HLZ XLB JFX ZDC. Contributed reagents/materials/analysis tools: XLB HMT FC. Wrote the paper: HLZ XLB.

References

1. Levy PE, Mobbs DC, Jones SK, Milne R, Campbell C, et al. (2007) Simulation of fluxes of greenhouse gases from European grasslands using the DNDC model. Agric Ecosyst Environ 121: 186–192.

2. Saggar S, Hedley CB, Giltrap DL, Lambie SM (2007) Measured and modelled estimates of nitrous oxide emission and methane consumption from a sheep-grazed pasture. Agric Ecosyst Environ 122: 357–365.

3. Hernandez-Ramirez G, Brouder SM, Smith DR, Van Scoyoc GE (2009) Greenhouse gas fluxes in an eastern corn belt soil: Weather, nitrogen source, and rotation. J Environ Qual 38: 841–854.

4. Wassmann R, Neue HU, Ladha JK, Aulakh MS (2004) Mitigating greenhouse gas emissions from rice-wheat cropping systems in Asia. Environ Devel Sustain 6: 65–90.

5. Forster P, Ramaswamy V, Artaxo P, Berntsen T, Betts R, et al. (2007) Changes in atmospheric constituents and in radiative forcing. In: Solomon S, Qin D, Manning M, Chen Z, Marquis M, et al. (Eds.) Climate Change 2007: The Physical Science Basis. Contribution of Working Group I to the Fourth Assessment Report of the Intergovernmental Panel on Climate Change, Cambridge University Press, Cambridge, United Kingdom and New York, NY, USA.

6. Lokupitiya E, Paustian K (2006) Agricultural soil greenhouse gas emissions: A review of national inventory methods. J Environ Qual 35: 1413–1427.

7. Verma A, Tyagi L, Yadav S, Singh SN (2006) Temporal changes in N$_2$O efflux from cropped and fallow agricultural fields. Agric Ecosyst Environ 116: 209–215.

8. Liu H, Zhao P, Lu P, Wang YS, Lin YB, et al. (2008) Greenhouse gas fluxes from soils of different land-use types in a hilly area of South China. Agric Ecosyst Environ 124: 125–135.

9. Tan Z, Liu S, Tieszen LL, Tachie-Obeng E (2009) Simulated dynamics of carbon stocks driven by changes in land use, management and climate in a tropical moist ecosystem of Ghana. Agric Ecosyst Environ 130: 171–176.

10. Wassmann R, Dobermann A (2006) Greenhouse Gas Emissions from Rice Fields: what do we know and where should we head for? Paper presented at: The 2nd Joint International Conference on "Sustainable Energy and Environment". Bangkok, Thailand. 21–23 November. Paper D-030 (O).

11. Pandey D, Agrawal M, Bohra JS (2012) Greenhouse gas emissions from rice crop with different tillage permutations in rice-wheat system. Agric Ecosyst Environ 159: 133–144.

12. Yagi K, Minami K (1990) Effect of organic matter application on methane emission from some Japanese paddy fields. Soil Sci Plant Nutr 36: 599–610.

13. Yagi K, Tsuruta H, Kanda KI, Minami K (1996) Effect of water management on methane emission from a Japanese rice paddy field: Automated methane monitoring. Global Biogeochem Cycles 10: 255–267.

14. Nishimura S, Sawamoto T, Akiyama H, Sudo S, Yagi K (2004) Methane and nitrous oxide emissions from a paddy field with Japanese conventional water management and fertilizer application. Global Biogeochem Cycles 18, GB2017, doi:10.1029/2003GB002207.

15. Al-Kaisi MM, Yin X (2005) Tillage and crop residue effects on soil carbon and carbon dioxide emission in corn-soybean rotations. J Environ Qual 34: 437–445.

16. Yao Z, Zheng X, Xie B, Mei B, Wang R, et al. (2009) Tillage and crop residue management significantly affects N-trace gas emissions during the non-rice season of a subtropical rice-wheat rotation. Soil Biol Biochem 41: 2131–2140.

17. Matthias AD, Blackmer AM, Bremmer JM (1980) A simple chamber technique for field measurements of emissions nitrous oxide from soils. J Environ Qual 9: 251–256.

18. Estavillo JM, Merino P, Pinto M, Yamulki S, Gebauer G, et al. (2002) Short term effect of ploughing a permanent pasture on N$_2$O production from nitrification and denitrification. Plant Soil 239: 253–265.

19. Dendooven L, Patiño-Zúñiga L, Verhulst N, Luna-Guido M, Marsch R, et al. (2012) Global warming potential of agricultural systems with contrasting tillage and residue management in the central highlands of Mexico. Agric Ecosyst Environ 152: 50–58.

20. Toma Y, Hatano R (2007) Effect of crop residue C: N ratio on N$_2$O emissions from Gray Lowland soil in Mikasa, Hokkaido, Japan. Soil Sci Plant Nutr 53: 198–205.

21. Lou Y, Ren L, Li Z, Zhang T, Inubushi K (2007) Effect of rice residues on carbon dioxide and nitrous oxide emissions from a paddy soil of subtropical China. Water Air Soil Pollut 178: 157–168.

22. Lu F, Wang X, Han B, Ouyang Z, Duan X, et al. (2010) Net mitigation potential of straw return to Chinese cropland: Estimation with a full greenhouse gas budget model. Ecol Appl 20: 634–647.

23. Sun W, Huang Y (2012) Synthetic fertilizer management for China's cereal crops has reduced N$_2$O emissions since the early 2000s. Environ Pollut 160: 24–27.

24. Leytem AB, Dungan RS, Bjorneberg DL, Koehn AC (2011) Emissions of ammonia, methane, carbon dioxide, and nitrous oxide from dairy cattle housing and manure management systems. J Environ Qual 40: 1383–1394.

25. Jiao Z, Hou A, Shi Y, Huang G, Wang Y, et al. (2006) Water management influencing methane and nitrous oxide emissions from rice field in relation to soil redox and microbial community. Commun Soil Sci Plant Anal 37: 1889–1903.

26. Li CF, Zhou DN, Kou ZK, Zhang ZS, Wang JP, et al. (2012) Effects of Tillage and Nitrogen Fertilizers on CH$_4$ and CO$_2$ Emissions and Soil Organic Carbon in Paddy Fields of Central China. PLoS ONE 7(5): e34642. doi:10.1371/journal.pone.0034642.

27. Zhong HP, Yue YZ, Fan JW (2003) Characteristics of crop straw resources in China and its Utilization. Res Sci 25(4): 62–67. (In Chinese).

28. Xiao T, He C, Ling X, Jin C. Wu C, et al. (2010) Comprehensive utilization, situation and countermeasure of crop straw resources in China. World Agric 12: 31–33. (In Chinese).

29. Hanaki M, Ito T, Saigusa M (2002) Effect of no-till rice (Oryza sativa L.) cultivation on methane emission in three paddy fields of different soil types with rice straw application. Jpn J Soil Sci Plant Nutr 73: 135–143.

30. Xiong YM, Huang GQ, Cao KW, Liu LW (2003) Review on Developing Ryegrass–Rice Rotation System in Double-cropping Rice Area of Middle and Lower Reaches of Changjiang River. Acta Agric Jiangxi 15: 47–51. (In Chinese).

31. Lapitan RL, Wanninkhof R, Mosier AR (1999) Methods for stable gas flux determination in aquatic and terrestrial systems. Develop Atmos Sci 24: 29–66.

32. Nakano T, Sawamoto T, Morishita T, Inoue G, Hatano R (2004) A comparison of regression methods for estimating soil-atmosphere diffusion gas fluxes by a closed-chamber technique. Soil Biol Biochem 36: 107–113.

33. Zheng X, Wang M, Wang Y, Shen R, Li J, et al. (1998) Comparison of manual and automatic methods for measurement of methane emission from rice paddy fields. Adv Atmos Sci 15: 569–579.

34. IPCC (1996) Climate Change 1995: The Science of Climate Change. In: Houghtom JT, Meira Filho LG, Callander BA, Harris N, Kattenberg A, et al. (Eds.) Intergovernmental Panel on Climate Change, Cambridge University Press, Cambridge, United Kingdom.

35. IPCC (2007) Climate change 2007: The physical science basis. Contribution of working group I to the fourth assessment report of the intergovernmental panel on climate change. Cambridge University Press, Cambridge, United Kingdom and New York, NY, USA.

36. Zhang JK, Jiang CS, Hao QJ, Tang QW, Cheng BH, et al. (2012) Effects of tillage-cropping systems on methane and nitrous oxide emissions from agro-ecosystems in a purple paddy soil. Environ Sci 33(6): 1980–1986. (In Chinese).

37. Gregorich EG, Rochette P, Hopkins DW, McKim UF, St-Georges P (2006) Tillage-induced environmental conditions in soil and substrate limitation determine biogenic gas production. Soil Biol Biochem 38: 2614–2628.

38. Wang M, Li J, Zhen X (1998) Methane Emission and Mechanisms of Methane Production, Oxidation, Transportation in the Rice Fields. Sci Atmos Sin 22: 600–612. (In Chinese).

39. Regina K, Pihlatie M, Esala M, Alakukku L (2007) Methane fluxes on boreal arable soils. Agric Ecosyst Environ 119: 346–352.

40. Yu K, Böhme F, Rinkleb J, Neue HU, DeLaune RD (2007) Major biogeochemical processes in soils: A microcosm incubation from reducing to oxidizing conditions. Soil Sci Soc Am J 71: 1406–1417.

41. Khalil MAK, Rasmussen RA, Shearer MJ, Chen ZL, Yao H, et al. (1998) Emissions of methane, nitrous oxide, and other trace gases from rice fields in China. J Geophys Res 103: 25241–25250.

42. Zhu R, Liu Y, Sun L, Xu H (2007) Methane emissions from two tundra wetlands in eastern Antarctica. Atmos Environ 41: 4711–4722.

43. Bossio DA, Horwath WR, Mutters RG, Van Kessel C (1999) Methane pool and flux dynamics in a rice field following straw incorporation. Soil Biol Biochem 31: 1313–1322.

44. Whalen SC, Reeburgh WS (1996) Moisture and temperature sensitivity of CH$_4$ oxidation in boreal soils. Soil Biol Biochem 28: 1271–1281.

45. Zona D, Janssens IA, Verlinden MS, Broeckx LS, Cools J, et al. (2011) Impact of extreme precipitation and water table change on N$_2$O fluxed in a bio-energy poplar plantation. Biogeosci Discuss 8: 2057–2092.

46. Xu W, Liu G, Liu W (2002) Effects of precipitation and soil moisture on N$_2$O emissions from upland soils in Guizhou. Chin J Appl Ecol 13(1): 67–70. (in Chinese).

47. Hao X, Chang C, Carefoot JM, Janzen HH, Ellert BH (2001) Nitrous oxide emissions from an irrigated soil as affected by fertilizer and straw management. Nutr Cy Agroecosyst 60: 1–8.

48. Kay BD, VandenBygaart AJ (2002) Conservation tillage and depth stratification of porosity and soil organic matter. Soil Till Res 66: 107–118.

49. Weier KL, Doran JW, Power JF, Walters DT (1993) Denitrification and the dinitrogen/nitrous oxide ratio as affected by soil water, available carbon, and nitrate. Soil Sci Soc Am J 57: 66–72.

50. Steinbach HS, Alvarez R (2006) Changes in soil organic carbon contents and nitrous oxide emissions after introduction of no-till in Pampean agroecosystems. J Environ Qual 35: 3–13.

8

Increasing Cropping System Diversity Balances Productivity, Profitability and Environmental Health

Adam S. Davis[1]*, Jason D. Hill[2], Craig A. Chase[3], Ann M. Johanns[4], Matt Liebman[5]

1 United States Department of Agriculture/Agricultural Research Service, Global Change and Photosynthesis Research Unit, Urbana, Illinois, United States of America, **2** Department of Bioproducts and Biosystems Engineering, University of Minnesota, St. Paul, Minnesota, United States of America, **3** Leopold Center for Sustainable Agriculture, Iowa State University, Ames, Iowa, United States of America, **4** Department of Economics, Iowa State University Extension and Outreach, Osage, Iowa, United States of America, **5** Department of Agronomy, Iowa State University, Ames, Iowa, United States of America

Abstract

Balancing productivity, profitability, and environmental health is a key challenge for agricultural sustainability. Most crop production systems in the United States are characterized by low species and management diversity, high use of fossil energy and agrichemicals, and large negative impacts on the environment. We hypothesized that cropping system diversification would promote ecosystem services that would supplement, and eventually displace, synthetic external inputs used to maintain crop productivity. To test this, we conducted a field study from 2003–2011 in Iowa that included three contrasting systems varying in length of crop sequence and inputs. We compared a conventionally managed 2-yr rotation (maize-soybean) that received fertilizers and herbicides at rates comparable to those used on nearby farms with two more diverse cropping systems: a 3-yr rotation (maize-soybean-small grain + red clover) and a 4-yr rotation (maize-soybean-small grain + alfalfa-alfalfa) managed with lower synthetic N fertilizer and herbicide inputs and periodic applications of cattle manure. Grain yields, mass of harvested products, and profit in the more diverse systems were similar to, or greater than, those in the conventional system, despite reductions of agrichemical inputs. Weeds were suppressed effectively in all systems, but freshwater toxicity of the more diverse systems was two orders of magnitude lower than in the conventional system. Results of our study indicate that more diverse cropping systems can use small amounts of synthetic agrichemical inputs as powerful tools with which to tune, rather than drive, agroecosystem performance, while meeting or exceeding the performance of less diverse systems.

Editor: John P. Hart, New York State Museum, United States of America

Funding: Funding for the study was provided by the US Department of Agriculture National Research Initiative (Projects 2002-35320-12175 and 2006-35320-16548), the Leopold Center for Sustainable Agriculture (Projects 2004-E06, 2007-E09, and 2010-E02), the Iowa Soybean Association, and the Organic Center. The funders had no role in study design, data collection and analysis, decision to publish, or preparation of the manuscript.

Competing Interests: The authors have declared that no competing interests exist.

* E-mail: adam.davis@ars.usda.gov

Introduction

One of the key challenges of the 21st century is developing ways of producing sufficient amounts of food while protecting both environmental quality and the economic well-being of rural communities [1,2]. Over the last half century, conventional approaches to crop production have relied heavily on manufactured fertilizers and pesticides to increase yields, but they have also degraded water quality and posed threats to human health and wildlife [3–6]. Consequently, attention is now being directed toward the development of crop production systems with improved resource use efficiencies and more benign effects on the environment [1,7]. Less attention has been paid to developing better methods of pest management, especially for weeds. Here we explore the potential benefits of diversifying cropping systems as a means of controlling weed population dynamics while simultaneously enhancing other desirable agroecosystem processes [8]. We focus on crop rotation, an approach to cropping system diversification whereby different species are placed in the same field at different times.

Rotation systems have been used for millennia to maintain soil fertility and productivity and to suppress pests, and can increase yields even in situations where substantial amounts of fertilizers and pesticides are applied [9,10]. Rotation systems also foster spatial diversity, since different crops within the rotation sequence are typically grown in different fields on a farm in the same year. Diversification through crop rotation can be an especially useful strategy in farming systems that integrate crop and livestock production. The addition of forage crops, including turnips and clovers, to cereal-based systems in northwestern Europe and England in the 1600s and 1700s enhanced nitrogen supply through fixation by legumes, and increased nutrient cycling due to greater livestock density and manure production. These changes allowed the intensification of both crop and livestock production and increased yields substantially [11,12]. Integrated crop–livestock systems remained widespread in northern Europe, England, and much of the humid, temperate regions of North America until the 1950s and 1960s, when increased availability of relatively low-cost synthetic fertilizers made mixed farming and nutrient recycling biologically unnecessary and specialized crop and livestock production more economically attractive. In recent years, there has been interest in reintegrating crop and livestock systems as a strategy for reducing reliance on fossil fuels, minimizing the use of increasingly expensive fertilizers, and limiting water pollution by nutrients, pathogens, and antibiotics [13,14].

Weeds are a ubiquitous and recurrent problem in essentially all crop production systems, and chemicals applied for weed control dominate the world market for pesticides [15]. With the introduction of crop genotypes engineered to tolerate herbicides, especially glyphosate, and with the continuing availability of older, relatively inexpensive herbicides, such as atrazine, successful weed management in conventional crop production systems has been largely taken for granted since the mid-1990s. Now, however, with expanded recognition of herbicides as environmental contaminants [4] and the increasing prevalence of herbicide resistant weeds [16], there is an important need to develop weed management strategies that are less reliant on herbicides and that subject weeds to a wide range of stress and mortality factors [17]. We believe that cropping system diversification may play an important role in the development of such strategies.

Here, we report the results of a large-scale, long-term experiment examining the consequences of cropping system diversification on agronomic, economic, and environmental measures of system performance. The experiment was conducted during 2003–2011 in Boone County, Iowa, within the central U.S. maize production region, and comprised three contrasting cropping systems varying in length of crop sequence, levels of chemical inputs, and use of manure. We compared a conventionally managed 2-yr rotation (maize-soybean) that received fertilizers and herbicides at rates comparable to those used on surrounding commercial farms with two more diverse cropping systems: a 3-yr rotation (maize-soybean-small grain + red clover) and a 4-yr rotation (maize-soybean-small grain + alfalfa-alfalfa) managed with reduced N fertilizer and herbicide inputs and periodic applications of composted cattle manure. Triticale was used as the small grain crop in 2003–2005; oat was used in 2006–2011. The 2-yr rotation is typical of cash grain farming systems in the region, whereas the 3-yr and 4-yr rotations are representative of farming systems in the region that include livestock. Details of the experimental site, management practices, sampling procedures, and data analyses are provided in the online SI section (Text S1, Figure S1, Tables S1-S4).

A central hypothesis framing our study was that cropping system diversification would result in the development of ecosystem services over time that would supplement, or eventually displace, the role of synthetic external inputs in maintaining crop productivity and profitability. Based on this hypothesis, we predicted that input requirements of the more diverse systems would initially be similar to that of the less diverse system, but would increasingly diverge from the less diverse system over time as the systems matured. We also predicted that crop yields, weed suppression, and economic performance of the three systems would be similar throughout the study. Finally, we predicted that reduced requirements for external synthetic inputs for pest management would result in a lower toxicological profile of the more diverse systems compared to the less diverse system.

Results

Crop Yields and Net Profitability

Cropping system diversification enhanced yields of maize and soybean grain and system-level harvested crop mass (grain, straw, and hay) while maintaining economic returns. The most parsimonious linear statistical models for each of these measures of system performance contained terms for main effects of *year* and *system*, but no interaction term ($AIC_{with\ interaction} = 319$; $AIC_{no\ interaction} = 315$). Over the 2003 to 2011 period, maize grain yield was on average 4% greater in the 3-yr and 4-yr rotations than in the 2-yr rotation (means for the 2-yr, 3-yr and 4-yr rotations are hereafter referred to as μ_2, μ_3 and μ_4, respectively; $\mu_2 = 12.3 \pm 0.1$ Mg ha^{-1}; $\mu_3 = 12.7 \pm 0.2$ Mg ha^{-1}; $\mu_4 = 12.9 \pm 0.2$ Mg ha^{-1}; pre-planned 1 d.f. contrast of *system*: $F_{1,7} = 8$, $P = 0.03$), and similar in the 3-yr and 4-yr rotations (Fig. 1a). Soybean grain yield during the same period was on average 9% greater in the 3-yr and 4-yr rotations than in the 2-yr rotation ($\mu_2 = 3.4 \pm 0.07$ Mg ha^{-1}; $\mu_3 = 3.8 \pm 0.08$ Mg ha^{-1}; $\mu_4 = 3.8 \pm 0.08$ Mg ha^{-1}; $F_{1,7} = 11.3$, $P = 0.01$) and similar in the 3-yr and 4-yr rotations (Fig. 1b). Harvested crop mass, averaged over the various crop phases comprising each cropping system, followed a similar pattern to maize and soybean grain yields. Mean crop biomass for 2003 to 2011 was 8% greater in the 3-yr and 4-yr rotations than in the 2-yr rotation ($\mu_2 = 7.9 \pm 0.08$ Mg ha^{-1}; $\mu_3 = 8.5 \pm 0.1$ Mg ha^{-1}; $\mu_4 = 8.6 \pm 0.2$ Mg ha^{-1}; *system*: $t_6 = 5.1$, $P = 0.002$), and similar in the 3-yr and 4-yr rotations (Fig. 1c).

We examined system profitability by calculating net returns to land and management, which represent profits to a farm operation without accounting for costs of land (e.g., rent or mortgage payments), management time (e.g., marketing), and federal subsidies. Profitability was analyzed for two temporal periods. From 2003 to 2005, considered the "startup" phase for the study, there were no differences among cropping systems in net profit, either through an analysis of main effects of *system* ($\mu_2 = \$448 \pm 17$ ha^{-1}; $\mu_3 = \$402 \pm 17$ ha^{-1}; $\mu_4 = \$457 \pm 15$ ha^{-1}; $F_{2,6} = 0.12$, $P = 0.89$) or by pre-planned 1-d.f. contrasts (2-yr vs. 3-yr and 4-yr rotations: $F_{1,7} = 0.10$, $P = 0.77$) (Fig. 1d). From 2006 to 2011, the "established" phase of the study, there were again no differences among systems, either through main effects of *system* ($\mu_2 = \$953 \pm 36$ ha^{-1}; $\mu_3 = \$965 \pm 34$ ha^{-1}; $\mu_4 = \$913 \pm 26$ ha^{-1}; $F_{2,6} = 0.62$, $P = 0.57$) or by pre-planned 1-d.f. contrasts (2-yr vs. 3-yr and 4-yr rotations: $F_{1,7} = 0.03$, $P = 0.86$).

Stability of system performance over time, as measured through a comparison of variances for the various products of the system, was similar for maize grain yield ($F_{2,6} = 2.4$, $P = 0.17$), soybean grain yield ($F_{2,6} = 0.95$, $P = 0.44$) and net returns to land and management during the startup phase of the study, 2003 to 2005 ($F_{2,6} = 0.05$, $P = 0.95$). Two system products, harvested crop mass from 2003 to 2011 and profit during the established phase of the study, 2006 to 2011, showed considerable differences in system stability over time, but in contrasting ways. Variance in mean harvested crop mass was greater in the 3-yr and 4-yr rotations than in the 2-yr rotation ($\sigma_2^2 = 0.27$; $\sigma_3^2 = 0.60$; $\sigma_4^2 = 0.95$; $F_{1,7} = 16$, $P = 0.005$). Conversely, cropping system diversification was associated with lower variance in profit during the established phase of the study. Variance in profit from 2006 to 2011 was lower in the 3-yr and 4-yr rotations than in the 2-yr rotation ($\sigma_2^2 = 1.5 \times 10^5$; $\sigma_3^2 = 8.1 \times 10^3$; $\sigma_4^2 = 6.3 \times 10^3$; $F_{1,7} = 16$, $P = 0.005$).

Agrichemical, Labor and Energy Inputs

Application rates of the primary agrichemicals used in this study, manufactured N fertilizer ($F_{2,14} = 117$, $P < 0.0001$) and herbicides ($F_{2,14} = 287$, $P < 0.0001$), both showed strong effects of cropping system. Manufactured N fertilizer applications were higher in the 2-yr rotation than in the 3-yr and 4-yr rotations ($\mu_2 = 80 \pm 3$ kg N ha^{-1}; $\mu_3 = 16 \pm 3$ kg N ha^{-1}; $\mu_4 = 11 \pm 2$ kg N ha^{-1}; $F_{1,17} = 16$, $P = 0.005$), with the difference between systems increasing over the course of the study ($F_{2,14} = 11.6$, $P = 0.001$) (Fig. 1e). Herbicide application rates followed a similar pattern, with greater amounts of herbicide applied in the 2-yr rotation than in the 3-yr and 4-yr rotations ($\mu_2 = 1.9 \pm 0.06$ kg a.i. ha^{-1};

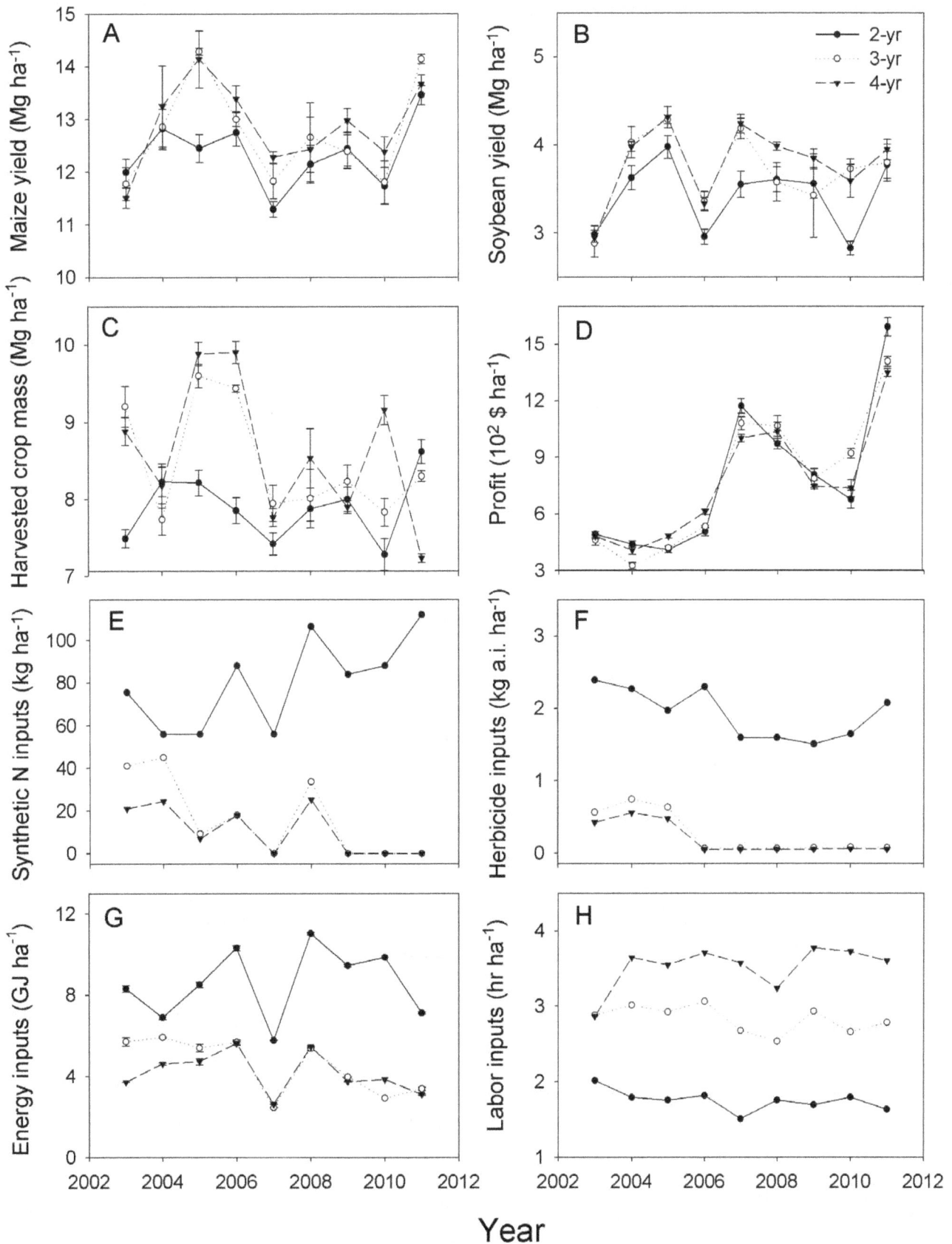

Figure 1. Cropping system performance over time. Annual performance of maize-soybean (2-yr), maize-soybean-small grain/red clover (3-yr), and maize-soybean-small grain/alfalfa-alfalfa (4-yr) cropping systems in Boone, IA, from 2003 to 2011. Performance metrics included: a) maize yield, b) soybean yield, c) rotation-level harvested crop mass, d) net returns to land and management, e) manufactured N fertilizer application rate, f) herbicide

application rate, g) fossil energy use, and h) labor requirements. Symbols represent the mean ± SEM of four replicate experimental blocks (N = 36 per cropping system).

$\mu_3 = 0.26 \pm 0.05$ kg a.i. ha^{-1}; $\mu_4 = 0.20 \pm 0.03$ kg a.i. ha^{-1}; $F_{1,17} = 610$, P<0.0001); differences among systems, however, did not increase over time (Fig. 1f).

Fossil energy use was strongly influenced by cropping system in both the startup ($F_{2,6} = 94$, P<0.0001) and established ($F_{2,6} = 116$, P<0.0001) phases of the study, with no difference in energy use between experimental phases ($F_{1,92} = 0.39$, P = 0.53) (Fig. 1g). From 2003 to 2011, inputs of energy were greater in the 2-yr rotation than in the 3-yr and 4-yr rotations ($\mu_2 = 8.6 \pm 0.1$ GJ ha^{-1}; $\mu_3 = 4.5 \pm 0.1$ GJ ha^{-1}; $\mu_4 = 4.2 \pm 0.04$ GJ ha^{-1}; $F_{1,7} = 55$, P = 0.0001). The partial correlations between energy use in a given cropping system and energy use in the maize phase of that rotation, taking into account the amount of N fertilizer applied to maize, were 0.94, 0.81 and 0.70 in the 2-yr, 3-yr and 4-yr systems, respectively (SI, Table S5). This indicated that synthetic N fertilizer use in the maize phase of the various cropping systems drove energy use within the maize phase, which in turn drove energy use by a given cropping system.

Demand for labor differed among the three cropping systems in both the startup ($F_{2,4} = 26$, P = 0.005) and established ($F_{2,10} = 299$, P<0.0001) study phases, but followed a contrasting pattern to energy requirements (Fig. 1h). Labor inputs were more than 33% lower in the 2-yr rotation than in the 3-yr and 4-yr rotations from 2003 to 2005 ($F_{1,5} = 35$, P = 0.002) and from 2006 to 2011 ($F_{1,11} = 59$, P<0.0001). Overall, there was a strong negative correlation (r = −0.79, P<0.0001) between fossil energy and labor inputs over time in the three cropping systems.

Divergent Weed Management Systems

Two lines of evidence indicate that weeds were managed effectively in all three cropping systems in both the 'startup' and 'established' phases, in spite of reducing herbicide use by 88% in the 3-yr and 4-yr rotations compared to the 2-yr rotation. First, weed seedbanks declined at an equal rate in all study systems (Fig. 2a). Selection among linear mixed effects regression models incorporating temporal autocorrelation among seedbank measurements over time supported different intercepts (*system*: $F_{2,6} = 16.8$, P = 0.0035) but did not support inclusion of a *year* by *system* interaction term ($AIC_s = 182$; $AIC_{s*y} = 185$), thus indicating a common slope ($b_1 = -0.18$). For all three systems, the time to decline to 95% of the weed seedbank levels in 2003 was 16.6 years. Declines in weed seedbanks reflected a focus of management attention on the timing of weed management activities and herbicide choices in all three systems, as well as the increased number and diversity of stress and mortality factors present in the 3-yr and 4-yr rotations [8,21]. Higher densities of weed seeds in the 3-yr and 4-yr rotations, as indicated by their greater intercept values than for the 2-yr rotation (Fig. 2a.), were the result of poorer weed control in the 3-yr and 4-yr rotations during the set-up of the experiment plots in 2002.

The second line of evidence concerns weed biomass, which was very low in all three cropping systems for the duration of the study (Fig. 2b), never exceeding 0.3% of harvested crop mass. Weed biomass was the same within a given crop phase, regardless of the cropping system in which it occurred (main effect of *system*: maize, $F_{2,6} = 1.47$; P = 0.30; soybean, $F_{2,6} = 0.88$; P = 0.46; small grain, $F_{1,3} = 1.24$; P = 0.31). There were differences in mean weed biomass among cropping systems ($\mu_2 = 0.0003 \pm 0.00007$ Mg ha^{-1}; $\mu_3 = 0.0076 \pm 0.0012$ Mg ha^{-1}; $\mu_4 = 0.009 \pm 0.001$ Mg ha^{-1}; $F_{2,6} = 12.7$; P<0.007). These differences arose mainly due

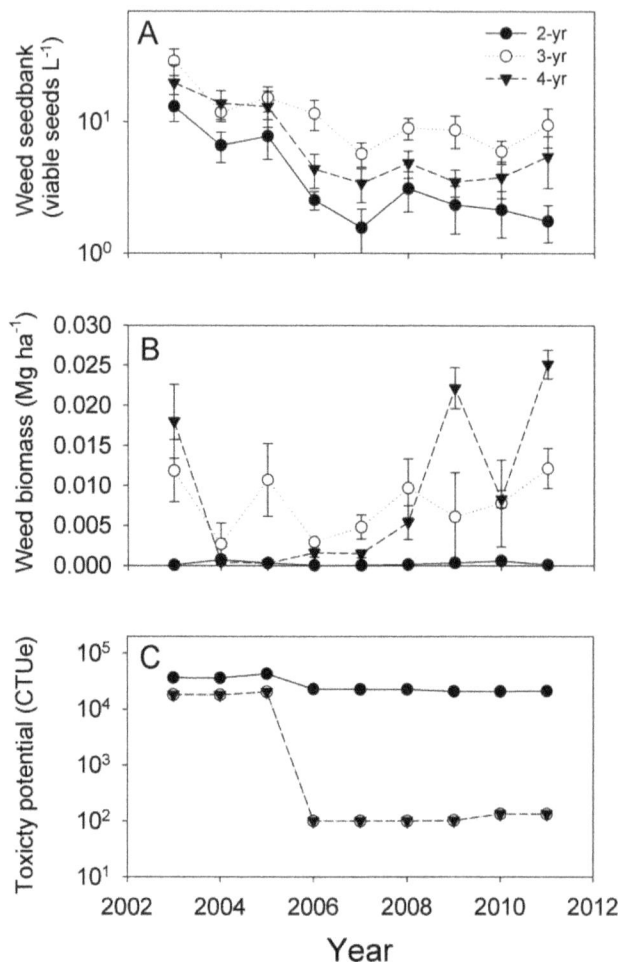

Figure 2. Divergent weed management systems. Weed management characteristics in maize-soybean (2-yr), maize-soybean-small grain/red clover (3-yr), and maize-soybean-small grain/alfalfa-alfalfa (4-yr) cropping systems in Boone, IA, from 2003 to 2011. Performance metrics included a) weed seed density in soil, b) weed aboveground biomass, and c) freshwater toxicity potential expressed in comparative toxic units (CTU_e). Symbols represent the mean ± SEM of four replicate experimental blocks (N = 36 per cropping system).

to the presence of a small grain phase in the 3-yr and 4-yr rotation crop sequences. Weed biomass did not differ between maize and soybean in any of the cropping systems ($F_{1,202} = 2.1$; P = 0.15), however weed biomass in the small grain phase of the 3-yr and 4-yr rotations was greater than weed biomass in the maize and soybean phases ($F_{1,206} = 174$; P<0.0001). In the 4-year system, weed biomass in alfalfa was intermediate between weed biomass levels in the maize/soybean and small grain phases.

Environmental toxicity, in relation to ecotoxicological profiles for herbicides used in this study (Fig. 2c), showed a strong effect of *system* ($F_{2,14} = 1673$, P<0.0001), with lower toxicity potential in the 3-yr and 4-yr rotations compared to the 2-yr rotation (*type*: $F_{1,17} = 2691$, P<0.0001). Ecotoxicity in the diversified and conventional systems diverged as the systems matured over time [*type* x *phase*: $F_{1,16} = 7.4$, P = 0.015], transitioning from a two-fold difference during 2003 to 2005 to a two hundred-fold difference in toxicity from 2006 to 2011 (Fig. 2c).

Discussion

Our results support the hypothesis that the development of ecosystem services over time in more diverse cropping rotations increasingly displaces the need for external synthetic inputs to maintain crop productivity. From 2003 to 2011, as predicted, the desired products (crop yield, weed suppression, and economic performance) of the more diverse and less diverse cropping rotations were similar, whereas external inputs and environmental impacts differed greatly among the systems (Fig. 3). Comparing these metrics of system performance by experimental phase (initial three years of system establishment versus the following six years) confirmed our prediction that system inputs and environmental impacts would diverge over time, whereas yield and profit would remain similar among more diverse and less diverse rotations. In the more diverse rotations, small amounts of synthetic agrichemical inputs thus served as powerful tools with which to tune, rather than drive, agroecosystem performance.

Grain production in the U.S. is dominated by short rotation systems designed to maximize grain yield and profit. These are important goals but represent only a portion of the many ecosystem services that managed lands may provide [18] and that should be considered when evaluating alternative production systems [1,19]. We believe that these functions are complementary, rather than competing, considerations for agroecosystem design. The results of this study demonstrate that more rotationally diverse cropping systems may be optimized in multiple dimensions, leveraging small agrichemical inputs with biological synergies arising from enhanced diversity of crop species and management tactics.

An example of the synergizing effects of cropping system diversification can be found in weed management in the 3-yr and 4-yr rotations. Weeds were suppressed as effectively in these systems as in the 2-yr rotation, with declining soil seedbanks and negligible weed biomass, yet herbicide inputs in the 3-yr and 4-yr rotation plots were 6 to 10 times lower, and freshwater toxicity 200 times lower, than in the 2-yr rotation. Improved efficiency and environmental sustainability of weed management in the 3-yr and 4-yr rotations resulted from integrating multiple, complementary tactics in an ecological weed management framework [8,20]. Mounting evidence for unintended effects resulting from heavy reliance on herbicides highlights the need to re-think the role of herbicides in weed management. Non-target impacts of herbicides include reproductive abnormalities and mortality in vertebrates [5,21–23] and potential for diminished non-crop nectar resources for key pollinator species [17,24,25]. Herbicide overuse has also resulted in widespread, accelerating evolution of weed genotypes resistant to one or more modes of herbicide action [26,27]. Our

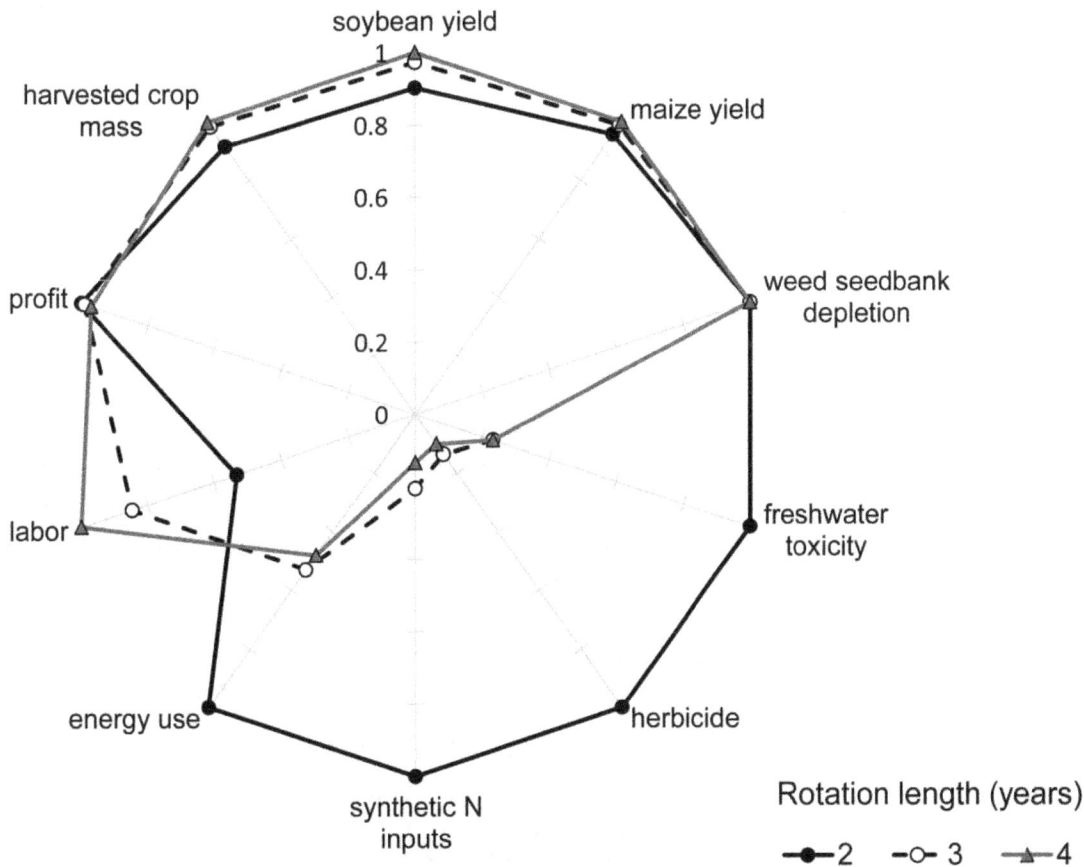

Figure 3. Multiple indicators of cropping system performance. Comparative long-term performance of maize-soybean (2-yr), maize-soybean-small grain/red clover (3-yr), and maize-soybean-small grain/alfalfa-alfalfa (4-yr) cropping systems in Boone, IA, averaged over the 2003–2011 study period. Variable means are normalized on a 0 to 1 scale, with 1 representing the cropping system with the largest absolute value for that variable (N = 36 per cropping system). Performance metrics included: maize and soybean yield, rotation-level harvested crop mass, net returns to land and management, manufactured N fertilizer and herbicide application rate, fossil energy use, labor requirements, freshwater toxicity potential and weed seedbank decline (measured as exponential decay constant).

data indicate that, in the context of a cropping system with weed suppressive characteristics, small herbicide inputs may contribute to a diverse suite of tactics that cumulatively provide effective, reliable, and more durable weed management.

The diversity-productivity-stability relationship has long been a key theme in ecology [28,29]. Recently, it has been applied in the context of bioenergy crop production to describe increases in biomass and ecosystem services, such as C sequestration, associated with increasing species diversity in polycultures of bioenergy feedstock crop species [30]. Our work supports the application of this concept to cropping systems more broadly. Future gains in more diverse systems may depend upon the application of ecological principles surrounding this relationship to cropping system design [31,32]. Cropping system diversification in this study included both crop species and management practices. In contrast to the 2-yr rotation, with two species, both of the 3-yr and 4-yr rotations included four crop species. In the 4-yr rotation, further temporal diversification was achieved by including a perennial-only crop phase (alfalfa hay) for one quarter of the rotation sequence. Our results showed productivity gains associated with greater diversity in system-level harvested crop mass and maize and soybean seed yields. We also observed increased stability of profit, with similar long-term means, in the 3-yr and 4-yr rotations compared to the 2-yr rotation.

Similar profits were attained through different pathways in the 3-yr and 4-yr rotations and the 2-yr rotation (Fig. 3). Increased labor, information intensive management and ecosystem services arising from increased biological N fixation (via the clover and alfalfa crops) and contrasting crop phenologies and competitive abilities were substituted in 3-yr and 4-yr rotations for the higher inputs of manufactured N, herbicides and energy from fossil fuels driving the 2-yr rotation. Energy use in maize drove differences among the cropping systems, and manufactured N inputs to maize contributed most strongly to energy balances for this crop. The high sensitivity of agricultural energy use to N fertilizer inputs provides a high-priority target for the redesign of cropping systems for increased sustainability.

Reintegration of crop and livestock production, as represented by the forage legumes and manure applications present in the more diverse systems, is not simply another aspect of cropping system diversification. Instead, it embodies an important principle in sustainable agriculture: system boundaries should be drawn to minimize externalities. Animal manure is produced regardless of whether feed grains are shipped to centralized concentrated animal feeding operations, or produced within integrated crop-livestock farming operations. In the former case, the manure may become a waste product and water pollutant if quantities exceed available land area for field application [33], whereas in the latter case, it contributes directly to crop nutrient requirements, improves soil quality, and reduces fossil fuel subsidies associated with grain transport and external N fertilizer inputs [14].

Substantial improvements in the environmental sustainability of agriculture are achievable now, without sacrificing food production or farmer livelihoods. When agrichemical inputs are completely eliminated, yield gaps may exist between conventional and alternative systems [19]. However, such yield gaps may be overcome through the strategic application of very low inputs of agrichemicals in the context of more diverse cropping systems. Although maize is grown less frequently in the 3-yr and 4-yr rotations than in the 2-yr rotation, this will not compromise the ability of such systems to contribute to the global food supply, given the relatively low contribution of maize and soybean production to direct human consumption and the ability of livestock to consume small grains and forages [34]. Through a

balanced portfolio approach to agricultural sustainability, cropping system performance can be optimized in multiple dimensions, including food and biomass production, profit, energy use, pest management, and environmental impacts.

Materials and Methods

Site Details and Agronomic Management

To investigate the relative performance of conventional and more diverse cropping systems, we conducted a 9-hectare experiment at the Iowa State University Marsden Farm (Figure S1), in Boone County, IA (42°01′ N; 93°47′ W; 333 m above sea level). The experiment site lies within a region of intensive rain-fed maize and soybean production and is surrounded by farms with high levels of productivity. Soils at the site are deep, fertile Mollisols. The experimental cropping system treatments included a conventionally managed 2-yr rotation (maize/soybean) that received agrichemicals at rates comparable to those used on commercial farms in the region, and more diverse cropping systems – a 3-yr rotation (maize/soybean/small grain + red clover green manure) and a 4-yr rotation (maize/soybean/small grain + alfalfa/alfalfa hay) – managed with reduced N fertilizer and herbicide inputs.

The entire site was planted with oat in 2001 and the cropping systems experiment was established in 2002 using a randomized complete block design with each crop phase of each rotation system present every year in four replicate blocks. Plots were 18 m x 85 m and managed with conventional farm machinery. Spring triticale was used as the small grain in 2003–2005, whereas oat was used in 2006–2010. Synthetic fertilizers were applied in the 2-yr rotation at conventional rates based on soil tests. In the 3-yr and 4-yr rotations, composted cattle manure was applied before maize production at a mean dry matter rate of 8.3 Mg ha^{-1} and substantial amounts of N were added through fixation by red clover and alfalfa [35,36,37]. Manure and legume N-fixation in the 3-yr and 4-yr rotations were supplemented with synthetic fertilizers based on soil tests, including the late-spring soil nitrate test for maize production [38]. Weed management in the 2-yr rotation was based largely on herbicides applied at conventional rates. In the 3-yr and 4-yr rotations, herbicides were applied in 38-cm-wide bands in maize and soybean and inter-row zones were cultivated; no herbicides were applied in small grain and forage legume crops. Choices of post-emergence herbicides used in each of the systems were made based on the identities, densities, and sizes of weed species observed in the plots. Other details of the farming practices used in the different cropping systems are described in Liebman et al. [39] and in the online SI materials (Text S1). Sampling procedures for determining crop yields, weed biomass and weed seed densities in soil are also described in the online SI materials (Text S1).

Energy and Economic Analyses

Energy inputs were divided into five categories: seed, fertilizer, pesticides, fuel for field operations, and propane and electricity used for drying maize grain after harvest. Data were obtained from logs describing all field operations, material inputs, and crop moisture characteristics for the experimental plots during the study period. Economic analyses measured performance characteristics of whole rotation systems under contrasting management strategies. We evaluated net returns to land and management on a unit land area basis, with land units divided in two equal portions for maize and soybean in the 2-yr rotation; three equal portions for maize, soybean, and small grains with red clover in the 3-yr rotation; and four equal portions for maize, soybean, small grains with alfalfa, and alfalfa in the 4-yr rotation. Net returns to land and management represented returns to a farm operation calculated without accounting for costs of land (e.g., rent or mortgage payments), management time (e.g., marketing), or possible federal subsidies. Data sources and assumptions for the energy and economic analyses are shown in the online SI materials.

Ecotoxicological Calculations

Freshwater ecotoxicity of pesticide use was estimated using the USEtox model [40–42]. Characterization factors (CFs) of ecotoxicity potential for active ingredients included transport to freshwater via surface water, soil, and air. CFs were available for eight of ten active ingredients applied in the three rotations. The two active ingredients for which CFs were unavailable are not of particular concern for freshwater ecotoxicity due either to their low toxicity (mesotrione) or low infiltration and persistence in freshwaters (lactofen) [43].

Statistical Analyses

The experiment was arranged in a randomized complete block design, with all entry points of the three crop rotations (i.e. all crops within each of the rotations) represented in four replicate blocks in each year of the study, for a total of 36 plots. Cropping system effects in time series data were analyzed using hierarchical linear mixed effects repeated measures models, modeling temporally correlated errors with an ARMA (auto-regressive moving average) correlation structure in the *nlme* package of R v.2.14.1 [44,45]. Fixed effects included *cropping system* and *experimental phase* (startup = 2003 to 2005; established = 2006 to 2011), and random effects included *replicate block* nested within *cropping system* and *year*. Partial correlations were estimated using the *corpcor* package in R v.2.14.1. In contrast to data for quantitative observations (e.g. crop yield or weed biomass) that varied by replicate block and year, data for input variables, such as synthetic fertilizer or herbicides and associated environmental toxicity metrics, did not vary among blocks for a particular rotation entry point in a given year, but did vary among years. Therefore, site-year was treated as the source of experimental replication for these latter variables in our statistical tests for effects of *cropping system* and *experimental phase*. This led to contrasting degrees of freedom in reported F-tests for these two data types. Finally, for variables with non-constant variance among cropping systems over time (crop biomass and profit), we used the 'varIdent' variance function within the *nlme* package to explicitly model differences in variances among cropping systems for these variables within our mixed effects models.

Supporting Information

Figure S1   Aerial view of Marsden Farm study, Boone IA. Crop abbreviations: m = maize, sb = soybean, g = small grain, a = alfalfa.

Table S1   Mean monthly air temperature and total monthly precipitation during the 2003–2011 growing seasons, and long-term temperature and precipitation averages. Data were collected about 1 km from the experimental site in Boone Co., IA.

Table S2   Crop identities and seeding rates in 2003–2011.

Table S3   Macronutrients applied in manufactured fertilizers, herbicide adjuvants, and manure in 2003–2011. Manufactured N, P, and K fertilizers were applied at rates that varied among years and rotations in response to soil test results. Manure was applied at a rate of 15.7 Mg ha^{-1} in maize phases of the 3-year and 4-year rotation systems, but moisture and nutrient concentrations varied among years, resulting in variable rates of macronutrient additions.

Table S4   Herbicide applications in 2003–2011 to maize and soybean in the three rotation systems. No herbicides were used for triticale, oat, red clover, and alfalfa grown within the 3-yr and 4-yr systems. Reported application rates reflect the effect of banding of herbicides over crop rows in the 3-yr and 4-yr systems.

Table S5   Simple and partial correlations between energy use within a given crop phase and mean rotation energy use and between energy use within a given crop phase and N fertilizer application rates.

Text S1   Detailed description of experimental site, management practices, scientific methods and statistical approach.

Acknowledgments

We are grateful to the many students, postdoctoral research associates and technicians who have made the Marsden Farm study possible.

Author Contributions

Conceived and designed the experiments: ASD ML. Performed the experiments: CAC AMJ ML. Analyzed the data: ASD JDH CAC AMJ ML. Contributed reagents/materials/analysis tools: ML. Wrote the paper: ASD JDH ML.

References

1. Foley JA, Ramankutty N, Brauman KA, Cassidy ES, Gerber JS, et al (2011) Solutions for a cultivated planet. Nature 478: 337–342.
2. Robertson GP, Swinton SM (2005) Reconciling agricultural productivity and environmental integrity: a grand challenge for agriculture. Front Ecol Environ 3: 38–46.
3. Dubrovsky NM, Burow KR, Clark GM, Gronber JM, Hamilton PA, et al. (2010) The quality of our nation's waters: Nutrients in the nation's streams and groundwater, 1992–2004. Circular 1350. Reston, VA: U.S. Geological Survey. http://pubs.usgs.gov/circ/1350/. Accessed 2012 Jul 10.
4. Gilliom RJ, Barbash JE, Crawford CG, Hamilton PA, Martin JD, et al. (2006) The quality of our nation's waters: pesticides in the nation's streams and ground water, 1992–2001. Circular 1291. Reston, VA: U.S. Geological Survey. http://pubs.usgs.gov/circ/2005/1291/. Accessed 2012 Jul 10.
5. Relyea RA (2005) The impact of insecticides and herbicides on the biodiversity and productivity of aquatic communities. Ecol Appl 15: 618–627.
6. Rohr JR, McCoy KA (2010) A qualitative meta-analysis reveals consistent effects of atrazine on freshwater fish and amphibians. Environ Health Persp 118: 20–32.
7. Tilman D, Cassman KG, Matson PA, Naylor R, Polasky S (2002) Agricultural sustainability and intensive production practices. Nature 418: 671–677.
8. Liebman M, Staver CP (2001) Crop diversification for weed management. In: Liebman M, Mohler CL, Staver CP, editors. Ecological management of agricultural weeds. Cambridge: Cambridge University Press. 322–374.
9. Bennett AJ, Bending GD, Chandler D, Hilton S, Mills P (2012) Meeting the demand for crop production: the challenge of yield decline in crops grown in short rotations. Biol Rev 87: 52–71.
10. Karlen DL, Varvel GE, Bullock DG, Cruse RM (1994) Crop rotations for the 21st century. Adv Agron 53: 1–45.
11. Grigg DB (1974) The agricultural systems of the world: An evolutionary approach. Cambridge: Cambridge University Press. 358 pp.

12. Grigg DB (1989) English agriculture: An historical perspective. Oxford: Basil Blackwell. 320 pp.
13. Magdoff F, Lanyon L, Liebhardt B (1997) Nutrient cycling, transformations, and flows: implications for a more sustainable agriculture. Adv Agron 60: 1–73.
14. Naylor R, Steinfeid H, Falcon W, Galloways J, Smil V, et al. (2005) Losing the links between livestock and land. Science 310: 1621–1622.
15. U.S. Environmental Protection Agency (EPA) (2011) Pesticides industry sales and usage: 2006 and 2007 market estimates. Washington, D.C.: U.S. Environmental Protection Agency. www.epa.gov/opp00001/pestsales/07pestsales/market_estimates2007.pdf. Accessed 2012 Jul 10.
16. Heap I (2012) International survey of herbicide resistant weeds. http://www.weedscience.org. Accessed 2012 Jul 10.
17. Mortensen DA, Egan JF, Maxwell BD, Ryan MR, Smith RG (2012) Navigating a critical juncture for sustainable weed management. BioSci 62: 75–84.
18. Jordan N, Warner KD (2010) Enhancing the multifunctionality of US agriculture. BioSci 60: 60–66.
19. Seufert V, Ramankutty N, Foley JA (2012) Comparing the yields of organic and conventional agriculture. Nature 485: 229–232.
20. Liebman M, Davis AS (2009) Managing weeds in organic farming systems: an ecological approach. In: Organic farming: The ecological system. Francis C, editor. Madison: American Society of Agronomy. 173–196.
21. Hayes TB, Collins A, Lee M, Mendoza M, Noriega N, et al. (2002) Hermaphroditic, demasculinized frogs after exposure to the herbicide atrazine at low, ecologically relevant doses. Proc Nat Acad Sci USA 99: 54376–55480.
22. Orton F, Lutz I, Kloas W, Rutledge EJ (2009) Endocrine disrupting effects of herbicides and pentachlorophenol: in vitro and in vivo evidence. Environ Sci Technol 43: 2144–2150.
23. Relyea RA (2005) The lethal impact of Roundup on aquatic and terrestrial amphibians. Ecol Appl 15: 1118–1124.
24. Egan JF, Mortensen DA (2012) Quantifying vapor drift of dicamba herbicides applied to soybean. Envir Toxicol Chem 31: 1023–1031.
25. Freemark K, Boutin C (1995) Impacts of agricultural herbicide use on terrestrial wildlife in temperate landscapes - a review with special reference to North America. Agric Ecosyst Environ 52: 67–91.
26. Powles SB, Yu Q (2010) Evolution in action: plants resistant to herbicides. Ann Rev Plant Biol 61: 317–347.
27. Tranel PJ, Riggins CW, Bell MS, Hager AG (2011) Herbicide resistances in Amaranthus tuberculatus: A call for new options. J Agric Food Chem 59: 5808–5812.
28. Chase JM, Leibold MA (2002) Spatial scale dictates the productivity-biodiversity relationship. Nature 416: 427–430.
29. Tilman D, Reich PB, Knops J, Wedin D, Mielke T, et al. (2001) Diversity and productivity in a long-term grassland experiment. Science 294: 843–845.
30. Tilman D, Hill J, Lehman C (2006) Carbon-negative biofuels from low-input high-diversity grassland biomass. Science 314: 1598–1600.
31. Smith RG, Gross KL, Robertson GP (2008) Effects of crop diversity on agroecosystem function: Crop yield response. Ecosyst 11: 355–366.
32. Smith RG, Mortensen DA, Ryan MR (2010) A new hypothesis for the functional role of diversity in mediating resource pools and weed-crop competition in agroecosystems. Weed Res 50: 185–185.
33. Jackson LL, Keeney DR, Gilbert EM (2000) Swine manure management plans in north-central Iowa: Nutrient loading and policy implications. J Soil Water Cons 55: 205–212.
34. Olmstead J (2011) Feeding the world? Twelve years later, U.S. grain exports are up, so too is hunger. Minneapolis, MN: Institute for Agriculture and Trade Policy. http://www.iatp. org/documents/feeding-the-world. Accessed 2012 Jul 10.
35. Heichel GH, Barnes DK, Vance CP, Henjum KI (1984) N₂ fixation, and N and dry matter partitioning during a 4-year alfalfa stand. Crop Sci 24: 811–815.
36. Heichel GH, Vance CP, Barnes DK, Henjum KI (1985) Dinitrogen fixation, and N and dry matter distribution during 4-year stands of birdsfoot trefoil and red clover. Crop Sci 25: 101–105.
37. Fox RH, Piekielek WP (1988) Fertilizer N equivalence of alfalfa, birdsfoot trefoil, and red clover for succeeding corn crops. J Prod Agric 1: 313–317.
38. Blackmer AM, Voss RD, Mallarino AP (1997) Nitrogen fertilizer recommendations for corn in Iowa. PM-1714. Ames, IA: Iowa State University Extension and Outreach. http://www.extension.iastate.edu/Publications/PM1714.pdf. Accessed 2012 Jul 10.
39. Liebman M, Gibson LR, Sundberg DN, Heggenstaller AH, Westerman PR, et al. (2008) Agronomic and economic performance characteristics of conventional and low-external-input cropping systems in the central corn belt. Agron. J. 100: 600–610.
40. Berthoud A, Maupu P, Huet C, Poupart A (2011) Assessing freshwater ecotoxicity of agricultural products in life cycle assessment (LCA): a case study of wheat using French agricultural practices databases and USEtox model. Int J Life Cycle Assess 16: 841–847.
41. Hauschild MZ, Huijbregts M, Jolliet O, MacLeod M, Margni M, et al. (2008) Building a model based on scientific consensus for life cycle impact assessment of chemicals: The search for harmony and parsimony. Environ Sci Technol 42: 7032–7037.
42. Rosenbaum RK, Bachmann TM, Gold LS, Huijbregts MAJ, Jolliet O, et al. (2008) USEtox-the UNEP-SETAC toxicity model: recommended characterisation factors for human toxicity and freshwater ecotoxicity in life cycle impact assessment. Int J Life Cycle Assess 13: 532–546.
43. Cornell PMEP (2008) Pesticide management education program. http://pmep.cce. cornell.edu/. Accessed 2012 Jul 10.
44. Crawley MJ (2007) The R Book. West Sussex: John Wiley and Sons. 942 pp.
45. R Development Core Team (2009) R: A language and environment for statistical computing (R Foundation for Statistical Computing, Vienna, Austria).

9

Effects of Nutrient Heterogeneity and Competition on Root Architecture of Spruce Seedlings: Implications for an Essential Feature of Root Foraging

Hongwei Nan, Qing Liu*, Jinsong Chen, Xinying Cheng, Huajun Yin, Chunying Yin, Chunzhang Zhao

Key Laboratory of Mountain Ecological Restoration and Bioresource Utilization & Ecological Restoration Biodiversity Conservation Key Laboratory of Sichuan Province, Institute of Biology, Chinese Academy of Sciences, Chengdu, China

Abstract

Background: We have limited understanding of root foraging responses when plants were simultaneously exposed to nutrient heterogeneity and competition, and our goal was to determine whether and how plants integrate information about nutrients and neighbors in root foraging processes.

Methodology/Principal Findings: The experiment was conducted in split-containers, wherein half of the roots of spruce (*Picea asperata*) seedlings were subjected to intraspecific root competition (the vegetated half), while the other half experienced no competition (the non-vegetated half). Experimental treatments included fertilization in the vegetated half (FV), the non-vegetated half (FNV), and both compartments (F), as well as no fertilization (NF). The root architecture indicators consisted of the number of root tips over the root surface (RTRS), the length percentage of diameter-based fine root subclasses to total fine root (SRLP), and the length percentage of each root order to total fine root (ROLP). The target plants used novel root foraging behaviors under different combinations of neighboring plant and localized fertilization. In addition, the significant increase in the RTRS of 0–0.2 mm fine roots after fertilization of the vegetated half alone and its significant decrease in fertilizer was applied throughout the plant clearly showed that plant root foraging behavior was regulated by local responses coupled with systemic control mechanisms.

Conclusions/Significance: We measured the root foraging ability for woody plants by means of root architecture indicators constructed by the roots possessing essential nutrient uptake ability (i.e., the first three root orders), and provided new evidence that plants integrate multiple forms of environmental information, such as nutrient status and neighboring competitors, in a non-additive manner during the root foraging process. The interplay between the responses of individual root modules (repetitive root units) to localized environmental signals and the systemic control of these responses may well account for the non-additive features of the root foraging process.

Editor: John Schiefelbein, University of Michigan, United States of America

Funding: This research was supported by grants from the Strategic Priority Research Program of the Chinese Academy of Sciences (XDA05050303) and the National Key Technology R & D Program (2011BAC09B04). The funders had no role in study design, data collection and analysis, decision to publish, or preparation of the manuscript.

Competing Interests: The authors have declared that no competing interests exist.

* E-mail: liuqing@cib.ac.cn

Introduction

Root foraging is one of the most important aspects of plant behavior because it can affect individual plant growth as well as plant fitness and community structure [1,2]. The said process can respond to the presence of neighboring competitor roots and the heterogeneous distribution of nutrients in the soil [3,4], particularly when the general levels of nutrient availability are low [5–7]. In nature, plants are simultaneously exposed to nutrient heterogeneity and the roots of neighbors. Recent studies reported that plant root growth could be an additive or a non-additive response to multiple forms of environmental information, which partially depends on the neighboring species or their competitive attributes [8–10]. Therefore, the incorporation of multiple simultaneous environmental conditions in root foraging studies may help to advance our understanding of the relationships between plant root systems and the environment.

In forest ecosystems, tree seedlings allow their roots to proliferate to acquire nutrients and water; seedlings typically contend with heterogeneous resources and competing neighbors, which both exert important effects on root foraging behavior [10,11]. Previous forestry studies on root competition have invested much effort toward investigating the effects of interspecific competition [3,12–14]; the importance of intraspecies interaction has received much less attention. Due to similarity in ecological characters, plants in intraspecies competition cannot avoid or alleviate adverse competition effect via niche complementarity. Accordingly, the thing missing from many studies of root competition is a detailed understanding of intraspecies interactions.

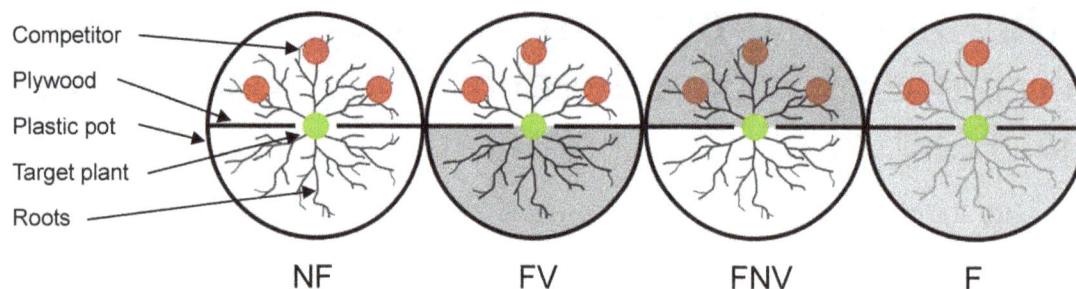

Figure 1. Schematic of the experimental treatments. The four treatments consisted of fertilization in the vegetated half (FV), the non-vegetated half (FNV), and both compartments (F), as well as no fertilization (NF).

Root architecture is defined as the spatial configuration of the root system, which has a key role in belowground resource acquisition [15,16]. Fitter et al. [17,18], as well as Farley and Fitter [19], demonstrated that a herringbone topology may be best for locating nutrient-rich patches in the soil, but a less herringbone topology is more suitable for exploiting these resources. Grime and Mackey reported that phenotypic plasticity for specific root architectural traits was significant in resource capture, as a result of nutrient heterogeneity in space and time [20]. In addition, root architecture was shown to be a primary factor affecting the degree of competition among roots of the same plant and/or neighboring plants [21–23]. More recently, Nord et al. found that the presence of a neighbor could lead to alterations in the root architecture, thereby keeping the root biomass stable [24]. To date, accumulating evidence indicated that root architecture was more sensitive to environmental stimuli than root biomass [24,25]. However, most studies addressing plant foraging ability have focused on root biomass but overlooked root architecture, which can contribute to a better understanding of the interactions between plant root systems and their environment.

Plant root foraging ability is closely related to root architecture, but none of the previous studies thus far have linked these aforementioned aspects of plant root systems. This oversight was probably because root functions, such as resource uptake and transport, were difficult to directly measure [26]. Previous studies mainly utilized lateral root attributes to assess the response of the root architecture to environmental stimuli; these attributes included descriptions of the morphological characteristics [27,28], spatial deployment pattern [29,30], and root-growth patterns [31–33]. However, all these measurements are unsuitable for precisely measuring the root foraging ability. In addition, the entire root system was traditionally divided into different parts based on size classes (e.g., 0–1 mm roots vs. 0–2 mm roots, based on their diameter), which did not provide information on the root system structure, function, and response to altered environmental conditions. This limitation is particularly true in woody plants because fine roots are complex branching structures composed of numerous individual root segments, which differ in their morphology and function. The position and form of individual roots on the branching fine root system are typically disregarded by the said classification modes [34–37]. Guo et al. examined the anatomy and mycorrhizal colonization of branch order in 23 Chinese temperate tree species, and demonstrated that active nutrient absorption was mainly achieved by the first three orders of the root system, particularly the first-order roots (tiny lateral branches at the very distal end of the root system) [37]. To effectively measure the root foraging ability, the first three root orders should collectively be taken into account, rather than the entire fine root system, when determining the root architecture

indicators for woody plants. To the best of our knowledge, none of the previous studies have employed such novel indirect assessment methods of root foraging.

Plants producing preferentially roots in nutrient-rich substrate patches were proposed to function as the primary root foraging mechanism by which plants cope with the naturally occurring heterogeneous nutrient supply in soil [5,38]. Several studies indicated that a plant in the presence of neighboring roots preferentially grows new roots in unoccupied soil before it does the same in a space already occupied by other species or conspecifics [21,39]. However, little information is available on how the foraging behavior of plant root systems responds to the simultaneous presence of nutrient heterogeneity and neighboring roots [8,10]. To obtain a more mechanistic understanding of plant root foraging response to neighbors and nutrients, we simultaneously manipulated nutrient heterogeneity and intraspecies competition conditions, investigated root foraging responses based on the root architecture, and assessed their influence on nutrient uptake in spruce (*Picea asperata*), the dominant tree species in the subalpine coniferous forests of western Sichuan, China.

Materials and Methods

Ethics Statement

The experiment was set up at an open field (31°25′N, 103°12′E, 2309 m, a.s.l.) in the Miyaluo natural reserve of Lixian County, Eastern Tibetan Plateau, in Sichuan, China. We obtained appropriate permissions from the Forestry Bureau of Lixian County, and from the forestry workers for field study. In present study, spruce (*P. asperata*) seedlings, the dominant tree species in natural reserve, were used as investigated subject, and we confirmed that our studies did not involve endangered or protected species. In addition, no specific permission was required for these locations because our study was the general pot experiment.

Experimental Design and Treatments

The experimental site had a montane monsoon climate, which was humid and rainy in summer but cold and dry in winter, with mean January and July temperatures of $-8°C$ and $12.6°C$, respectively. The mean annual precipitation ranged from 600 mm to 1100 mm, and the mean annual evaporation was from 1000 mm to 1900 mm. The soil was classified as mountain brown earth [40].

On April 2011, 32 large circular plastic pots (38 cm in diameter, 30 cm deep) were divided into two parts of equal volume using solid plywood planks (see Fig. 1). The pots were filled with sieved, root free soil (4.5 mm mesh) from the neighboring forest. The basic soil properties were as follows: pH, 5.85; soil organic C,

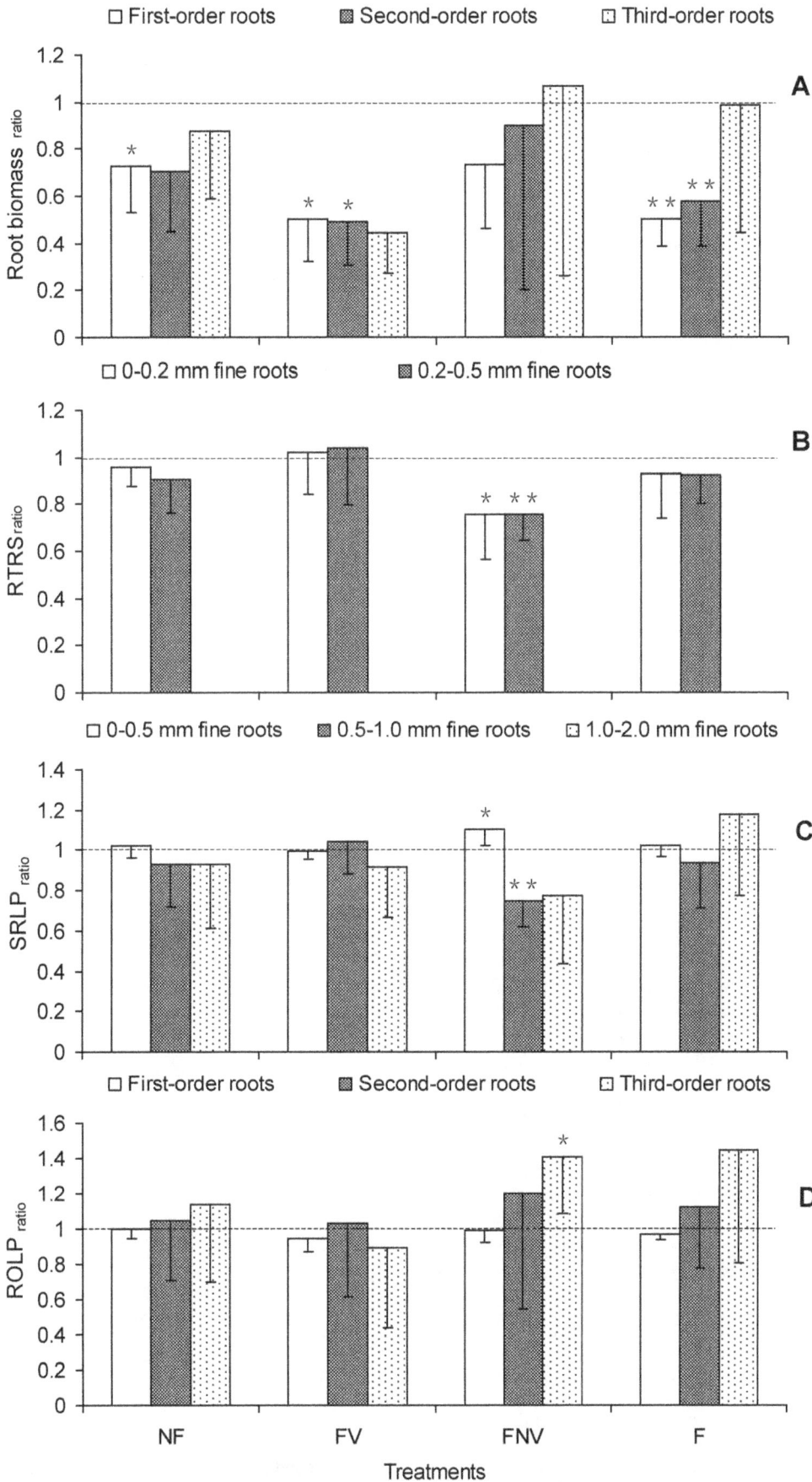

Figure 2. The ratio "vegetated half: non-vegetated half" in root system biomass and architecture. (A) root biomass ratio; (B) the number of root tips over the root surface ratio (RTRS $_{ratio}$); (C) the length percentage ratio of diameter-based fine root subclasses to the total fine root length (SRLP $_{ratio}$); (D) the length percentage ratio of each root order to the total fine root length (ROLP $_{ratio}$). Asymmetrical root biomass and architecture (i.e. ratios significantly different from 1.0) are indicated above the columns (**$P<0.01$, *$P<0.05$). Error bars represent one SE of the mean.

62.70 mg·g^{-1}; total N, 3.66 mg·g^{-1}; total P, 0.43 mg·g^{-1}; and total K, 7.92 mg·g^{-1}.

At the beginning of May, three-year-old spruce (*P. asperata*) saplings of similar sizes were randomly established in the pots; the root systems of these saplings had nearly homogeneous and symmetrical distribution around the stem axis. One sapling to be used as the target plant was carefully placed in the middle of each pot. The main root of this sapling was then inserted into a narrow (3 cm) gap carved into the plywood plank, whereas the lateral roots were equally arrayed into separate compartments. Three spruce saplings were planted in half of each pot (the "vegetated half") to function as competitors, whereas the other half (the "non-vegetated half") had no saplings (Fig. 1).

In this study, all the four treatments were established by applying fertilizer in different compartments or otherwise. These treatments included fertilization in the vegetated half (FV), non-vegetated half (FNV), and both compartments (F), as well as no fertilization (NF); each treatment had eight pots. The fertilizer contained NPK in a 15:1:1 ratio, based on Hoagland's hydroponic solution [41]. The fertilizer was applied from June to mid-September at 1.0 g N·m^{-2} every 10 days (a total of ten times throughout the growing season).

Root Measurements

In mid-September, all the target plant seedlings were carefully harvested by hand with the help of a watering hose, taking care to maintain the integrity of the root systems. Roots were then separated from each seedling and divided into two groups (without including the main root) based on the compartment where they were grown. All the root systems in each group were carefully washed free of soil. Their length, surface area, volume, and number of tips were measured using the WinRHIZO image analysis software (Régent instruments, Quebec, QC, Canada). In order to obtain more accurate morphological results, we scanned all the root systems, which were time- and energy-consuming, unlike previous studies that merely selected a few root samples per plant. Subsequently, three root samples per plant and compartment were chosen from the scanned roots. Each of the said root samples contained at least eight intact distal root segments, including more than three root orders. The samples were dissected to obtain the first three root orders using scalpel blades in large petri dish. The most distal root tips were classified as the first-order roots, whereas the second- and third-order roots were dissected according to the order of streams in geography [34]. The root morphologies of the first three root orders, such as the length and surface area, were assessed using the same image analysis software as mentioned above to determine the length and surface area ratio among the first three orders. Finally, all the root systems per plant

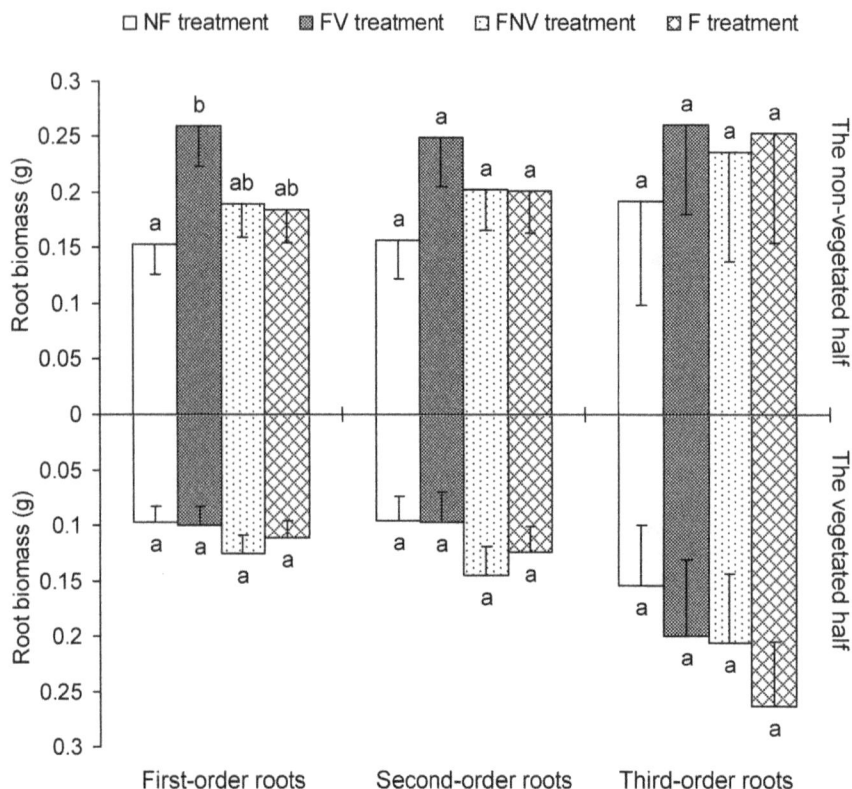

Figure 3. Root system biomass in the vegetated half and in the non-vegetated half. Letters indicate the same root order difference between treatments (LSD tests, following ANOVA). Error bars represent 1 SE of the mean.

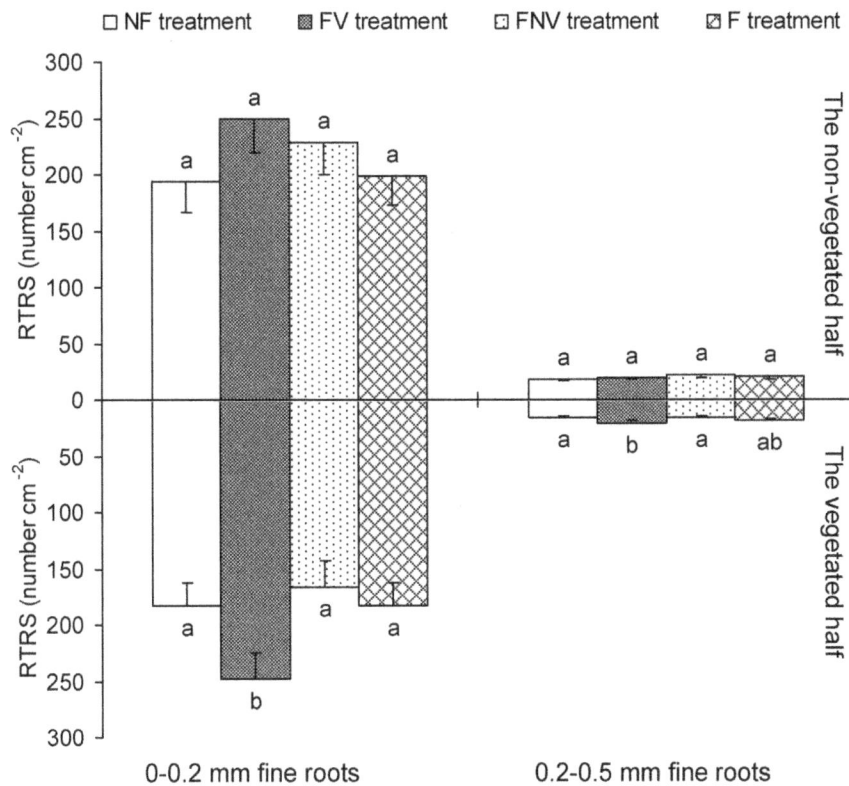

Figure 4. The number of root tips over root surface (RTRS), root architecture indicator, in the vegetated half and in the non-vegetated half. Letters indicate the same subclass (0–0.2 mm or 0.2–0.5 mm fine roots) difference between treatments (LSD tests, following ANOVA). Error bars represent 1 SE of the mean.

and compartment were divided into two groups according to their diameter (fine roots, ≤ 2 mm, other roots, >2 mm). Their biomass was measured using a digital balance after drying in an oven at 70°C for 48 h. When the total biomass of fine roots per plant and compartment was calculated, the root biomass of the first three orders was then added to obtain the final value. Furthermore, the main roots and shoots of each seedling were washed carefully, and their biomass was measured using a similar method to determine the whole plant biomass.

A root architecture indicator defined in our study is the number of root tips over root surface (RTRS). More than 96% of the root tips were located in 0–0.5 mm fine roots, as demonstrated in our previous work. Thus, this region of the root surface alone was used for calculating the RTRS values to avoid errors. To further investigate the root architecture, we divided 0–0.5 mm fine roots into two subclasses based on their diameter, namely, the 0–0.2 mm and the 0.2–0.5 mm root systems. The RTRS of these subclasses could be calculated using of the above mentioned root morphology measurements.

Another root architecture indicator was the length percentage of diameter-based fine root subclasses to the total fine root length (subclass root length percentage, SRLP). Given that the first three orders in the root systems were the primary parts involved in nutrient absorption [37] and constituted the main body of 0–0.5 mm fine roots (average root diameter of the third-root order was approximately 0.46 mm in our study), we divided the whole fine root (≤ 2 mm) into three subclasses based on their diameter: the 0–0.5 mm, 0.5–1.0 mm, and 1.0–2.0 mm subclasses [42]. A high SRLP value of the 0–0.5 mm root system indicated more

efficient root foraging ability, which could be calculated from root morphology measurements, as mentioned above.

To improve our understanding of the mechanisms involved, we determined the length percentage of each root order to the total fine root length as another indicator of the root architecture (root order length percentage, ROLP). The surface area and length of the first-order root were analyzed using the link analysis tool provided by WinRhizo™ 2009. According to Pregitzer's definition [34], the first-order roots consisted of the external-external and the external-internal links; the morphological parameters of the first-order roots were equal to the sum of both links [43]. The root morphology of the second- and third-order was calculated using the morphology ratio among the first three orders, as described above. Based on these results, the ROLP could be calculated.

The biomass of the first three root orders, as described above, was a suitable indicator to assess the root foraging ability, except for the root architecture. We had acquired the respective length and surface area of the first three orders in the preceding methods, but their volumes remained unknown. Based on the morphological measurements of the abovementioned root samples, the respective regression of the volume on surface area for the first three orders was established. The respective volume of the first three orders per plant and compartment were calculated. Given the strongly linear relationship between the fine root volume and its biomass [44], the respective biomass of the first three orders were calculated using Cheng's formula based on their volumetric percentage to that of the fine roots (≤ 2 mm) [42].

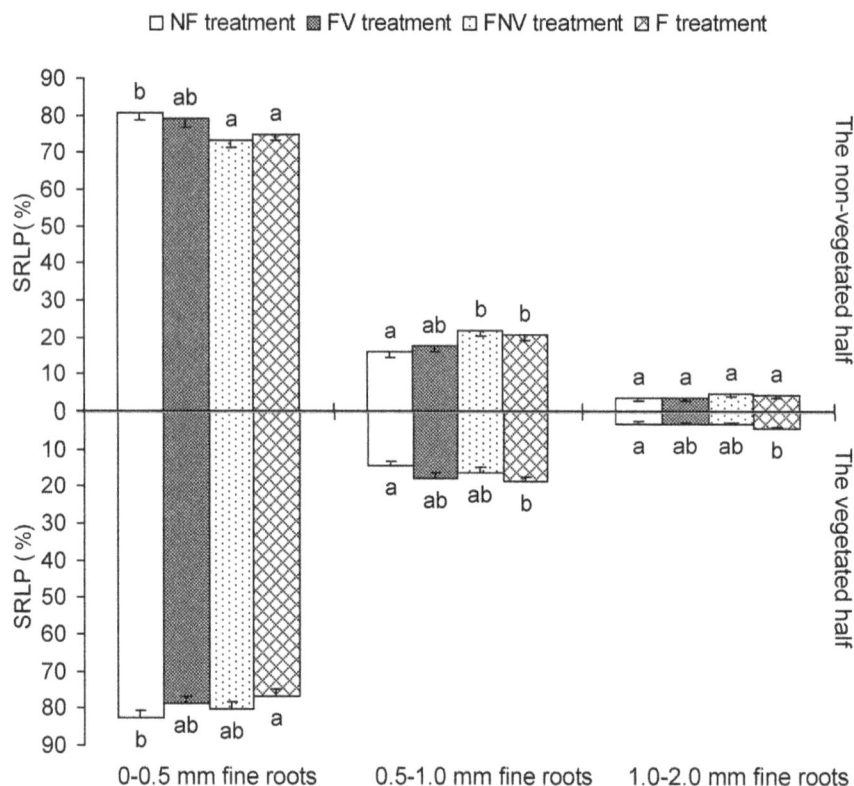

Figure 5. The length percentage of diameter-based fine root subclasses to the total fine root length (subclass root length percentage, SRLP), root architecture indicator, in the vegetated half and in the non-vegetated half. Letters indicate the same subclass (0–0.5 mm, 0.5–1.0 mm or 1.0–2.0 mm fine roots) difference between treatments (LSD tests, following ANOVA). Error bars represent 1 SE of the mean.

Growth Measurement

Each target plant seedling was tagged when they were planted in early May. The basal diameter of each seedling was measured and recorded. Another fifteen additional spruce seedlings with similar sizes to the planted target seedlings were selected. Their basal diameters and whole plant biomass weights were simultaneously measured to establish the regression model of the whole plant biomass according to the basal diameter. Based on the regression equation, the initial biomass for each of the target plant seedlings was calculated. At the end of the growing season, the final plant biomass harvested was measured, as described above.

The relative growth rate (RGR) for each plant was calculated using the formula $RGR = [\ln w_2 - \ln w_1]/T$ [45]; where w_2 and w_1 are the final and initial plant biomass, respectively, whereas T is the number of months between the initial and final measurements (i.e., 3.5 months).

Statistical Analyses

The root response was evaluated for each pot using the ratio between the root variable values in the vegetated and non-vegetated halves (e.g., $RTRS_{ratio} = RTRS_{vegetated\ half}/RTRS_{non-vegetated\ half}$). The values of the root variables were considered to be higher in the non-vegetated half when the ratio was significantly lower than 1, and lower when the ratio is higher than 1 (i.e., the ratio is equal to 1 for symmetrical root growth). This difference was analyzed using a paired-sample t-test. Furthermore, the effects of different treatments on the root architecture and biomass in the vegetated and non-vegetated regions as well as the relative growth rate (RGR) were examined using factorial ANOVA for a randomized block design, with treatments as the fixed factors.

The root biomass, architecture, and relative growth rate were recorded as dependent variables. The data were transformed when necessary using the natural logarithmic transformation to satisfy the normality and homogeneity of the variances. The overall data was statistically analyzed using the SPSS program (SPSS 13.0, Chicago).

Results

The First Three Order Root Biomass

The first three root orders were the most important sections of fine-root systems for nutrient and water acquisition. For woody plants with complicated branching order root systems, the fine root (≤2 mm) biomass was not suitable for measuring the root foraging ability. The first-order roots in the NF treatment, as well as the first- and second-order roots in both FV and F treatments, showed significantly lower root biomass ratios (i.e. ratios were significantly less than 1), whereas no significant differences were found for the third-order roots in all the four treatments, as well as in the first three root orders of the FNV plants (Fig. 2). These results indicated that except for the FNV treatment, root competition reduced the absorbing root biomass in the vegetated half of the target plant, which were mainly concentrated on the first two root orders. Furthermore, different root order responses were observed for various forms of root competition. The root biomass ratio in FNV treatment may have not been significantly different from 1 because the absorbing root biomass decreased as the soil resources were increased by the increased use of fertilizers in the non-vegetated half. In addition, the first-order root biomass significantly varied among the non-vegetated halves of the FV and

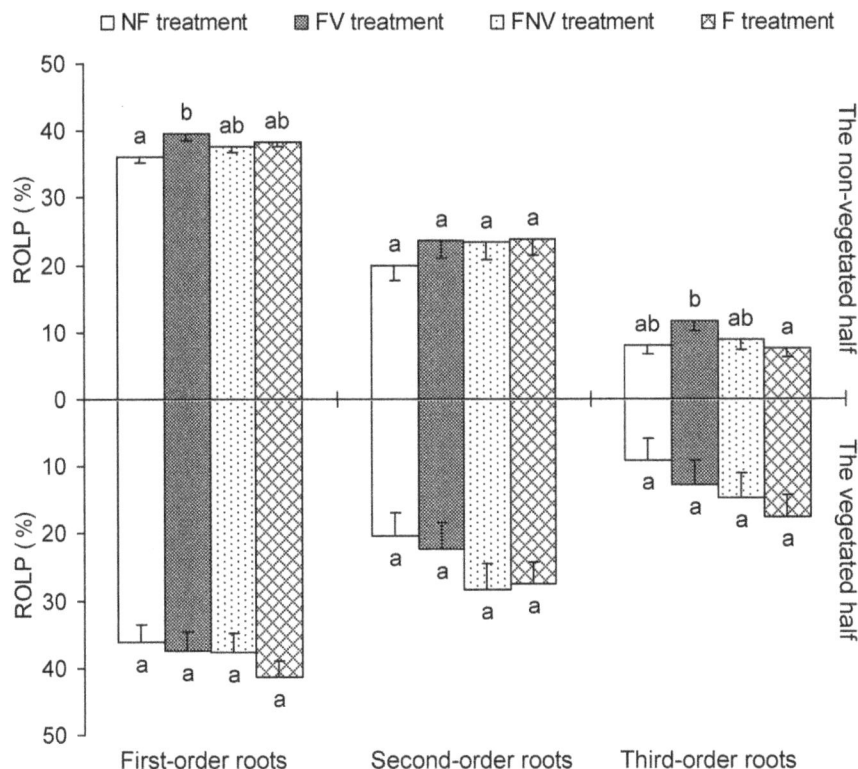

Figure 6. The length percentage of each root order to the total fine root length (root order length percentage, ROLP), root architecture indicator, in the vegetated half and in the non-vegetated half. Letters indicate the same root order difference between treatments (LSD tests, following ANOVA). Error bars represent 1 SE of the mean.

NF treatments. By contrast, the biomass was not significantly different between the second- and third-order roots from the non-vegetated halves of all four treatments. No significant differences were observed among treatments in the vegetated half (Fig. 3).

Root Architecture Indicator: RTRS

RTRS of both 0–0.2 mm and 0.2–0.5 mm fine roots in the non-vegetated half was shown to be significantly higher than that in the vegetated half for FNV treatment (i.e. ratio less than 1), and no difference was found in the other treatments (Fig. 2). In the fertilization of the non-vegetated half for FNV treatment, the

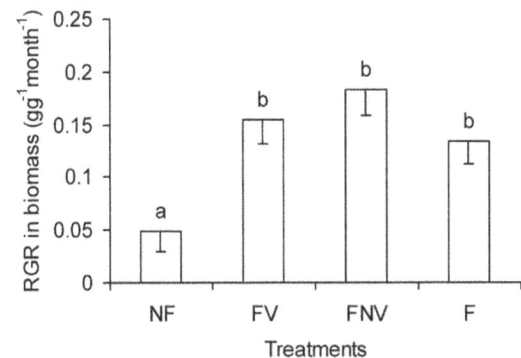

Figure 7. The relative growth rate (RGR) of target plant in different treatments. Letters indicate RGR differences between treatments (LSD tests, following ANOVA). Error bars represent 1 SE of the mean.

target plants increased spatial nutrient uptake by altering RTRS. In the vegetated half, RTRS of 0–0.2 mm fine roots for the FV treatment was significantly higher compared with the other three treatments, reaching a maximum of 247.7 cm^{-2}. RTRS of 0.2–0.5 mm fine roots for the FV treatment was also higher than those obtained in the NF and FNV treatments, with values of 20.3, 15.6, and 16.1 cm^{-2}, respectively (Fig. 4). Since the RTRS of 0–0.2 mm fine roots was much higher than that of 0.2–0.5 mm fine roots in all of the four treatments, RTRS of the latter had little effects on root foraging ability compared with the former. The RTRS of 0–0.2 mm fine roots in the vegetated half significantly increased from 182.8 cm^{-2} in the NF treatment to 247.7 cm^{-2} in the FV treatment, and significantly decreased to 182.6 cm^{-2} in the F treatment. In addition, there was no significant difference in the RTRS values of both fine root subclasses among all the four treatments in the vegetated half (Fig. 4).

Root Architecture Indicator: SRLP

The 0–0.5 mm root systems mainly consisted of the first three orders; the SRLP of which may reflect length proportion of the root systems being able to absorb nutrient and water in the soil to whole fine root. 0–0.5 mm fine root in the FNV treatment had significantly higher SRLP ratios (i.e. the ratio was significantly more than 1), 0.5–1.0 mm fine roots had lower SRLP ratios (i.e. less than 1), whereas no differences were found between the vegetated and non-vegetated halves in all the other three treatments (Fig. 2). The significantly higher SRLP ratio of 0–0.5 mm fine roots in the FNV treatment indicated the target plant's attempt to strengthen nutrient acquisition in the observed space. The SRLP of 0–0.5 mm fine roots in the F treatment was significantly lower in the vegetated and non-vegetated halves, as compared with that of the NF treatment. The opposite trend was found for the SRLP of 0.5–1.0 mm fine roots (Fig. 5). The lower SRLP of 0–0.5 mm fine roots helped reduce the absorbing root length density, thereby alleviating root competition intensity within the same plant root system.

Root Architecture Indicator: ROLP

The length percentages of the first three root orders against the total fine root contributed to the further analysis of the inner changes in the SRLP of 0–0.5 mm fine roots, as mentioned above. The third-order ROLP ratio in the FNV treatment was significantly more than one (Fig. 2), which explained the higher SRLP ratio of the 0–0.5 mm fine roots in the FNV treatment. The ROLP of the first-order roots in the non-vegetated half with the FV treatment was significantly higher than in those with the NF treatment. The ROLP in the non-vegetated half of the third-order roots in the FV treatment were likewise higher than in the F treatment (Fig. 3). However, no significant differences in the ROLP values of the three root orders were found among all the four treatments (Fig. 6).

The Relative Growth Rate (RGR)

The RGR of the FV and FNV treatments were significantly higher, as compared with that in NF treatment, but were not significantly different from the F treatment (Fig. 7). Given that the total amount of nutrients used in the FV and FNV treatment was half of that in the F treatment, the absence of significant differences between both treatments indicated that the target plants, which were simultaneously exposed to nutrient heterogeneity and the roots of the neighboring plants, had excellent nutrient uptake abilities.

Discussion

De Kroon et al. proposed that the interplay between local responses and systemic modifications of these responses was an essential feature of plant foraging. More specifically, plant foraging for resources was achieved through various processes acting in concert at the level of the repetitive units (modules), from which plant roots were constructed. These processes involved individual module responses to localized environmental signals and the systemic control of these responses. This systemic control can be achieved by the signals received from connected modules exposed to different conditions or by those reflecting the overall resource status of the plant [2]. However, past evidence of the interplay between local responses and systemic modifications in plant root foraging behavior was limited. In the present study, the significant increase in the RTRS of 0–0.2 mm fine roots in the vegetated half occurred from the NF treatment (without fertilizers) to the FV treatment (with fertilizers only in the vegetated half); the significant decrease in the RTRS of the 0–0.2 mm fine roots in the vegetated half was observed from the FV to the F treatments (i.e., nutrients are supplemented in both the vegetated and non-vegetated halves, as compared to FV treatment), with the conversion of the whole plant root resource status from the localized nutrient supply in the FV treatment (i.e., fertilizer only in vegetated half) to the overall nutrient supply in the F treatment (i.e., fertilizer in both halves). These results clearly showed that RTRS was regulated by the local responses and the systemic controlled mechanisms. For the first time, our findings provided new evidence based on root architecture for De Kroon's concept, which directly reflect the root foraging ability of woody plants.

Cahill et al. hypothesized that plants integrate information from both resource and neighbor-based cues in the environment in a non-additive manner [8]. However, they measured the horizontal spread of the roots, which was unsuitable for precisely exploring the root foraging ability, as compared with the root biomass or architecture. In our study, the 0–0.5 mm fine roots SRLP of both the vegetated and non-vegetated halves decreased with the increasing nutrient concentrations, based on the results of the NF and F treatments. Therefore, the target plant adopted strategies to ease the competition within the same plant root system as the nutrient status increased. The RTRS of the vegetated half and the ROLP of the first-order roots in the non-vegetated half were higher in the FV treatment than in other treatments. In addition, these indicators were significantly different between the vegetated and non-vegetated halves in the FNV treatment. Collectively, we were able to show that plants used novel root foraging behaviors under different combinations of environmental conditions, such as neighboring plants and localized fertilization. We took full advantage of the root architecture indicators to effectively measure foraging behaviors and provide pronounced evidence that woody plant root foraging behavior was a non-additive response to multiple forms of environmental information.

When grown in heterogeneous conditions, plants preferentially produce roots in nutrient-rich substrate patches, and enhance the uptake efficiency of these roots, as compared with other roots of the same plant outside the patch zone [38,46]. The differences between the NF and FV treatments indicated that the target plants increased their nutrient uptake in nutrient-rich patches by altering the root architecture (RTRS) under the conditions of constant absorbing root biomass. Despite the intense competition in the same patches, root competition did not affect the attempts of plants to absorb resources in nutrient-rich patches. In addition, the RTRS ratio in the FNV treatment was less than 1, which reflected the attempt of the target plants to strengthen the nutrient intake in nutrient-rich patches. Mommer et al. suggested that the root response to nutrient distribution in a competitive environment depended on the competitive strength of the neighboring species; in their study, competition with a superior competitor led the inferior *Agrostis stolonifera* to increase relative root investment in the nutrient-poor patch instead of the nutrient-rich patch [10]. Under similar competitive strength conditions by neighboring species (i.e., intraspecific competition), the target plants in the present study still had enhanced nutrient uptake in the nutrient-rich patches, which showed that plants seemed to prefer nutrient intake in nutrient-rich patches than in the nutrient-poor counterparts unless forced by enormous environmental stress, such as competition with more superior competitor (with larger competitive advantage). Therefore, the unit cost of soil resource acquisition was lower in the nutrient-rich patches than in the nutrient-poor ones.

Some plants may engage in a game of "Tragedy of the Commons" when competing for soil resources. Thus, a plant in the presence of neighboring roots should preferentially place new roots in unoccupied soil instead of the space containing roots of other species or conspecifics [21,39]. The target plant in the FV treatment had a higher ROLP and biomass for the first root order in the non-vegetated half, as compared with the NF treatment; higher ROLP was observed in the third-order roots of the non-vegetated half with the FV treatment, as compared with the F treatment. Despite the lower soil resource concentration in the non-vegetated half than in the vegetated one, the plant still attempted to increase the nutrient intake in this space. Furthermore, the plants intensified nutrient uptake in the non-vegetated half by altering the RTRS in FNV treatment, as described above. Therefore, plants simultaneously exposed to nutrient heterogeneity and neighboring plants still attempted to increase nutrient uptake in the space free of other plant roots, regardless of the distribution of resources.

The non-additive root growth response under the combined nutrients and neighbors environments (i.e. interactions occur) may be due to the interplay between local responses and systemic modifications of the response. When intense competitive signals were received from the connected modules (i.e., roots in the vegetated half) in the FV treatment, the target plants increased their nutrient uptake in the non-vegetated half by investing more first-order root biomass and increasing the ROLP of first-order roots in the non-vegetated half, as compared with the NF treatment, and by increasing the ROLP of the third-order roots in the FV treatment, as compared with the F treatment. Because the fraction of nutrients obtained from the non-vegetated half to nutrients the whole plant desired was increased, the intense competition in the vegetated half was alleviated in the FV treatment. In other words, target plants increasing their nutrient uptake in the non-vegetated half helped decrease the fraction of nutrients obtained from the vegetated half. The interaction between roots in the different halves (modules) triggered potential nutrient uptake ability of whole plant root system, with more powerful nutrient uptake observed in both non-vegetated and vegetated halves. Although facilitators of soil resource acquisition were present in the non-vegetated half, as well as higher nutrient concentrations and the absence of interspecific root competition, the target plants in the FNV treatment still increased their nutrient uptake in the vegetated half than in the non-vegetated one, with higher SRLP in the 0–0.5 mm fine roots and higher ROLP of the third-order roots. Therefore, competition was strengthened in the vegetated half, based on the interplay between the local responses and systemic controls. This response was necessary for late-succession trees to be established in fully occupied belowground environments to ensure long-term success of the said tree population. Given the similar nutrient concentration between two halves in the NF and F treatments, induction of root growth in nutrient-rich patches was lost and root competition became the most important environmental stimulus. That is, our study indicated that under the combinations of homogeneous nutrients and root competition, target plants adopted the strategies of deceasing SRLP in 0–0.5 mm fine roots, either in the non-vegetated or vegetated halves, to alleviate inter- and intra-plant

root competition with the increasing nutrient concentration. The lower SRLP in 0–0.5 mm fine roots (the significant region in nutrient absorption) contributed to mitigate intra-plant root competition because competition among roots of the same plant was three- to five-times greater than competition among roots of neighbouring plants [47]. Collectively, the interplay between the local responses and the systemic response modifications in root foraging was far more complicated under a combination of neighboring competitors and nutrient heterogeneity than that of neighboring competitors and homogeneous nutrient conditions. The sophisticated interaction between local response and systemic control originated from the existing nutrient differences and neighboring plant roots, which triggered the potential root foraging ability under a combination of neighboring competitors and nutrient heterogeneity. This phenomenon may account for the similar relative growth rate (RGR) among the plants in the FV, FNV, and F treatments.

In this study, contrary to the small biomass difference in the first three root orders between different treatments, root architecture indicators that originated from these root systems were greatly varied. Therefore, the root architecture responded to environmental stimuli more sensitively than the root biomass. Moreover, the plant's attempt to increase nutrient uptake was reflected by the altered root architecture but with constant biomass. Given that the roots possessing essential nutrient uptake ability represent only a portion of the entire root system for woody plants, the root architecture indicators constructed by these roots (i.e., the first three root orders or the 0–0.5 mm roots in diameter) in our study were more precisely measured the root foraging ability, as compared with the methods used in previous investigations. These root architecture indicators provided us with a novel and effective means to explore woody plant root foraging behavior.

Acknowledgments

We thank Liangchun Gong and Tiangui Si for their help with fieldwork.

Author Contributions

Conceived and designed the experiments: HN QL JC. Performed the experiments: HN HY CY CZ. Analyzed the data: HN JC. Contributed reagents/materials/analysis tools: XC. Wrote the paper: HN QL.

References

1. Kembel SW, De Kroon H, Cahill JF, Mommer L (2008) Improving the scale and precision of hypotheses to explain root foraging ability. Ann Bot 101: 1295–1301.
2. De Kroon H, Visser EJW, Huber H, Mommer L, Hutchings MJ (2009) A modular concept of plant foraging behaviour: the interplay between local responses and systemic control. Plant Cell Environ 32: 704–712.
3. Messier C, Coll L, Poitras-Lariviere A, Belanger N, Brisson J (2009) Resource and non-resource root competition effects of grasses on early- versus late-successional trees. J Ecol 97: 548–554.
4. Coomes DA, Grubb PJ (2000) Impacts of root competition in forests and woodlands: a theoretical framework and review of experiments. Ecol Monogr 70: 171–207.
5. Farley RA, Fitter AH (1999) Temporal and spatial variation in soil resources in a deciduous woodland. J Ecol 87: 688–696.
6. Hodge A (2004) The plastic plant: root responses to heterogeneous supplies of nutrients. New Phytol 162: 9–24.
7. Hodge A (2009) Root decisions. Plant Cell Environ 32: 628–640.
8. Cahill JF, McNickle GG, Haag JJ, Lamb EG, Nyanumba SM, et al. (2010) Plants integrate information about nutrients and neighbors. Science 328: 1657–1657.
9. Fang S, Gao X, Deng Y, Chen X, Liao H (2011) Crop root behavior coordinates phosphorus status and neighbors: from field studies to three-dimensional in situ reconstruction of root system architecture. Plant Physiol 155: 1277–1285.
10. Mommer L, van Ruijven J, Jansen C, van de Steeg HM, de Kroon H (2012) Interactive effects of nutrient heterogeneity and competition: implications for root foraging theory? Funct Ecol 26: 66–73.
11. Blair BC (1998) Root foraging and soil nutrient heterogeneity: Presence and effecton competition in tropical forests. Biotropica 30: 21–21.
12. Leuschner C, Hertel D, Coners H, Buttner V (2001) Root competition between beech and oak: a hypothesis. Oecologia 126: 276–284.
13. Gunaratne AMTA, Gunatilleke CVS, Gunatilleke IAUN, Weerasinghe HMSPM, Burslem DFRP (2011) Release from root competition promotes tree seedling survival and growth following transplantation into human-induced grasslands in Sri Lanka. Forest Ecol Manag 262: 229–236.
14. Fletcher EH, Thetford M, Sharma J, Jose S (2012) Effect of root competition and shade on survival and growth of nine woody plant taxa within a pecan [Carya illinoinensis (Wangenh.) C. Koch] alley cropping system. Agroforest Syst 86: 49–60.
15. Fitter AH (1987) An architectural approach to the comparative ecology of plant-root systems. New Phytol 106: 61–77.
16. Lynch J (1995) Root architecture and plant productivity. Plant Physiol 109: 7–13.
17. Fitter AH, Stickland TR (1991) Architectural analysis of plant-root systems.2. Influence of nutrient supply on architecture in contrasting plant-species. New Phyt 118: 383–389.
18. Fitter AH, Stickland TR, Harvey ML, Wilson GW (1991) Architectural analysis of plant-root systems.1. Architectural correlates of exploitation efficiency. New Phyt 118: 375–382.
19. Farley RA, Fitter AH (1999) The responses of seven co-occurring woodland herbaceous perennials to localized nutrient-rich patches. J Ecol 87: 849–859.
20. Grime JP, Mackey JML (2002) The role of plasticity in resource capture by plants. Evol Ecol 16: 299–307.

21. Gersani M, Brown JS, O'Brien EE, Maina GM, Abramsky Z (2001) Tragedy of the commons as a result of root competition. J Ecol 89: 660–669.

22. Rubio G, Liao H, Yan XL, Lynch JP (2003) Topsoil foraging and its role in plant competitiveness for phosphorus in common bean. Crop Sci 43: 598–607.

23. Lynch JP, Ho MD (2005) Rhizoeconomics: Carbon costs of phosphorus acquisition. Plant Soil 269: 45–56.

24. Nord EA, Zhang C, Lynch JP (2011) Root responses to neighbouring plants in common bean are mediated by nutrient concentration rather than self/non-self recognition. Funct Plant Biol 38: 941–952.

25. Bolte A, Villanueva I (2006) Interspecific competition impacts on the morphology and distribution of fine roots in European beech (*Fagus sylvatica L.*) and Norway spruce *(Picea abies (L.) Karst.)*. Eur J Forest Res 125: 15–26.

26. Lucash MS, Eissenstat DM, Joslin JD, McFarlane KJ, Yanai RD (2007) Estimating nutrient uptake by mature tree roots under field conditions: challenges and opportunities. Trees-Struct Funct 21: 593–603.

27. Malamy JE (2005) Intrinsic and environmental response pathways that regulate root system architecture. Plant Cell Environ 28: 67–77.

28. Postma JA, Lynch JP (2012) Complementarity in root architecture for nutrient uptake in ancient maize/bean and maize/bean/squash polycultures. Ann Bot 110: 521–534.

29. Comas LH, Eissenstat DM (2004) Linking fine root traits to maximum potential growth rate among 11 mature temperate tree species. Funct Ecol 18: 388–397.

30. Hartnett DC, Potgieter AF, Wilson GWT (2004) Fire effects on mycorrhizal symbiosis and root system architecture in southern African savanna grasses. Afr J Ecol 42: 328–337.

31. Fujita H, Syono K (1996) Genetic analysis of the effects of polar auxin transport inhibitors on root growth in Arabidopsis thaliana. Plant Cell Physiol 37: 1094–1101.

32. Marchant A, Kargul J, May ST, Muller P, Delbarre A, et al. (1999) AUX1 regulates root gravitropism in Arabidopsis by facilitating auxin uptake within root apical tissues. Embo J 18: 2066–2073.

33. Grabov A, Ashley MK, Rigas S, Hatzopoulos P, Dolan L, et al. (2005) Morphometric analysis of root shape. New Phytol 165: 641–651.

34. Pregitzer KS, DeForest JL, Burton AJ, Allen MF, Ruess RW, et al. (2002) Fine root architecture of nine North American trees. Ecol Monogr 72: 293–309.

35. Wells CE, Glenn DM, Eissenstat DM (2002) Changes in the risk of fine-root mortality with age: A case study in peach, Prunus persica (Rosaceae). Am J Bot 89: 79–87.

36. Guo DL, Mitchell RJ, Hendricks JJ (2004) Fine root branch orders respond differentially to carbon source-sink manipulations in a longleaf pine forest. Oecologia 140: 450–457.

37. Guo D, Xia M, Wei X, Chang W, Liu Y, et al. (2008) Anatomical traits associated with absorption and mycorrhizal colonization are linked to root branch order in twenty-three Chinese temperate tree species. New Phytol 180: 673–683.

38. Day KJ, John EA, Hutchings MJ (2003) The effects of spatially heterogeneous nutrient supply on yield, intensity of competition and root placement patterns in *Briza media* and *Festuca ovina*. Funct Ecol 17: 454–463.

39. O'Brien EE, Brown JS (2008) Games roots play: effects of soil volume and nutrients. J Ecol 96: 438–446.

40. Hu R, Lin B, Liu Q (2011) Effects of forest gaps and litter on the early regeneration of picea asperata plantations. Scientia Silvae Sinicae 47: 23–29.

41. Garcia AL, Franco JA, Nicolas N, Vicente RM (2002) Influence of amino acids in the hydroponic medium on the growth of tomato plants. J Plant Nutr 29: 2093–2104.

42. Cheng S, Widden P, Messier C (2005) Light and tree size influence belowground development in yellow birch and sugar maple. Plant Soil 270: 321–330.

43. Lei P, Scherer-Lorenzen M, Bauhus J (2012) Belowground facilitation and competition in young tree species mixtures. Forest Ecol Manag 265: 191–200.

44. Ozier-Lafontaine H, Lecompte F, Sillon JF (1999) Fractal analysis of the root architecture of Gliricidia sepium for the spatial prediction of root branching, size and mass: model development and evaluation in agroforestry. Plant Soil 209: 167–180.

45. Zheng M, Lai L, Jiang L, An P, Yu Y, et al. (2012) Moderate water supply and partial sand burial increase relative growth rate of two Artemisia species in an inland sandy land. J Arid Environ 85: 105–113.

46. Zhou J, Dong BC, Alpert P, Li HL, Zhang MX, et al. (2012) Effects of soil nutrient heterogeneity on intraspecific competition in the invasive, clonal plant Alternanthera philoxeroides. Ann Bot 109: 813–818.

47. Rubio G, Walk T, Ge ZY, Yan XL, Liao H, et al. (2001) Root gravitropism and below-ground competition among neighbouring plants: A modelling approach. Ann Bot 88: 929–940.

10

Nutrient Presses and Pulses Differentially Impact Plants, Herbivores, Detritivores and Their Natural Enemies

Shannon M. Murphy[1]*, Gina M. Wimp[2], Danny Lewis[2], Robert F. Denno[3]†

1 Department of Biological Sciences, University of Denver, Denver, Colorado, United States of America, **2** Biology Department, Georgetown University, Washington, DC, United States of America, **3** Department of Entomology, University of Maryland, College Park, Maryland, United States of America

Abstract

Anthropogenic nutrient inputs into native ecosystems cause fluctuations in resources that normally limit plant growth, which has important consequences for associated food webs. Such inputs from agricultural and urban habitats into nearby natural systems are increasing globally and can be highly variable, spanning the range from sporadic to continuous. Despite the global increase in anthropogenically-derived nutrient inputs into native ecosystems, the consequences of variation in subsidy duration on native plants and their associated food webs are poorly known. Specifically, while some studies have examined the effects of nutrient subsidies on native ecosystems for a single year (a nutrient pulse), repeated introductions of nutrients across multiple years (a nutrient press) better reflect the persistent nature of anthropogenic nutrient enrichment. We therefore contrasted the effects of a one-year nutrient pulse with a four-year nutrient press on arthropod consumers in two salt marshes. Salt marshes represent an ideal system to address the differential impacts of nutrient pulses and presses on ecosystem and community dynamics because human development and other anthropogenic activities lead to recurrent introductions of nutrients into these natural systems. We found that plant biomass and %N as well as arthropod density fell after the nutrient pulse ended but remained elevated throughout the nutrient press. Notably, higher trophic levels responded more strongly than lower trophic levels to fertilization, and the predator/prey ratio increased each year of the nutrient press, demonstrating that food web responses to anthropogenic nutrient enrichment can take years to fully manifest themselves. Vegetation at the two marshes also exhibited an apparent tradeoff between increasing %N and biomass in response to fertilization. Our research emphasizes the need for long-term, spatially diverse studies of nutrient enrichment in order to understand how variation in the duration of anthropogenic nutrient subsidies affects native ecosystems.

Editor: James F. Cahill, University of Alberta, Canada

Funding: This research was funded by the National Parks Ecological Research Fellowship Program, a partnership between the National Park Service, the Ecological Society of America and the National Park Foundation that was funded by the Andrew W. Mellon Foundation. The funders had no role in study design, data collection and analysis, decision to publish, or preparation of the manuscript.

Competing Interests: The authors have declared that no competing interests exist.

* E-mail: Shannon.M.Murphy@du.edu

† Deceased.

Introduction

Natural and anthropogenic inputs of nutrients into native ecosystems often promote fluctuations in the availability of resources that normally limit plant growth [1,2,3,4]. Such inputs promote changes in primary productivity and plant diversity that in turn can have important, community-wide consequences for associated food webs [4,5,6,7,8]. Even a short-term increase in a resource that is limiting (a resource pulse) can have extended effects on community structure, trophic interactions and ecosystem function [9]. Although a growing number of studies document the widespread effects of sporadic nutrient or basal-resource pulses on plant productivity and food web structure [5,7,9,10,11,12,13,14,15,16,17,18], few have examined how long-term nutrient loading (a resource press) impacts recipient communities [3,4,19,20], and fewer yet have contrasted the food web effects of a nutrient pulse with a press in the same system [3,21]. In previous work, we found that increased primary production via nitrogen fertilization alters arthropod community structure and composition in *Spartina* marshes; species richness of

herbivores, predators, parasitoids and detritivores all increased in response to nitrogen addition [18]. That study was the first to examine how food web structure is altered through trophic dynamics that extend solely from enhanced plant production, and not from changes in plant community composition, but it only examined the food web response to a nutrient pulse within a single season. What happens to arthropod food webs when wetlands receive a nutrient press over several years is unknown yet of critical importance given the increasing amount of nitrogen runoff into salt marshes. There are a few notable examples of ongoing nutrient press studies [22,23,24,25,26,27] but these studies focus primarily on plant responses to nutrient subsidies and the responses of multiple trophic levels to such presses remains poorly understood. Here we extend our previous research on food web responses to nutrient subsidies to compare the effects of a resource pulse with a resource press on the arthropod food web of two mid-Atlantic salt marshes in North America.

Inputs of limiting nutrients (e.g., nitrogen) frequently have important effects on plant species richness, plant community composition, primary productivity, and plant tissue quality

[3,4,28]. For example, long-term nutrient loading can lead to the simplification of both plant and associated arthropod communities due in part to a strong correlation between plant and insect species diversity [4]. As a result, determining the direct effects of enrichment on arthropod communities, as opposed to indirect effects mediated by plant composition, can be experimentally daunting [3,4,19,21]. However, by restricting our research to the natural monocultures of *Spartina alterniflora* (hereafter *Spartina*) found in salt marshes, our study is uniquely able to focus on altered food web structure and dynamics that extend solely from enhanced plant productivity and not from compositional changes in the plant community [13].

Understanding how the duration of anthropogenically-derived nutrient subsidies affects natural food webs is important from a conservation perspective in light of the nearly seven-fold increase in agricultural nitrogen fertilization and an extreme global increase of nutrient runoff into natural systems [29]. In particular, land development and agriculture are jeopardizing coastal wetlands at an alarming rate, and one of the major threats is nutrient pollution from neighboring anthropogenic sources, which alters vegetation dynamics by increasing nitrogen availability [30,31]. An estimated 50% of the variation in nitrogen availability in *Spartina* marshes is explained by shoreline development, such as housing developments and agriculture [30]. Annual nitrogen inputs from fertilizer exceed 1000 kg/km^2 and are expected to double in the near future [32,33,34]. In addition to fertilizer-derived nitrate, animal waste is estimated to add more than 2000 kg/km^2 and waste water from urban areas contributes an additional 100–500 kg/km^2 to the annual nitrogen-load in aqueous runoff [35,36]. As terrestrially-derived nitrate flows downstream, about one quarter is intercepted by coastal wetlands (e.g. *Spartina* marshes) before reaching open waters [37]. Nitrogen that is retained in the marsh is incorporated into plant biomass, denitrified or buried in marsh sediments [13,37,38]. Because *Spartina* is N-limited, nitrogen subsidies result in dramatic increases in biomass, plant nitrogen content and detritus [13,14,30,39,40]. Stable isotope analyses confirm that allochthonous nitrogen is taken up by *Spartina* [41] and is transported directly up the food chain from producers to primary consumers [42]. Thus, nutrient runoff from anthropogenic sources has direct consequences for *Spartina* and its associated consumers. Our research on how nutrient pulses and presses alter food web structure is particularly relevant because inputs of nutrients from agricultural and urban habitats into nearby natural systems can be highly variable and span the range from sporadic (e.g. nutrient pulse) to continuous (e.g. nutrient press) [32,35,37,43]. Yet, how nitrogen pulses and presses differentially affect the recipient community of consumers is poorly known.

Some arthropod species may respond to a nutrient press by retaining the density achieved during the first year of enrichment, but others are likely to exhibit more complex responses. Enrichment affects each arthropod species indirectly via one or more paths through the complex marsh interaction web. Feedbacks and time lags may mean that the effect of enrichment on a species may continue to change over many years. For example, increased live plant biomass can be expected to produce increased thatch (dead) biomass after a time lag, which is in turn expected to decrease intraguild predation and cannibalism [44,45]. An increase in predator population growth rate may follow, with subsequent effects on prey density. Such interactions among responses may continue to reverberate over long time frames and are not predictable from short-term experiments.

Here, we present the results of an experiment in which we tested the responses of the arthropod food web to a nutrient pulse and a nutrient press. We replicated our experiment at two field sites, Tuckerton, NJ (TUCK) and Cape Hatteras National Seashore, NC (CHNS), which are located on opposite sides of a biogeographic break in VA where *Spartina* switches from annual to perennial aboveground growth [46]. We chose to work in a higher-latitude marsh and a lower-latitude marsh to investigate site-to-site variation in pulse and press dynamics. Pennings *et al.* [47] showed that *Spartina* from marshes at higher latitudes in North America is more palatable to herbivores than in marshes at lower latitudes. Recently, McCall and Pennings [48] demonstrated that latitude and tidal range explain much of the geographic variation in biotic and abiotic variables among marshes at higher and lower latitudes. Notably, the two marshes examined in this study vary greatly in tidal range (longer tidal inundation at CHNS relative to TUCK), which is positively correlated with *Spartina* height [49]. The differences in our study marshes may play an important role in how readily plants use nutrient subsidies and the degree to which the arthropod community responds.

We expected our nutrient pulse to produce results similar to those of earlier marsh pulse fertilization experiments [13,14,15,18]. In pulse plots, we predicted that 1) species at all trophic levels would respond positively to enrichment, although the increase in detritus and detritivores would be delayed, 2) effects would be greater among higher trophic levels, and 3) all responses would gradually return to control levels after fertilization ceased (Fig. 1). In press plots, which received fertilizer over multiple years, we predicted that herbivores and predators might exhibit complex extended responses (indicated by '?' in Fig. 1b,e). Notably, we expected that a nutrient press would not be a simple extension of a nutrient pulse due to effects on higher trophic levels that could cascade to affect herbivorous prey. Reproductive responses by univoltine spiders to increased prey could lead to increased predator density during the second year, and could be further amplified by an increase in thatch, which reduces predator interference [45]. These increased predator densities were expected to depress herbivore densities [50], but if suppression was relatively mild, herbivore densities could remain elevated and allow predator densities to continue to rise. In contrast, more intense predation could lead to a severe decline in herbivores, followed by a decline in predators. Similar feedbacks and delayed responses may occur among plants, detritus and detritivores, but since those dynamics have been studied less, we predicted only that these groups would retain the levels reached during the first year of fertilization (Fig. 1a,b,d). Finally, we predicted that some responses to fertilization would differ between the two marshes. We predicted that herbivores and predators would respond more strongly to fertilization at TUCK, the high-latitude marsh, because of expected higher grass palatability [47]. We also predicted that *Spartina* height would increase more at CHNS, where tidal inundation was greater.

Materials and Methods

Study Sites and Organisms

We conducted our study in two salt marshes dominated by natural monocultures of *Spartina* along the east coast of the United States, TUCK (39° 31.6′N, 74° 19.2′W) and CHNS (35° 47.6′N, 75° 32.8′W). All necessary permits were obtained for the described field studies (permits for TUCK were obtained from K. Able at the Rutgers University Marine Station and permits for CHNS were obtained from B. Commins at the National Park Service Research Permit and Reporting System and J. Ebert, M. Lyons, T. Broili, M. Carfioli, and S. Strickland at CHNS). We focused on a reduced food web composed of the numerically-dominant herbivores,

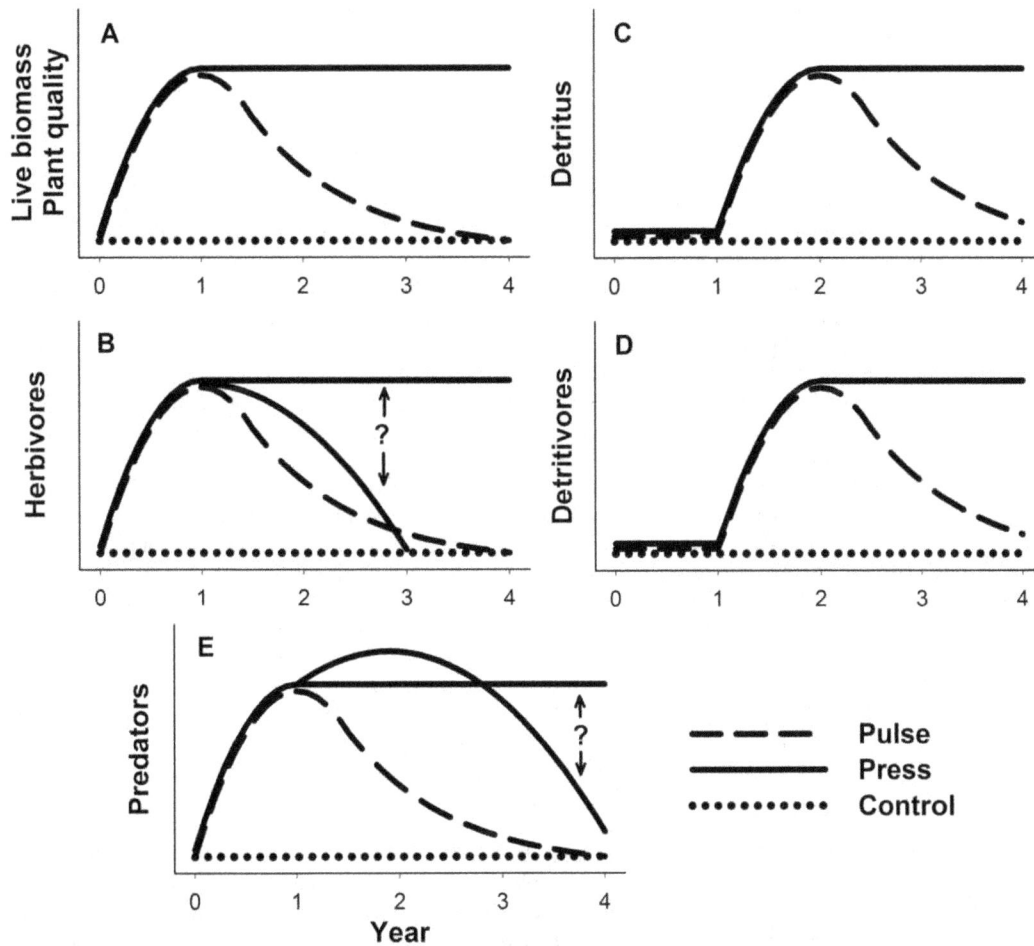

Figure 1. Conceptual model of possible inter-annual responses of *Spartina* plant parameters and consumer densities to press and pulse nitrogen subsidies. Plant characteristics (A, C) and arthropod densities (B, D, E) are expected to vary over time due to seasonal and stochastic events not related to the treatment effect. Here we show the predicted responses of the pulse and press treatments in relation to the control treatment, which is held constant over time. Question mark (?) in B represents possible different responses by herbivores depending on level of predator suppression. Question mark (?) in E represents possible different responses by predators depending on whether predators over-exploit their prey.

detritivores, and predators that are common at both field sites [18]. *Spartina* serves as the only host plant for a variety of insect herbivores [13]. Planthoppers (*Prokelisia dolus* and *P. marginata*) are most abundant (~80% of herbivore biomass); remaining herbivores are rare relative to *Prokelisia* and consist of other planthoppers (e.g., *Delphacodes penedetecta*) and true bugs (e.g., *Trigonotylus uhleri*). Detritivores feed on *Spartina* detritus and epiphyton associated with *Spartina* on the marsh surface (e.g. amphipod *Orchestia grillus*, isopod *Venezillo parvus*). Natural enemies, including invertebrate predators and parasitoids, attack herbivores and detritivores associated with *Spartina*, but predators are the most important source of mortality [50]; these include omnivores that feed on *Spartina* and other herbivores (e.g. katydids *Conocephalus spartinae* and *Orchelimum fidicinium*), generalist predators (e.g. web-building spider *Grammonota trivittata*), and specialist predators (e.g. the mirid *Tytthus vagus* attacks planthopper eggs) [44]. Top carnivores (e.g. hunting spiders *Pardosa littoralis*, *Clubiona* sp.) feed on herbivores, detritivores, specialist predators and sometimes each other [51]. All organisms are hereafter referred to by their genera or feeding guild.

Nutrient Manipulations

To investigate the differential effects of a nutrient pulse versus press on the arthropod food web, we manipulated nutrient subsidies with a one-way design. Year 1 (2005) was the only year that we fertilized pulse treatment plots. In years 2–4 (2006–2008), we continued to fertilize press plots in the same manner as during year 1 (see Table S1 for a complete list of fertilizer addition and sample dates). At each site, we established 10 blocks, each with three 2×2 m treatment plots, and assigned plots randomly to one of three treatments: control (no fertilization), pulse (fertilization during year 1 only) and press (fertilization during all years). Our plots were necessarily 2×2 m to accommodate restrictions associated with working in a protected National Seashore (CHNS), but previous work shows that the population dynamics of the major herbivores and predators as well as treatment effects on trophic composition in our plots scale up to the dynamics that prevail in larger plots (>100 m²) [13,50]. We fertilized each plot with 60 grams/m² of a 3:1 mixture of granular ammonium nitrate (N-P-K: 34-0-0) and triple phosphate (0-45-0) three times during the season, for a total of 180 grams/m² per year (Table S1). The elevated N-content achieved by our fertilization treatment is comparable to that for plants in *Spartina* marshes that experience high nutrient loading from nearby coastal developments [30]. Previous work has demonstrated that N additions after peak biomass is attained have little impact on *Spartina* growth [52], thus we applied N only at the beginning of each growing-season for all fertilization treatments.

Plant and Arthropod Samples

We measured *Spartina* biomass and height before the initiation of fertilization treatments (the first sample collected in May 2005) and subsequently during peak biomass each season (Table S1). On each collection date, we harvested all of the plant biomass within a 0.047-m^2 quadrat from each plot. We sorted quadrat samples into live and dead (thatch) plant material, measured the height of living culms and counted the number of tillers. We washed the plant material with deionized water, dried it in a drying oven at 60°C for three days and weighed it. To measure the treatment effects on the N-and C-content of *Spartina*, we collected plant snips (5–10 *Spartina* culms per plot) that were processed as described above, ground in a Wiley mill, and sent to the Cornell Stable Isotope Laboratory for analysis. Plant snips and quadrats were collected several times during each growing season (Table S1). To ensure that our fertilization treatments remained in the appropriate treatment plot and did not spread into the adjacent matrix or neighboring plots, we collected plant snips from 1 m outside each plot in 2005 and 2006 (Table S1).

To measure the treatment effects on the arthropod food web, we sampled arthropods with a D-vac suction sampler with a restricted suction head (0.036 m^2), which we placed in 5 different locations within each plot for 3-seconds. At both sites, we sampled the arthropod community 5 times in 2005 and 2006, and 4 times in 2007 and 2008 (Table S1). After collection, we stored the arthropod samples in ethanol and later sorted, counted and identified individuals to genus and species.

We expected plant characteristics and arthropod densities to vary over time, even in control plots; therefore, to isolate the effects of fertilization from other types of variation, we calculated pulse and press treatment effects in each block, based on the treatment/ control ratio. More precisely, treatment effect = ln((treatment value +1)/(control value +1)), where "treatment value" was the value from the press or pulse plot in the block, and "control value" was the value from the control plot in the same block (units in g/ m^2 for plant characteristics and individuals/m^2 for arthropods). A positive treatment effect therefore means that the value in the pulse or press plot was higher than the value in the control plot in the same block.

Herbivore Damage

To assess the amount of damage inflicted by herbivores, we haphazardly chose 10 *Spartina* culms from each quadrat from CHNS on August 17, 2006. We were able to distinguish damage by three different herbivores: snail, katydid and *Trigonotylus*. For each leaf, we measured length and the amount of leaf that was damaged to the nearest 0.5 cm; for snails we measured the length of each radulation, for katydids we measured the length of each chew mark and for *Trigonotylus* we measured their distinctive 'spotting' damage. Snail damage was minimal so we did not include it in our analyses. We then divided each type of damage by the total length of all leaves in the culm to get average damage per cm of leaf.

Statistical Analyses

To test for pre-treatment differences, we performed ANOVA on plant and arthropod characteristics in the May 2005 pre-treatment sample, using the treatment that a plot would later receive (press, pulse or control) as the explanatory variable (SAS proc anova). For subsequent statistical analyses, we calculated the effect of fertilization treatment on each *Spartina* characteristic and on the density of each arthropod taxon in each block on each sample date. We used treatment effect as the response variable in a repeated measures ANOVA with explanatory variables time, marsh (CHNS or TUCK), treatment (press or pulse) and all possible interactions (SAS proc mixed). Block within marsh was treated as a random effect. We treated pre-fertilization data collected in May 2005 as a separate time category from the post-fertilization data collected later the same year. In order to determine the appropriate variance-covariance structure for our repeated measures ANOVA, we first explored temporal autocorrelation among data. We used an autoregressive structure when correlation decayed over time (AR(1) in proc mixed), compound symmetry (CS in proc mixed) when correlation showed no temporal trend, and a simple variance components matrix (VC in proc mixed) when autocorrelation was absent. When variance changed over time, we used a heterogeneous structure (ARH(1) or CSH in proc mixed). When there was a significant year by treatment interaction, we used pre-planned t-tests to compare the effects of press and pulse treatments on plant characteristics and arthropod densities at both marshes during each of the four years and during the pre-treatment collection. A false error rate correction for multiple tests was applied to the results of those tests (SAS proc multtest). To test whether fertilizer remained in the appropriate treatment plot and did not spread into the adjacent matrix or neighboring plots, we performed ANOVA on live *Spartina* biomass and culm height 1 m outside plots in 2005 and 2006; treatment was the fixed effect and block and marsh were random effects. We calculated herbivore load for each plot on each sample date as (herbivore density +1)/(*Spartina* live biomass). Herbivore density was the density of all planthoppers, katydids and *Trigonotylus*. We then calculated the effect of press and pulse treatments on herbivore load in each block on each sample date as ln(treatment load/control load). In a similar manner, we calculated predator/prey ratio for each plot as (predator density +1)/(prey density +1), and then calculated effect sizes as above. Predators consisted of all spiders and prey consisted of the herbivores listed above. As a measure of total nitrogen uptake by *Spartina*, we used grams nitrogen in live biomass per square meter of marsh surface, calculated as %N multiplied by live biomass per square meter. We tested for differences in this N density between marshes and fertilization treatments with repeated measures ANOVA (SAS proc mixed) using marsh, fertilization treatment and their interaction as fixed factors and block within marsh as a random effect. For herbivore damage, we analyzed *Trigonotylus* damage using repeated measures ANOVA with block as a random effect and the ten culms treated as repeated measures from the same plot (SAS proc mixed). Katydid-damaged area had so many zero values that parametric analysis was not possible, so we used the randomization test of Ruxton et al. [53], modified to account for repeated measures [54]. Separate tests were performed on all treatment pairs and P-values were adjusted for multiple tests (SAS proc multtest). We deposited our data in the Dryad Repository:

Results

With some notable exceptions, the 19 plant and arthropod response variables followed similar trajectories over the course of the experiment. In plots that received the one-year fertilizer pulse, responses were significantly greater than controls during the first year, after which they gradually declined, and by year 4 they were

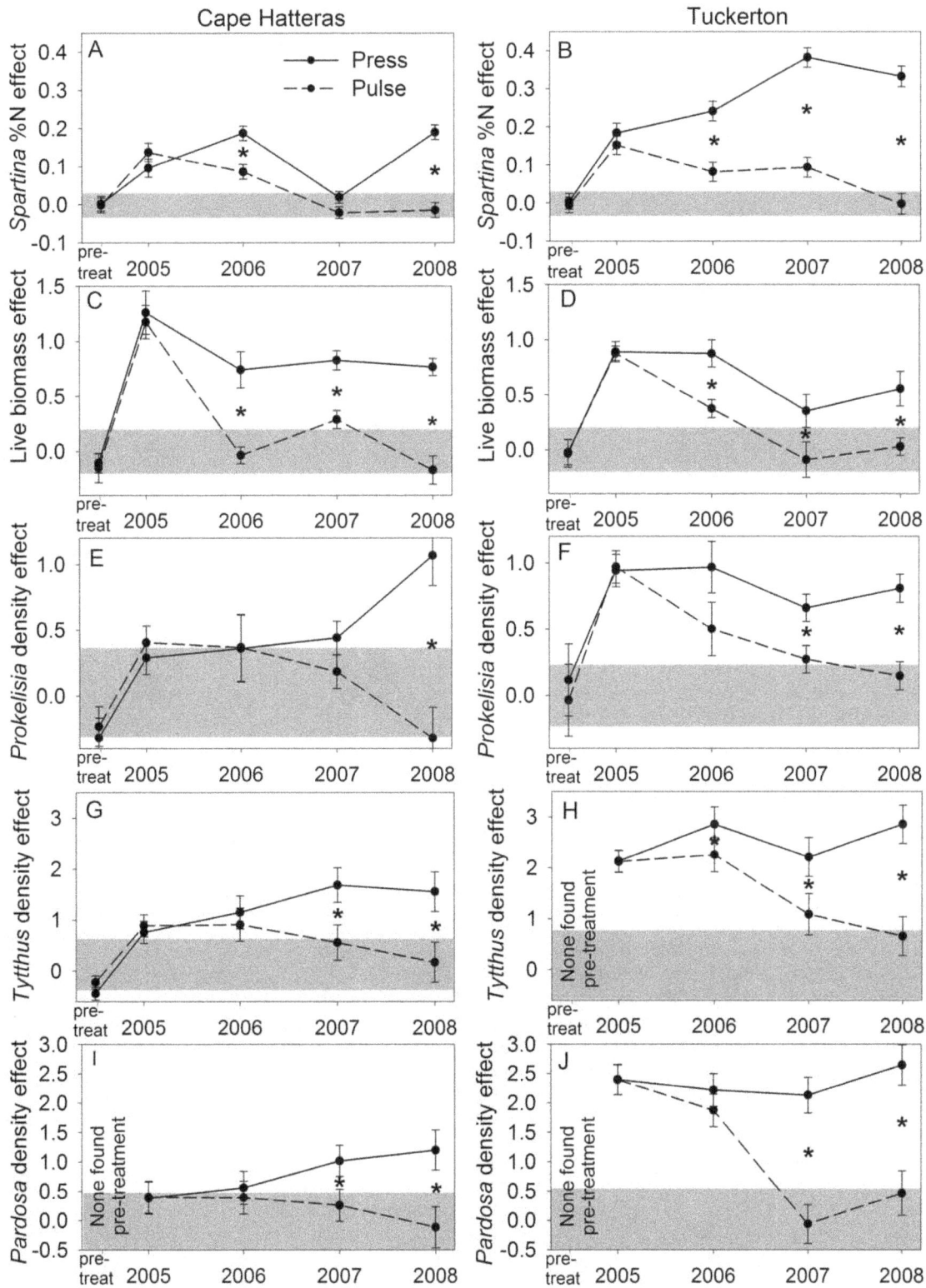

Figure 2. Plant and arthropod responses to nutrient manipulations. Effects of press (solid line) and pulse (dashed line) treatments on *Spartina* %N (A, B), *Spartina* live biomass density (C, D), density of *Prokelisia* planthoppers (E, F), density of the predatory mirid *Tytthus* (G, H), and density of the top intraguild predator *Pardosa* (I, J) (results from CHNS on left, TUCK on right; means ±se). Effect means within gray bands were not significantly different from zero, meaning that treatment and control values did not differ. Asterisks indicate that pulse and press treatments had significantly different effects in a given year ($\alpha = 0.05$). Asterisks are displayed only when the year by treatment interaction was significant. See methods for information on how treatment effects were calculated.

no different from controls (Figs. 2, S1, S2, S3). These results replicated those of earlier pulse experiments. In plots that received press fertilization, on the other hand, responses remained higher than controls throughout the four years of the experiment.

Pre-treatment Differences and Fertilization Effects Outside Study Plots

In our pre-treatment sample, treatments did not differ significantly from each other for any measures, except at CHNS where *Tytthus* densities were lower in plots that would later receive nutrient presses than in control plots ($F_{2,27} = 3.80$, $P = 0.03$). We detected no effect of fertilization treatment on *Spartina* biomass (year 1: $F_{2,46.1} = 0.32$, $P = 0.73$; year 2: $F_{2,46.3} = 0.51$, $P = 0.60$), or on culm height 1 m outside of treatment plots (year 1: $F_{2,47} = 0.21$, $P = 0.81$; year 2: $F_{2,46.3} = 0.56$, $P = 0.58$).

Plant Responses to Nutrient Manipulations

Spartina %N increased in response to fertilization, but it responded much more strongly at TUCK than at CHNS (marsh*treatment interaction $F_{1,28.7} = 40.37$, $P < 0.0001$) (Fig. 2a,b). In contrast, live *Spartina* biomass (measured as dry weight) responded more strongly at CHNS (marsh*treatment interaction $F_{1,65.5} = 12.48$, $P = 0.0008$) (Fig. 2c,d). These responses exaggerated already-existing differences between control plots at the two marshes, where percent nitrogen was higher at TUCK ($t_{29.9} = 17.50$, $P < 0.0001$), and live biomass was higher at CHNS ($t_{6.31} = 2.69$, $P = 0.03$). As a result, the two marshes became even more different from one another when fertilized (Fig. 3a). Despite these significant differences between marshes, *Spartina* nitrogen density (grams N in live Spartina/m2) in control plots was very similar at the two marshes ($t_{80.4} = 0.64$, $P = 0.52$) and increased by virtually the same, very large, amount when fertilized (Fig. 3b). Nitrogen density in press treatments did not differ between marshes ($t_{76.6} = 0.45$, $P = 0.65$), but was higher than control plots ($t_{88.1} = 16.14$, $P < 0.0001$). At CHNS this result was accomplished largely through increased biomass whereas at TUCK this happened largely through increased %N with a relatively small increase in biomass.

The response of thatch to fertilization was delayed by a year, but starting in year 2 it followed the usual trajectory: thatch in pulse plots peaked in year 2 and then declined, whereas thatch in press plots remained elevated through year 4 (Fig. S1e). However, it is notable that thatch levels in press plots declined in years 3 and 4 even though they remained above control levels.

Contrary to our prediction, culm length responded more strongly at TUCK than at CHNS ($F_{1,19.6} = 6.40$, $P = 0.02$) (Fig. S1a), despite the fact that tidal inundation was greater at CHNS. Culm density exhibited a positive response to fertilization only during year 1 (Fig. S1b). Tiller density was not affected by fertilization at either marsh ($F_{1,102} = 0.09$, $P = 0.77$) (Fig. S1c).

Arthropod Responses to Nutrient Manipulations

Densities of arthropods exhibited positive responses to pulse and press fertilization with few exceptions (Figs. 2, S2, S3). Those exceptions included densities of the amphipod *Orchestia*, which responded erratically to fertilization, increasing and decreasing in different years (Fig. S2c). *Prokelisia* planthopper density responded positively at TUCK, but did not become significantly greater than controls in CHNS press plots until year 3 (Fig. 2e,f). Finally, in press plots at TUCK, densities of *Trigonotylus* (Fig. S2b) and hunting spiders (Fig. S3c) were not significantly greater than controls in year 3.

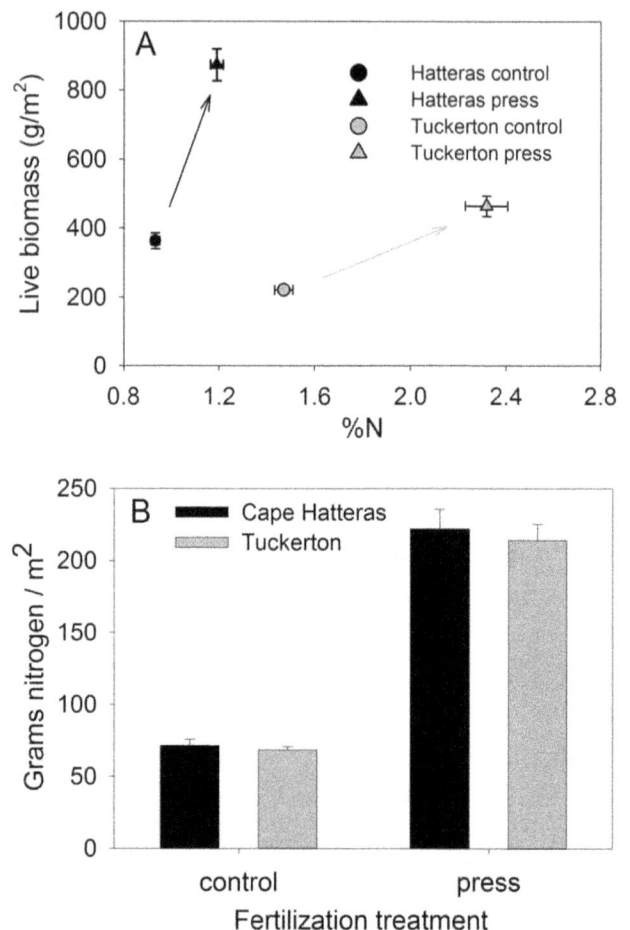

Figure 3. Similarities and differences in the *Spartina* response to fertilization at CHNS and TUCK (means ±se). Responses of *Spartina* live biomass and %N to press fertilization differed significantly between the two marshes when averaged over the duration of the study (A). Arrows indicate the change caused by fertilization at each marsh. In contrast, fertilization increased nitrogen density (gN/m² in live *Spartina*) by the same amount at the two marshes when averaged over the duration of the study (B).

Under press fertilization, most predator densities remained elevated (Figs. 2g–j, S3b,c) and the predator/prey ratio increased throughout the experiment at both marshes (Fig. 4c,d). However, the resulting increase in predation pressure was not sufficient to depress herbivore densities, all of which were greater in the press than control during the last year of the experiment at both marshes (Figs. 2e,f, S2a,b).

Five of the nine arthropod response variables that were measured at both marshes differed significantly between marshes, and all five responded more strongly at TUCK than at CHNS, supporting our prediction of stronger responses at the high-latitude marsh, possibly due to more palatable grass. Those five responses were densities of *Prokelisia* planthoppers ($F_{1,2.2} = 9.64$, $P = 0.0051$) (Fig. 2e,f), *Delphacodes* planthoppers ($F_{1,20.4} = 33.26$, $P < 0.0001$) (Fig. S2a), *Tytthus* egg predators ($F_{1,44.1} = 24.78$, $P < 0.0001$) (Fig. 2g,h), *Pardosa* wolf spiders ($F_{1,20.4} = 35.13$, $P < 0.0001$) (Fig. 2i,j), and herbivore load ($F_{1,9.92} = 9.25$, $P = 0.013$) (Fig. 4a,b).

Although herbivore load was significantly higher at TUCK than at CHNS, fertilization had a relatively weak effect on herbivore load even at TUCK ($t_{15.1} = 1.76$, $P = 0.099$) (Fig. 4a,b), indicating that herbivore density increased by roughly the same percentage

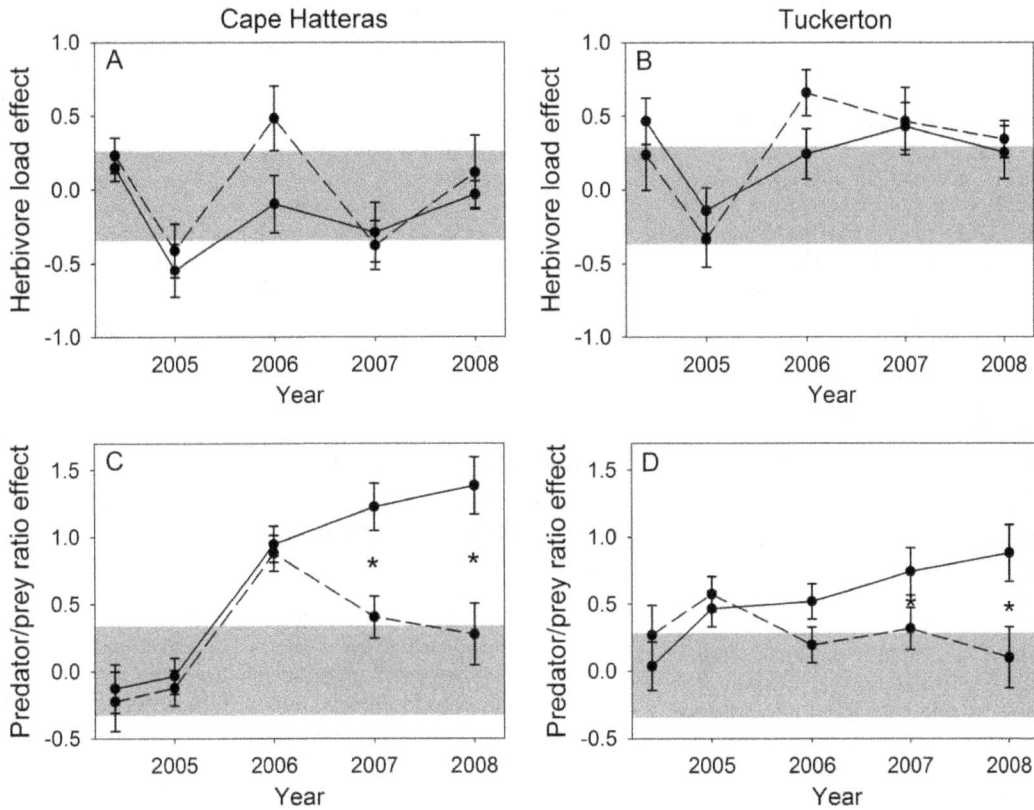

Figure 4. Response of herbivore load and predator/prey ratio to nutrient manipulations. Effects of press (solid line) and pulse (dashed line) treatments on herbivore load (A, B) and predator/prey ratio (C, D) at the CHNS and TUCK (means ±se). Effects within gray bars were not significantly different from zero, meaning that treatment and control plots did not differ. Herbivores consisted of all planthopper adults and nymphs plus *Trigonotylus*; predators consisted of all spiders. Asterisks indicate that pulse and press treatments had significantly different effects in a given year (α = 0.05).

as *Spartina* biomass in response to fertilization. The fraction of *Spartina* tissue damaged by katydids increased with fertilization ($P < 0.0001$; Fig. S4a), and damage in the press treatment was greater than the pulse treatment ($P = 0.02$; Fig. S4a). In contrast, fertilization did not reliably increase damage by *Trigonotylus*; pulse plots had significantly higher levels of damage than controls ($t_{22.9} = 3.56$, $P = 0.0017$), but press plots did not ($t_{22.9} = 0.03$, $P = 0.98$) (Fig. S4b).

Discussion

As in earlier studies of short-term nutrient pulses [14], we found that plants and arthropods on both marshes responded positively to a nutrient pulse during the first year and then gradually returned to control levels over the next four years. When nutrient addition was maintained over multiple years as a nutrient press, many responses simply maintained the level reached during the first year of fertilization, but others exhibited more complex, extended responses. As predicted, the responses by higher trophic levels to the extended fertilization of a nutrient press meant that results from a pulse study could not have predicted food web responses to a nutrient press. One key food web characteristic, the predator-to-prey ratio, continued to increase in press plots over the course of the experiment at both marshes (Fig. 4c,d). Thatch biomass, which has important effects on predator interference and prey suppression [45], showed the opposite trend. Thatch peaked during the second year of fertilization and then declined at both marshes even under continued fertilization (Fig. S1e). The fact

that these extended responses occurred in a natural monoculture shows that long-term enrichment can affect predator-prey interactions without the mediation of plant species turnover, and that long-term experiments are valuable in determining extended effects.

Predator densities generally responded more robustly to enrichment than herbivore densities, increasing the predator/prey ratio in agreement with theory and earlier pulse fertilization experiments [13,15,56]. However, the continued rise of the predator/prey ratio in press treatments relative to controls contrasts with earlier pulse studies and may have been caused by slow-developing predator responses, including reproductive responses, and their response to increased thatch and *Spartina* biomass. Web-building spiders in particular may have benefitted from more space for web construction. Marsh predators, especially *Tytthus* and *Pardosa*, are capable of suppressing herbivore densities [45], and our predictions included the possibility that increased predator abundance would lead to over-exploitation of prey and subsequent decline in predator densities (Fig. 1b,e). However, it appears that improved *Spartina* quantity and quality allowed herbivore abundance to increase despite increased predation pressure (Figs. 2e,f, S2a,b). The decrease in thatch biomass in press plots relative to controls during years 3 and 4, after a peak in year 2 (Fig. S1e) may have been caused by a delayed response of decomposers to increased thatch quality [57]. Because higher levels of thatch decrease cannibalism, this trend toward lower thatch levels may eventually lead to an increase in intraguild predation and cannibalism, potentially reducing herbivore suppression and increasing herbivore damage. Such feedback loops highlight the importance of conducting nutrient press studies because the effects of nutrient addition on higher trophic levels, and thereby trophic cascades, may take years to be fully manifested.

In addition to differences between nutrient pulses and presses, the contrast in how *Spartina* responded to a nutrient press at the two marshes was striking. At TUCK, plant quality (%N) increased with a minimal increase in biomass, while at CHNS biomass increased with only a small increase in quality (Fig. 3). Our results demonstrate that vegetation at different sites can respond to the same degree of enrichment in very different ways, perhaps due to a tradeoff between increases in plant quality and biomass. We found that the arthropod food web had a greater response to nutrient subsidies at TUCK than at CHNS, perhaps because consumers responded more readily to increases in plant quality than biomass. The modest increase in plant quality at CHNS may explain why herbivores took longer to respond than at TUCK and why there was a much weaker effect on higher trophic levels. Our prediction that detritus (thatch) would not accumulate until the second year of the experiment, was supported at both sites (Fig. S1e). However, detritivores did not always track the detrital signal. *Orchestia* demonstrated positive, negative, or neutral responses to fertilization depending on year (Fig. S2c). *Venezillo* responded positively to fertilization, with a stronger response to the nutrient press (Fig. S2d). Although the overall response of *Venezillo* met predictions, densities were not lagged as predicted. Both of these species have been recorded in the literature as feeding on *Spartina* detritus [58,59], but may also feed heavily on live *Spartina* or algae [60,61], which may explain why their responses did not correspond with our predictions for 'true' detritivore species.

Herbivore responses varied spatially between marshes; the magnitude and duration of planthopper responses were both greater at TUCK (Figs. 2e,f, S2a), but the herbivore *Trigonotylus* had a greater response to the press treatment at CHNS (Fig. S2b). Although herbivore densities increased in pulse and press plots as predicted, our prediction that we would also observe greater levels of herbivore damage on *Spartina* plants in those plots was mixed. Katydid densities and damage were greater in pulse and press plots relative to controls (Figs. S3a, S4a), but greater *Trigonotylus* densities did not translate into increased damage in pulse and press treatments (Figs. S2b, S4b).

Our study demonstrates that nutrient subsidies can have very different impacts on the arthropod food web depending on subsidy duration. In general, plant and arthropod measures returned to ambient conditions within 3–4 years after a nutrient pulse, but remained elevated during the entirety of a nutrient press. Several observed responses displayed consistent trends as the press continued, including increases in the predator/prey ratio, densities of *Pardosa* and web-building spiders, and a decrease in thatch biomass after the spike in year 2. These long-term trends imply that the ultimate effect of enrichment on these marshes remains unknown. Nutrient pulses and presses can also have very distinct responses in different salt marshes; in the high-latitude marsh, plants responded by increasing plant quality while in the low-latitude marsh, plants increased biomass. Arthropod response was more consistent between marshes, but the magnitude of response was much greater at TUCK, the high-latitude marsh, perhaps due to individual taxa responses to higher plant quality (%N) at that marsh.

Recently there has been a call for long-term studies on resource pulses and their direct and indirect effects on the recipient plant and arthropod communities [62,63]. Our research suggests that investigations into how resource pulses and presses differ is also necessary, especially in systems where input regimes of nutrients from anthropogenic sources into natural systems, such as salt marshes, can be highly variable. As agricultural production and nitrogen application continues to intensify, natural ecosystems will experience a long-term press in nitrogen loading from anthropogenic sources. Our study demonstrates that persistent nitrogen addition has the potential to reshape food web interactions by differentially impacting higher trophic levels. While previous nutrient press studies have primarily focused on the impacts of nutrient addition on plants and herbivores, here we demonstrate that nutrient additions may lead to feedback loops that impact prey suppression and ultimately change rates of plant production and decomposition via increasingly greater impacts on natural enemies.

Supporting Information

Figure S1 Effects of press and pulse treatments on additional *Spartina* characteristics not included in Figure 1. Effects for press (solid line) and pulse (dashed line) treatments are displayed for Cape Hatteras National Seashore, NC (panels on left), and Tuckerton, NJ (panels on right). Error bars indicate standard errors of the means. Effect means within gray bands were not significantly different from zero, meaning that treatment and control values did not differ. Asterisks indicate that pulse and press treatments had significantly different effects in a given year (alpha = 0.05). Treatment effect was calculated as ln((treatment value +1)/(control value +1)), where "treatment value" was the value from the press or pulse plot in a block, and "control value" was the value from the control plot in that same block. A) Effect of press (solid line) and pulse (dashed line) treatments on the average length of *Spartina* culms (plant height). B) Effect of press (solid line) and pulse (dashed line) treatments on the density of *Spartina* culms per square meter. Neither fertilization treatment had a consistent affect on culm density at either marsh. C) Effect of press (solid line) and pulse (dashed line) treatments on the number of *Spartina* tillers per square meter. Neither fertilization treatment significantly affected tiller density at either marsh. D) Effect of press (solid line) and pulse (dashed line) treatments on grams of nitrogen per square meter of marsh surface in live *Spartina* biomass. E) Effect of press (solid line) and pulse (dashed line) treatments on grams of *Spartina* thatch per square meter.

Figure S2 Effects of press and pulse treatments on additional herbivores and algivores not included in Figure 1. Effects for press (solid line) and pulse (dashed line) treatments are displayed for Cape Hatteras National Seashore, NC (panels on left), and Tuckerton, NJ (panels on right). Error bars indicate standard errors of the means. Effect means within gray bands were not significantly different from zero, meaning that treatment and control values did not differ. Asterisks indicate that pulse and press treatments had significantly different effects in a given year (alpha = 0.05). Treatment effect was calculated as ln((treatment value +1)/(control value +1)), where "treatment value" was the value from the press or pulse plot in a block, and "control value" was the value from the control plot in that same block. A) Effect of press (solid line) and pulse (dashed line) treatments on the density of the planthopper *Delphacodes penedetecta*. The pre-treatment press effect at Cape Hatteras was not significant despite its high mean because of high variance among blocks. B) Effect of press (solid line) and pulse (dashed line) treatments on the density of the mirid herbivore *Trigonotylus uhleri*. C) Effect of press (solid line) and pulse (dashed line) treatments on the density of the amphipod *Orchestia grillus*. Very few amphipods were collected at the Tuckerton marsh. D) Effect of press (solid line) and pulse (dashed line) treatments on the density of the isopod *Venezillo parvus*. Virtually no isopods were collected at Cape Hatteras.

Figure S3 Effects of press and pulse treatments on additional omnivore and predator densities not included in Figure 1. Effects for press (solid line) and pulse (dashed line) treatments are displayed for Cape Hatteras National Seashore, NC (panels on left), and Tuckerton, NJ (panels on right). Error bars indicate standard errors of the means. Effect means within gray bands were not significantly different from zero, meaning that treatment and control values did not differ. Asterisks indicate that pulse and press treatments had significantly different effects in a given year (alpha = 0.05). Treatment effect was calculated as ln((treatment value +1)/(control value +1)), where "treatment value" was the value from the press or pulse plot in a block, and "control value" was the value from the control plot in that same block. A) Effect of press (solid line) and pulse (dashed line) treatments on densities of katydids per square meter. Katydids belonged to the genera *Conocephalus* and *Orchelimum*. Katydid densities at Tuckerton were too low to calculate reliable treatment effects. B) Effect of press (solid line) and pulse (dashed line) treatments on densities of web-building spiders per square meter. The most common families of web-building spiders at both marshes were Linyphiidae and Dictynidae. C) Effect of press (solid line) and pulse (dashed line) treatments on densities of hunting spiders other than *Pardosa* per square meter. Effects of fertilization on *Pardosa* density are displayed in figures 1I and 1J. During the pre-treatment collection, hunting spiders were found in only three blocks at Tuckerton.

Figure S4 Effects of fertilization treatment on degree of herbivore damage to *Spartina* plants at CHNS. A) Fraction of leaf with damage caused by katydids (*Conocephalus spartinae* and *Orchelimum fidicinium*). B) Fraction of leaf with damage caused by the herbivorous mirid *Trigonotylus uhleri*.

Table S1 Compilation of dates that experimental manipulations were initiated or maintained and that plant and arthropod samples were collected. At both of our field sites (Tuckerton, NJ [TUCK] and Cape Hatteras National Seashore, NC [CHNS]), fertilization treatments were initiated in 2005 and maintained from 2006–2008. Below we list the dates that we fertilized treatment plots and collected samples to measure plant biomass, plant %N (both within and 1 m outside study plots) and arthropod abundance.

Acknowledgments

Our co-author, Bob Denno, unfortunately passed away before this article could be published. His contributions to the conceptualization and early stages of this work were critical to its success. We miss him dearly. We thank C. Finke, M. Douglas, B. Crawford, E. Parilla, D. McCaskill, R. Pearson, H. Martinson and J. Hines for help in the field and lab, everyone at TUCK (especially K. Able, Rutgers University Marine Station) and CHNS (J. Ebert, B. Commins, M. Lyons, T. Broili, M. Carfioli, S. Strickland) for facilitating permits. We thank the DC PIG (a plant–insect discussion group), Steve Pennings and an anonymous reviewer for helpful comments on earlier drafts of this manuscript.

Author Contributions

Conceived and designed the experiments: SM RD. Performed the experiments: SM GW DL RD. Analyzed the data: DL. Contributed reagents/materials/analysis tools: SM GW DL RD. Wrote the paper: SM GW DL.

References

1. Robinson D (1994) The responses of plants and their roots to non-uniform supplies of nutrients. New Phytologist 127: 635–647.
2. Inouye R, Tilman D (1995) Convergence and divergence of old-field vegetation after 11 years of nitrogen addition. Ecology 76: 1872–1887.
3. Siemann E (1998) Experimental tests of effects of plant productivity and diversity on grassland arthropod diversity. Ecology 79: 2057–2070.
4. Haddad NM, Haarstad J, Tilman D (2000) The effects of long-term nitrogen loading on grassland insect communities. Oecologia 124: 73–84.
5. Polis GA, Hurd SD (1996) Linking marine and terrestrial food webs: allochthonous input from the ocean supports high secondary productivity on small islands and coastal land communities. American Naturalist 147: 396–423.
6. Polis GA, Anderson WB, Holt RD (1997) Toward an integration of landscape and food web ecology: the dynamics of spatially subsidized food webs. Annual Review of Ecology and Systematics 28: 289–316.
7. Polis GA, Hurd SD, Jackson CT, Sanchez-Pinero F (1998) Multifactor population limitation: variable spatial and temporal control of spiders on Gulf of California islands. Ecology 79: 490–502.
8. Huxel GR, McCann K (1998) Food web stability: the influence of trophic flows across habitats. American Naturalist 152: 460–469.
9. Ostfeld RS, Keesing F (2000) Pulsed resources and community dynamics of consumers in terrestrial ecosystems. Trends in Ecology and Evolution 15: 232–237.
10. Hurd LE, Wolf LL (1974) Stability in relation to nutrient enrichment in arthropod consumers of old-field successional ecosystems. Ecological Monographs 44: 465–482.
11. Bakelaar GR, Odum EP (1978) Community and population level responses to fertilization in an old-field ecosystem. Ecology 59: 660–665.
12. Strauss S (1987) Direct and indirect effects of host-plant fertilization on an insect community. Ecology 68: 1670–1678.
13. Denno RF, Gratton C, Peterson MA, Langellotto GA, Finke DL, et al. (2002) Bottom-up forces mediate natural-enemy impact in a phytophagous insect community. Ecology 83: 1443–1458.
14. Gratton C, Denno RF (2003) Inter-year carryover effects of a nutrient pulse on Spartina plants, herbivores, and natural enemies. Ecology 84: 2692–2707.
15. Stiling P, Rossi AM (1997) Experimental manipulations of top-down and bottom-up factors in a tri-trophic system. Ecology 78: 1602–1606.
16. Yang LH (2004) Periodical cicadas as resource pulses in North American forests. Science 306: 1565–1567.
17. McFarlin CR, Brewer JS, Buck TL, Pennings SC (2008) Impact of fertilization on a salt marsh food web in Georgia. Estuaries and Coasts 31: 313–325.
18. Wimp GM, Murphy SM, Finke DL, Huberty AF, Denno RF (2010) Increased primary production shifts the structure and composition of a terrestrial arthropod community. Ecology 91: 3303–3311.
19. Kirchner TB (1977) The effects of resource enrichment on the diversity of plants and arthropods in a shortgrass prairie. Ecology 58: 1334–1344.
20. Gruner DS, Taylor AD (2006) Richness and species composition of arboreal arthropods affected by nutrients and predators: a press experiment. Oecologia 147: 714–724.
21. Hoekman D, Dreyer J, Jackson RD, Townsend PA, Gratton C (2011) Lake to land subsidies: Experimental addition of aquatic insects increases terrestrial arthropod densities. Ecology 92: 2063–2072.
22. Valiela I, Teal JM, Persson NY (1976) Production and dynamics of experimentally enriched salt marsh vegetation: belowground biomass. Limnology and Oceanography 21: 245–252.
23. Crawley MJ, Johnston AE, Silvertown J, Dodd M, de Mazancourt C, et al. (2005) Determinants of species richness in the park grass experiment. American Naturalist 165: 179–192.
24. Gough L, Osenberg CW, Gross KL, Collins SL (2000) Fertilization effects on species density and primary productivity in herbaceous plant communities. Oikos 89: 428–439.
25. Pennings SC, Clark CM, Cleland EE, Collins SL, Gough L, et al. (2005) Do individual plant species show predictable responses to nitrogen addition across multiple experiments? Oikos 110: 547–555.
26. Suding KN, Collins SL, Gough L, Clark C, Cleland EE, et al. (2005) Functional- and abundance-based mechanisms explain diversity loss due to N fertilization. Proceedings of the National Academy of Sciences 102: 4387–4392.
27. Clark CM, Cleland EE, Collins SL, Fargione JE, Gough L, et al. (2007) Environmental and plant community determinants of species loss following nitrogen enrichment. Ecology Letters 10: 596–607.
28. Tilman D (1987) Secondary succession and the pattern of plant dominance along experimental nitrogen gradients. Ecological Monographs 57: 189–214.

29. Tilman D (1999) Global environmental impacts of agricultural expansion: The need for sustainable and efficient practices. Proceedings of the National Academy of Sciences 96: 5995–6000.

30. Bertness MD, Ewanchuk PJ, Silliman BR (2002) Anthropogenic modification of New England salt marsh landscapes. Proceedings of the National Academy of Sciences 99: 1395–1398.

31. Bertness MD, Silliman BR, Jefferies R (2004) Salt marshes under siege. American Scientist 92: 54–61.

32. Boyer EW, Goodale CL, Jaworski NA, Howarth RW (2002) Anthropogenic nitrogen sources and relationships to riverine nitrogen export in the northeastern U.S.A. Biogeochemistry 57/58: 137–169.

33. Howarth RW, Billen G, Swaney D, Townsend A, Jaworski N, et al. (1996) Regional nitrogen budgets and riverine N & P fluxes for the drainages to the North Atlantic Ocean: natural and human influences. Biogeochemistry 35: 75–139.

34. Howarth RW, Boyer EW, Pabich WJ, Galloway JN (2002) Nitrogen use in the United States from 1961–2000 and potential future trends. Ambio 31: 88–96.

35. Mayer B, Boyer EW, Goodale CL, Jaworski NA, van Breemen N, et al. (2002) Sources of nitrate in rivers draining sixteen watersheds in the northeastern U.S.: Isotopic constraints. Biogeochemistry 57/58: 171–197.

36. Valiela I, Bowen JL (2002) Nitrogen sources to watersheds and estuaries: role of land cover mosaics and losses within watersheds. Environmental Pollution 118: 239–248.

37. Valiela I, Cole ML (2002) Comparative evidence that salt marshes and mangroves may protect seagrass meadows from land-derived nitrogen loads. Ecosystems 5: 92–102.

38. Valiela I, Teal JM (1979) The nitrogen budget of a salt marsh ecosystem. Nature 280: 652–656.

39. Mendelssohn IA (1979) Nitrogen metabolism in the height forms of *Spartina alterniflora* in North Carolina. Ecology 60: 574–584.

40. Mendelssohn IA (1979) The influence of nitrogen level, form, and application method on the growth response of *Spartina alterniflora* in North Carolina. Estuaries 2: 106–112.

41. Cole ML, Valiela I, Kroeger KD, Tomasky GL, Cebrian J, et al. (2004) Assessment of a ∂^{15}N isotopic method to indicate anthropogenic eutrophication in aquatic ecosystems. Journal of Environmental Quality 33: 124–132.

42. Carmichael RH, Annett B, Valiela I (2004) Nitrogen loading to Pleasant Bay, Cape Cod: application of models and stable isotopes to detect incipient nutrient enrichment of estuaries. Marine Pollution Bulletin 48: 137–143.

43. Vitousek PM, Mooney HA, Lubchenco J, Melillo JM (1997) Human domination of Earth's ecosystems. Science 277: 494–499.

44. Denno RF, Gratton C, Döbel H, Finke DL (2003) Predation risk affects relative strength of top-down and bottom-up impacts on insect herbivores. Ecology 84: 1032–1044.

45. Finke DL, Denno RF (2002) Intraguild predation diminished in complex-structured vegetation: implications for prey suppression. Ecology 83: 643–652.

46. Turner RE (1976) Geographic variations in salt marsh macrophyte production: a review. Contributions in Marine Science 20: 47–68.

47. Pennings SC, Siska EL, Bertness MD (2001) Latitudinal differences in plant palatability in Atlantic coast salt marshes. Ecology 82: 1344–1359.

48. McCall BD, Pennings SC (In Press) Geographic variation in salt marsh structure and function. Oecologia.

49. Valiela I, Teal JM, Deuser WG (1978) The nature of growth forms in the salt marsh grass *Spartina alterniflora*. American Naturalist 112: 461–470.

50. Döbel HG, Denno RF (1994) Predator planthopper interactions. In: Denno RF, Perfect TJ, editors. Planthoppers: Their Ecology and Management. New York, NY: Chapman and Hall. 325–399.

51. Matsumura M, Trafelet-Smith GM, Gratton C, Finke DL, Fagan WF, et al. (2004) Does intraguild predation enhance predator performance? A stoichiometric perspective. Ecology 89: 2601–2615.

52. Silvanima JVC, Strong DR (1991) Is host-plant quality responsible for the populational pulses of salt-marsh planthoppers (Homoptera, Delphacidae) in northwestern Florida. Ecological Entomology 16: 221–232.

53. Ruxton GD, Rey D, Neuhauser M (2010) Comparing samples with large numbers of zeros. Animal Behaviour 80: 937–940.

54. Anderson MJ, ter Braak CJF (2003) Permutation tests for multi-factorial analysis of variance. Journal of Statistical Computation and Simulation 73: 85–113.

55. Murphy SM, Wimp GM, Lewis D, Denno RF (2012) Data from: Nutrient presses and pulses differentially impact plants, herbivores, detritivores and their natural enemies. Dryad Digital Repository. http://dx.doi.org/10.5061/dryad.fb006.

56. Hunter MD, Price PW (1992) Playing chutes and ladders: heterogeneity and the relative roles of bottom-up and top-down forces in natural communities. Ecology 73: 724–732.

57. Melillo JM, Aber JD, Muratore JF (1982) Nitrogen and lignin control of hardwood leaf litter decomposition dynamics. Ecology 63: 621–626.

58. Agnew AM, Shull DH, Buchsbaum R (2003) Growth of a salt marsh invertebrate on several species of marsh grass detritus. Biological Bulletin 205: 238–239.

59. Zimmer M, Pennings SC, Buck TL, Carefoot TH (2004) Marsh litter and detritivores: a closer look at redundancy. Estuaries 27: 753–769.

60. Galvan K (2008) The diet of saltmarsh consumers [PhD Dissertation]: Louisiana State University. 238 p.

61. Wimp GM, Murphy SM, Lewis D, Douglas MR, Ambikapathi R, et al. (In Press) Predator hunting mode influences patterns of prey use from grazing and epigeic food webs. Oecologia.

62. Yang LH, Bastow JL, Spence KO, Wright AN (2008) What can we learn from resource pulses? Ecology 89: 621–634.

63. Nowlin WH, Vanni MJ, Yang LH (2008) Comparing resource pulses in aquatic and terrestrial ecosystems. Ecology 89: 647–659.

11

Grassland Resistance and Resilience after Drought Depends on Management Intensity and Species Richness

Anja Vogel[1]*, Michael Scherer-Lorenzen[2], Alexandra Weigelt[3]

1 Institute of Ecology, Friedrich Schiller University Jena, Jena, Germany, **2** Michael Scherer-Lorenzen, Faculty of Biology – Geobotany, University of Freiburg, Freiburg, Germany, **3** Alexandra Weigelt, Institute of Biology, University of Leipzig, Leipzig, Germany

Abstract

The degree to which biodiversity may promote the stability of grasslands in the light of climatic variability, such as prolonged summer drought, has attracted considerable interest. Studies so far yielded inconsistent results and in addition, the effect of different grassland management practices on their response to drought remains an open question. We experimentally combined the manipulation of prolonged summer drought (sheltered vs. unsheltered sites), plant species loss (6 levels of 60 down to 1 species) and management intensity (4 levels varying in mowing frequency and amount of fertilizer application). Stability was measured as resistance and resilience of aboveground biomass production in grasslands against decreased summer precipitation, where resistance is the difference between drought treatments directly after drought induction and resilience is the difference between drought treatments in spring of the following year. We hypothesized that (i) management intensification amplifies biomass decrease under drought, (ii) resistance decreases with increasing species richness and with management intensification and (iii) resilience increases with increasing species richness and with management intensification.

We found that resistance and resilience of grasslands to summer drought are highly dependent on management intensity and partly on species richness. Frequent mowing reduced the resistance of grasslands against drought and increasing species richness decreased resistance in one of our two study years. Resilience was positively related to species richness only under the highest management treatment. We conclude that low mowing frequency is more important for high resistance against drought than species richness. Nevertheless, species richness increased aboveground productivity in all management treatments both under drought and ambient conditions and should therefore be maintained under future climates.

Editor: Jon Moen, Umea University, Sweden

Funding: The University of Jena funded the drought experiment within the ProChance program (www.uni-jena.de, grant to AW) and the University of Zurich gave additional support (www.uzh.de). The management experiment was supported by the Friedrich Schiller University of Jena and the ETH Zurich (funds from N. Buchmann). The Jena Experiment was funded by the Deutsche Forschungsgemeinschaft (FOR 456,www.dfg.de) with additional support from the Friedrich Schiller University of Jena, the Max Planck Society, the University of Zurich and the Swiss National Science Foundation (3100AQ-107531 to B. Schmid). The funders had no role in study design, data collection and analysis, decision to publish, or preparation of the manuscript.

* E-mail: anja.vogel@uni-jena.de

Introduction

There is agreement that the world's ecosystems will likely have to cope with future climatic changes, such as increased mean temperatures, a higher frequency of extreme weather events as well as changes in wind and precipitation patterns [1]. Among the different scenarios are projected decreases in summer precipitation and increases in autumn, winter and spring precipitation in subtropical and temperate regions (see also [2] that shows the same for the region of our study site). Along with the present and ongoing climate change, biodiversity is challenged by land-use changes to meet the growing demand for ecosystem services [3]. The consequences of those changes for ecosystem functioning, ecosystem services and human wellbeing have been the focus of research in the last few decades. It has been found that plant species diversity can have positive effects on multiple ecosystem processes [4] and if many times, places, functions and environmental changes were considered [5]. Aboveground plant biomass production, the most-studied process in biodiversity research, has

been consistently found to rise in response to plant diversity in grasslands [6,7,8,9,10], an important finding for agricultural management. This positive relationship of species richness and productivity even holds under nutrient-rich conditions [11,12] and perturbations such as intense livestock grazing [13].

Beyond productivity itself, the temporal stability of biomass production has also been found to relate positively with species richness [14,15,16,17], meaning that temporal variability in productivity is lower in species-rich compared to species-poor communities. Besides the consistent results of diversity effects on temporal stability, the relationship between diversity and other aspects of stability, like resistance and resilience after perturbations (such as climatic changes) are mixed. Resistance, the degree of change after perturbations [18], is usually calculated as the difference of some performance measure between perturbed and unperturbed conditions and reflects the extent to which the mean of an ecosystem property changes after a single perturbation event. Resilience (in the sense of engineering resilience [19]), the time after perturbation until pre-perturbation levels are regained [18], is usually expressed as the rate of return of a variable at a given time after perturbation [20,21,22]. Both, resistance and resilience after perturbations were expected to increase with species richness [14]. In experimental studies it has been found to be true for resistance of biomass production against parasitism [23] and herbivore attack [24], but also neutral relationships between herbivore resistance and species richness have been reported [13,25]. The experimental results on resistance of grassland biomass production against drought show mostly negative or neutral relationships with species richness [21,22,26], depending on the level of productivity. It has been documented that more productive grasslands have a lower ability to withstand perturbation (lower resistance) than less productive communities, indicated by the higher loss of absolute aboveground biomass after a drought perturbation [22]. Thus, depending on the slope of the biodiversity-productivity relationship under unperturbed conditions, resistance either decreases with increasing species richness (negative slope) [21,22], increases with species richness (positive slope, which was only documented for a grassland experiment, where species richness was not independently manipulated) [26] or does not change (no significant relationship), when species richness has no effect on productivity like in the studies by Kahmen et al. [27] and Wang et al. [28]. It can thus be hypothesized that high species richness has a positive effect on the productivity of ecosystems but also results in a lower ability to resist future climatic changes. Besides the important ecosystem property to resist perturbations, a fast return to preperturbation levels (high resilience) would help to maintain ecosystem functioning. In very early studies it had been suggested that resistance and resilience are inversely correlated [29]. MacGillivray et al. [30] supported this prediction by measuring drought resistance and resilience in semi-natural grassland communities. It is also consistent with the work by van Ruijven et al. [22] who reported about experimental grasslands with decreasing resistance with species richness and increasing resilience. Although they confirmed that resistance rather depends on initial productivity of grasslands than on species richness per se (proportional resistance did not decrease with species richness), they found species richness to be a stronger predictor for recovery than productivity (recovery and proportional recovery increased with species richness).

Despite these findings, it remains unclear whether the idea that species richness promotes the productivity of grasslands at the consequence of lower resistance but higher resilience against perturbation, is applicable to a broad range of grasslands. For example, Gilgen and Buchmann [31] found that drought effects on biomass production vary between different grassland types along an altitudinal gradient. However, the explanatory factors of that study, namely environmental conditions and management strategies, were not independent and their individual effects could thus not be separated. In managed grasslands, management intensity has profound consequences on species richness and productivity. Fertilized grasslands, for example, have low species richness but a high productivity due to the high resource availability. Because of a low resistance with high productivity, one would expect that fertilization decreases the resistance of grasslands to perturbation. Indeed, Grime et al. [32] suggested that fertile grasslands would be less resistant compared to extensively managed grasslands because they contain more species with a high relative growth rate, which are highly susceptible to drought stress. However, higher growth rates should lead to faster regrowth after perturbation and we would expect a high resilience in fertilized grasslands.

We experimentally manipulated plant species richness, management intensity (amount of fertilizer and mowing frequencies, Figure 1) and prolonged summer drought separately to investigate the effects of biodiversity on resistance and resilience against drought perturbation in grasslands differing in their management intensity. We hypothesize that (i) biomass decreases under drought perturbation, especially under high management intensity due to the higher productivity under fertilization, (ii) resistance decreases with increasing species richness and with management intensification but proportional resistance (measure of resistance that is corrected for initial biomass productivity) does not change with species richness and (iii) resilience as well as proportional resilience increases with increasing species richness as well as with management intensity.

Results

The harvest at the end of the induced drought period showed significant increases in aboveground biomass with sown species richness in both years (Table 1, Figure 2). This was true for all management and drought treatments. Aboveground biomass was significantly lower under drought only in the frequently mown grasslands (M4F100, M4F200, Figure 2), i.e., the drought response of grasslands was affected by management intensity (Table 1).

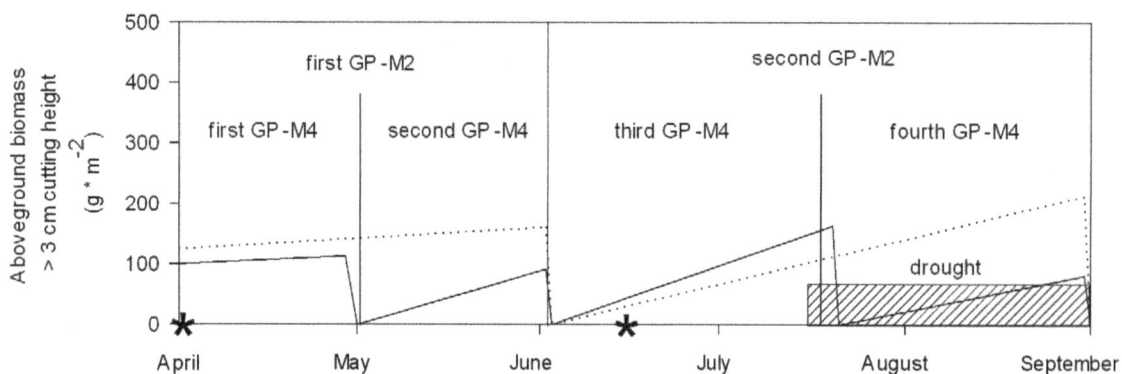

Figure 1. Time course of the growing season, including management and drought interventions of our study system. Stars indicate fertilization dates, vertical lines show mowing dates and dashed area represents our drought period. Frequently mown grassland types (M4, solid line) had four cuts per year and therefore four growth periods (first to fourth GP) previous to every cut. Normal mown grassland types (M2, dotted line) had two cuts per year and therefore two growth periods. Aboveground biomass is plotted as mean standing biomass above cutting height for each mowing grassland type.

Table 1. Summary of mixed effects models for aboveground biomass in August 2008 and 2009 to test for effects of diversity (numbers of initially sown species and functional group richness), management and drought treatments.

	Biomass August 2008				Biomass August 2009			
	df	AIC	L ratio	p	AIC	L ratio	p	
Nullmodel	11	1882.024			2058.232			
Block	14	1885.305	2.719	0.4370	2061.191	3.041	0.3854	
Species Richness (log-scale) = SR	15	1843.338	43.967	<0.0001 ***	2026.597	36.595	<0.0001	***
Number of Functional Groups = FG	16	1845.331	0.007	0.9336	2026.316	2.280	0.1310	
Management	19	1815.300	36.031	<0.0001 ***	1941.106	91.210	<0.0001	***
Drought	20	1789.755	27.545	<0.0001 ***	1927.831	15.275	0.0001	***
Management×Drought	23	1786.773	8.982	0.0295 *	1930.126	3.705	0.2951	
Management×SR	26	1789.410	3.363	0.3390	1935.555	0.572	0.9029	
Management×FG	29	1793.380	2.030	0.5662	1937.716	3.838	0.2795	
Drought×SR	30	1793.947	1.433	0.2313	1939.645	0.072	0.7890	
Drought×FG	31	1794.484	1.463	0.2264	1941.613	0.032	0.8579	
SR×FG	32	1795.201	1.283	0.2574	1942.251	1.362	0.2431	

Models were fitted by stepwise inclusion of variables and likelihood ratio tests (L ratio) were applied to assess statistical significance of variables (p-values). Significance is given with * = $p<0.05$. ** = $p<0.01$. *** = $p<0.001$; df = degrees of freedom.

Resistance slightly decreased with increasing species richness in 2008 (Table 2, Figure 3 left) and with increasing mowing frequency in both years (Table 2, Figure 3). Proportional resistance was not affected by species richness but still decreased with increasing mowing frequency in 2008 (Table 2, Figure 3). We found no effect of the interaction of species diversity and management treatment (mowing frequency or amount of fertilizer) on resistance or proportional resistance. Resilience was positively related to species richness only in the most intensively managed grassland (significant interaction of SR and F in Table 3, Figure 4). Proportional resilience was not affected by species richness (Table 3).

Discussion

In our study we independently manipulated biodiversity loss, management intensity and drought. We were therefore able to distinguish the effects of all single treatments from their interactions on aboveground biomass. We found that the response of experimental grasslands to drought depends on management intensity. Aboveground biomass decreased after induced summer drought only in grasslands with frequent mowing (four times per year), not in grasslands with only two cuts per year. Differences in growth status due to mowing may explain our findings. The low canopy height after mowing generally increases soil surface evaporation through increased wind speed at ground level and low plant cover. Hu et al. [33] reported increased evaporation with decreased canopy density, measured as leaf area index (LAI), especially at LAI values lower than 2 m^2 * m^{-2}. After mowing, LAI in our experiment was close to 0 m^2 * m^{-2} (data not shown). In addition to reduced precipitation this would mean a further decrease of soil moisture in the frequently mown (M4-) treatment compared to the normal mown (M2-) treatment. Consequently, a decrease of aboveground biomass only occurs if drought hits the communities at an early growth status, when soil is not sufficiently covered by plants. Thus, grasslands with high mowing frequency and hence frequently low LAI have to be considered as more sensitive to drought. In contrast, extensively mown grasslands would only suffer from drought if it occurs in the regrowth phase after mowing. This was not the case in both of the study years for

our grasslands mown only twice a year (M2). They were well-advanced in height growth at the beginning of the drought treatment. The same might have been the case in the study of Jentsch et al. [34], where no effect on aboveground productivity could be detected when drought was induced during peak growing season in June [34]. In contrast Pfisterer and Schmid [21] found negative drought effects in their grassland with the same management as in our experiment (two cuts per year and drought induction before late cut). The longer drought period (and therefore shorter time for normal regrowth) compared to our experiment could be one explanation for contrasting results but also site specific (climatic conditions) and treatment differences have to be taken into account. This study site has mean annual precipitation amounts twice as high as in our site (Table 4) and rain shelters were adjusted close to vegetation, thus a stronger heat effect could be assumed.

Along with our second hypothesis we wanted to test, whether the observed relationships of resistance and species richness [22] change with management intensity. We only found slightly decreasing resistance with species richness and this effect of species richness did not change with management intensity. Instead we found a strong effect of management intensity itself on resistance because high mowing frequency decreased the resistance of grasslands against drought. High mowing frequency resulted in both lower absolute biomass accumulation (resistance) as well as lower relative biomass accumulation under drought (proportional resistance) compared to normal mown stands. Our results support the idea that species richness affects resistance due to increasing productivity with species number, and not due to number of species *per se* [22], because proportional resistance was not related to species richness. De Boeck *et al.* [35] found that more productive and species-rich communities have a higher evapotranspiration and water use efficiency compared to monocultures. They concluded that decreased aboveground biomass is one potential mechanism for saving water, because it reduces the transpirational surface of the canopy.

After the second drought in the following year, resistance was constant across the plant diversity gradient, meaning that there

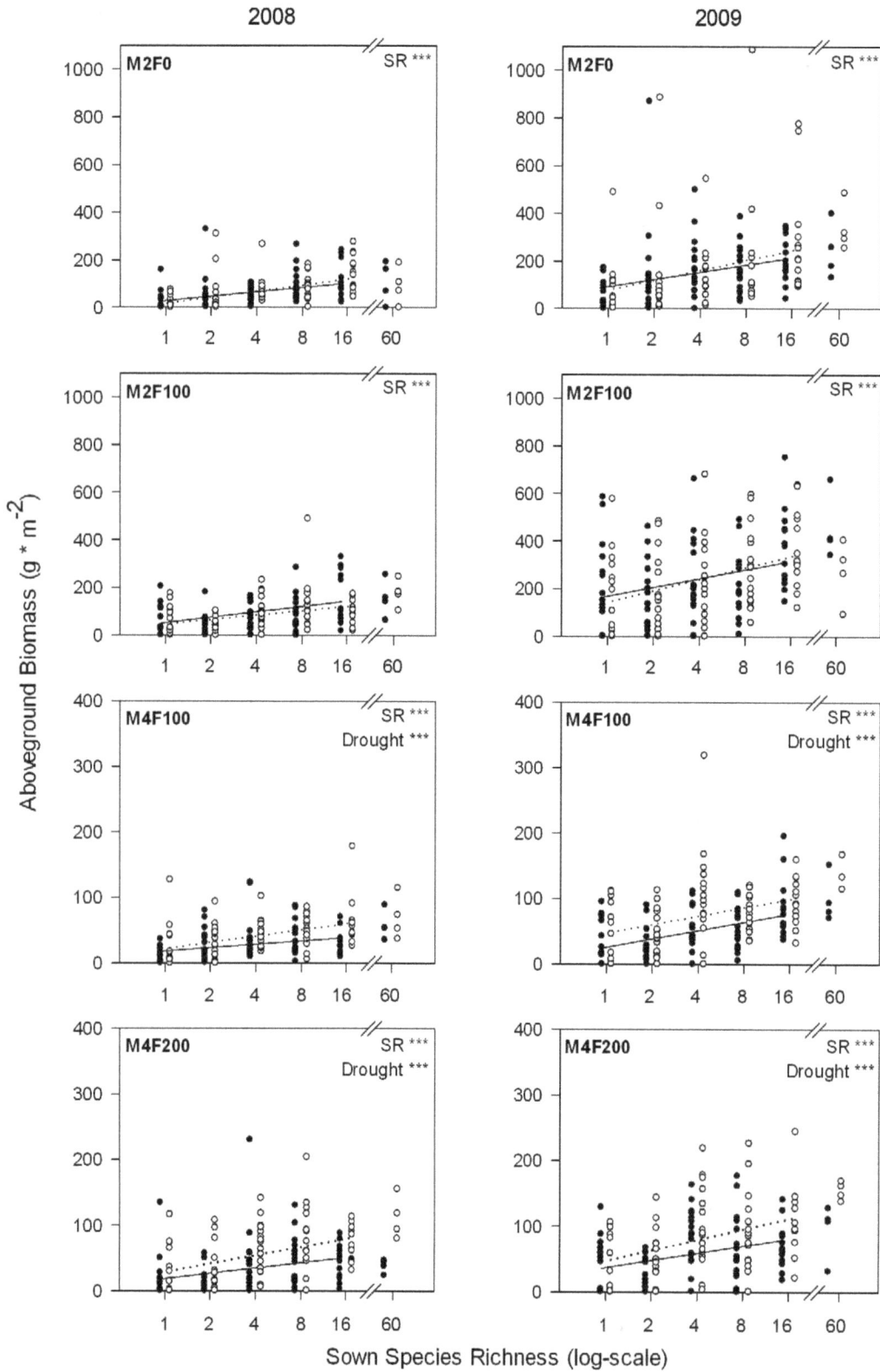

Figure 2. Aboveground biomass across sown species richness gradient for each management treatment. Aboveground biomass at the end of the induced drought period in August 2008 (left column) and 2009 (right column) measured regrowth since the last cut (M2-types: June; M4-types: July). The ambient treatment is given in open circles (dotted regression line) and drought treatment in closed circles (solid regression line). Significant effects obtained from mixed models for every single management treatment per year: SR = effect of sown species richness (linear), drought = difference in drought and ambient treatments, * = p<0.05, ** = p<0.01, *** = p<0.001.

Table 2. Summary of mixed effects models for resistance and proportional resistance of aboveground biomass after drought in August 2008 and 2009 to test for effects of diversity (numbers of initially sown species and functional group richness), management (separated into mowing and fertilizer amounts) and drought treatments.

	df	Resistance 2008				Resistance 2009				df	proportional Resistance 2008				proportional Resistance 2009		
		AIC	L ratio	p		AIC	L ratio	p			AIC	L ratio	p		AIC	L ratio	p
Nullmodel	3	1026.667				1158.463				6	914.183				1025.004		
Block	6	1028.312	4.355	0.2256		1157.890	6.573	0.0868		9	917.457	2.726	0.4358		1028.257	2.746	0.4324
Species richness (log-scale) = SR	7	1025.371	4.941	0.0262	*	1157.731	2.598	0.1417		10	918.179	1.278	0.2582		1030.178	0.080	0.7778
Number of functional groups = FG	8	1025.530	1.840	0.1749		1159.167	0.563	0.4529		11	919.132	1.047	0.3062		1032.016	0.161	0.6880
Mowing = M	9	1007.267	20.263	<0.0001 ***		1145.509	15.658	0.0001 ***		12	909.735	11.397	0.0007 ***		1030.359	3.657	0.0558
M×SR	10	1007.980	1.287	0.2565		1147.369	0.141	0.7077		13	911.734	0.001	0.9757		1032.358	0.001	0.9788
M×FG	11	1008.535	1.445	0.2293		1149.119	0.250	0.6169		14	910.415	3.320	0.0685		1034.324	0.034	0.8534
Fertilizer amount = F	12	1010.209	0.325	0.5685		1150.820	0.298	0.5851		15	911.540	0.874	0.3498		1036.322	0.002	0.9626
F×SR	13	1011.601	0.608	0.4354		1151.880	0.940	0.3322		16	913.419	0.122	0.7272		1038.275	0.047	0.8290
F×FG	14	1010.700	2.901	0.0885		1152.796	1.084	0.2978		17	912.595	2.824	0.0929		1039.751	0.524	0.4689

Models were fitted by stepwise inclusion of variables and likelihood ratio tests (L ratio) were applied to assess statistical significance of variables (p-values). Significance is given with * = p<0.05, ** = p<0.01, *** = p<0.001; df = degrees of freedom.

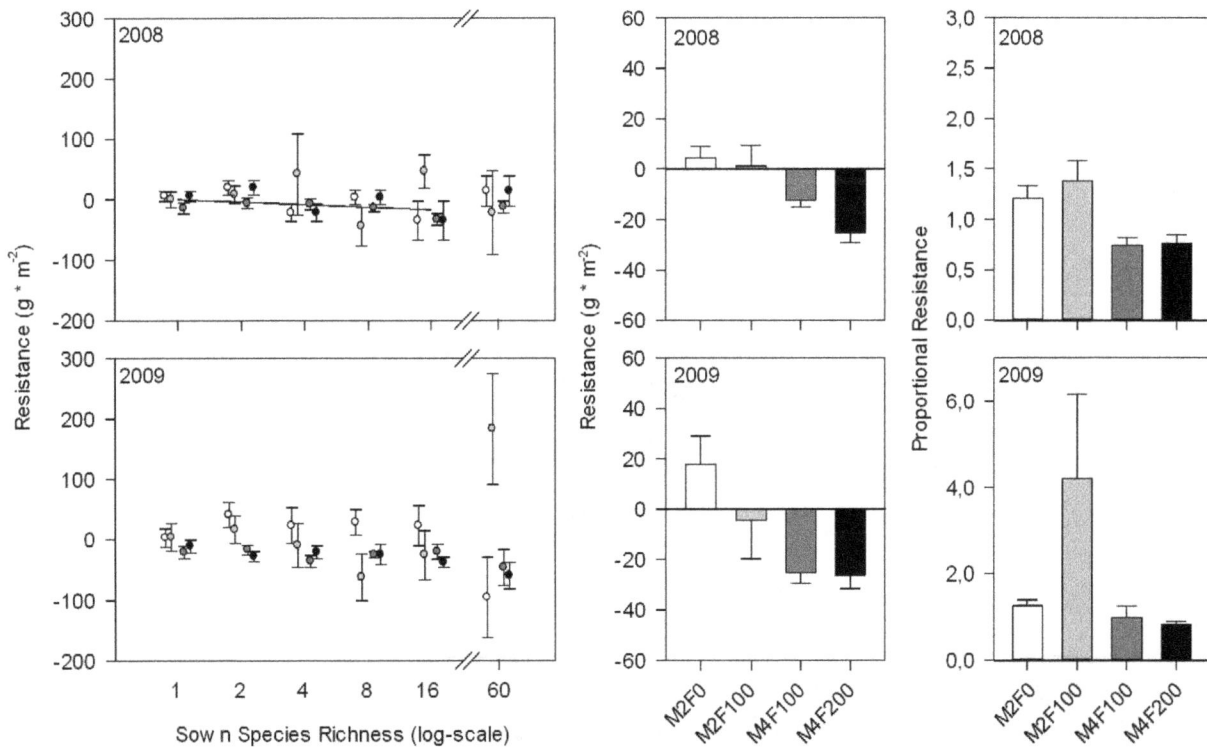

Figure 3. Resistance of biomass production for 2008 and 2009. Resistance was calculated as the difference of drought and corresponding ambient treatments at the end of the drought period in August and was plotted against species richness (left) and management treatments (middle). Regression lines are given for significant effects of species richness. Proportional resistance (ratio of drought to ambient treatment) was plotted against management treatment (right). Management treatments are shown in white (M2F0), gray (M2F100), dark gray (M4F100) and black (M4F200).

Table 3. Summary of mixed effects models for resilience and proportional resilience of aboveground biomass of the first cut in spring 2009 to test for effects of diversity (numbers of initially sown species and functional group richness), management (separated into mowing and fertilizer amounts) and drought treatments.

	df	Resilience Spring 2009				proportional Resilience 2009		
		AIC	L ratio	p		AIC	L ratio	p
Nullmodel	6	3898,242				1024,902		
Block	9	3903,177	1,065	0.7855		1026,706	4,196	0.2410
Species richness (log-scale) = SR	10	3903,771	1,406	0.2357		1026,624	2,081	0.1491
Number of functional groups = FG	11	3905,674	0,097	0.7557		1028,398	0,226	0.6345
Mowing = M	12	3905,225	2,449	0.1176		1029,770	0,628	0.4280
M×SR	13	3907,219	0,006	0.9395		1030,613	1,157	0.2821
M×FG	14	3909,208	0,011	0.9161		1032,446	0,167	0.6826
Fertilizer amount = F	15	3907,540	3,668	0.0555		1032,955	1,491	0.2221
F×SR	16	3904,600	4,939	0.0263	*	1033,658	1,297	0.2548
F×FG	17	3905,130	1,470	0.2253		1033,873	1,784	0.1816

Models were fitted by stepwise inclusion of variables and likelihood ratio tests (L ratio) were applied to assess statistical significance of variables (p-values). Significance is given with * = $p<0.05$, ** = $p<0.01$, *** = $p<0.001$; df = degrees of freedom.

was no stronger decrease in absolute amounts of aboveground biomass in the species rich compared to the species poor communities due to drought. Different precipitation patterns might explain this year-to-year changes, since rainfall just before the induced drought period was different in both years (Table 4).

The main rain events of 2008 occurred in the very wet April, whereas May and June were unusually dry. Soil moisture (volumetric water content measured in 8 cm depth of an unsheltered reference area) decreased to an average of 17.0% in summer. In contrast in 2009, rain events were regularly distributed

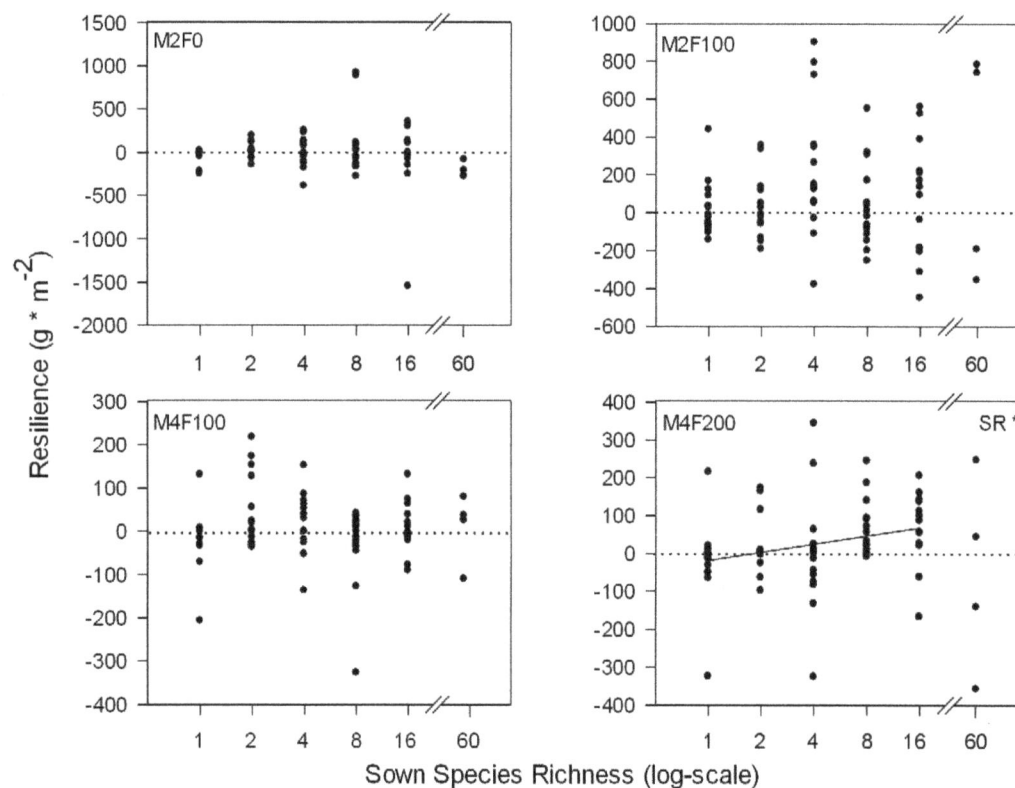

Figure 4. Resilience of biomass over sown species richness for each management treatment. Resilience of biomass was calculated as the difference of drought and corresponding ambient treatments for the first harvest in spring 2009 (M2-types: June, M4-types: April). Regression lines are given for significant effects obtained from linear models for every single management treatment per year: SR = effect of sown species richness (linear), drought = difference in drought and ambient treatments, * = $p<0.05$, ** = $p<0.01$, *** = $p<0.001$.

Table 4. Climatic parameters measured on field site during the two study years 2008 and 2009 with reference period 1961–1990 measured by the German Weather Service DWD in Jena, city center.

Month	Air temperature (°C)			Precipitation (mm)			Soil moisture (Vol%)	
	1961–90	2008	2009	1961–90	2008	2009	2008	2009
J	0.40	5.00	−3.09	37.00	24.50	9.00	37.29	22.11
F	1.40	3.76	1.15	34.00	20.40	33.70	37.15	33.31
M	4.80	5.11	5.04	43.00	55.80	42.50	37.97	37.01
A	8.60	7.90	11.58	57.00	91.80	73.70	37.35	31.39
M	13.40	14.05	13.89	62.00	22.00	62.60	25.62	31.05
J	16.70	17.13	15.01	75.00	54.40	52.90	21.74	28.58
J	18.20	18.51	18.34	52.00	40.60	85.10	17.75	31.29
A	17.40	17.90	18.59	63.00	58.60	14.60	16.61	22.34
S	14.20	12.05	14.56	42.00	50.00	53.60	21.71	23.67
O	9.80	9.13	8.42	39.00	55.30	47.30	26.30	28.70
N	5.00	5.60	8.06	41.00	19.90	68.30	28.16	34.00
D	1.70	1.28	0.64	42.00	38.60	80.00	30.59	36.22
Year	9.30	11.75	11.48	587.00	531.90	623.30	28.19	29.97

Values represent monthly means (temperature, soil moisture) or sums (precipitation).

and average soil moisture was at 25.3% in summer. A higher soil moisture at the beginning of the drought treatment in 2009 may have stimulated a better growth of the plants compared to the year before and might have more rapidly lowered the loss of soil water through evaporation. Species-rich communities are thought to be more water efficient [35,36], i.e. they produce more biomass per unit of water. Such communities could therefore benefit from higher soil moisture compared to species-poor communities. In consequence, they could be proportionally less affected from drought stress than species poor communities. Whether annual precipitation patterns would explain the different resistance-species richness patterns between years, can only be underpinned by long-term data.

Resilience increased with species richness as previously reported [22] but only under highest management intensity. In contrast, we found that species richness only affected absolute resilience, whereas the proportion of biomass increase in previously dried subplots compared to their ambient conditions (proportional resilience) did not change with species richness. Furthermore, resilience and species richness were only positively related in the very intensively managed grasslands (frequently mown and high fertilized). The positive relationship of resilience and species richness in the M4F200 management treatment was weak but significant and might be strongly due to the positive responses of the 8-species-mixtures. It is known that species richness increases shoot-root-ratios indicating a better resource use with species richness [37]. We can only speculate why this was more effective in the previously dried subplots compared to the corresponding ambient treatment under highest management intensity. It may be explained in part by higher fertilizer amounts affecting below-ground processes. It has been reported that drought did not necessarily increase root growth [38,39,40]. Together with a decreased aboveground biomass, drought may have decreased shoot-root-ratios especially in more diverse mixtures. The corresponding ambient treatments might have a much higher shoot-root-ratio due to the higher biomass and the well-known fact that fertilization decreases root growth relative to shoot growth. After drought, this lower shoot-root-ratio of the dried subplots may

be a prerequisite for better aboveground biomass allocation in comparison to the ambient in the intensively fertilized treatment. Furthermore, the plasticity and intensity of growth responses of roots and shoots under drought is highly species-specific [39,41], and thus plant species composition is an important determinant of community root and shoot growth [31]. This could explain that especially the 8-species-mixtures had a high resilience, when intensively managed.

Our results indicate that management intensity affects the resistance of grasslands after drought, with growth phenology being the underlying cause: grasslands at the regrowth stage are more sensitive to decreased precipitation and loose more biomass, than grasslands at later stages with a more fully developed canopy. As a consequence, low mowing frequency enhances drought resistance because of a lower probability to face reduced precipitation during the regrowth stage. Nevertheless, species richness and aboveground biomass were positively related even under drought conditions, which shows that biomass yield is higher the more diverse a community is, no matter under which management intensity and climatic conditions the community grew. Thus high plant species diversity should be maintained under future climates.

Materials and Methods

Study site and experimental treatments

We used the gradient of plant species richness established in the Jena biodiversity experiment and superimposed a gradient of management intensity and a drought treatment. The field site is located in the floodplain of the river Saale in Jena (Thuringia, Germany, 50°55′N, 11°35′E, 130 m above sea level) with a mean annual air temperature of 9.3°C and precipitation amount of 587 mm measured during 1961–1990. The study site was used as a highly fertilized arable field before [42]. The soils are loamy Eutric Fluvisols. In 2002, 80 grassland plots of different plant species mixtures were established from a pool of 60 mesophilic grassland species from Molinio-Arrhenateretea meadows typical for the regional alluvial plains.

The gradient in plant species richness (1, 2, 4, 8, 16 and 60 species) in the Jena experiment is combined with a gradient in the number of functional groups (1, 2, 3 or 4 functional groups namely grasses, small herbs, tall herbs and legumes) with about four replicates per species richness×functional groups combination. Mixtures were arranged in a randomized block design to account for edaphic variations with increasing distance to the river Saale. Experimental plots were maintained by weeding blockwise in two annual weeding campaigns. For further details see [42].

The gradient in management intensity was established in 2006 with four subplots on every plot of the 80 plant mixtures. Management varied in mowing regime (M2: two cuts, M4: four cuts per year) and the amount of NPK-fertilizer application (F0: no fertilizer; F100: 100.0 kg N ha^{-1} a^{-1}, 43.6 kg P ha^{-1} a^{-1}, 83.0 kg K ha^{-1} a^{-1}; F200: 200 kg N ha^{-1} a^{-1}, 87.2 kg P ha^{-1} a^{-1}, 166.0 kg K ha^{-1} a^{-1}) and was combined as follows: M2F0, M2F100, M4F100, M4F200 [11]. All three fertilizer treatments were arranged randomly on an area of each 1.6 m×4 m within the main plots of 20×20 m, while M2F0 treatment was always located in the central core area of the plots, representing the standard management of the whole field site. Fertilization was done twice a year (31 March and 23 June 2008, 31 March and 16 June 2009) and mowing was done in spring (end of April, only M4-subplots), in early summer (beginning of June, all subplots), end of July (M4-subplots) and in late summer at the beginning of September (all subplots).

Drought was induced in 2008 and 2009 using transparent rain shelters during six weeks in summer previous to the last annual cut (25 July to 2 September 2008 and 16 July to 1 September 2009, Figure 1). Rain shelters were made of LDPE greenhouse film (www.dm-folien.com) in 2008 and of PVC sheets (www.paruschke-kunststoffe.de, product code: PVCSPK7018K10) in 2009 because of its higher durability. Rain shelters were inclined in a height of 1.3 to 1.5 m to enable ventilation and runoff of rain water in one direction 1 m away from our core area. Control subplots remained unsheltered and received ambient precipitation. We established one sheltered (hereafter named "drought" treatment) and one unsheltered subplot (named "ambient") of 1.6 m×2 m size for each management treatment in each of the 80 plots covering the whole diversity gradient. Measurements of the soil water content revealed a soil moisture decrease of about 17% in the drought treatment compared to the ambient treatment. The open-side construction of the rain shelters could not prevent a temperature increase of about 1.5–2.2°C on soil surface in the drought treatments, but no warming was detected at 20 cm height. The PVC sheets reduced photosynthetically active radiation (PAR) by 28% maximum. In 2009 we established an additional roof control in all plots, e.g. a sheltered subplot where we added collected rain water, to measure the pure roof effect (heat, altered light conditions) on our response variable. We found that the results of the roof control were similar to those of the ambient treatment (data not shown).

Since the Jena Experiment was established, climatic conditions were measured by a weather station directly on field site so that we were able to document weather data during our experimental phase 2008–2009 (Table 4). Rain shelters excluded 59.5 mm precipitation in 2008 (reduction of 40% compared to unroofed subplots during the summer July–September) and 53.7 mm in 2009 (reduction of 35%).

Data collection

All measurements were restricted to a central area of 1 m×1 m on every subplot to minimize edge effects (precipitation, varying soil nutrients, different height of neighboring vegetation). We clipped aboveground biomass of a 20 cm×50 cm area at 3 cm height above soil surface two days prior to every mowing event of a subplot. Biomass was sorted into sown species, unsown weeds and dead plant material, dried until constant weight (70°C, 48 h) and weighed. Here we present the results of the biomass of sown species.

Statistical analysis

We calculated resistance and resilience from our biomass data according to van Ruijven and Berendse [22]. In contrast to that study, which compared perturbed and unperturbed plots in two consecutive years, we were able to compare data from each drought treatment with its corresponding ambient treatment at the same time. Resistance of biomass production was calculated as the difference of biomass under perturbed and unperturbed conditions (drought - ambient) at the end of the drought period in August. Proportional resistance calculated as the ratio of drought to ambient treatment biomass was determined to account for productivity effects on resistance. Resilience determines the change in biomass production after perturbation and was calculated as difference of post-drought biomass and the corresponding ambient treatment from the first harvest after drought (M4-subplots: April 2009, M2-subplots: June 2009). Proportional resilience was calculated as the ratio of post-drought biomass and the corresponding ambient treatment from the first harvest after drought. The proportional values indicate, whether the ratio of biomass decrease or increase due to drought change.

We analyzed the data with mixed effects models using the nlme-package of R 2.8.1. to account for the nested design of our experiment (drought/ambient nested within management nested within plots of different diversity levels). Because we were not interested in effects of each species combination, we used plots as random factors in the model, as well as management and our drought treatment. We fitted a series of models by stepwise inclusion of fixed effects. First, we included block in our fixed term to account for all edaphic variation in the field and the blockwise management and data sampling. Then we included sown species richness and functional group richness as diversity factors and our experimental treatments (management, drought) and their interaction with diversity treatments stepwise in the fixed term of the models with the maximum likelihood method. We applied likelihood ratio tests for model comparison and estimating the significance of the fixed effects. We are aware that sown diversity and realized diversity can vary between the management treatments, because management intensification is expected to reduce species diversity [43]. We therefore fitted additional models with realized species richness and realized functional group richness instead of sown species and functional group richness. These models presented the same conclusions as with design variables. For better comparison with other experimental results, we present results of sown diversity effects in the paper and realized diversity effects in the (Table S1, S2, S3, Figure S1, S2, S3 and Methods S1 for information on data acquisition of realized species richness). To meet the assumptions of mixed effects models (normally distributed within group errors and random effects), biomass and resistance were log transformed. When data were heteroscedastic (in case of resilience), variance functions were included [44]. The 60 species mixtures merely serve as reference plots and are excluded from analysis as they are not fully compatible with the experimental design of the experiment.

Supporting Information

Methods S1 Realized species richness was recorded during drought period in August 2008 and 2009 in every subplot. We recorded presence and absence of every single sown species in 10 squares of 1 dm^2 size along one transect and repeated this three times within our study area of 1 m^2. The realized species number was the sum of all species that were present in at least one out of the 30 squares.

Figure S1 Aboveground biomass across realized species richness gradient for each management treatment. Aboveground biomass at the end of the induced drought period in August 2008 (left) and 2009 (right) for each management treatment since the last cut (M2-types: June; M4-types: July). The ambient treatment is given in open circles (dotted regression line) and drought treatment in closed circles (solid regression line). Significant effects obtained from mixed models for every single management treatment per year: SR = effect of realized species richness (linear), drought = difference in drought and ambient treatments, * = p<0.05, ** = p<0.01, *** = p<0.001.

Figure S2 Resistance in biomass production over realized species richness. Resistance was calculated as the difference of drought and corresponding ambient treatments. Realized species richness represents the mean of realized species numbers of drought and ambient treatment. Management treatments are shown in white (M2F0), gray (M2F100), dark gray (M4F100), black (M4F200).

Figure S3 Resilience of biomass over realized species richness for each management treatment. Resilience was calculated as the difference of drought and corresponding ambient treatments for the first harvest in spring 2009 (M2-types: June, M4-types: April). Realized species richness represents the mean of realized species numbers of drought and ambient treatment. Regression lines are given for significant effects obtained from linear models for every single management treatment per year: SR = effect of realized species richness (linear), drought = difference in drought and ambient treatments, * = p<0.05, ** = p<0.01, *** = p<0.001.

Table S1 Summary of mixed effects models for aboveground biomass in August 2008 and 2009 to test for effects of management, drought and diversity (realized numbers of species and functional groups) treatments.

Table S2 Summary of mixed effects models for resistance and proportional resistance of aboveground biomass after drought in August 2008 and 2009 to test for effects of diversity (realized numbers of species and functional groups) and management treatments (separated into mowing and fertilizer amounts).

Table S3 Summary of mixed effects models for resilience computed as the difference between previously drought and ambient treatment in aboveground biomass as well as for proportional resilience of the first cut in spring 2009 to test for effects of management (separated into mowing and fertilizer amounts) and diversity (realized numbers of species and functional groups) treatments.

Acknowledgments

We thank numerous student helpers, the gardeners Steffen Eismann, Steffen Ferber, Silke Hengelhaupt, Ute Köber, Katja Kunze, Heike Scheffler as well as Gerlinde Kratzsch, Victoria Stabrey, Stefan Lorenz, Kymbat Dikambaeva and interns Kyle Siefers and Kathrin Erfurt for their help in maintaining the field site and help during data collection. We especially thank Victor Malakhov for his help during installation and maintaining the rain shelters as well as help during biomass harvests. We also want to thank Bernhard Schmid for his support throughout the experiment as well as Dan Flynn and two anonymous reviewers for helpful comments on the manuscript.

Author Contributions

Conceived and designed the experiments: AW MSL. Performed the experiments: AV AW. Analyzed the data: AV. Contributed reagents/materials/analysis tools: AW MSL. Wrote the paper: AV AW MSL. Interpretation of results: AV AW MSL.

References

1. IPCC (2007) Climate Change 2007: Synthesis Report Intergovernmental Panel on Climate Change.
2. Jacob D, Göttel H, Kotlarski S, Lorenz P, Sieck K (2008) Klimaauswirkungen und Anpassung in Deutschland-Phase 1: Erstellung regionaler Klimaszenarien für Deutschland. In:, , Umweltbundesamt, editor (2008) Climate Change 11-08. Dessau-Roßlau: Umweltbundesamt.
3. Millennium Ecosystem Assessment M, editor (2005) Ecosystems and Human Well-being: Synthesis. Washington, DC: Island Press.
4. Hector A, Bagchi R (2007) Biodiversity and ecosystem multifunctionality. Nature 448: 188–191.
5. Isbell F, Calcagno V, Hector A, Connolly J, Harpole WS, et al. (2011) High plant diversity is needed to maintain ecosystem services. Nature 477: 199–U196.
6. Hector A, Schmid B, Beierkuhnlein C, Caldeira MC, Diemer M, et al. (1999) Plant diversity and productivity experiments in European grasslands. Science 286: 1123–1127.
7. Tilman D, Reich PB, Knops J, Wedin D, Mielke T, et al. (2001) Diversity and productivity in a long-term grassland experiment. Science 294: 843–845.
8. van Ruijven J, Berendse F (2003) Positive effects of plant species diversity on productivity in the absence of legumes. Ecology Letters 6: 170–175.
9. Marquard E, Weigelt A, Temperton VM, Roscher C, Schumacher J, et al. (2009) Plant species richness and functional composition drive overyielding in a six-year grassland experiment. Ecology 90: 3290–3302.
10. Reich PB, Knops J, Tilman D, Craine J, Ellsworth D, et al. (2001) Plant diversity enhances ecosystem responses to elevated CO2 and nitrogen deposition (vol 410, pg 809, 2001). Nature 411: 824–+.
11. Weigelt A, Weisser WW, Buchmann N, Scherer-Lorenzen M (2009) Biodiversity for multifunctional grasslands: equal productivity in high-diversity low-input and low-diversity high-input systems. Biogeosciences 6: 1695–1706.

12. Kirwan L, Luescher A, Sebastia MT, Finn JA, Collins RP, et al. (2007) Evenness drives consistent diversity effects in intensive grassland systems across 28 European sites. Journal of Ecology 95: 530–539.
13. Isbell FI, Wilsey BJ (2011) Increasing native, but not exotic, biodiversity increases aboveground productivity in ungrazed and intensely grazed grasslands. Oecologia 165: 771–781.
14. Yachi S, Loreau M (1999) Biodiversity and ecosystem productivity in a fluctuating environment: The insurance hypothesis. Proceedings of the National Academy of Sciences of the United States of America 96: 1463–1468.
15. Tilman D, Reich PB, Knops JMH (2006) Biodiversity and ecosystem stability in a decade-long grassland experiment. Nature 441: 629–632.
16. van Ruijven J, Berendse F (2007) Contrasting effects of diversity on the temporal stability of plant populations. Oikos 116: 1323–1330.
17. Eisenhauer N, Milcu A, Allan E, Nitschke N, Scherber C, et al. (2011) Impact of above- and below-ground invertebrates on temporal and spatial stability of grassland of different diversity. Journal of Ecology 99: 572–582.
18. Pimm SL (1984) The complexity and stability of ecosystems. Nature 307: 321–326.
19. Gunderson LH (2000) Ecological resilience - in theory and application. Annual Review of Ecology and Systematics 31: 425–439.
20. Tilman D (1996) Biodiversity: Population versus ecosystem stability. Ecology 77: 350–363.
21. Pfisterer AB, Schmid B (2002) Diversity-dependent production can decrease the stability of ecosystem functioning. Nature 416: 84–86.
22. van Ruijven J, Berendse F (2010) Diversity enhances community recovery, but not resistance, after drought. Journal of Ecology 98: 81–86.
23. Joshi J, Matthies D, Schmid B (2000) Root hemiparasites and plant diversity in experimental grassland communities. Journal of Ecology 88: 634–644.

24. Pfisterer AB, Diemer M, Schmid B (2003) Dietary shift and lowered biomass gain of a generalist herbivore in species-poor experimental plant communities. Oecologia 135: 234–241.

25. Scherber C, Heimann J, Kohler G, Mitschunas N, Weisser WW (2010) Functional identity versus species richness: herbivory resistance in plant communities. Oecologia 163: 707–717.

26. Tilman D, Downing JA (1994) Biodiversity and Stability in Grasslands. Nature 367: 363–365.

27. Kahmen A, Perner J, Buchmann N (2005) Diversity-dependent productivity in semi-natural grasslands following climate perturbations. Functional Ecology 19: 594–601.

28. Wang YF, Yu SX, Wang J (2007) Biomass-dependent susceptibility to drought in experimental grassland communities. Ecology Letters 10: 401–410.

29. Leps J, Osbornovakosinova J, Rejmanek M (1982) Community stability, complexity and species life-history strategies. Vegetatio 50: 53–63.

30. Macgillivray CW, Grime JP, Band SR, Booth RE, Campbell B, et al. (1995) Testing predictions of the resistance and resilience of vegetation subjected to extreme events. Functional Ecology 9: 640–649.

31. Gilgen AK, Buchmann N (2009) Response of temperate grasslands at different altitudes to simulated summer drought differed but scaled with annual precipitation. Biogeosciences 6: 2525–2539.

32. Grime JP, Brown VK, Thompson K, Masters GJ, Hillier SH, et al. (2000) The response of two contrasting limestone grasslands to simulated climate change. Science 289: 762–765.

33. Hu ZM, Yu GR, Zhou YL, Sun XM, Li YN, et al. (2009) Partitioning of evapotranspiration and its controls in four grassland ecosystems: Application of a two-source model. Agricultural and Forest Meteorology 149: 1410–1420.

34. Jentsch A, Kreyling J, Elmer M, Gellesch E, Glaser B, et al. (2011) Climate extremes initiate ecosystem-regulating functions while maintaining productivity. Journal of Ecology 99: 689–702.

35. De Boeck HJ, Lemmens C, Bossuyt H, Malchair S, Carnol M, et al. (2006) How do climate warming and plant species richness affect water use in experimental grasslands? Plant and Soil 288: 249–261.

36. Van Peer L, Nijs I, Reheul D, De Cauwer B (2004) Species richness and susceptibility to heat and drought extremes in synthesized grassland ecosystems: compositional vs physiological effects. Functional Ecology 18: 769–778.

37. Bessler H, Temperton VM, Roscher C, Buchmann N, Schmid B, et al. (2009) Aboveground overyielding in grassland mixtures is associated with reduced biomass partitioning to belowground organs. Ecology 90: 1520–1530.

38. Kreyling J, Beierkuhnlein C, Elmer M, Pritsch K, Radovski M, et al. (2008) Soil biotic processes remain remarkably stable after 100-year extreme weather events in experimental grassland and heath. Plant and Soil 308: 175–188.

39. Weisshuhn K, Auge H, Prati D (2011) Geographic variation in the response to drought in nine grassland species. Basic and Applied Ecology 12: 21–28.

40. Molyneux DE, Davies WJ (1983) Rooting pattern and water relations of 3 pasture grasses growing in drying soil. Oecologia 58: 220–224.

41. Foulds W (1978) Response to soil-moisture supply in 3 leguminous species.1. growth, reproduction and mortality. New Phytologist 80: 535–545.

42. Roscher C, Schumacher J, Baade J, Wilcke W, Gleixner G, et al. (2004) The role of biodiversity for element cycling and trophic interactions: an experimental approach in a grassland community. Basic and Applied Ecology 5: 107–121.

43. Cop J, Vidrih M, Hacin J (2009) Influence of cutting regime and fertilizer application on the botanical composition, yield and nutritive value of herbage of wet grasslands in Central Europe. Grass and Forage Science 64: 454–465.

44. Pinheiro JC, Bates DM, eds. Mixed-Effects Models in S and S-PLUS. New York: Springer Verlag. 528 p.

12

Timing of Favorable Conditions, Competition and Fertility Interact to Govern Recruitment of Invasive Chinese Tallow Tree in Stressful Environments

Christopher A. Gabler[1,2]*, Evan Siemann[1]

1 Department of Ecology and Evolutionary Biology, Rice University, Houston, Texas, United States of America, **2** Department of Biology and Biochemistry, University of Houston, Houston, Texas, United States of America

Abstract

The rate of new exotic recruitment following removal of adult invaders (reinvasion pressure) influences restoration outcomes and costs but is highly variable and poorly understood. We hypothesize that broad variation in average reinvasion pressure of *Triadica sebifera* (Chinese tallow tree, a major invader) arises from differences among habitats in spatiotemporal availability of realized recruitment windows. These windows are periods of variable duration long enough to permit establishment given local environmental conditions. We tested this hypothesis via a greenhouse mesocosm experiment that quantified how the duration of favorable moisture conditions prior to flood or drought stress (window duration), competition and nutrient availability influenced *Triadica* success in high stress environments. Window duration influenced pre-stress seedling abundance and size, growth during stress and final abundance; it interacted with other factors to affect final biomass and germination during stress. Stress type and competition impacted final size and biomass, plus germination, mortality and changes in size during stress. Final abundance also depended on competition and the interaction of window duration, stress type and competition. Fertilization interacted with competition and stress to influence biomass and changes in height, respectively, but did not affect *Triadica* abundance. Overall, longer window durations promoted *Triadica* establishment, competition and drought (relative to flood) suppressed establishment, and fertilization had weak effects. Interactions among factors frequently produced different effects in specific contexts. Results support our 'outgrow the stress' hypothesis and show that temporal availability of abiotic windows and factors that influence growth rates govern *Triadica* recruitment in stressful environments. These findings suggest that native seed addition can effectively suppress superior competitors in stressful environments. We also describe environmental scenarios where specific management methods may be more or less effective. Our results enable better niche-based estimates of local reinvasion pressure, which can improve restoration efficacy and efficiency by informing site selection and optimal management.

Editor: Sebastien Lavergne, CNRS/Université Joseph-Fourier, France

Funding: This work was supported by the US National Science Foundation (DDIG, DEB-0910361), Garden Club of America (Fellowship in Ecological Restoration), Society of Wetland Scientists (Student Research Grant), Native Plant Society of Texas (A.M. Gonzalez Grant), Wintermann Research Fund, University of Houston Coastal Center and Lodieska Stockbridge Vaughn, Houston Livestock Show and Rodeo and Wray-Todd Fellowships. The funders had no role in study design, data collection and analysis, decision to publish, or preparation of the manuscript.

Competing Interests: The authors have declared that no competing interests exist.

* E-mail: cagabler@uh.edu

Introduction

Reinvasion pressure, or the rate of new exotic recruitment following removal of mature conspecifics, varies broadly among similarly invaded habitats and is crucial to restoration outcomes and costs but is poorly understood and difficult to predict [1–3]. When restoring habitats dominated by an exotic plant, invader density governs strength of impacts on communities and ecosystem functions [4] and influences required management methods, which can have diverse impacts on non-target species [5] and vary widely in cost [6]. Accurately estimating reinvasion pressure can improve restoration efficacy and efficiency by informing site selection and optimal management strategies [2], but the mechanisms driving its variation are poorly understood despite their importance to restoration and exotic plant control [3].

Triadica sebifera (Chinese tallow tree) is a major invader in the southeastern United States that exhibits broad variation in average reinvasion pressure during restorations of habitats it previously dominated [7] (Gabler & Siemann, unpublished data). We hypothesize that this variation arises predominantly because differences among invaded habitats in their temporal availability of moisture conditions suitable for *Triadica* recruitment drive differences in average *Triadica* recruitment success [2]. Differences in average reinvasion pressure can become masked over time by *Triadica* dominance because ontogenetic niche expansions (increases in niche breadth during development) enable *Triadica* to persist in moisture conditions unsuitable for its recruitment [8].

Our 'outgrow the stress' hypothesis further posits that short-term reinvasion pressure depends on propagule abundance and spatiotemporal availability of realized recruitment windows, which are akin to 'safe sites' [9] but emphasize ontogenetic niche expansions [2]. Realized recruitment windows are periods of variable duration that permit exotics with expanding niches to become established and are determined by abiotic conditions and interspecific interactions with recipient communities. This hypothesis stresses factors that influence seedling growth during temporary periods of suitable environmental conditions and may thus influence establishment success, which Holmgren *et al.* [10] demonstrate can influence vegetation structure on a landscape scale and are central to the present work.

Existing hypotheses explaining recruitment or invasion success have long emphasized spatial and/or temporal availability of conditions suitable for germination and establishment. Harper [9] defined 'safe sites' as sites free of specific hazards, e.g. intolerable moisture conditions, and argued that all colonization occurs as a function of their availability. Grubb [11] described similar 'regeneration niches' and Johnstone [12] elaborated safe sites to consider their dispersion in time. These suggest strict limits in resource availability or climatic tolerances define recruitment opportunities, and indeed many studies demonstrate that temporary variation across distinct abiotic thresholds can permit or preclude plant establishment in stressful habitats [13,14].

More recent spatiotemporal niche-based invasion hypotheses consider a broader range of biotic and abiotic factors and place greater emphasis on their interactions, e.g. competition with resident species for fluctuating resources [15]. Such hypotheses recognize the importance of stochastic disruptions to communities, like earlier theories, but accentuate historical contingencies arising from these and other irregular events, e.g. 'niche opportunity' of Shea and Chesson [16]. They also identify "grayer" abiotic thresholds resulting from stress mediating effects of certain biotic interactions [17] or life history strategies, e.g. the storage effect [18]. Our 'outgrow the stress' hypothesis takes spatiotemporal niche-based invasion hypotheses one step further – albeit strictly in the context of reinvasion – by considering not only fluctuations in

the biotic and abiotic environment but also ontogenetic changes in invader environmental tolerances and ultimately the impacts of environment on ontogenetic development of invaders and thus their individual tolerances through time [2].

Availability of soil resources, including water, and competition for these and other resources are fundamental factors limiting plant distributions [19,20], so we expect they are principally important in defining realized recruitment windows. Nutrient availability has been show to have strong effects on invasion success [21–23]. Fertilization increased *Triadica* invasion pressure in coastal prairies by increasing seedling survival, height and biomass [24], but increased *Triadica* survival in coastal prairies with nutrient addition was not always observed and performance benefits were nutrient specific [25].

Water regime is crucial to *Triadica*'s local distribution and can vary considerably on small spatial scales in its introduced range [26]. Though established *Triadica* seedlings have relatively broad moisture tolerances [27,28], moisture requirements for germination and survival and growth of young seedlings are relatively narrow [8]. In other systems, interannual variation in precipitation can influence establishment success among years [29] and cause episodic recruitment [30]. Preliminary results from experimental restorations of eleven sites dominated by *Triadica* suggest reinvasion pressure correlates with soil moisture, and that addition of native seeds may decrease *Triadica* recruitment success in favorable moisture conditions but, at least early in restoration, increase recruitment success in more stressful conditions (Gabler & Siemann, unpublished data). Interactions between these fundamental factors remain unclear, especially in high stress environments where they may be most important to *Triadica* recruitment success.

We began validating our 'outgrow the stress' hypothesis by demonstrating that *Triadica* undergoes rapid ontogenetic moisture niche expansions, which enable seedlings to tolerate conditions

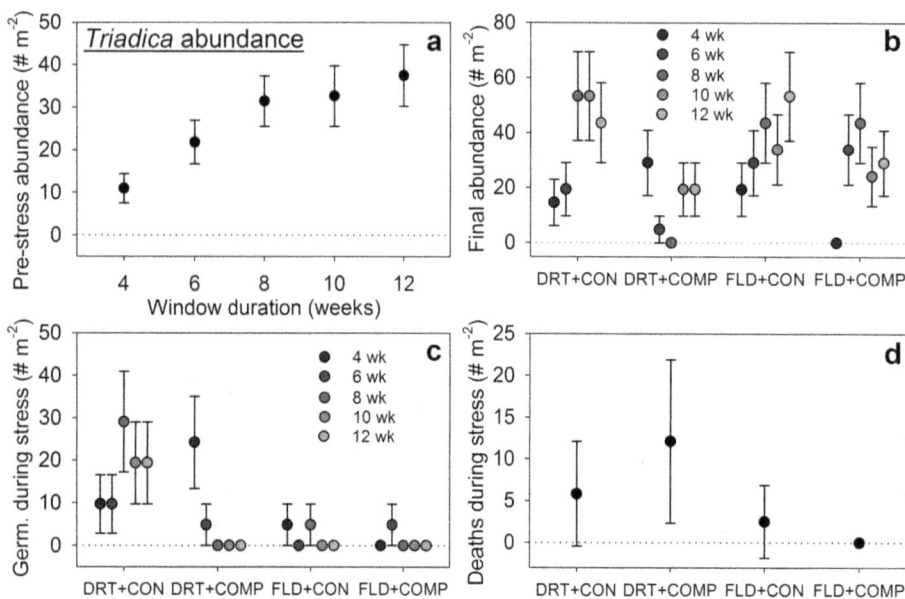

Figure 1. Effects of experimental treatments on four metrics of *Triadica* abundance. Panels (a–d) represent abundance metrics (means ±1 SE) broken down by treatments that significantly affected that metric. Legend: stress type – drought (DRT) or flood (FLD); competition – natives added (COMP) or control (CON). (a) Pre-stress *Triadica* abundance was higher in longer window duration treatments. (b) Final abundance was higher with longer window durations and generally lower with drought stress or competition. (c) Flood or competition generally reduced germination during stress; more germinants were observed after shorter windows but only with competition. (d) More seedlings died during stress when subject to drought or competition.

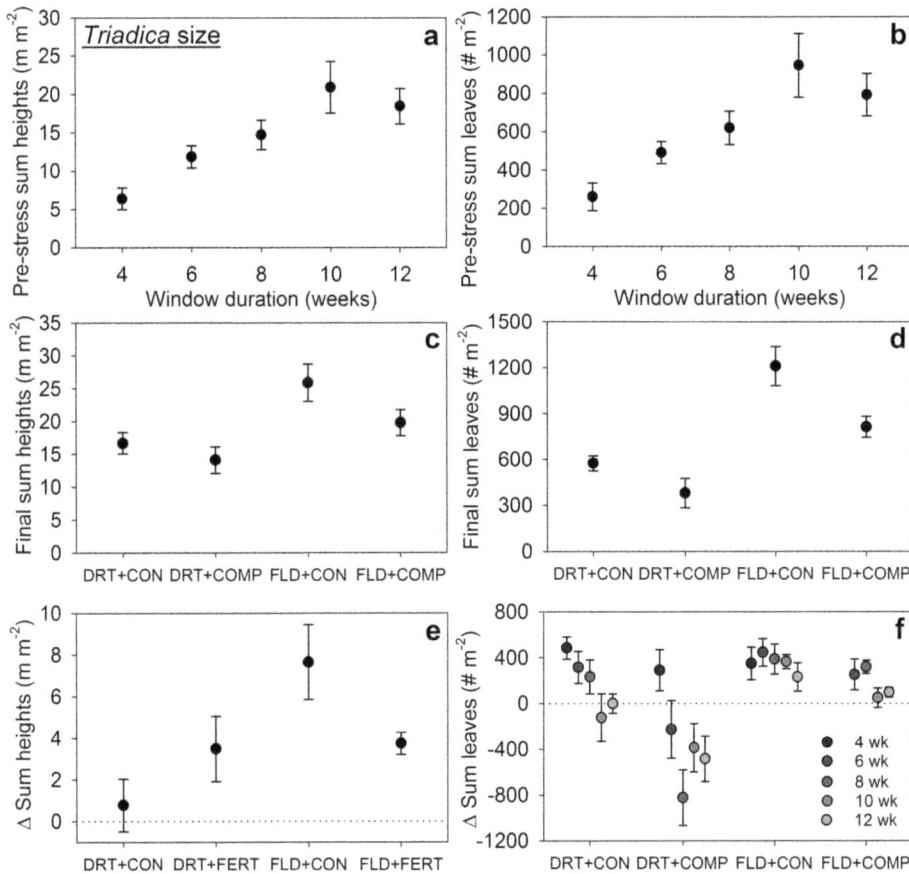

Figure 2. Effects of experimental treatments on six metrics of *Triadica* size. Panels (a–f) represent size metrics (means ±1 SE) broken down by treatments that significantly affected that metric. Legend: stress type – drought (DRT) or flood (FLD); competition – natives added (COMP) or control (CON); fertilization – NPK added (FERT) or control (CON). Pre-stress pot sums of *Triadica* seedling heights (a) and leaf abundances (b) were higher in longer window duration treatments. Final pot sums of seedling heights (c) and leaves (d) were lower with drought or competition. Absolute changes (Δ) during stress in pot sums of *Triadica* seedling heights (e) and leaf abundances (f) were lower and sometimes negative in drought stress and lower with competition. Fertilization increased changes in height in drought but decreased them in flood (e). Changes in leaves decreased with window duration (f), and decreases in changes in leaves with competition were greater in drought (f).

that do not permit germination (i.e. continuous flooding and short-term drought) within two months of germination [8]. In this work, we continue vetting this hypothesis by investigating realized recruitment windows. Here we quantify how the duration of favorable moisture conditions prior to flood or drought stress (window duration), competition and nutrient availability influence *Triadica* recruitment within highly stressful but variable environments.

Greater understanding of how temporal moisture fluctuations, competition and nutrient availability influence *Triadica* survival and performance would improve estimates of average reinvasion pressure within particular habitats, and would enhance our ability to predict short-term reinvasion pressure based on climate forecasts [2,31]. Knowledge of context-dependent effects of native seed addition or fertilization on *Triadica* recruitment would promote management plans better fit to local circumstances and more able to mitigate or exploit stochastic events such as extreme weather or nutrient or seed pulses. Both provide valuable decision-making tools for restoration and *Triadica* management, and these approaches are applicable in other invaded systems.

We investigated how window duration and key ecological factors influence *Triadica* recruitment in stressful environments by performing a mesocosm experiment manipulating window dura-

tion, stress type, competition and fertilization. If availability of realized recruitment windows governs recruitment in stressful habitats, longer window durations prior to stress should increase *Triadica* abundance and performance once stress resumes. If size confers tolerance in plants with ontogenetic niche expansions [32], factors influencing growth rates should affect recruitment during finite windows of opportunity [2], thus fertilization and competition should increase and decrease *Triadica* success, respectively. To better understand how temporal availability of realized recruitment windows influence exotic recruitment and key biotic and abiotic factors shape realized recruitment windows, we ask: (i) How do window duration, competition and fertilization interact to influence *Triadica* seedling abundance and performance? (ii) How does the nature of water stress influence *Triadica* success and/or alter the effects of other factors?

Methods

Focal Species

Chinese tallow tree [*Triadica sebifera* (L.) Small, Euphorbiaceae; synonym *Sapium sebiferum*; '*Triadica*' throughout] is a major invasive species in the southeastern United States naturalized from Texas to Arkansas and eastward from Florida to North Carolina and in

Table 1. Results of ANODEVs testing effects of experimental treatments on *Triadica* abundance.

Factor	d.f.	χ^2	p	d.f.	χ^2	p	χ^2	p	d.f.	χ^2	p
		pre-stress abundance			final abundance		germinants during stress			deaths during stress	
Window	4	15.7	**0.0035**	4	10.4	**0.0349**	2.7	0.61	4	0.5	0.97
Stress				1	1.0	0.31	18.6	**<0.0001**	1	16.2	**<0.0001**
Comp	1	1.5	0.22	1	9.4	**0.0021**	6.5	**0.0106**	1	5.7	**0.0165**
Fert	1	0.7	0.39	1	1.4	0.23	0.3	0.56	1	0.0	0.85
W*S				4	7.2	0.13	2.6	0.63	4	6.0	0.20
W*C	4	0.3	0.99	4	1.6	0.81	14.8	**0.0052**	4	1.2	0.87
W*F	4	1.3	0.87	4	2.5	0.64	4.0	0.41	4	0.6	0.99
S*C				1	2.8	0.09	0.0	0.84	1	2.6	0.09
S*F				1	0.4	0.54	0.4	0.55	1	1.9	0.17
C*F	1	0.9	0.34	1	0.0	0.92	0.1	0.71	1	0.0	0.88
W*S*C				4	17.2	**0.0018**	3.9	0.42			
W*S*F				4	4.9	0.30	2.0	0.73			
W*C*F	4	3.6	0.47	4	2.0	0.73	1.6	0.80			
S*C*F				1	0.0	0.92	0.0	1.00			
W*S*C*F				4	2.7	0.60	0.0	1.00			

Experimental treatments include window duration (W), stress type (S), competition (C), fertilization (F) and their interactions. Pre-stress and final abundances are the numbers of live *Triadica* seedlings observed before and after 28 days of water stress, respectively. Germinants and deaths during stress are abundances of those instances observed during this stress period.

California [26,33]. *Triadica* aggressively displaces native plants in grasslands (e.g. imperiled coastal prairies), wetlands and forests and can form monocultures in only two decades [26,34]. *Triadica* is a superior competitor due to a combination of high growth rates [35], prolific seed production [36], broad abiotic tolerances [27,37] and low herbivore loads in its introduced range [38].

Triadica seeds exhibit dormancy and can remain viable in seed banks for 5+ years [26]. Seeds require specific abiotic conditions to cue germination, namely widely oscillating day-night temperatures, which are characteristic of exposed soil and promote *Triadica* germination in disturbed conditions [7,39,40], and moist but unsaturated soils, which promote germination in moisture conditions optimal for seedling survival and growth [8]. Established *Triadica* juveniles have broad moisture tolerances [27,28], but the moisture requirements of newly germinated *Triadica* seedlings are relatively narrow and rapidly broaden in the first months of development to enable persistence in conditions ranging from constant flooding to short-term drought (an ontogenetic moisture niche expansion); survival in flooded conditions depends strongly on plant size, specifically whether seedlings have any emergent leaves [8]. We expect rapid moisture niche expansions early in ontogeny are crucial to *Triadica* establishment success during brief windows of favorable conditions in temporally variable environments. Furthermore, we hypothesize that size confers moisture tolerance and thus factors such as competition and nutrient availability also influence minimum establishment time and recruitment success during abiotic windows of opportunity [2].

Greenhouse Mesocosm Experiment

We manipulated duration of favorable moisture conditions prior to water stress (window duration), competition and nutrient availability in mesocosms and quantified *Triadica* abundance, survival and performance through a period of water stress. Our balanced full factorial design used 2.8 L pots with five window duration, two competition, two fertilization and two stress type treatments with 10 replicates per treatment combination (n = 400 pots). In July 2008 we filled 2.8 L tapered square plastic Treepots (36 cm tall, 6–10 cm diameter; Stuewe & Sons, Oregon, USA) with ~2 L field soil collected from Justin Hurst Wildlife Management Area (JHWMA) in southeast Texas and randomly assigned treatments to each. Soils collected near 28.959502 N, −95.461348 W were expansive Pledger (85%) and Brazoria Clay (10%) vertisols (very-fine, smectitic, hyperthermic Typic Hapluderts) with 60–80% clay content. JHWMA is limited-access public land owned and managed by the Texas Parks & Wildlife Department, whose staff granted us access and permission to collect soil. We added 10 washed *Triadica* seeds to each pot from a well-mixed batch collected in 2007 from source trees in southeast Texas. This provided relatively dense seed banks (~1000 seeds/m²) typical of habitats dominated by *Triadica*. We housed pots in a climate controlled greenhouse under natural light with day temperatures of 29–31°C and night temperatures of 19–21°C, which approximates spring in southeast Texas. See Gabler and Siemann [8] for additional site description and seed preparation protocols.

We established five window duration treatments by exposing pots to identical well-drained and well-watered conditions for 4, 6, 8, 10 or 12 weeks before imposing water stress. We established two competition treatments by adding nothing (control, CON) or 0.5 g each of *Schizachyrium scoparium* (Michx.) Nash (little bluestem) and *Leersia oryzoides* (L.) Sw. (rice cutgrass) seeds at the time of *Triadica* seed addition (competition, COMP). We chose these species to ensure that natives were present and alive to compete with *Triadica* in both types of water stress. Both have relatively broad moisture tolerances, but *Schizachyrium* tolerates substantial drought whereas *Leersia* tolerates persistent flooding. We established two fertilization treatments by adding water (control, CON) or 4 g/m² nitrogen,

Table 2. Results of ANOVAs testing effects of experimental treatments on *Triadica* performance.

Factor	d.f.	pre-stress sum of heights F_{91}	p	pre-stress sum of leaves F_{91}	p	d.f.	final sum of heights F_{94}	p	final sum of leaves F_{94}	p	stem biomass F_{94}	p	leaf biomass F_{94}	p	aboveground biomass F_{94}	p
Window	4	4.5	**0.0026**	4.8	**0.0017**	4	2.0	0.11	0.2	0.92	1.5	0.21	0.5	0.71	0.8	0.55
Stress	1	2.9	0.09	2.3	0.13	1	14.2	**0.0003**	47.1	**<0.0001**	30.8	**<0.0001**	39.1	**<0.0001**	36.5	**<0.0001**
Comp	1	0.5	0.46	2.8	0.10	1	6.1	**0.0157**	17.3	**<0.0001**	8.7	**0.0043**	11.0	**0.0014**	10.0	**0.0022**
Fert	1	1.7	0.20	0.6	0.46	1	0.5	0.50	0.2	0.69	1.1	0.29	0.5	0.49	0.7	0.40
W*S	4	0.4	0.82	0.5	0.75	4	0.3	0.87	2.2	0.08	1.0	0.42	1.8	0.14	1.3	0.29
W*C	4	0.9	0.49	1.6	0.18	4	1.9	0.11	1.7	0.15	2.3	0.07	3.2	**0.0177**	2.9	**0.0264**
W*F	4	1.1	0.38	1.2	0.32	4	2.3	0.06	1.6	0.19	1.1	0.36	1.3	0.26	1.1	0.35
S*C	1	0.3	0.61	1.2	0.28	1	1.3	0.25	1.3	0.26	3.8	0.06	5.0	**0.0290**	4.8	**0.0314**
S*F	1	0.1	0.73	0.0	0.84	1	0.1	0.73	0.1	0.71	2.7	0.11	3.3	0.07	3.6	0.06
C*F	1	0.9	0.36	2.1	0.16	1	0.1	0.70	0.4	0.52	5.3	**0.0245**	4.0	**0.0489**	5.4	**0.0231**

Experimental treatments include window duration (W), stress type (S), competition (C), fertilization (F) and their interactions. Pre-stress and final sums of heights and leaves are summed totals within individual mesocosm pots of *Triadica* seedling heights and leaf abundances observed before and after 28 days of water stress, respectively. Stem, leaf and aboveground biomasses are sums of dry tissue masses collected from individual pots after 28 days of water stress. All values were square root transformed for analyses.

Table 3. Results of ANOVAs testing effects of experimental treatments on absolute changes (Δ) in *Triadica* performance.

factor	d.f.	Δ sum of heights F_{121}	p	Δ sum of leaves F_{121}	p
Window	4	2.4	0.05	4.3	**0.0028**
Stress	1	12.7	**0.0006**	17.9	**<0.0001**
Comp	1	24.2	**<0.0001**	15.4	**0.0002**
Fert	1	0.1	0.71	0.2	0.66
W*S	4	0.4	0.84	0.5	0.75
W*C	4	1.3	0.29	0.3	0.85
W*F	4	0.5	0.75	0.3	0.85
S*C	1	3.9	0.05	4.3	**0.0402**
S*F	1	4.8	**0.0307**	1.6	0.21
C*F	1	1.6	0.21	2.4	0.12

Treatments include window duration (W), stress type (S), competition (C), fertilization (F) and their interactions. Absolute changes (final – initial; Δ) in sums of *Triadica* seedling heights and leaf abundances for each pot were untransformed for analyses. Initial values were measured immediately before we initiated water stress (pre-stress). Final values were measured after 56 days of either drought or flood stress (post-stress). Analyses included pots that had at least one live *Triadica* seedling at the initial or final survey (n = 122).

1.3 g/m^2 phosphorus, 2.7 g/m^2 potassium and micronutrients (Ultra Turf fertilizer; Vigoro Corp., Illinois, USA) dissolved/suspended in water at seed addition and 8 weeks later (fertilized, FERT). We established treatments for type of water stress by either blocking drainage and "topping off" pots at watering so 8–10 cm of standing water persisted (flood, FLD) or discontinuing watering altogether (drought, DRT) after the designated period of favorable conditions ended. We watered according to treatments and weeded pots not subject to competition thrice weekly. In each pot we quantified *Triadica* abundance, height and leaf count of individual *Triadica* seedlings, and percent cover (visual estimate) and maximum height of native plants at the onset of water stress and 14 and 28 days later. After final surveys we harvested living aboveground biomass of all *Triadica* seedlings from all treatments, native plants from competition treatments, and root biomass of living *Triadica* seedlings from 12-week window treatments. Biomass samples were oven-dried at 70°C for 48 h and weighed.

Analyses

To evaluate effects of experimental treatments on *Triadica* abundance, we fit abundance (count) data using generalized linear models (GLMs; 'glm' in R 2.13; R Foundation for Statistical Computing, Vienna, Austria) with Poisson probability distributions. We used analysis of deviance (ANODEV, a form of likelihood ratio testing; 'anova' in R) with chi-squared tests to determine whether window duration, competition, fertilization, stress type and/or their interactions influenced final seedling abundance or the number of germinants or deaths during stress. We analyzed pre-stress seedling abundance the same way but excluded stress type as a model term because stress had not yet been imposed. For abundances of seedlings and germinants we considered all pots (n = 400). For deaths we considered pots that had at least one *Triadica* seedling at at least one survey (n = 111). Importantly, there were zero *Triadica* seedlings in ~3/4 of the pots. High proportions of zeroes are often suggestive of over-dispersed and zero-inflated data.

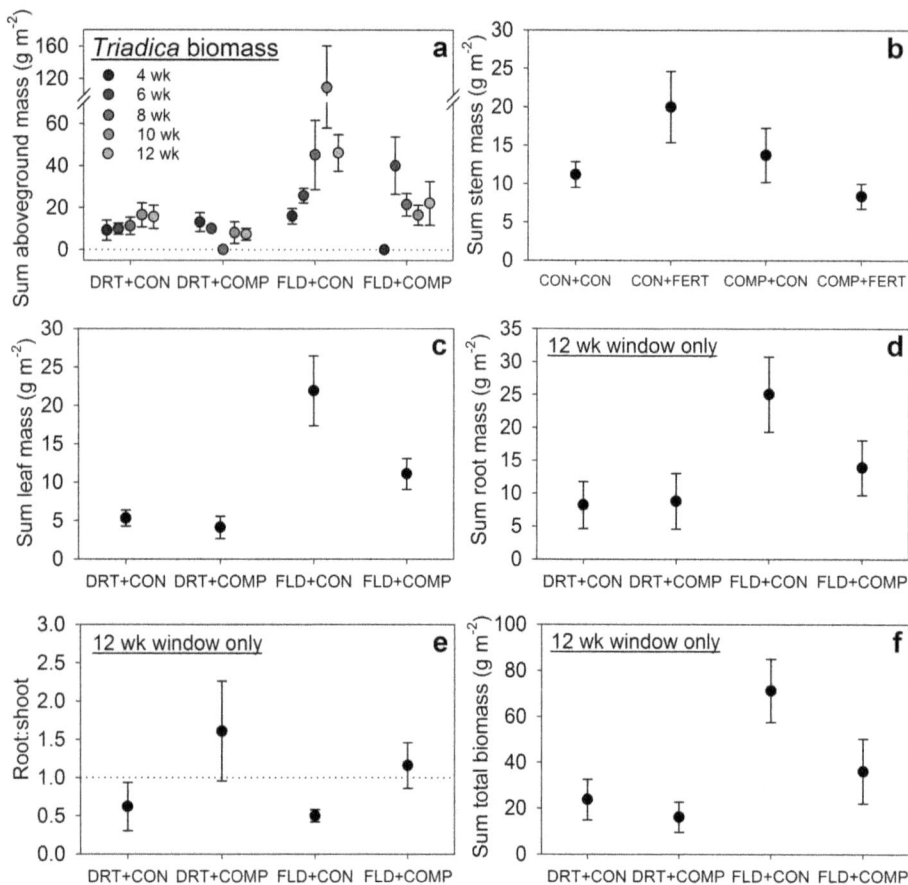

Figure 3. Effects of experimental treatments on six metrics of *Triadica* biomass. Panels (a–f) represent biomass metrics (means ±1 SE) broken down by treatments that significantly affected that metric (except in C where window effects are not show and E-G where insignificant trends are shown). Legend: stress type – drought (DRT) or flood (FLD); competition – natives added (COMP) or control (CON); fertilization – NPK added (FERT) or control (CON). Pot sums of aboveground biomass (a), stem biomass (b) and leaf biomass (c) were reduced in drought or competition treatments, and increased with fertilization without competition but decreased with fertilization with competition (e.g. panel b). Aboveground and leaf biomasses increased with window duration, but these relationships were inconsistent with competition (e.g. panel a). Competition reduced aboveground and leaf biomasses in flood only, but they were lowest in drought overall (e.g. panel c). Sums of root (d) and whole plant biomass (f) were lower in drought treatments. (f) Competition increased root:shoot ratios.

To explore this possibility, we fit abundance data using zero-inflated regression models for count data ('zeroinfl' in R, package 'pscl') and compared models using negative binomial versus Poisson probability distributions. For each response variable, the dispersion parameter (theta) in the negative binomial model was highly insignificant ($p > 0.94$), and z-scores and p-values for all model terms were highly similar between corresponding models. The same held true if we ignored zero-inflation and compared negative binomial GLMs ('glm.nb' in R, package 'MASS') with our initial Poisson GLMs. Lastly, we performed likelihood ratio tests for over-dispersion in count data using 'odTest' in R (package 'pscl'). For each response variable we failed to reject the null hypothesis that data are not over-dispersed (initial abundance: $p = 0.31$, final abundance: $p = 0.42$, interval germination: $p = 0.45$). We also found no evidence that zero-inflation (a type of overdispersion) was a factor in our analyses. First, 'odTest' should detect zero-inflation but did not. Second, none of the zero-inflation model terms in any of our zero-inflated models were significant. These findings provide strong evidence that our count data are neither over-dispersed nor zero-inflated, and they support our use of Poisson GLMs for abundance data.

We also used logistic regression to directly address whether absence of *Triadica* seedlings depended on treatments and to further investigate the reliability of our Poisson GLMs. We performed logistic regressions by fitting binary *Triadica* presence/absence data using 'glm' in R with a binomial distribution and the 'logit' link function. We then used ANODEVs with chi-squared tests to evaluate whether treatments influenced pre-stress or final seedling presence or presence of germinants. Results showed that exactly the same set of factors and interactions that significantly influenced seedling abundances also significantly predict presence/absence of *Triadica* seedlings. This further validates the results of our Poisson GLMs. Our zero-inflated and logistic regression analyses considered all pots (n = 400).

This abundance of zeroes did complicate our analyses of seedling performance. Only 92 pots had at least one *Triadica* seedling at the pre-stress survey, and 95 pots did at the final survey. Crucially, the number of pots with seedlings present depended on treatments and resulted in unbalanced sample sizes. Therefore, we performed ANOVAs utilizing Type II Sums of Squares ('Anova' in R, package 'car') to test whether treatments and/or their interactions influenced pre-stress and final sums of *Triadica* seedling heights and leaf abundances, sums of *Triadica* stem, leaf, root, aboveground and total biomasses, absolute changes (final − initial) in seedling heights and leaf abundances; and root:shoot. Response variables incorporating root biomass include the 12 week window treatment only and were log_e transformed, absolute changes were untransformed, and all other *Triadica* performance metrics were square root transformed for analyses. Due to unbalanced sample sizes, degrees of freedom were sparse and missing values produced unreliable covariance structure among higher order interactions for some response variables. Therefore, we simplified the most complex models by removing all third and fourth order interaction terms. Doing so rectified these issues, resulting in constant degrees of freedom across response variables and consistent output among alternative models. The order of terms can have particularly strong effects on model output with unbalanced designs, thus we used Type II SSs because they do not depend on order of terms. Models using Type III SSs are also suitable and produced equivalent results.

We used Holm-Bonferroni adjusted pairwise t-tests to identify differences between treatment means. We used pot sums instead of averages because we are more interested in population level effects and because averages are confounded by uncontrolled variation in *Triadica* seedling ages.

Results

Main Effects

Window duration influenced pre-stress and final *Triadica* seedling abundance (Table 1) and presence (Table S1), pre-stress sums of *Triadica* heights and leaves (Table 2), and changes in leaf abundance during stress (Table 3). All increased as window duration increased (Figs. 1a–b, 2a–b), except changes in leaf abundance which decreased (Fig. 2f). Final *Triadica* abundances in 8 and 12 weeks window treatments were 2.2-fold higher than the 4 weeks treatment ($p = 0.026$ and 0.018, respectively; $p = 0.051$ for 10 versus 4 weeks treatments). Final heights and leaves and stem, leaf and aboveground biomasses were 62%, 24% and 2.4-, 2.0- and 2.2-fold greater, respectively, in longer windows than in the shortest, but these differences were not significant.

Stress type affected abundances of *Triadica* germinants and deaths during stress (Table 1), presence of germinants during stress (Table S1), final heights and leaves, *Triadica* stem, leaf, root, aboveground and total biomasses (Table 2), and changes in heights and leaves during stress (Table 3). Germinants and deaths were rare during flood stress but ~10-fold more frequent during drought (Fig. 1c). Otherwise, drought typically reduced *Triadica* performance; flood treatments demonstrated 46% greater final height, 2-fold greater final leaf abundance, 2.6-fold greater changes in height, 10-fold greater changes in leaf abundance (Figs. 2c–f), 2.8-fold more stem biomass, 3.5-fold more leaf biomass, and 3.1-fold more aboveground biomass (Figs. 3a–c). In 12 week window treatments, flood treatments increased *Triadica* root biomass 2.5-fold ($F_{1,23} = 5.2$, $p = 0.0364$) and total biomass 2.8-fold versus drought ($F_{1,23} = 6.3$, $p = 0.0231$; Fig. 3d–f).

Competition with native grasses impacted final *Triadica* seedling abundance and numbers of germinants and deaths during stress (Table 1), final presence of *Triadica* seedlings and presence of germinants (Table S1), final heights and leaves, *Triadica* stem, leaf and aboveground biomasses (Table 2), root:shoot, and changes in heights and leaves (Table 3). Competition decreased final seedling abundance 44% by reducing germinants 60% and increasing deaths 2.8-fold (Fig. 1b–d). Final presence of seedlings and presence of germinants decreased 44% and 68% with competition, respectively. Competition reduced final *Triadica* heights 16%, leaf abundance 26%, and stem, leaf and aboveground biomass by 33–

36% (Fig. 2c–d, 3a–c). *Triadica* growth during stress was essentially arrested with competition. Competition reduced increases in heights 99% (approximately zero change) and reduced increases in leaf abundance 116% (leaves were lost; Fig. 2e–f). Among 12 week window treatments, competition increased root:shoot 2.4-fold ($F_{1,23} = 7.2$, $P = 0.0165$; Fig. 3e).

Fertilization had no effect on abundance or presence of *Triadica* seedlings, germinants or deaths (Tables 1, S1). *Triadica* size and mass were independent of main effects of fertilization (Table 2), but depended on interactions with fertilization discussed below.

Interactions

The interaction of window duration and competition influenced abundance of *Triadica* germinants during stress (Table 1) and leaf and aboveground biomasses (Table 2). Germinant abundance was generally consistent across window treatments without competition, but it decreased as window duration increased with competition (Fig. 1c). Leaf and aboveground biomasses increased with window duration without competition, but not with competition (Fig. 2a). The 3-way interaction of window, stress type and competition affected final *Triadica* abundance and presence (Tables 1, S1). Without competition, abundances were greater in longer window treatments regardless of stress type. With competition, abundances in drought pots were highest in the shortest window and lowest in intermediate windows, whereas the opposite was true in flood pots (lowest in short, highest in intermediate; Fig. 1b). Final *Triadica* presence followed the same general pattern.

The interaction of stress type and competition affected *Triadica* leaf and aboveground biomasses and changes in leaf abundance (Tables 2, 3). In drought treatments, these biomasses were lowest and competition insignificantly reduced both. In flood treatments, biomasses were >3-fold higher overall and competition significantly reduced them 46–50% (Fig. 3a,c). Changes in leaf abundance were insignificantly lower with competition in flood, but significantly lower and negative with competition in drought (Fig. 2f). Changes in sums of heights depended on the interaction of stress and fertilization (Table 3). Fertilization increased vertical growth 4.5-fold during drought, but reduced increases 51% during flooding. Increases in height were similar in drought and flood treatments with fertilization.

The interaction of competition and fertilization influenced final *Triadica* stem, leaf and aboveground biomasses (Table 2). Fertilization increased these 79%, 57% and 69%, respectively, without competition but reduced them 39%, 35% and 37%, respectively, with competition.

Experimental treatments also influenced performance of native plants. Analyses and results pertaining to native plants are presented in the supporting information (Appendix S1, Fig. S1, Table S2).

Discussion

Reinvasion pressure is vital to restoration outcomes and costs, but it can vary widely among equivalently invaded habitats and is difficult to predict [1–3]. Invasive *Triadica sebifera* (Chinese tallow tree) exhibits rapid ontogenetic niche expansions in its moisture tolerance early in life [8]. This likely contributes to broad variation in average reinvasion pressure among restorations of *Triadica*-dominated ecosystems [7,8]. For plants with expanding abiotic niches like *Triadica*, our 'outgrow the stress' hypothesis holds that reinvasion pressure is determined by spatiotemporal availability of realized recruitment windows when exotic propagules are abundant [2]. This study tested two basic tenants of this hypothesis. (1) There is a minimum establishment time for exotics wherein they must germinate and grow to a stage and/or size capable of tolerating subsequent conditions. (2) Factors influencing growth can influence individual attainment of required tolerances and thus permit or preclude recruitment. Our results were not this black and white, but they clearly validate the core predictions of the 'outgrow the stress' hypothesis below.

Prediction 1

Recruitment success will scale with temporal availability of abiotic windows. Our results clearly show that longer periods of favorable conditions prior to water stress (i.e. greater abiotic window availability) increased *Triadica* abundance before and after subsequent stress periods. Overall, longer windows also increased final *Triadica* size, but benefits were less straightforward because longer windows had insignificant or negative impacts on some response variables in specific environmental contexts. This increased variance across window treatments and resulted in window duration having few significant main effects on size. Few studies have considered ontogenetic niche expansions. To our knowledge there are no other direct experimental tests of whether longer abiotic windows increase recruitment in stressful environments. However, observational studies often link longer periods between stressful events to increased recruitment. For example, Stokes [41] observed enhanced *Salix nigra* recruitment in areas subject to less frequent inundation. Manipulation of abiotic window durations in the field poses significant logistical challenges, but is necessary to experimentally test this prediction in a more natural setting. Window frequency is another aspect of window availability that should affect reinvasion in different ways and merits investigation.

Two metrics of *Triadica* success were reduced among longer window treatments, but only in specific contexts. There were fewer germinants during stress in longer windows with competition (Table 1, Fig. 1c), and seedlings gained fewer or lost more leaves among longer windows during drought (Table 2, Fig. 2f). Window duration also had insignificant or inconsistent effects on leaf and aboveground biomass in drought or competition treatments, respectively (Table 2, Fig. 3a). We expect the mechanism here reflects resource limitation and varying levels of demand. Water was most limiting in drought, and adding natives produced greater demand and competition for water. Longer windows exacerbated water limitation by increasing pre-stress size and abundance of *Triadica* seedlings and native plants (Figs. 1a, 2a–b, S1a–b), which produced higher total water demand. We observed similarly decreased success among the largest seedlings in the driest moisture treatments when investigating *Triadica*'s moisture niche expansions [8]. Reduced *Triadica* growth during flood stress with competition likely reflects reduced light availability (Figs. 2c–d, 3a–c). Fertilization may augment this effect during flood (Figs. 2e, 3b). Notably, we previously observed peak flood-induced mortality among *Triadica* seedlings shorter than sustained flood depths [8]. This likely explains mortality and final seedling abundances of zero observed in 4 week window treatments with flooding, which may reflect a strict minimum establishment time.

Prediction 2

If size confers tolerance [32], factors that increase or decrease growth rates will have similar effects on recruitment of plants with expanding niches during finite windows of opportunity in stressful environments. Our results clearly show that competition with native plants decreased *Triadica* size and mass, as well as abundance. Fertilization increased *Triadica* performance in some contexts, but did not affect abundance. This may be because field

soils utilized were relatively fertile. Total N content in soils near our collection site was 0.26% of dry mass in 2009 (Gabler and Siemann, unpublished data). Stress type treatments produced differences in size and mass often larger than between competition treatments, and it had strong effects on germinant and death abundances. Yet, stress type only affected final seedling abundance in interaction with window duration and competition. Decreases in *Triadica* abundance in drought treatments are unlikely to have been an effect of enhanced performance of the drought-tolerant native because native performance also decreased in drought (Appendix S1, Figure S1, Table S2). Rather, what were likely relatively weaker native competitors had a stronger overall competitive effect, presumably due to the scarcity of water.

Our findings largely support this prediction of the 'outgrow the stress' hypothesis. It holds that changes in growth rate affect recruitment by altering minimum establishment times, but it acknowledges that differences in required establishment time do not mandate differences in recruitment. Recruitment ultimately depends on whether minimum establishment times exceed the duration of favorable conditions available locally. Environmental factors should only influence recruit abundance when factors shorten or extend required establishment times across the critical threshold of local duration of suitable conditions. Size and mass are meaningful aspects of recruitment success and reinvasion pressure in their own rights, and may affect future survival in manifold ways.

Implications for Restoration and Management

We quantified baseline establishment times for *Triadica* across a variety of realistic environmental scenarios that are consistent with its documented ontogenetic niche expansions [8] and observed restoration outcomes in previously *Triadica*-dominated habitats [7] (Gabler and Siemann, unpublished data). We quantified how key factors interact to influence *Triadica* establishment and growth. These results increase our understanding of the mechanisms of reinvasion in *Triadica* and generally, and may improve predictions of exotic recruitment based on local conditions [2,31]. Enhanced understanding and predictions can inform management of *Triadica* and other exotics exhibiting niche expansions.

Our findings concern some phenomena beyond our control (e.g. drought). However, managers can take advantage of these events to increase management efficacy or efficiency. Native addition is commonly recommended to suppress invaders [3,42]. However, the literature disagrees over whether this is effective, especially for superior competitors like *Triadica* [3]. Adding natives decreased *Triadica* establishment and growth in this study, but not in all contexts. Results suggest fertilization can reduce *Triadica* performance in some contexts, but we do not recommend this because fertilization can have negative effects. The reinvasion pressure framework validated here can inform site selection and exotic management strategies during restoration or control efforts, especially where exotics exhibit ontogenetic niche expansions.

Supporting Information

Figure S1   Effects of experimental treatments on native plants.

Table S1   Results of ANODEVs using logistic regression models to test effects of experimental treatments on *Triadica* presence.

Table S2 Results of ANOVAs testing effects of experimental treatments on native plant abundance and performance.

Table S3 Experimental data used in this study.

Appendix S1 Analyses and results of experimental treatments on native plants.

Acknowledgments

We thank Amy Dunham, Volker Rudolf and an anonymous reviewer for valuable comments on earlier versions of this manuscript.

Author Contributions

Conceived and designed the experiments: CAG ES. Performed the experiments: CAG. Analyzed the data: CAG. Contributed reagents/materials/analysis tools: CAG ES. Wrote the paper: CAG. Provided statistical advice and key comments and revisions on earlier versions of this manuscript: ES.

References

1. Buckley YM, Bolker BM, Rees M (2007) Disturbance, invasion and re-invasion: managing the weed-shaped hole in disturbed ecosystems. Ecology Letters 10: 809–817.
2. Gabler CA, Siemann E (2012) Environmental variability and ontogenetic niche shifts in exotic plants may govern reinvasion pressure in restorations of invaded ecosystems. Restoration Ecology 20: 545–550.
3. Kettenring KM, Adams CR (2011) Lessons learned from invasive plant control experiments: a systematic review and meta-analysis. Journal of Applied Ecology 48: 970–979.
4. Grime JP (1998) Benefits of plant diversity to ecosystems: immediate, filter and founder effects. Journal of Ecology 86: 902–910.
5. Rinella MJ, Maxwell BD, Fay PK, Weaver T, Sheley RL (2009) Control effort exacerbates invasive-species problem. Ecological Applications 19: 155–162.
6. Epanchin-Niell RS, Hastings A (2010) Controlling established invaders: integrating economics and spread dynamics to determine optimal management. Ecology Letters 13: 528–541.
7. Donahue C, Rogers WE, Siemann E (2006) Restoring an invaded prairie by mulching live *Sapium sebiferum* (Chinese tallow trees): Effects of mulch on sapium seed germination. Natural Areas Journal 26: 244–253.
8. Gabler CA, Siemann E (2013) Rapid ontogenetic niche expansions in invasive Chinese tallow tree permit establishment in unfavourable but variable environments and can be exploited to streamline restoration. Journal of Applied Ecology DOI: 10.1111/1365-2664.12071.
9. Harper JL (1977) Population Biology of Plants. San Diego: Academic Press.
10. Holmgren M, Lopez BC, Gutierrez JR, Squeo FA (2006) Herbivory and plant growth rate determine the success of El Nino Southern Oscillation-driven tree establishment in semiarid South America. Global Change Biology 12: 2263–2271.
11. Grubb PJ (1977) The maintenance of species-richness in plant communities: the importance of the regeneration niche. Biological Reviews 52: 107–145.
12. Johnstone IM (1986) Plant invasion windows: A time-based classification of invasion potential Biological Reviews 61: 369–394.
13. Balke T, Bouma TJ, Horstman EM, Webb EL, Erftemeijer PLA, et al. (2011) Windows of opportunity: thresholds to mangrove seedling establishment on tidal flats. Marine Ecology-Progress Series 440: 1–9.
14. Peringer A, Rosenthal G (2011) Establishment patterns in a secondary tree line ecotone. Ecological Modelling 222: 3120–3131.
15. Davis MA, Grime JP, Thompson K (2000) Fluctuating resources in plant communities: a general theory of invasibility. Journal of Ecology 88: 528–534.
16. Shea K, Chesson P (2002) Community ecology theory as a framework for biological invasions. Trends in Ecology & Evolution 17: 170–176.
17. Arredondo-Nunez A, Badano EI, Bustamante RO (2009) How beneficial are nurse plants? A meta-analysis of the effects of cushion plants on high-Andean plant communities. Community Ecology 10: 1–6.
18. Chesson P (2000) Mechanisms of maintenance of species diversity. Annual Review of Ecology and Systematics 31: 343–366.
19. Vitousek PM, Aber JD, Howarth RW, Likens GE, Matson PA, et al. (1997) Human alteration of the global nitrogen cycle: Sources and consequences. Ecological Applications 7: 737–750.
20. Casper BB, Jackson RB (1997) Plant competition underground. Annual Review of Ecology and Systematics 28: 545–570.
21. Brewer JS, Cralle SP (2003) Phosphorus addition reduces invasion of a longleaf pine savanna (Southeastern USA) by a non-indigenous grass (*Imperata cylindrica*). Plant Ecology 167: 237–245.
22. Busey P (2003) Cultural management of weeds in turfgrass: A review. Crop Science 43: 1899–1911.

23. Tomassen HBM, Smolders AJP, Limpens J, Lamers LPM, Roelofs JGM (2004) Expansion of invasive species on ombrotrophic bogs: desiccation or high N deposition? Journal of Applied Ecology 41: 139–150.
24. Siemann E, Rogers WE, Grace JB (2007) Effects of nutrient loading and extreme rainfall events on coastal tallgrass prairies: invasion intensity, vegetation responses, and carbon and nitrogen distribution. Global Change Biology 13: 2184–2192.
25. Siemann E, Rogers WE (2007) The role of soil resources in an exotic tree invasion in Texas coastal prairie. Journal of Ecology 95: 689–697.
26. Bruce KA, Cameron GN, Harcombe PA, Jubinsky G (1997) Introduction, impact on native habitats, and management of a woody invader, the Chinese tallow tree, *Sapium sebiferum* (L) Roxb. Natural Areas Journal 17: 255–260.
27. Butterfield BJ, Rogers WE, Siemann E (2004) Growth of Chinese tallow tree (*Sapium sebiferum*) and four native trees under varying water regimes. Texas Journal of Science 56: 335–346.
28. Hall RBW, Harcombe PA (1998) Flooding alters apparent position of floodplain saplings on a light gradient. Ecology 79: 847–855.
29. Bartha S, Meiners SJ, Pickett STA, Cadenasso ML (2003) Plant colonization windows in a mesic old field succession. Applied Vegetation Science 6: 205–212.
30. Crawley MJ (1990) Rabbit grazing, plant competition and seedling recruitment in acid grassland. Journal of Applied Ecology 27: 803–820.
31. Young TP, Petersen DA, Clary JJ (2005) The ecology of restoration: historical links, emerging issues and unexplored realms. Ecology Letters 8: 662–673.
32. Kunstler G, Coomes DA, Canham CD (2009) Size-dependence of growth and mortality influence the shade tolerance of trees in a lowland temperate rain forest. Journal of Ecology 97: 685–695.
33. Aslan CE (2011) Implications of newly-formed seed-dispersal mutualisms between birds and introduced plants in northern California, USA. Biological Invasions 13: 2829–2845.
34. Harcombe PA, Hall RBW, Glitzenstein JS, Cook ES, Krusic P, et al. (1999) Sensitivity of Gulf Coast forests to climate change. LafayetteLA: US Department of the Interior, USGS, National Wetlands Research Center. 45–66 p.
35. Lin J, Harcombe PA, Fulton MR, Hall RW (2004) Sapling growth and survivorship as affected by light and flooding in a river floodplain forest of southeast Texas. Oecologia 139: 399–407.
36. Renne IJ, Gauthreaux SA, Gresham CA (2000) Seed dispersal of the Chinese tallow tree (*Sapium sebiferum* (L.) Roxb.) by birds in coastal South Carolina. American Midland Naturalist 144: 202–215.
37. Jones RH, McLeod KW (1989) Shade tolerance in seedlings of Chinese tallow tree, American sycamore, and cherrybark oak. Bulletin of the Torrey Botanical Club 116: 371–377.
38. Siemann E, Rogers WE (2003) Herbivory, disease, recruitment limitation, and success of alien and native tree species. Ecology 84: 1489–1505.
39. Donahue C, Rogers WE, Siemann E (2004) Effects of temperature and mulch depth on Chinese tallow tree (*Sapium sebiferum*) seed germination. Texas Journal of Science 56: 347–356.
40. Nijjer S, Lankau RA, Rogers WE, Siemann E (2002) Effects of temperature and light on Chinese tallow (*Sapium sebiferum*) and Texas sugarberry (*Celtis laevigata*) seed germination. Texas Journal of Science 54: 63–68.
41. Stokes KE (2008) Exotic invasive black willow (*Salix nigra*) in Australia: influence of hydrological regimes on population dynamics. Plant Ecology 197: 91–105.
42. Firn J, House APN, Buckley YM (2010) Alternative states models provide an effective framework for invasive species control and restoration of native communities. Journal of Applied Ecology 47: 96–105.

Effect of Different Fertilizer Application on the Soil Fertility of Paddy Soils in Red Soil Region of Southern China

Wenyi Dong[1], **Xinyu Zhang**[1]*, **Huimin Wang**[1], **Xiaoqin Dai**[1], **Xiaomin Sun**[1], **Weiwen Qiu**[2], **Fengting Yang**[1]

1 Key Laboratory of Ecosystem Network Observation and Modeling, Institute of Geographic Sciences and Natural Resources Research, Chinese Academy of Sciences, Beijing, People's Republic of China, **2** The New Zealand Institute for Plant and Food Research Limited, Christchurch, New Zealand

Abstract

Appropriate fertilizer application is an important management practice to improve soil fertility and quality in the red soil regions of China. In the present study, we examined the effects of five fertilization treatments [these were: no fertilizer (CK), rice straw return (SR), chemical fertilizer (NPK), organic manure (OM) and green manure (GM)] on soil pH, soil organic carbon (SOC), total nitrogen (TN), C/N ratio and available nutrients (AN, AP and AK) contents in the plowed layer (0–20 cm) of paddy soil from 1998 to 2009 in Jiangxi Province, southern China. Results showed that the soil pH was the lowest with an average of 5.33 units in CK and was significantly higher in NPK (5.89 units) and OM (5.63 units) treatments ($P<0.05$). The application of fertilizers have remarkably improved SOC and TN values compared with the CK, Specifically, the OM treatment resulted in the highest SOC and TN concentrations (72.5% and 51.2% higher than CK) and NPK treatment increased the SOC and TN contents by 22.0% and 17.8% compared with CK. The average amounts of C/N ratio ranged from 9.66 to 10.98 in different treatments, and reached the highest in OM treatment ($P<0.05$). During the experimental period, the average AN and AP contents were highest in OM treatment (about 1.6 and 29.6 times of that in the CK, respectively) and second highest in NPK treatment (about 1.2 and 20.3 times of that in the CK). Unlike AN and AP, the highest value of AK content was observed in NPK treatments with 38.10 mg·kg^{-1}. Thus, these indicated that organic manure should be recommended to improve soil fertility in this region and K fertilizer should be simultaneously applied considering the soil K contents. Considering the long-term fertilizer efficiency, our results also suggest that annual straw returning application could improve soil fertility in this trial region.

Editor: Peter Shaw, Roehampton University, United Kingdom

Funding: This work was supported by the Knowledge Innovation Program of the Chinese Academy of Sciences (KZCX2-EW-310) and National Natural Science Foundation of China (No. 41171153, 41001179). The authors declare that no additional external funding was received for this study. The funders had no role in study design, data collection and analysis, decision to publish, or preparation of the manuscript.

Competing Interests: The authors have declared that no competing interests exist.

* E-mail: zhangxy@igsnrr.ac.cn

Introduction

Red soils, which can be classified as Ultisols in the Soil Taxonomy System of the USA and Acrisols and Ferralsols in the FAO legend [1], occupy approximately 2.04 million km^2 in tropical and subtropical regions of China [1–3]. In these red soil regions, Rice (*Oryza sativa* L.) is the main cereal crop, contributing 19% and 29% of the world rice area and rice production, respectively [4]. Paddy soil, formed under interchange between drying and wetting rice field conditions, is considered to be the most important soil resource for the food security of China [5]. In recent years, due to the rapid population growth and a continuous decline in the amount of cultivated land area, the rate of fertilizer application keeps on rising in these regions in order to obtain high crop production in agriculture [6]. Nevertheless, instead of improving the soil structure and fertility, the long-term inappropriate fertilization has caused severe degradation of red soils, characterized by high acidity, low nutrients and a disturbed, unbalanced ecosystem [7]. Therefore, how to ameliorate degraded paddy soils in the red soil region and maintain the region's sustainable development of agricultural production has become an urgent problem.

Recently, soil quality has gained attention as a result of environmental issues related to soil degradation and production sustainability under different farming systems [8]. It has been considered by previous researches that the concentrations of soil nutrients (e.g., organic C, N, P, and K) are good indicators of soil quality and productivity because of their favorable effects on the physical, chemical, and biological properties of soil [9]. Soil pH affects the chemical reactions in soil [10]. Extremes of pH in soils, for example, will lead to a rapid increase in net negative surface charge and thus increases the soil's affinity for metal ions [11,12]. Soil organic components, such as soil organic carbon (SOC) or total N (TN) are the most critical indices of paddy soil fertility [13]. Dynamics of SOC and TN storage in agricultural soils drives microbial activity and nutrient cycles, promotes soil physical properties and water retention capacity, and reduces erosion [14]. Moreover, it has been recognized that soil available nutrients (including N, P and K), coming from mineralization and available components of fertilizer, can be directly absorbed by plants, contributing greatly to the soil fertility [15].

With the development of agricultural production, fertilization has been widely used as a common management practice to maintain soil fertility and crop yields [16]. Long-term field experiments (LTFEs) using different agronomic management can provide direct observations of changes in soil quality and fertility and can be predictions of future soil productivity and soil environment interactions [17,18]. Over past decades, a great number of long-term experiments were initiated to examine the effects of fertilization on soil fertility in the world [19–22]. Some studies have documented that the use of fertilizers was necessary, and that continuous fertilizer application increased the concentrations of SOC, TN and other nutrients in plough layers compared with the initial value at the beginning of the experiment [23–25]. Manure amendments markedly increased the contents of SOC, TN, and other available nutrients, and reduced soil acidification [13,26]. However, other studies have shown that the continued use of fertilizers may result in the decline of soil quality and productivity [27,28]. Long-term application of fertilizer was inadequate to maintain levels of nutrients, the SOC and TN significantly decreasing under the fertilizer treatment and the available N (AN), available P (AP) and available K (AK) did not show clear changes with time or between treatments despite some variation [29,30]. Thus, there's still some debate over the effect of different fertilization treatments on soil fertility.

Qianyanzhou experimental station, which was founded in 1982 by Chinese Academy of Sciences (CAS), is noted for its studies on the integrated development and management of natural resources in red earth hilly regions. It has been identified as the red soil hilly system of international experimental demonstration research station in UNESCO' s programme on Man and the Biosphere (MAB), and now has become an essential station of the Chinese Ecosystem Research Network (CERN) [31]. In paddy soil of southern China, some LTFEs have been initiated [32,33]. These experiments have provided basic data for research into paddy soil. However, most of them dealt with different chemical fertilizer rates in long-term experiments but few studies focused on different types of fertilizers (such as green manure and rice straw return) affecting the soil fertility. A thorough understanding of how these fertilizers and varying management practices affect the long-term soil fertility of conventional cropping systems is still lacking in Qianyanzhou region, one of the most important typical red earth hill regions in China. In this study, five fertilization treatments (no fertilizer, rice straw return, chemical fertilizer, organic manure and green manure) were applied. Our objectives were to (i) assess the changes of soil fertility parameters in Qianyanzhou from 1998 to 2009 (ii) evaluate the effects of different fertilization treatments on soil fertility parameters, and (iii) put forward suggestions to improve soil fertility in agricultural regions of southern China.

Materials and Methods

Experimental Site

Qianyanzhou Experimental Station (115°03′29.2″E, 26°44′29.1″N) of CAS is situated on the typical red earth hill region in the mid-subtropical monsoon landscape zone of Taihe County, Jiangxi Province, China. The average elevation is approximately 100 m, and relative relief is 20–50 m. Qianyanzhou Experimental Station has a subtropical monsoon climate. According to the statistics of the meteorological data, the mean annual temperature at this site is 17.8°C, and the annual active accumulated temperatures (above 0 and 10 degrees Celsius) are 6543.8 and 5948.2 degrees Celsius respectively. The annual precipitation and evaporation are 1471.2 mm and 259.9 mm

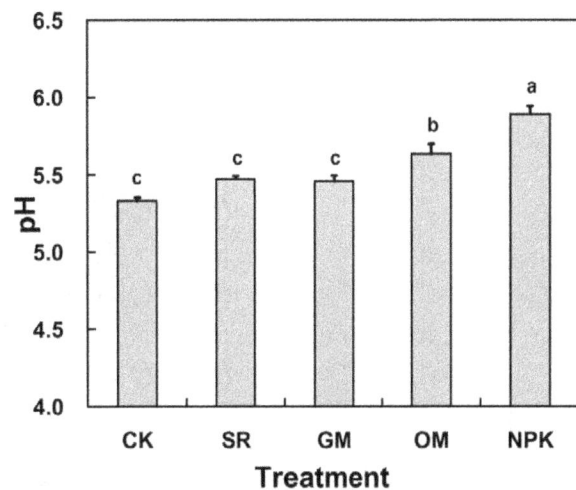

Figure 1. Average soil pH in the different fertilizer treatments. CK: no fertilizer; SR: straw returning rice field; NPK: chemical fertilizer; OM: organic manure and GM: green manure.

respectively, with the mean relative humidity of 83%. The frost-free period is 290 d and global radiation is 4223 $MJ \cdot m^{-2}$. Our experimental field is located in the flat floodplain where the soil-forming parent material consists of red sandstone and sandy conglomerate. Based on the investigation and analysis before our experiment, it can be concluded that paddy soil is the main soil type with bulk density of 1.50 $g \cdot cm^{-3}$ (0–20 cm), pH of 5.97, soil organic carbon of 9.71 $g \cdot kg^{-1}$, total N content of 1.02 $g \cdot kg^{-1}$, available P content of 1.56 $mg \cdot kg^{-1}$ and available K content of 17.61 $mg \cdot kg^{-1}$.

Experimental Design

A long-term fertilization experiment was conducted initially in 1998 under a double rice cropping system (rice-rice-winter fallow) which is one of the most common cropping systems in the region. Summer rice was sown at the end of April and harvested in July. Winter rice was sown at the end of July and harvested in November. During the growing season, hand weeding was done to control weeds.

There were five treatments in total: no fertilizer (CK), straw returning rice field (SR), chemical fertilizer (NPK), organic manure (OM) and green manure (GM). All treatments were arranged in a randomized block design with three replications [25,34,35], totalling 15 plots. Each plot was 15 m^2 (3 m×5 m) and was isolated by concrete walls (50 cm depth and 15 cm above the soil surface). These fertilization systems were chosen based on several common fertilization experiences from local farmers. In SR treatment, all the aboveground rice residues were returned to the soil after harvest, about 4500 $kg \cdot hm^{-2}$ on dry weight. In NPK treatments, inorganic fertilizers were applied at the rates of N-P_2O_5-K_2O at 225-135-225 $kg \cdot hm^{-2}$ by using urea, calcium-magnesium phosphate and potassium chloride. Before sowing, 60% of N, P and K fertilizer were applied as base fertilizers and the remaining fertilizer was applied as top-dressing. In OM treatment, organic manure which came from the faeces of pigs was applied at the rate of 4100 $kg \cdot hm^{-2}$ fresh weight. All pig manure used in our experiment came from a pig farm in Taihe County, where the composting-process was conducted at high temperatures and a good organic fertilizer was obtained after a few months of fermentation by sterilization, deodorization, and so on [36,37]. In GM treatment, fresh Chinese milk vetch (*Astragalus sinicus* L.) was

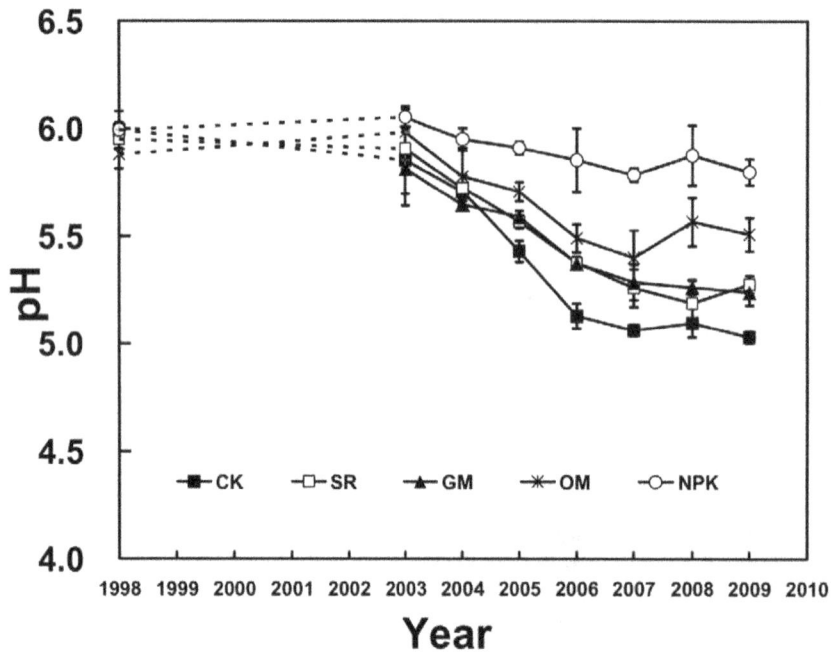

Figure 2. Dynamics of soil pH in the different fertilizer treatments during 1998–2009. Soil samples for 1999–2002 were not analyzed (dashed lines).

applied at the rate of 22500 kg·hm^{-2} fresh weight according to the local conventional green manure application rate.

Soil Sampling and Analysis

Soil samples in the 15 plots were collected annually during 2003–2009 at 7–10 days after the harvest of the late rice. To reflect the real effect of long-term fertilization on the soil fertility, the data in the first 5 years of this experiment were not obtained to analyze in our paper. In each plot, soils were sampled with an auger with 5 cm internal diameter in the plough layer (0–20 cm) at five randomly selected locations and then mixed as one sample [35,38,39]. All fresh soil samples were air-dried and sieved through

a 2.0 mm and 0.25 mm sieve and stored for nutrient analysis [38,39].

Soil physical and chemical properties were measured using the methods described by Bao [40]. Soil pH was measured with glass electrode in a 1:2.5 soil/water suspension. SOC was measured by a K_2CrO_7-H_2SO_4 oxidation procedure and TN by the Kjeldahl method. Soil C/N values were calculated as the ratio SOC to TN. AN was determined by using a micro-diffusion technique after alkaline hydrolysis. AP was determined by the Olsen method. AK was measured by flame photometry after NH_4OAc neutral extraction.

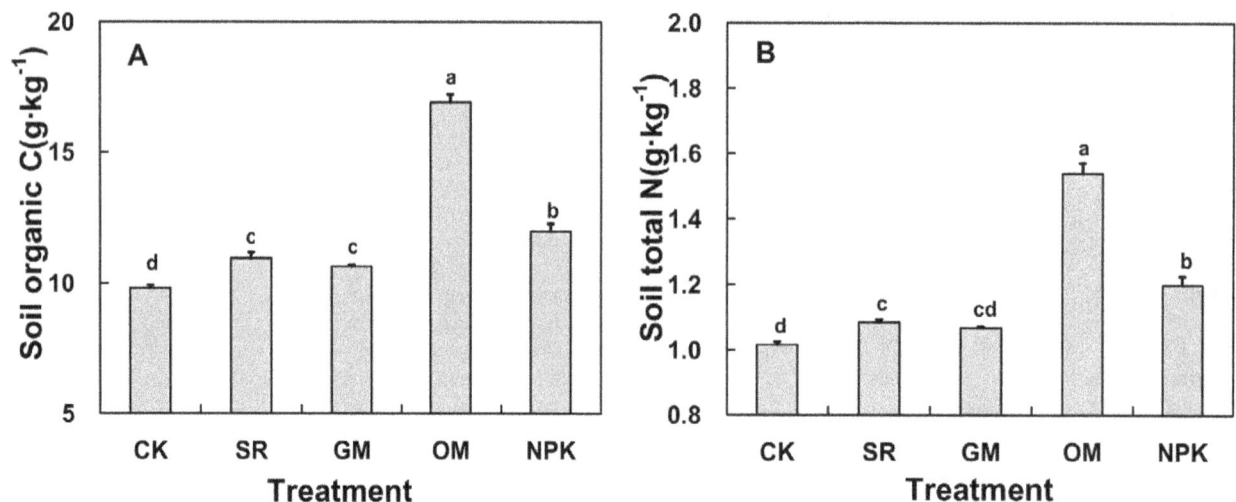

Figure 3. Average SOC and TN in the different fertilizer treatments. CK: no fertilizer; SR: straw returning rice field; NPK: chemical fertilizer; OM: organic manure and GM: green manure.

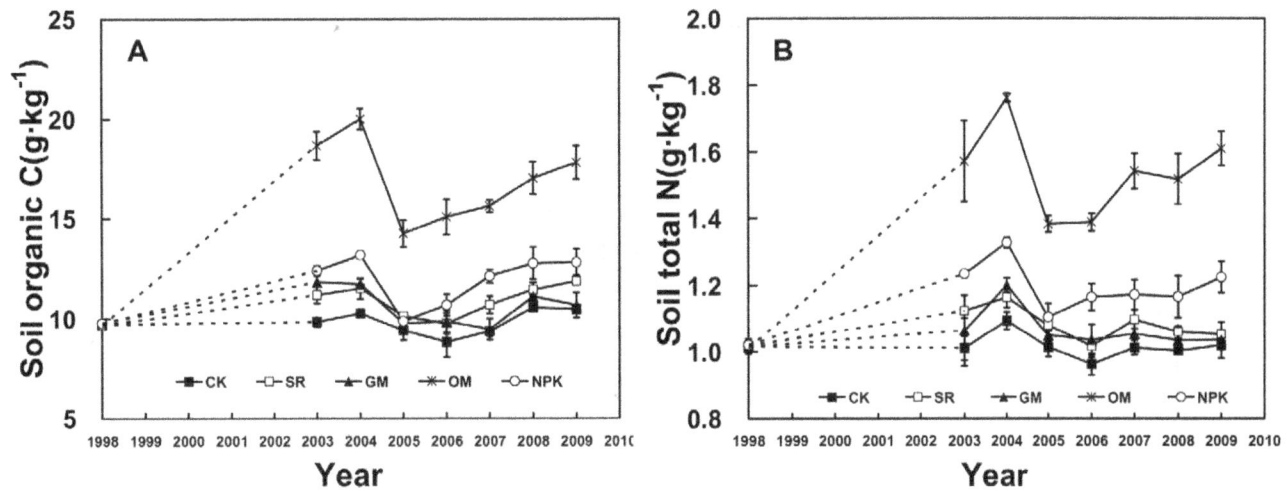

Figure 4. Dynamics of SOC and TN in the different fertilizer treatments during 1998–2009. Soil samples for 1999–2002 were not analyzed (dashed lines).

Data Analyses

All results were reported as means ± standard error (SE) for three replicates. One-way of variance (ANOVA) and Duncan's multiple comparisons were performed to determine the differences among the fertilizer treatments in terms of the long-term soil nutrient contents means (during 2003–2009). Then the annual values under different treatments were used to investigate the dynamics of nutrient contents during the whole period. All statistical analyses were performed using the SPSS software package (version 15.0) (Statistical Graphics Crop, Princeton, USA). A difference at $P<0.05$ level was considered as statistically significant.

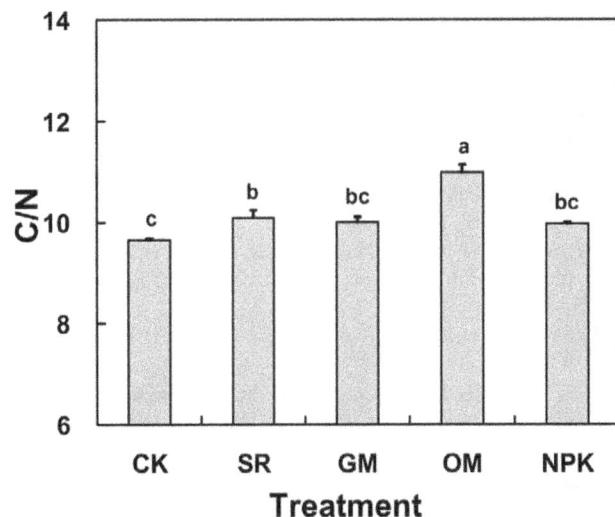

Figure 5. Average C/N ratios in the different fertilizer treatments. CK: no fertilizer; SR: straw returning rice field; NPK: chemical fertilizer; OM: organic manure and GM: green manure.

Results

Effects of Different Fertilizer Treatments on Soil pH

The average soil pH was shown in Fig.1. Statistical analysis revealed that fertilization treatments led to a significant increase in soil pH compared with the CK treatment ($P<0.05$). The soil pH was the lowest in CK with an average of 5.33 units and highest in NPK treatment with 5.89 units. In OM. treatment, the soil pH was relatively higher than CK (reaching 5.63 units).

During 1998 to 2009, the soil pH in NPK treatment appeared relatively stable despite some slight drop with the time (Fig.2). However, the values in other treatments showed a clear decline trend with time despite some variations. In the beginning years after fertilization, there was no evident difference in pH among all treatments, but eventually the soil pH in CK reduced dramatically and declined sharply from 5.71 to 5.03 (0.68 units lower). In SR, GM and OM treatments, the soil pH values declined by 0.57, 0.57 and 0.27 units respectively during the experimental period.

Effects of Different Fertilizer Treatments on Soil Organic C and Total N

The SOC and TN contents showed statistically significant differences among the five treatments (Fig.3). We observed that the application of fertilizers (especially OM and NPK fertilizers) had remarkably improved SOC and TN values compared with the CK. Specifically, the OM treatment resulted in the highest SOC and TN concentrations (16.93 and 1.54 g·kg^{-1}, respectively), which was 72.5%. and 51.2%[1] higher than that of CK. The SOC and TN in NPK treatment were significantly higher than CK, reaching 11.97 and 1.20 g·kg^{-1}, respectively. While in SR and GM treatment, the SOC was remarkably higher than CK, reaching 10.94 and 10.64 g·kg^{-1} respectively, but significant differences in TN contents between GM and CK were not observed.

The SOC in different treatments had a similar trend over time (Fig.4A). From 1998 to 2004, the SOC showed a clear increase with time due to fertilization, rising from initial 9.65–9.78 g·kg^{-1} to 11.51–20.00 g·kg^{-1} in 2004, respectively. Then SOC content dropped sharply but quickly reached at stable level. It was also obtained that the SOC content in OM was obviously higher than

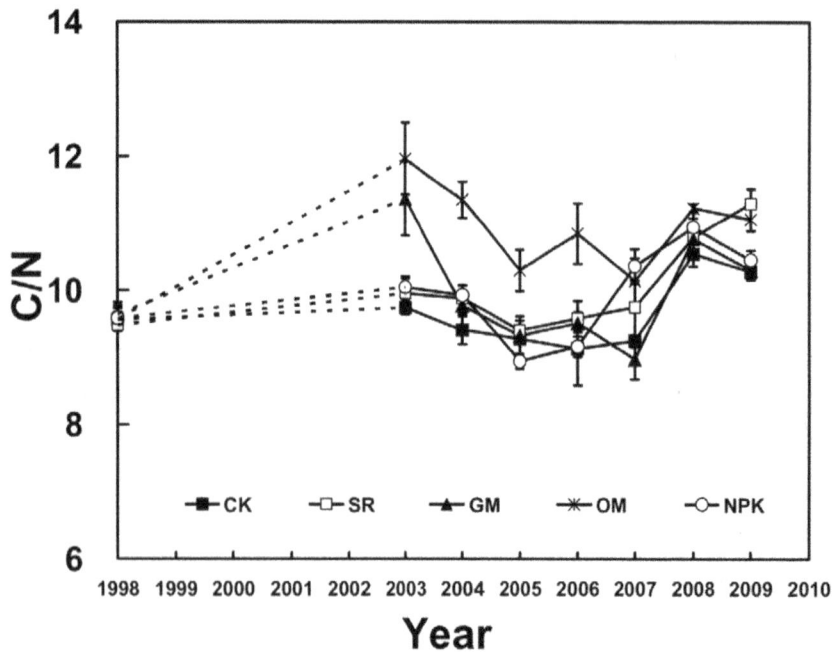

Figure 6. Dynamics of soil C/N ratios in the different fertilizer treatments during 1998–2009. Soil samples for 1999–2002 were not analyzed (dashed lines).

the other treatments during the experiment period, whereas that in CK remained relative stable (about 10 g·kg^{-1}).

The dynamics of TN content in the five treatments followed similar patterns with SOC during 1998–2009 (Fig.4B). In the first few years, TN content tended to increase rapidly in the OM treatment (from 1.01 to 1.76 g·kg^{-1}), followed by NPK treatment (from 1.02 to 1.33 g·kg^{-1}). Thereafter, both of them declined and then maintained a certain level. Meanwhile, the soil TN contents in SR, GM and CK treatments were relatively steady, at approximately 1.05 g·kg^{-1}.

Effects of Different Fertilizer Treatments on Soil C/N Ratio

There were marked differences in soil C/N ratio among different treatments due to fertilizer application (Figure5). The average C/N ratio in the OM treatment (10.98) was obviously higher than the other treatments ($P<0.05$, Fig.5). Similarly, in SR treatment, the C/N ratio was significantly higher than CK. Nevertheless, there were no significant differences of C/N ratios in the CK, SR, GM and NPK treatments, ranging from 9.66 to 10.00.

Dynamics of soil C/N ratios during 1998–2009 are shown in Fig.6. From 1998 to 2003, the C/N ratios increased sharply in OM and GM treatments (25.0%and 17.9% higher than the initial amount), and then the values declined slowly and constantly till 2007 (reaching 10.14 and 8.98 respectively). However, the other treatments including CK, SR and NPK fluctuated at a stable level (approximately 10.0) from 1998 to 2007. In the last two years of the experiment period, all the five treatments displayed similar trends without a significant difference varying slightly between 8.98 and 11.29.

Effects of Different Fertilizer Treatments on Soil Available Nutrients

A comparison of available nutrients among the treatments indicated that the fertilizer had a notable influence on soil AN, AP

and AK ($P<0.05$, Fig.7A–C). During the experiment period, the average AN and AP contents in OM were highest (about 1.6 and 29.6 times of the CK, respectively) and second was NPK treatment (about 1.2 and 20.3 times of the CK). However, there were no obvious differences of AN and AP between SR, GM and CK treatments. Unlike AN and AP, the highest value of AK content was found in NPK treatment, which was 38.10 mg·kg^{-1} (about 2.2 times of the CK), and there were no obvious differences among the other four treatments (Fig.7C).

AN in the OM treatment obviously increased with time due to fertilization at the beginning, and then the value tended to rise with a slight fluctuation before remaining at the highest level in comparison to the other treatments. We also found a similar trend in the NPK treatment but lower than OM and did not find the significant differences between SR, GM and CK (Fig.8A).

During the entire experiment period, AP concentrations in CK, SR and GM treatments remained at an extremely low level (approximately 1.80 mg·kg^{-1}) and were almost the same as the initial values. On the contrary, AP in both OM and NPK treatments displayed similar changes over time. Specifically, the value rose sharply in the first few years of fertilization (63.59 and 45.54 mg·kg^{-1} respectively in 2003) then progressively reduced till 2007 (25.37 and 16.68 mg·kg^{-1} respectively) and later increased slightly (Fig.8B).

The NPK treatment significantly increased AK content, especially from 2007 to 2009. However, the significant changes of AK in other treatments with time were not observed, maintaining at a stable and low level. (Fig.8C).

Discussion

Many experiments have been conducted on the relationship between fertilization and soil pH [35,41]. Some studies demonstrated that the soil pH was decreased to a certain extent with different fertilizer treatments [6]. In our study, the soil pH tended to drop in different treatments with time (Fig.2), and the decline

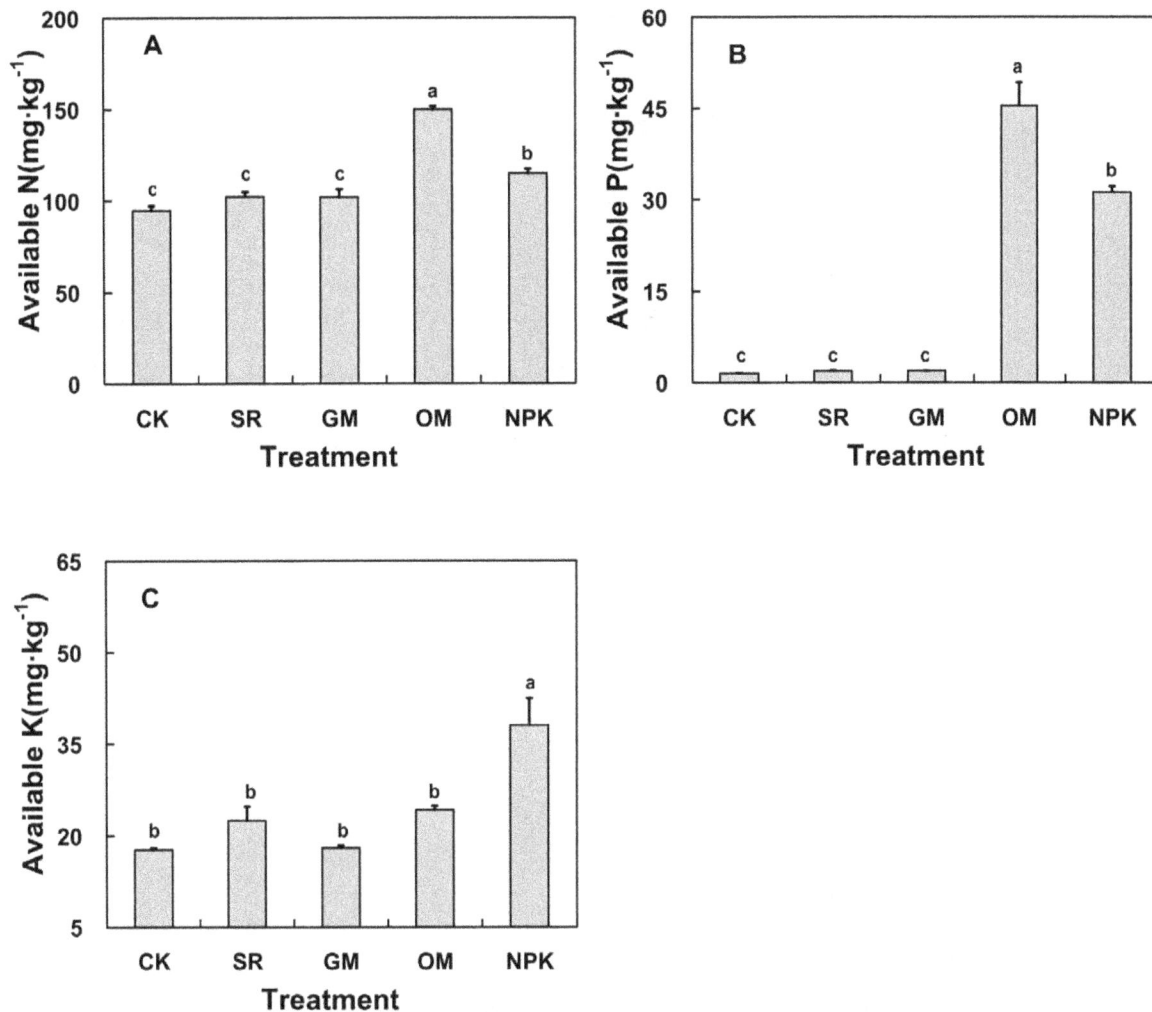

Figure 7. Average soil AN, AP and AK in the different fertilizer treatments. CK: no fertilizer; SR: straw returning rice field; NPK: chemical fertilizer; OM: organic manure and GM: green manure.

rate in the NPK and OM treatments were relatively lower than CK, but producing higher pH values in the NPK and OM treatments than in the CK (Fig.1), which suggested that chemical fertilizer and organic manure could alleviate soil acidification to some extent. It has been reported that the application of alkaline fertilizer (e.g. calcium magnesium phosphate fertilizer) would return some alkaline substance to soils and thus increase the soil pH [42]. In addition, the application of organic manure could improve soil acidity by increasing the soil organic matter, promoting the soil maturation, improving the soil structure, and enhancing the soil base saturation percentage, which is in line with Zhang [35] and Li [43]. Moreover, studies showed that the soil pH in CK was lower than the initial value, which indicated that the acid deposition could have a great influence on the soil acidification in this trial region [44]. Since a too high or too low pH is harmful to the crop growth, it might be a practicable measure to establish the proper range of soil pH through fertilizer use.

Soil organic matter is a key contributor to soil due to its capacity to affect plant growth indirectly and directly [45]. As SOC and TN constitute heterogeneous mixtures of organic substances, they are widely used as the main parameters for evaluating soil fertility [46]. Meanwhile, human activities such as fertilizer practices and

cropping systems play a key role in the regulation of C and N contents in agricultural soils [47]. In our experiments, SOC and TN contents increased considerably in the fertilization treatments compared with CK, especially in OM and NPK treatments (Fig.3), suggesting that organic and chemical fertilizer are beneficial to the accumulation of soil organic matter and thus improves soil fertility. This may be because both the application of organic manure and chemical fertilizer can improve soil aggregation, soil water retention, and reduce bulk density of the soil in the plough layer, promoting crop growth and the return of more root residues to the soil [48]. Under the SR application, SOC content was increased gradually with time, suggesting that the continuous SR supply had a positive effect in sustaining SOC, in accordance with the finding of Nie [49]. In farmland ecosystems, pig manure is easy to accumulate in soil and has lower ammonia losses than other fertilizers [33], so the TN concentration was significantly increased by continual annual OM applications compared to the other treatments (Fig.4B).

Previous studies have pointed out that the soil C/N ratio plays a key role in mineralization and accumulation of SOC and TN contents and that a high C/N ratio generally slows the mineralization process [47,50].As found by Zhang [51], the C/N ratio was obviously higher in OM applications than the other

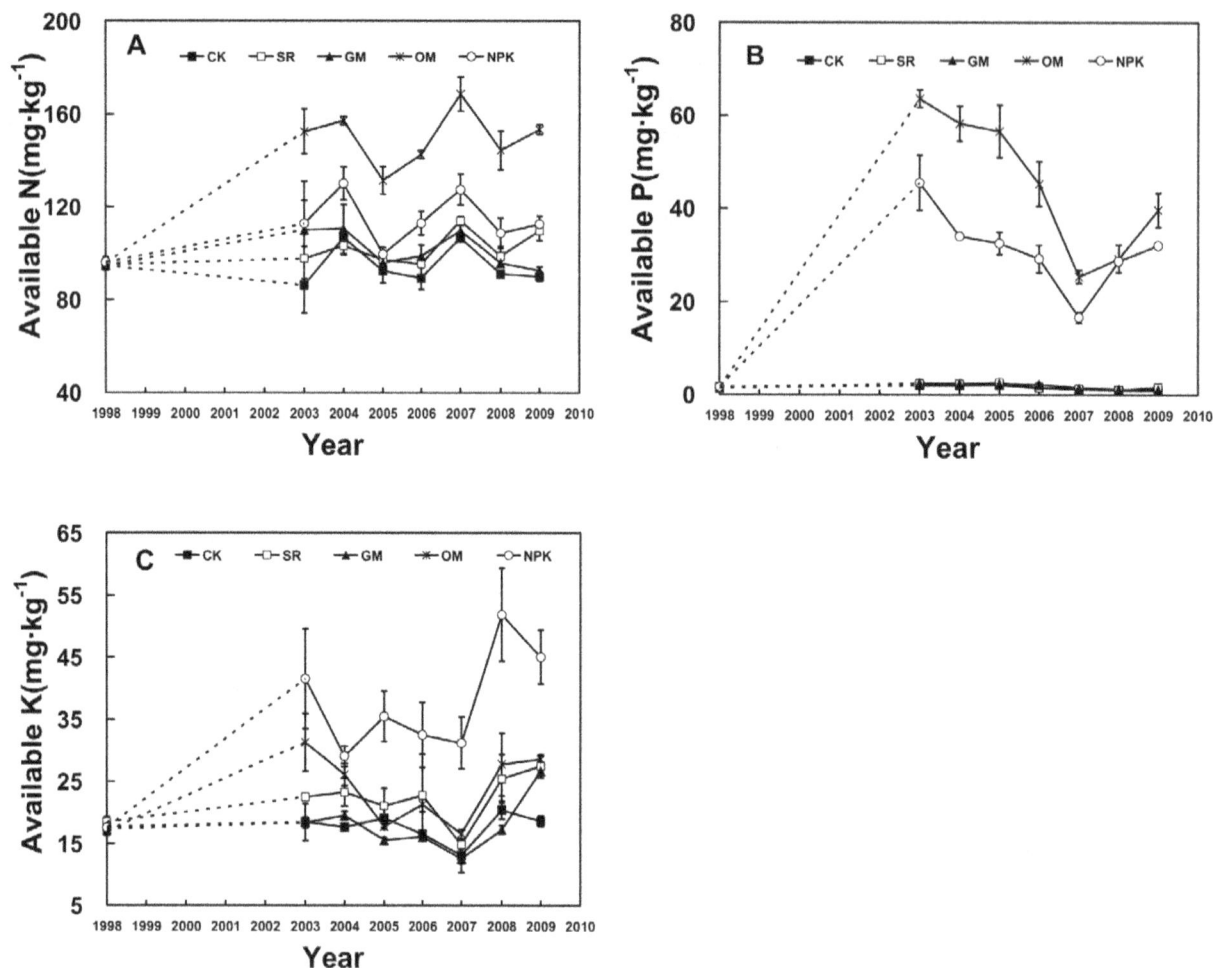

Figure 8. Dynamics of soil AN, AP and AK in the different fertilizer treatments during 1998–2009. soil samples for 1999–2002 were not analyzed (dashed lines).

treatments attributed to the SOC was more markedly enriched relative to TN. Moreover, we observed that there was a slight rise in soil C/N ratios with time in SR and NPK treatments (Fig.6), suggesting both SOC and TN accumulated gradually with time and the increase in SOC buildup was more rapid than TN over the past years in SR and NPK treatments [52].

It is known that fertilization is crucial for maintaining soil available nutrient levels, because fertilization ensures the largely constant presence of active microorganisms and the regular dynamic of biomass carbon [53]. Our research also showed that soil available nutrient contents were significantly affected by different fertilization treatments. Long-term OM application led to significantly higher values of soil AN and AP, compared to the other fertilization treatments (Fig.8A and 8B). It has also been reported by Huang [25] that significant AN and AP increases were observed in the manure-applied treatments. In addition, the AN and AP contents were maintained at a very higher level in the NPK treatments (Fig.8C) as a result of the long-term high inputs of N and P fertilizers [54], whereas AN and AP did not show statistically significant differences between CK, SR and GM treatments. Unlike AN and AP, AK was the highest in NPK treatments in our results, followed by the OM treatments, demonstrating that the K supplement from chemical and manure fertilizers are important and this may be the most advantageous

way to solve the problem in China, where K resources are quite limited. Previous studies have shown that application of rice straw significantly increased available K while increasing organic matter contents [54,55]. Similarly, the AK in SR treatment was higher than CK in our study. It was noted that the SOC, TN, C/N ratios and available nutrients had no significant difference under GM treatment in our study. Previous study had described that the decomposition process of green manure (as a fresh organic matter) was very slow and complicated, affected by soil temperature, moisture, plough back time and so on [56]. Hence, the real mechanism of the nutrient release of green manure deserves further study in order to make better use of green manure and increase its fertilizer efficiency in our study region.

Considering the dynamic changes of soil available nutrient content, we found that AN significantly increased with time from 1998 to 2009 in OM and NPK treatments, suggesting that the long-term soil organic matter played a major role in releasing soil AN. The evident increase of AK in NPK treatment over time in our study, suggests that continuously applying K fertilizer would dramatically improve the soil AK supply, In addition, AP values in both OM and NPK treatments were increased greatly compared to the other three treatments during the whole experiment period. This is consistent with many previous studies showing the accumulation of P is the most obvious and long-term manure or inorganic fertilizer application can significantly alter the amounts and proportion of labile and stable soil P pools [57–59]. That the OM treatment resulted in even higher AP value than NPK, also support the view that organic fertilizer is much more conducive to soil P availability rather than commercial P fertilizers in cropping systems that receive predominantly organic P amendments [59].

In conclusion, significant differences in soil fertility of paddy soils in the red soil region of southern China among different fertilization treatments were found in our study. Application of OM and NPK resulted in a substantial increase of SOC, TN, C/N ratios, AN and AP contents relative to the other fertilization treatments. Thus, it is likely that the OM and NPK application improves soil fertility. Meanwhile, the application of NPK would increase soil AK, leading to the highest AK contents. Continuous application of SR also had a positive effect in sustaining SOC, TN and C/N ratios. However, the effect of GM application on soil fertility was not remarkable compared to CK. Hence, organic manure should be recommended to improve soil fertility in this region and K fertilizer should be simultaneously applied considering the soil K contents. Moreover, in terms of the long-term

fertilizer efficiency, annual straw returning application year by year could be adopted in this trail region.

Supporting Information

Table S1 Data statistic and analysis for samples in this study. The detailed data of soil nutrient contents in different fertilization treatments for each samples in this study is shown.

Acknowledgments

We thank Jingdong Zou in the Qianyanzhou experimental Station of Chinese Ecosystem Research Network (CERN) for many sampling and analysis work.

Author Contributions

Conceived and designed the experiments: WD XZ XS. Performed the experiments: WD XZ XS. Analyzed the data: WD XZ. Contributed reagents/materials/analysis tools: HW XD FY. Wrote the paper: WD XZ XS WQ.

References

1. FAO-Unesco(1974)Soil Map of the World 1: 5000000. Legend, Volume 1. Unesco, Paris.
2. Li CK (1983) Red Soils of China. Science Press, Beijing.
3. Zhang MK, Xu JM (2005) Restoration of surface soil fertility of an eroded red soil in southern China. Soil and Tillage Research 80: 13–21.
4. Sun W, Huang Y (2011) Global warming over the period 1961–2008 did not increase high-temperature stress but did reduce low-temperature stress in irrigated rice across China. Agricultural and Forest Meteorology 151: 1193–1201.
5. Ma L, Yang LZ, Ci E, Cheng YQ, Wang Y, et al. (2009) Effects of long-term fertilization on distribution and mineralization of organic carbon in paddy soil. Acta Pedologica Sinica 46: 1050–1058.
6. Wang BR, Cai ZJ, Li DC (2010) Effect of different long-term fertilization on the fertility of red upland soil. Journal of Soil and Water Conservation 24: 85–88.
7. Chen WC, Wang KR, Xie XL (2009) Effects on distributions of carbon and nitrogen in a reddish paddy soil under long-term different fertilization treatments. Chinese Journal of Soil Science 40: 523–528.
8. Galantini J, Rosell R (2006) Long-term fertilization effects on soil organic matter quality and dynamics under different production systems in semiarid Pampean soils. Soil and Tillage Research 87: 72–79.
9. Cao C, Jiang S, Ying Z, Zhang F, Han X (2011) Spatial variability of soil nutrients and microbiological properties after the establishment of leguminous shrub Caragana microphylla Lam. plantation on sand dune in the Horqin Sandy Land of Northeast China. Ecological Engineering 37: 1467–1475.
10. Zhao J, Dong Y, Xie X, Li X, Zhang X, et al. (2011) Effect of annual variation in soil pH on available soil nutrients in pear orchards. Acta Ecologica Sinica 31: 212–216.
11. Wu Z, Gu Z, Wang X, Evans L, Guo H (2003) Effects of organic acids on adsorption of lead onto montmorillonite, goethite and humic acid. Environmental Pollution 121: 469–475.
12. Yang JY, Yang XE, He ZL, Li TQ, Shentu JL, et al. (2006) Effects of pH, organic acids, and inorganic ions on lead desorption from soils. Environmental Pollution 143: 9–15.
13. Liu M, Li ZP, Zhang TL, Jiang CY, Che YP (2011) Discrepancy in response of rice yield and soil fertility to long-term chemical fertilization and organic amendments in paddy soils cultivated from Infertile upland in subtropical China. Agricultural Sciences in China 10: 259–266.
14. Manna M, Swarup A, Wanjari R, Mishra B, Shahi D (2007) Long-term fertilization, manure and liming effects on soil organic matter and crop yields. Soil and Tillage Research 94: 397–409.
15. Vogeler I, Rogasik J, Funder U, Panten K, Schnug E (2009) Effect of tillage systems and P-fertilization on soil physical and chemical properties, crop yield and nutrient uptake. Soil and Tillage Research 103: 137–143.
16. Shen JP, Zhang LM, Guo JF, Ray JL, He JZ (2010) Impact of long-term fertilization practices on the abundance and composition of soil bacterial communities in Northeast China. Applied Soil Ecology 46: 119–124.
17. Blair N, Faulkner R, Till A, Poulton P (2006) Long-term management impacts on soil C, N and physical fertility: Part I: Broadbalk experiment. Soil and Tillage Research 91: 30–38.
18. Li BY, Huang SM, Wei MB, Zhang HL, Shen AL, et al. (2010) Dynamics of soil and grain micronutrients as affected by long-term fertilization in an aquic Inceptisol. Pedosphere 20: 725–735.
19. Mitchell CC, Westerman RL, Brown JR, Peck TR (1991) Overview of long-term agronomic research. Agronomy Journal 83: 24–25.
20. Bhandari A, Sood A, Sharma K, Rana D (1992) Integrated nutrient management in a rice-wheat system. Journal of the Indian Society of Soil Science 40: 742–747.
21. Dawe D, Dobermann A, Moya P, Abdulrachman S, Singh B, et al. (2000) How widespread are yield declines in long-term rice experiments in Asia? Field Crops Research 66: 175–193.
22. Ladha J, Dawe D, Pathak H, Padre A, Yadav R, et al. (2003) How extensive are yield declines in long-term rice-wheat experiments in Asia? Field Crops Research 81: 159–180.
23. Whitbread A, Blair G, Konboon Y, Lefroy R, Naklang K (2003) Managing crop residues, fertilizers and leaf litters to improve soil C, nutrient balances, and the grain yield of rice and wheat cropping systems in Thailand and Australia. Agriculture, Ecosystems & Environment 100: 251–263.
24. Bi L, Zhang B, Liu G, Li Z, Liu Y, et al. (2009) Long-term effects of organic amendments on the rice yields for double rice cropping systems in subtropical China. Agriculture, Ecosystems & Environment 129: 534–541.
25. Huang S, Zhang W, Yu X, Huang Q (2010) Effects of long-term fertilization on corn productivity and its sustainability in an Ultisol of southern China. Agriculture, Ecosystems & Environment 138: 44–50.
26. Gu YF, Zhang XP, Tu SH, Lindström K (2009) Soil microbial biomass, crop yields, and bacterial community structure as affected by long-term fertilizer treatments under wheat-rice cropping. European Journal of Soil Biology 45: 239–246.
27. Kumar A, Yadav DS (2001) Long-term effects of fertilizers on the soil fertility and productivity of a rice–wheat System. Journal of Agronomy and Crop science 186: 47–54.
28. Yang S (2006) Effect of long-term fertilization on soil productivity and nitrate accumulation in Gansu oasis. Agricultural Sciences in China 5: 57–67.
29. Shen J, Li R, Zhang F, Fan J, Tang C, et al. (2004) Crop yields, soil fertility and phosphorus fractions in response to long-term fertilization under the rice monoculture system on a calcareous soil. Field Crops Research 86: 225–238.
30. Su YZ, Wang F, Suo DR, Zhang ZH, Du MW (2006) Long-term effect of fertilizer and manure application on soil-carbon sequestration and soil fertility under the wheat–wheat–maize cropping system in northwest China. Nutrient Cycling in Agroecosystems 75: 285–295.
31. Li J, Liu Y, Yang X (2006) Studies on water-vapor flux characteristic and the relationship with environmental factors over a planted coniferous forest in Qianyanzhou Station. Acta Ecologica Sinica 26: 2449–2456.
32. Yuan YH, Li HX, Huang QR, Hu F, Pan GX, et al. (2008) Effects of long-term fertilization on dynamics of soil organic carbon in red paddy soil. Soils 40: 237–242.
33. Chen AL, Xie XL, Wen WY, Wang W, Tong CL (2010) Effect of long term fertilization on soil profile nitrogen storage in a reddish paddy soil. Acta Ecologica Sinica 30: 5059–5065.
34. Sikka R, Kansal BD (1995) Effect of fly-ash application on yield and nutrient composition of rice, wheat on pH and available nutrient status of soil. Bioresource Technology 51: 199–203.
35. Zhang HM, Wang BR, Xu MG, Fan TL (2009) Crop yield and soil responses to long-term fertilization on a red soil in southern China. Pedosphere 19: 199–207.
36. Bhamidimarri SMR, Pandey SP (1996) Aerobic thermophilic composting of piggery solid wastes. Water Science and Technology 33: 89–94.
37. Imbeah M (1998) Composting piggery waste: a review. Bioresource Technology 63: 197–203.

38. Wang YC, Wang EL, Wang DL, Huang SM, Ma YB, et al. (2010) Crop productivity and nutrient use efficiency as affected by long-term fertilization in North China Plain. Nutrient Cycling in Agroecosystems 86: 105–119.

39. Kapkiyai JJ, Karanja NK, Qureshi JN, Smithson PC, Woomer PL (1999) Soil organic matter and nutrient dynamics in a Kenyan nitisol under long-term fertilizer and organic input management. Soil Biology and Biochemistry 31: 1773–1782.

40. Bao SD (2005) Soil and Agricultural Chemistry Analysis. Agriculture Press, Beijing. (In Chinese).

41. Daugelene N, Butkute R (2008) Changes in phosphorus and potassium contents in soddy-podzolic soil under pasture at the long-term surface application of mineral fertilizers. Eurasian Soil Science 41: 638–647.

42. Wu XC, Li ZP, Zhang TL (2008) Long-term effect of fertilization on organic carbon and nutrients content of paddy soils in red soil region. Ecology and Environment 17: 2019–2023. (In Chinese).

43. Li BY, Huang SM, Wei MB, Zhang HL, Shen AL, et al. (2010) Dynamics of soil and grain micronutrients as affected by long-term fertilization in an aquic Inceptisol. Pedosphere 20: 725–735.

44. Liu KH, Fang YT, Yu FM, Liu Q, Li FR, et al. (2010) Soil acidification in response to acid deposition in three subtropical forests of subtropical China. Pedosphere 20: 399–408.

45. Lee SB, Lee CH, Jung KY, Park KD, Lee D, et al. (2009) Changes of soil organic carbon and its fractions in relation to soil physical properties in a long-term fertilized paddy. Soil and Tillage Research 104: 227–232.

46. Huang QR, Hu F, Huang S, Li HX, Yuan YH, et al. (2009) Effect of long-term fertilization on organic carbon and nitrogen in a subtropical paddy soil. Pedosphere 19: 727–734.

47. Tong C, Xiao H, Tang G, Wang H, Huang T, et al. (2009) Long-term fertilizer effects on organic carbon and total nitrogen and coupling relationships of C and N in paddy soils in subtropical China. Soil and Tillage Research 106: 8–14.

48. Hyvönen R, Persson T, Andersson S, Olsson B, Ågren GI, et al. (2008) Impact of long-term nitrogen addition on carbon stocks in trees and soils in northern Europe. Biogeochemistry 89: 121–137.

49. Nie J, Zhou J, Wang H, Chen X, Du C (2007) Effect of long-term rice straw return on soil glomalin, carbon and nitrogen. Pedosphere 17: 295–302.

50. Khalil M, Hossain M, Schmidhalter U (2005) Carbon and nitrogen mineralization in different upland soils of the subtropics treated with organic materials. Soil Biology and Biochemistry 37: 1507–1518.

51. Zhang M, He Z (2004) Long-term changes in organic carbon and nutrients of an Ultisol under rice cropping in southeast China. Geoderma 118: 167–179.

52. Darilek JL, Huang B, Wang Z, Qi Y, Zhao Y, et al. (2009) Changes in soil fertility parameters and the environmental effects in a rapidly developing region of China. Agriculture, Ecosystems & Environment 129: 286–292.

53. Nardi S, Morari F, Berti A, Tosoni M, Giardini L (2004) Soil organic matter properties after 40 years of different use of organic and mineral fertilisers. European Journal of Agronomy 21: 357–367.

54. Li Z, Zhang T, Chen B (2006) Changes in organic carbon and nutrient contents of highly productive paddy soils in Yujiang county of Jiangxi province, China and their environmental application. Agricultural Sciences in China 5: 522–529.

55. Chen XW, Li BL (2003) Change in soil carbon and nutrient storage after human disturbance of a primary Korean pine forest in Northeast China. Forest Ecology and Management 186: 197–206.

56. Tejada M, Gonzalez JL, García-Martínez AM, Parrado J (2008) Application of a green manure and green manure composted with beet vinasse on soil restoration: Effects on soil properties. Bioresource technology 99: 4949–4957.

57. Lwkin LY, Kosilova AN, Dubanina GV (1994) The effect of long-term application of fertilizers on soil fertility and winter hardiness and productivity of winter wheat on typical chernozen. Agrokhimiya, 1: 38–43.

58. Qu JF, Dai JJ, Xu MG, Li JM (2009) Advances on effects of long-term fertilization on soil phosphorus. Chinese Journal of Tropical Agriculture, 29: 75–80.

59. Motavalli PP, Miles RJ (2002) Soil phosphorus fractions after 111 years of animal manure and fertilizer application. Biology and Fertility of Soils, 36: 35–42.

14

Salt Marsh as a Coastal Filter for the Oceans: Changes in Function with Experimental Increases in Nitrogen Loading and Sea-Level Rise

Joanna L. Nelson*, Erika S. Zavaleta

Environmental Studies Department, University of California Santa Cruz, Santa Cruz, California, United States of America

Abstract

Coastal salt marshes are among Earth's most productive ecosystems and provide a number of ecosystem services, including interception of watershed-derived nitrogen (N) before it reaches nearshore oceans. Nitrogen pollution and climate change are two dominant drivers of global-change impacts on ecosystems, yet their interacting effects at the land-sea interface are poorly understood. We addressed how sea-level rise and anthropogenic N additions affect the salt marsh ecosystem process of nitrogen uptake using a field-based, manipulative experiment. We crossed simulated sea-level change and ammonium-nitrate (NH_4NO_3)-addition treatments in a fully factorial design to examine their potentially interacting effects on emergent marsh plants in a central California estuary. We measured above- and belowground biomass and tissue nutrient concentrations seasonally and found that N-addition had a significant, positive effect on a) aboveground biomass, b) plant tissue N concentrations, c) N stock sequestered in plants, and d) shoot:root ratios in summer. Relative sea-level rise did not significantly affect biomass, with the exception of the most extreme sea-level-rise simulation, in which all plants died by the summer of the second year. Although there was a strong response to N-addition treatments, salt marsh responses varied by season. Our results suggest that in our site at Coyote Marsh, Elkhorn Slough, coastal salt marsh plants serve as a robust N trap and coastal filter; this function is not saturated by high background annual N inputs from upstream agriculture. However, if the marsh is drowned by rising seas, as in our most extreme sea-level rise treatment, marsh plants will no longer provide the ecosystem service of buffering the coastal ocean from eutrophication.

Editor: Just Cebrian, MESC; University of South Alabama, United States of America

Funding: The following sources of funding awarded to J.L. Nelson supported the research: National Oceanic and Atmospheric Administration/ National Estuarine Research Reserve Graduate Research Fellowship (grant number NA08NOS4200268); Phi Beta Kappa; CA Environmental Quality Initiative; Garden Club of America Coastal Wetlands scholarship; and from the University of California Santa Cruz: STEPS (Science, Technology, Engineering, Policy and Society "Institute, at University of California, Santa Cruz), STARS (Services for Transfer and Re-entry Students" at University of California, Santa Cruz), the Marilyn C. Davis Scholarship, the Center for Dynamics at the Land-Sea Interface, and the Environmental Studies department. The funders had no role in study design, data collection and analysis, decision to publish, or preparation of the manuscript.

Competing Interests: The authors have declared that no competing interests exist.

* E-mail: jnelson@stanfordalumni.org

Introduction

Human activity has altered biotic and abiotic environmental controls at rates, scales, and in combinations that are unprecedented: the hydrologic cycle, biodiversity, land cover, the use of biological productivity, water quality, and the cycling of nitrogen (N) have all changed at global scales [1,2,3]. Multiple global environmental changes converge in particular at the land-sea interface, with anthropogenic disturbances originating from both the marine and terrestrial realms, making this an important place to study interactions.

Sea-level rise (due to climate change) and N pollution are two dominant drivers of global change affecting ecosystems; although both are recognized threats to coastal salt marshes, their interacting effects are unknown. Coastal salt marshes are highly productive ecosystems [4,5] that provide a number of ecosystem services, including interception of watershed-derived nitrogen (N) and other pollutants before they reach the ocean [6,7,8]. Salt marsh is a threatened habitat in California, having lost 75 to 90 percent of its historic extent [9,10].

Sea-level rise is changing the character and location of the land-sea interface and therefore the existence, distribution, and potential migration of salt marshes [11,12]. Salt marsh existence depends on the relative elevation difference between sea level and the marsh platform, determined not only by sea level, but by marsh subsidence, erosion, and rate of sediment delivery or organic-matter accretion. A recent study projects a global sea-level rise by 2100 of 0.5 to 1.4 meters above the 1990 level [13], which exceeds the 2007 IPCC maximum estimate of 0.6 meters. Sea level affects marsh distribution and density through the mechanisms of waterlogging and salinity stress [14,15]. Resilience of salt marsh to sea-level rise depends on the ability of a) halophytes to migrate upland, or b) the marsh platform to rise at a similar pace due to sediment accretion or organic-matter accretion. Whether coastal marshes can keep pace with accelerating sea-level rise is an open question [12]. At our study site, Elkhorn Slough, marshes are unlikely to keep pace with sea-level rise because the Slough has physical, hard shoreline barriers – levees, tide gates, and rip-rap – that will likely obstruct marsh migration towards the uplands. Secondly, paleoecological research in Elkhorn Slough indicates the rate of sediment accretion on the marsh platform has been 2–5 mm/yr for the past 50 years, and 1–2 mm/yr in the 200 years before that [16], which is lower than the predicted rate of 5–7 mm/yr of sea-level rise [2]. In this estuary, marsh-platform building is dominated by sediment accretion [16]. Elkhorn Slough marshes have been stable over the past five years, as sedimentation of 3–4 mm/yr has been closely matched by subsidence [17].

Nitrogen pollution, often due to run-off from agricultural and urban lands, has increased exponentially in recent decades [18] and poses one of the greatest threats to estuarine ecological function [19,20,21]. The leading sources of added nitrogen are the application of synthetic fertilizer in agriculture and human population growth rates in coastal areas, with associated runoff [22,23]. Nitrogen supply in salt marshes affects plant productivity and biomass, and plant physiology, such as resource allocation and tissue N content [24,25,26].

Nutrient enrichment of coastal and estuarine systems can lead to altered biogeochemical cycles, disruptive or harmful blooms of phytoplankton and macroalgae, changes in food webs and biodiversity [21], and hypoxic or anoxic ocean regions, also called "dead zones" [27,28,29]. In the USA, three quarters of all major estuaries have hypoxic "Dead Zones" [30]. Pathways of nitrogen

interception in the coastal environment include plant uptake into tissue, denitrification by microbial communities, and burial in sediments [31,32]. In the present study, we focus on plant uptake by emergent marsh plants and quantify the ecosystem service of the "coastal filter" (e.g., [33,34]) represented by N sequestered.

In our study, we address the question: 1) How do sea-level rise and anthropogenic nitrogen additions affect the salt marsh ecosystem process of nitrogen uptake? This is the first study we are aware of to investigate the presence and type of interactions between the two stressors in an empirical, controlled experiment in temperate salt marsh. Salt marsh plant zonation has been clearly described, including the observation that increased waterlogging through relative sea-level rise detrimentally affects marsh plant growth and survival [35,36]. Our novel contribution is to measure the responses of plant growth and N sequestration during simultaneous changes to inundation and N exposure, in order to quantify potential changes to salt marsh ecosystem services. Our objective was to examine changes in the salt marsh's ability to serve as a coastal filter with increases in sea-level rise and nitrogen loading.

In any sea-level-rise scenario, salt marsh plants will experience increased inundation depths and times. We expected the dominant

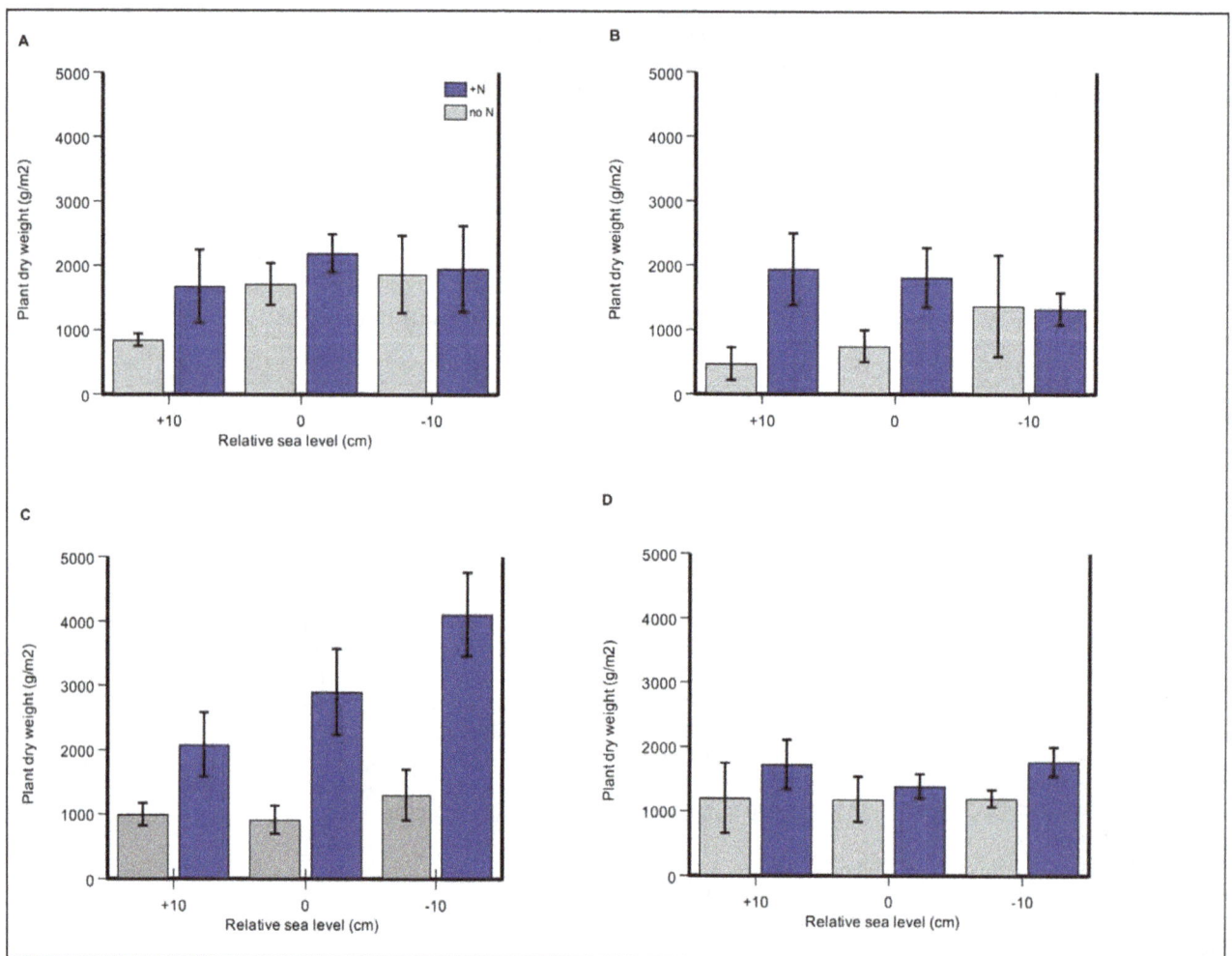

Figure 1. Nitrogen addition increased aboveground marsh biomass. Aboveground salt marsh plant biomass (g m^{-2}) in a) July 2008; b) Nov 2008; c) July 2009; and d) Nov 2009 harvests. Salt marsh plant species are the dominant *Sarcocornia pacifica*, as well as *Jaumea carnosa*, *Frankenia salina*, and *Distichlis spicata*. Four out of five harvests are shown: April 2009 was very similar to November of each year. Control treatment (no N) is shown in grey, and N-addition treatment (+N) in blue. Error bars depict standard error of the mean.

A

B

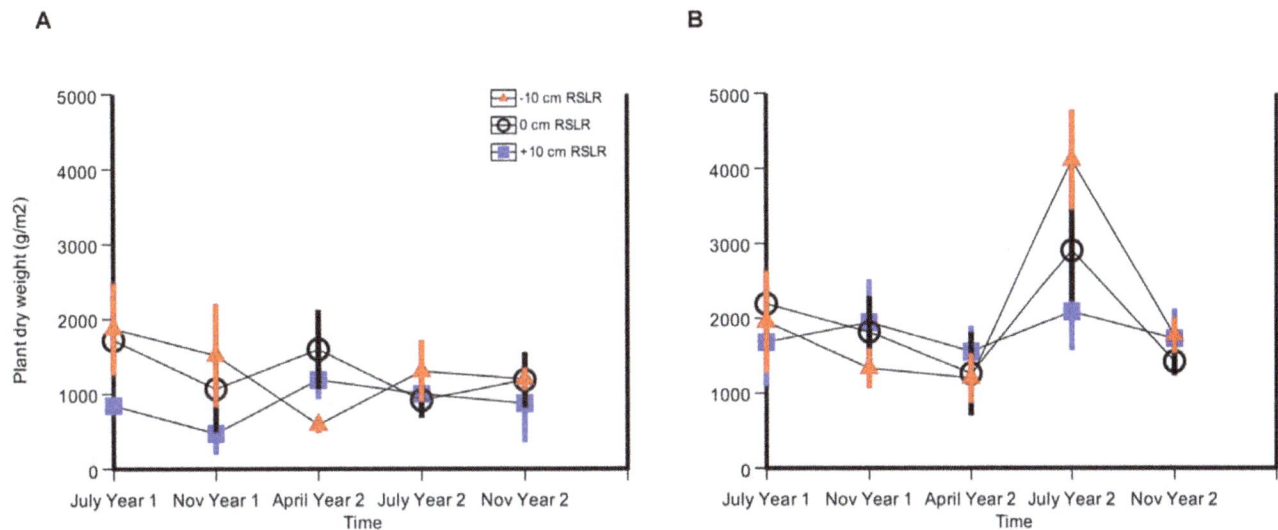

Figure 2. Nitrogen addition increased aboveground marsh biomass most strongly in the second growing season. Aboveground salt marsh plant biomass (g m^{-2}) in a time-series depiction of a) control treatment (no N) and b) N-addition treatment (+N). Error bars depict standard error of the mean.

plant, *Sarcocornia pacifica* (Standley) (pickleweed), to decrease in both abundance (biomass) and extent (experimental sea levels where the plants survived) due to ecological drowning. We expected diminished nutrient uptake as plants were physiologically stressed and dying. We anticipated that experimentally raising the marsh platform – analogous to a transplant experiment to higher elevation – (i.e., reducing the frequency of inundation) would improve halophytes' ability to take up nitrogen. Finally, we had a general expectation that nitrogen addition above background levels would increase marsh plant growth, providing antagonistic effects to marsh drowning in the field (e.g., [37,38]) – but a threshold might exist, where chronic nutrient addition contributed to toxic effects or no longer contributed to growth.

Finally, nitrogen incorporated into plant tissue will continue to cycle when the plant dies or senesces, and decomposes, raising the question of whether plant-bound nutrients have truly been "intercepted" from the ocean. The slower turnover time of nitrogen bound in organic form is generally considered beneficial in buffering the rates and amounts of available-N delivery [24]. How long N is intercepted in standing biomass depends on the lifespan of pickleweed; the plant senesces some succulent tissue annually, but the average lifespan of the perennial plant is unknown. In other N-cycling studies of halophytes – with a focus on temporal dynamics – decomposition is positively influenced by N in tissues, negatively affected by C:N ratios, and occurs in autumn and winter despite lower temperatures [39]; maximum N accumulation in a *Spartina* marsh in Georgia occurred in aboveground tissues in summer and belowground tissues in winter [40]; in a US Northeast *Spartina* marsh, seasonal differences in total N pools of herbaceous species were most strongly influenced by belowground fine root matter and dead macro-organic matter fluxes [41]. The timing of nutrient delivery and plant uptake is important to the efficacy of marsh as a coastal filter [24]. *Sarcocornia* is most productive (with green, succulent, new tissue) in the summer months and dormant (with woody stems) in the winter [37]. Thus, there is a potential "mismatch" in timing in Pacific Coast marshes, where maximum plant production occurs in summer and peak nutrient runoff arrives with winter rains. This timing mismatch could mean that a heightened winter nutrient

pulse has relatively greater effect on belowground growth than it does on then-dormant-aboveground marsh plants, as well as that dormant winter plants have a weaker influence on winter nitrogen movement through the marsh and to the coastal ocean. To capture the dynamics of this potential mismatch, we explored marsh response to N addition and sea-level rise simulations by harvesting plants in the months of April, July, and Nov/Dec (spring, summer, and winter).

Results

Above- and belowground biomass production

Nitrogen addition increased aboveground marsh biomass (N-level F = 11.08, p = 0.006) (Fig. 1 and 2, Table 1). Nitrogen-addition effects were strongest in Year Two of treatments, particularly in July during the summer growing season (Fig. 1). For example, in July 2009 at −10 cm relative sea level, fertilized plots and unfertilized plots had mean biomass of 4.1 (±0.67) kg m^{-2} and 1.3 (±0.41) kg m^{-2}, respectively – a three-fold difference.

In contrast, relative sea-level rise had no significant effect on biomass (RSL F = 1.04, p = 0.39) and did not influence the N response (N-level x RSL F = 0.90, p = 0.43) (Table 1). The only harvest in which both treatments had any type of interactive or synergistic effect was the summer (July) of Year Two, where effects were additive: in the presence of N-addition, biomass decreased linearly with relative sea-level rise (Fig. 1). This pattern differed from the first year of the experiment, where in the absence of N-addition (ambient conditions), biomass decreased linearly with relative sea-level rise in both July and November (Fig. 1).

Within each year, plant growth increased with N-addition most strongly in the summer, and biomass was highest in July of Year Two (Fig. 1 and 2). Pairwise comparisons of the significant N effect on biomass indicated that July of the first year was significantly different than all three of the second-year harvests (factor = season, p = 0.04, p<0.001 and p<0.001). In the second year, April biomass was lower than July of that same year (p = 0.04). The significant N effect on biomass was also apparent in a main-effects

Table 1. Results of statistical analyses (General Linear Model).

Model and response variable	Source	df	F	p
General Linear Model: repeated measures. Data transformed ln(x+1). Excludes 30cm RSLR				
Aboveground biomass				
	N level	1	11.08	**0.006**
	Elevation	2	1.04	0.39
	N level * Elev	2	0.90	0.43
	Error	12		
Aboveground tissue [N]				
	N level	1	35.81	**<0.001**
	Elevation	2	0.08	0.92
	N level * Elev	2	0.47	0.64
	Error	11		
	Within subjects: Month	4	5.04	0.002
N stored succulent pickleweed				
	N level	1	13.88	**0.003**
	Elevation	2	2.99	0.09
	N level * Elev	2	0.58	0.57
	Error	12		
	Within subjects: Month	4	17.52	**<0.001**
	Within: Month* N level	4	4.10	**0.006**
N stored all species				
	N level	1	64.48	**<0.0001**
	Elevation	2	1.26	0.32
	N level * Elev	2	0.48	0.63
	Error	12		
	Within subjects: Month	3	26.06	**<0.0001**
General Linear Model				
Root biomass data transformed ln(x)				
July 2009 (Nov 2009)	N level	1	0.55 (3.48)	0.47(0.09)
	Elevation	2	0.47(0.09)	0.64 (0.92)
	N level * Elev	2	1.64(0.42)	0.23(0.67)
	Error	12		
Shoot:root ratio data needed no transformation				
July 2009	N level	1	12.32	**0.004**
	Elevation	2	2.14	0.16
	N level * Elev	2	0.51	0.61
	Error	12		
Nov 2009	Block	2	2.239	0.223
	N level	1	0.226	0.659
	Elevation	2	0.252	0.789
	N level * Elev	2	0.334	0.735

Table 2. Results of statistical analyses (paired t-test).

Main effects of N	df	t	p
Paired t-test of no N and +N treatments			
Aboveground biomass	44	−4.81	**<0.0001**
N stored all species	44	−6.81	**<0.0001**

test (averaged over all elevations and seasons) (t = −4.81, p<0.001) (Table 2).

Root biomass tended to increase with nitrogen addition in November (N level F = 3.48, p = 0.09), but relative sea-level rise did not have a discernible effect (RSL July F = 0.47, p = 0.635; RSL Nov, F = 0.09, p = 0.915) (Fig. 3 and Table 1). Root biomass in November 2009 was almost double that of July 2009: November's fertilized root biomass at −10 cm relative sea level

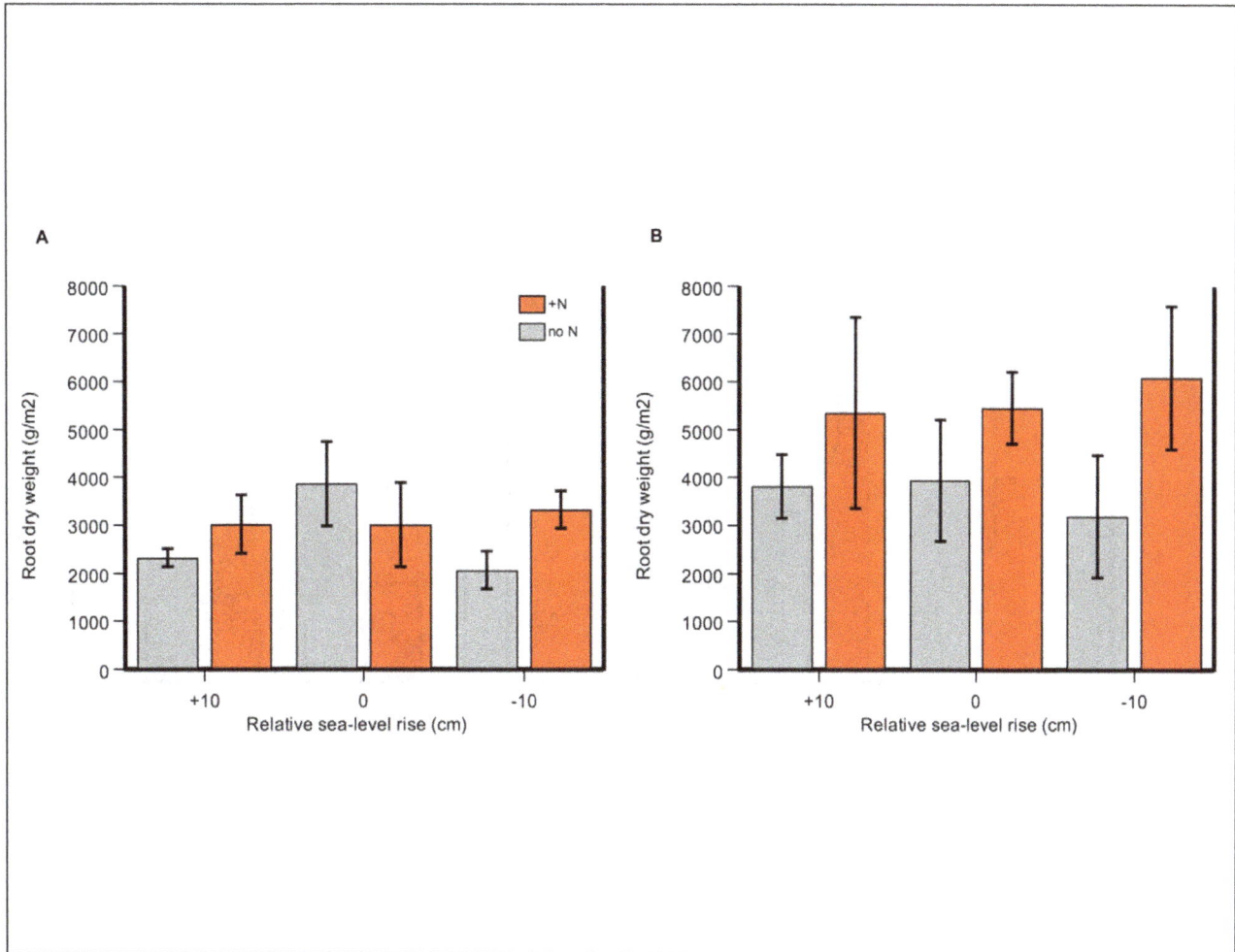

Figure 3. Marsh root biomass was higher in winter than summer. Salt marsh root biomass is almost twice as high in the dormant season of winter as in summer. Root biomass (g m^{-2}) in a) July 2009, and b) November 2009. Control treatment (no N) is shown in grey, and N-addition treatment (+N) in red. Error bars depict standard error of the mean.

averaged 6096 g (± 1527), compared to 3330 g (± 419) in July, a 183% increase.

Because of the strong effect of N increasing aboveground biomass in July and modest effect of increasing belowground biomass in November, N strongly increased shoot:root ratios in July (N level F = 12.31, p = 0.004) (Fig. 4). Changing relative sea level did not exert a significant effect on shoot:root ratios (RSL F = 2.14, p = 0.16) or influence the N treatment (N-level x RSL F = 0.51, p = 0.61).

There was very little evidence for spatial variation in marsh growth, in that a test for a block effect was non-significant in all analyses. Although there was a strong and interpretable overall response to treatments, salt marsh responses varied temporally, by season.

Extreme sea-level rise treatment. In the highest simulated sea-level rise of 30 cm, all salt marsh plants died in Year Two of the experiment, between spring and summer. N-addition led to greater biomass in only one of three harvests with living plants, in winter of the first year (Fig. S1).

Plant tissue nitrogen

Nitrogen concentration. Nitrogen concentration (mg N g^{-1}) in aboveground plant tissue increased strongly in plots with N-addition (N-level F = 35.81; p<0.001) (Fig. S2 and Table 1). Similar to results for biomass, simulated sea-level rise did not have an effect (RSL F = 0.08, p = 0.92), and there was no interaction between the treatments (F = 0.47, p = 0.64) (Table 1). There were significant within-subject (within-plot) effects of season (F = 5.04, p = 0.002), leading to an exploration of temporal variation: N concentration in July of the first year was significantly different than N concentrations in July and November of the second year (factor = season, p<0.001 for each comparison), and N concentration in April of the second year was significantly different than July or November of the same year (factor = season p<0.001 and p = 0.002).

Treatment effects on plant N concentration were most apparent in Year Two, as with biomass, but in the dormant season of November rather than the growing season of July. At a maximum – November 2009 in the +10 cm sea-level rise plots –pickleweed (*S. pacifica*) succulent tissue had a concentration of 37.17 (±23.9) mg N g^{-1} plant tissue when fertilized compared to 9.06 (±0.27) mg N g^{-1} plant tissue in controls, a 410% difference (Fig. S2).

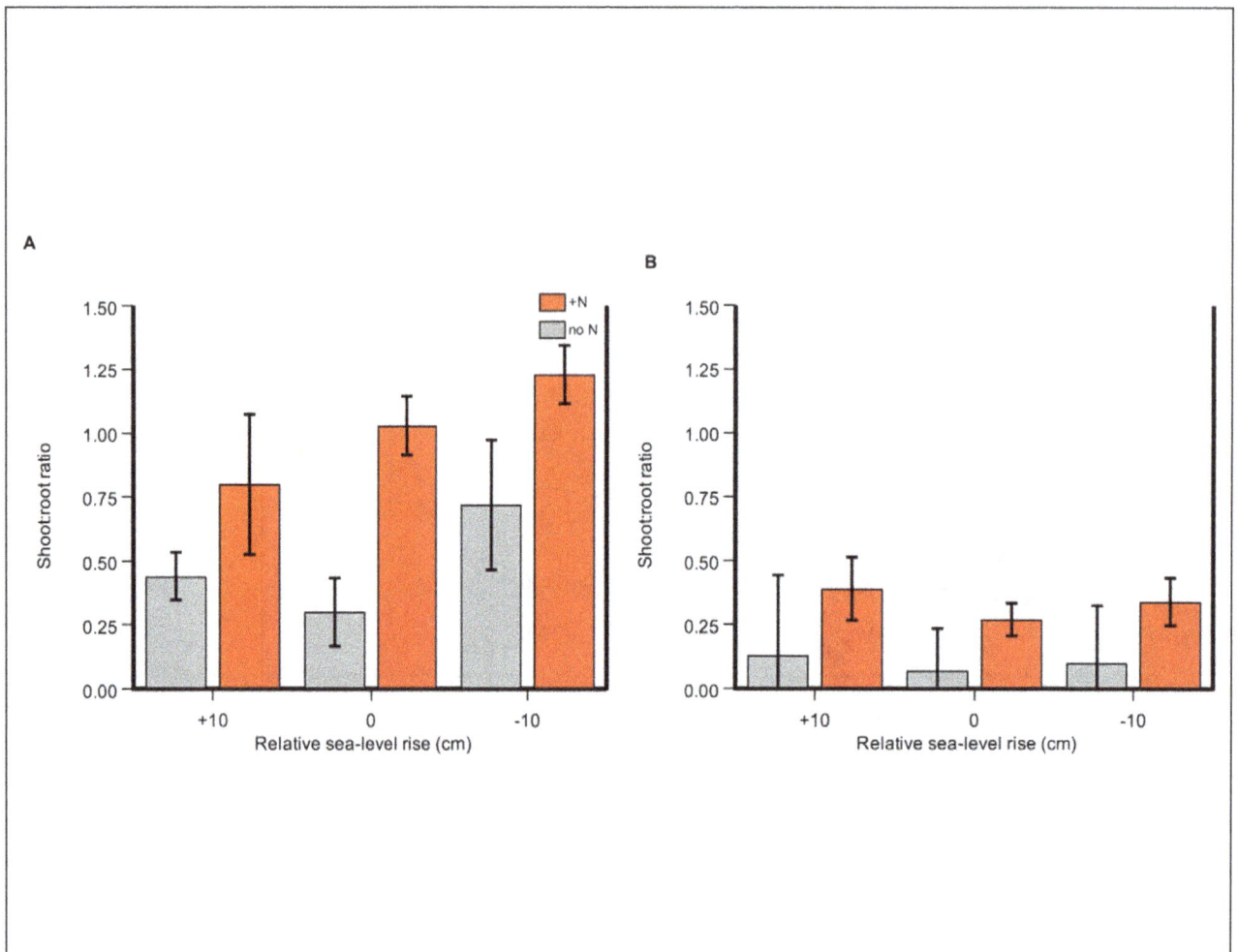

Figure 4. Nitrogen addition strongly increased shoot: Root ratios in the summer growing season, but not the winter. Because of the strong effect of N increasing aboveground biomass in July and modest effect of increasing belowground biomass in November, N strongly increased shoot:root ratios in July. Shoot:root ratios (unitless) in a) July 2009, and b) November 2009. Control treatment (no N) is shown in grey, and N-addition treatment (+N) in green. Error bars depict standard error of the mean.

N addition significantly increased root-N concentration in coarse roots only, in November (N-level $F = 25.32$, $p < 0.001$; RSL $F = 1.50$, $p = 0.26$). There were no discernible treatment effects on fine roots (N-level $F = 0.002$, $p = 0.96$; RSL $F = 0.50$, $p = 0.62$).

Extreme sea-level rise. In the highest simulated sea-level rise of 30 cm, N concentration in aboveground tissues increased significantly with added inorganic N only in April of the second year (N-level $F = 20.41$, $p = 0.01$) (Fig. S1).

Plant nitrogen sequestration. Total nitrogen sequestered in all halophyte species and tissue types – a product of nitrogen concentration and biomass of all species – increased strongly in response to N addition (N level $F = 64.48$, $p < 0.0001$) (Fig. 5 and Table 1). N sequestered in succulent pickleweed (gN m^{-2}) only increased strongly in response to N addition (N level $F = 13.88$, $p = 0.003$) (Fig. 6 and 7). Relative sea level did not have a significant effect (RSL $F = 2.99$, $p = 0.09$), with no interaction between treatments (N-level x RSL $F = 0.581$, $p = 0.57$). Pickleweed sequestered more N in the summer seasons (season $F = 17.53$, $p < 0.001$) (Fig. 6), although a significant interaction between season and N-level makes this difficult to interpret (season-by-N level $F = 4.10$, $p = 0.006$). At a maximum, fertilized plants stored more than four times as much nitrogen as controls: in

July 2009 at -10 cm relative sea-level rise, plants sequestered 22.8 (± 5.6) gN m^{-2} compared to no-N plots with 4.8 (± 1.6) gN m^{-2}, a difference of 475 percent (Fig. 6 and 7). At that same time and plot elevation, biomass increased at a lower rate of 316 percent (4107 g m^{-2} average fertilized biomass vs. 1300 g m^{-2} average unfertilized biomass) (Fig. 1).

Extreme sea-level rise treatment. There was no significant effect of N addition on N stored in plots with 30 cm of simulated sea-level rise ($F = 0.69$, $p = 0.45$). However, there was a within-plot effect of season ($F = 4.48$, $p = 0.05$) (Fig. S1).

Discussion

Important ecosystem functions and services provided by temperate salt marsh are at risk of being diminished by directional, ecological change. Multiple global changes can interact to dampen, amplify, or add to each other's effects, adding complexity that is important to address in our understanding of ecological processes. In this study, the global changes of sea-level rise and nitrogen pollution have strong effects on salt marsh productivity and nutrient cycling. First, salt marshes buffer terrestrial N pollution through plant uptake; however, sea-level rise quickly

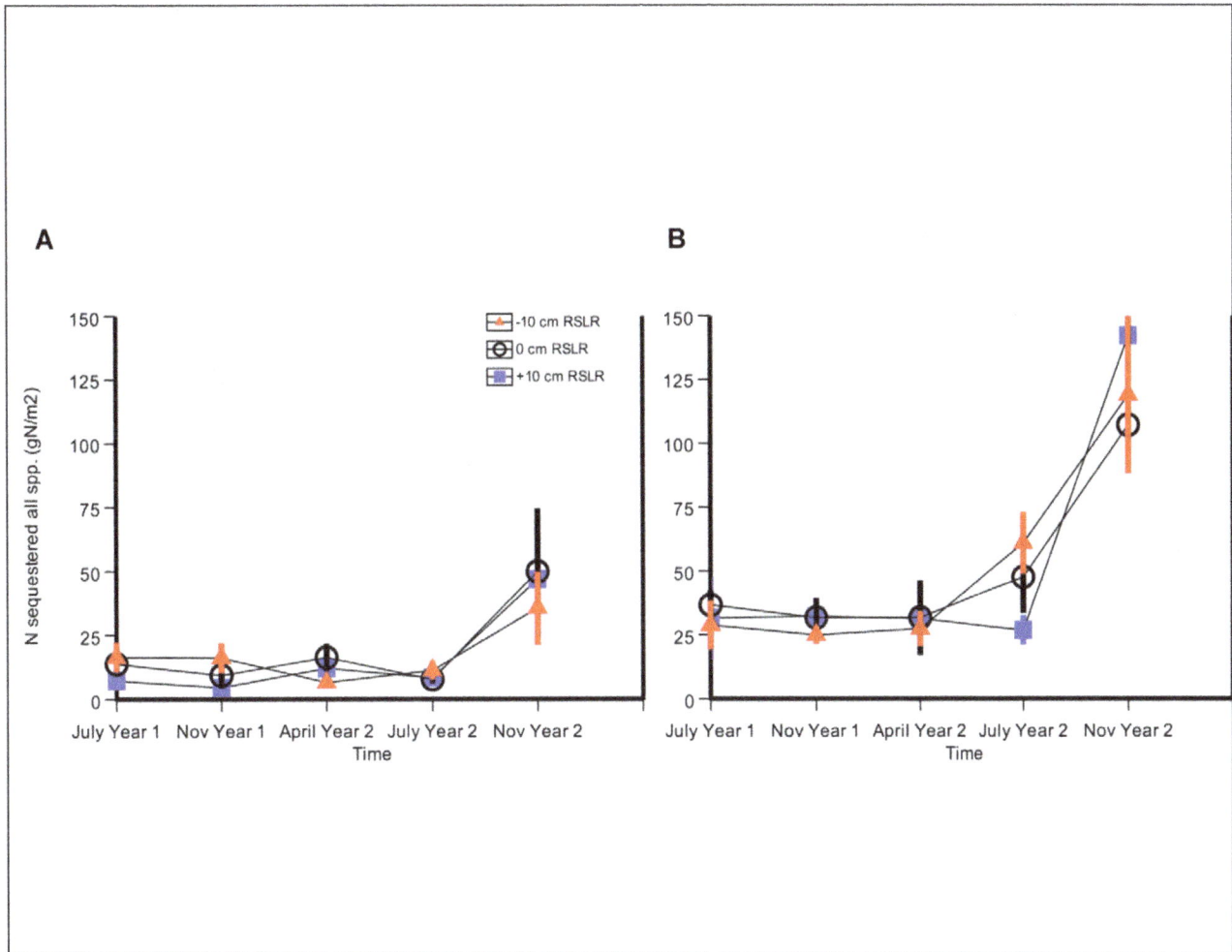

Figure 5. Salt marsh functions as a coastal filter, with a winter maximum across all halophyte species. Nitrogen sequestered (gN m^{-2}) in a time-series depiction of a) control treatment (no N) and b) N-addition treatment (+N). Error bars depict standard error of the mean.

diminishes salt marsh extent. Second, halophytes sequester more nitrogen – through growth and increased tissue N concentrations – with inorganic N addition. Third, there are seasonal variations in response to treatments – where plants grow more, sequester more nitrogen in succulent tissue, and generally respond more strongly to the combination of treatments in the summer growing season rather than the winter dormant season – with the exception of N sequestration across all species, which was highest in November of the second year of the study.

This is the first study to examine the interaction of nitrogen pollution and sea-level rise on the capacity of temperate salt marshes to intercept land-derived N to protect ocean functioning. Our results suggest the plants serve as a robust N trap, or coastal filter. Additionally, in the case of Coyote Marsh, Elkhorn Slough, this function is not saturated. However, if the marsh is drowned by rising seas – as it was in the most extreme sea-level rise simulation – the plants will no longer provide the ecosystem service of buffering the ocean from eutrophication.

Interacting effects of N addition and sea-level rise on salt marsh

In our investigation of salt marsh as a coastal filter, simulated sea-level rise reduced marsh resilience to N loading, and additions of inorganic N led to more N uptake into plants. Nitrogen had the dominant effect on plant growth and N sequestration; simulated sea-level rise only had a significant effect on plant growth and N uptake when waterlogging killed plants; and in the peak growing season of the second year (summer), the effects of the two perturbations were additive: in the presence of N-addition, plant biomass and N sequestration decreased linearly with relative sea-level rise.

Marshes buffer estuarine waters from N loading through plant uptake. In response to N addition, N concentrations increased in succulent, annual tissue of the dominant marsh plant, pickleweed; growth and shoot:root ratios of all four marsh species increased, with a larger proportion of N-rich shoots relative to lower-N roots. Together, these three factors drove the magnitude of N sequestered on a per area basis, which was four times higher in fertilized plots.

It is notable that in an estuarine environment with *high* concentrations of nitrate in the main channel water (up to 250–300 µM NO$_3$-N in winter [42,43]), salt marsh plants continued to be N-limited (indicated by increased biomass with N addition). High nitrate concentrations in the main channel of the Slough fuel productivity that can be categorized as beyond "eutrophic" to "hypertrophic" [44]. Therefore, although "control" plots are

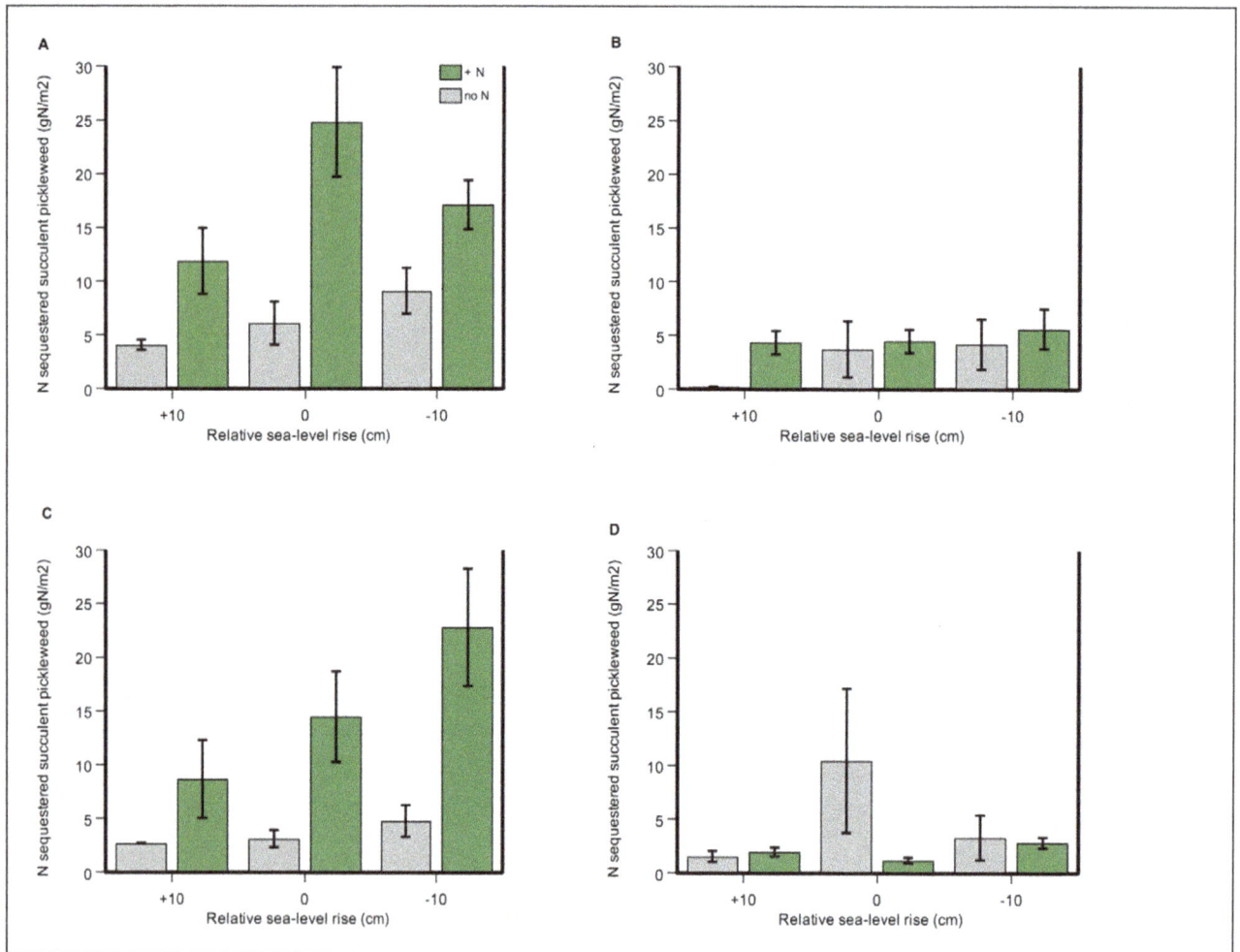

Figure 6. Salt marsh functions as a coastal filter in new-growth, succulent pickleweed, especially in summers. At a maximum, plants with added inorganic nitrogen sequestered more than four times as much as N as controls. Total N sequestered in *S. pacifica* new-growth tissue (gN m^{-2}) in a) July 2008; b) November 2008; c) July 2009; d) November 2009. Four out of five harvests are shown: April 2009 was very similar to November of each year. Error bars depict standard error of the mean.

bathed in high concentrations of nitrogen during tidal inundation, "+N" treatment plots show still higher growth and continued uptake.

Salt marsh plants in Coyote Marsh were vulnerable to the extreme sea-level rise simulation, in that all plants died by the middle of the second year of the experiment and ceased to provide the filtering function of N uptake. This result is consistent with the estuarine literature [35,36,45]; however, our novel contribution was to look at N sequestration as plants were inundated, physiologically stressed, and dying. In the simulated sea-level rise of +30 cm, plants did not have significantly higher N concentrations in their tissue than other elevation treatments, so there was no compensatory effect of more N sequestration per biomass. Therefore, a decrease in biomass indicates a proportional decrease in N uptake, implying that sea-level rise will severely diminish the buffering function of salt marsh.

Plants exposed to the less extreme simulation of sea-level rise (+10 cm) survived throughout the two-year experiment, and their biomass did not differ significantly from that of the ambient marsh platform. These results suggest plant growth was not adversely affected by the 10-cm sea-level rise treatment, and that – unsurprisingly – the rate and magnitude of tidal inundation

matter in terms of species' responses and survival. However, plant biomass reached a one-time maximum in plots with the higher-elevation treatment (summer of the second year), supporting the idea that a marsh platform at a higher elevation in the range of MHW to MHHW could promote marsh plant growth. Although the simulated sea-level rises we imposed were sudden, rather than the gradual rate predicted (5–7 mm/year eustatic rise in 50 years [2]), the total amount of rise is on par with IPCC 2007 predictions (25–35 cm in the next 50 years).

In contrast to aboveground measures, root biomass did not respond as strongly to N-addition in our two-year study, either showing no response or tending to increase. There is disagreement in the literature about the vulnerability of salt marshes to eutrophication, which centers on belowground responses. Some results indicate that nutrient-enriched sediments, such as treated sewage sediments, have no growth-retarding effects on marsh plants [46], while other studies show relatively lower root growth in marshes with nutrient addition [47,48] which can contribute to subsidence of the marsh platform [48,49]. Halophyte roots' potential contribution to building marsh platform elevation, and therefore marsh sustainability, points to the importance of measuring both above- and belowground biomass, and we

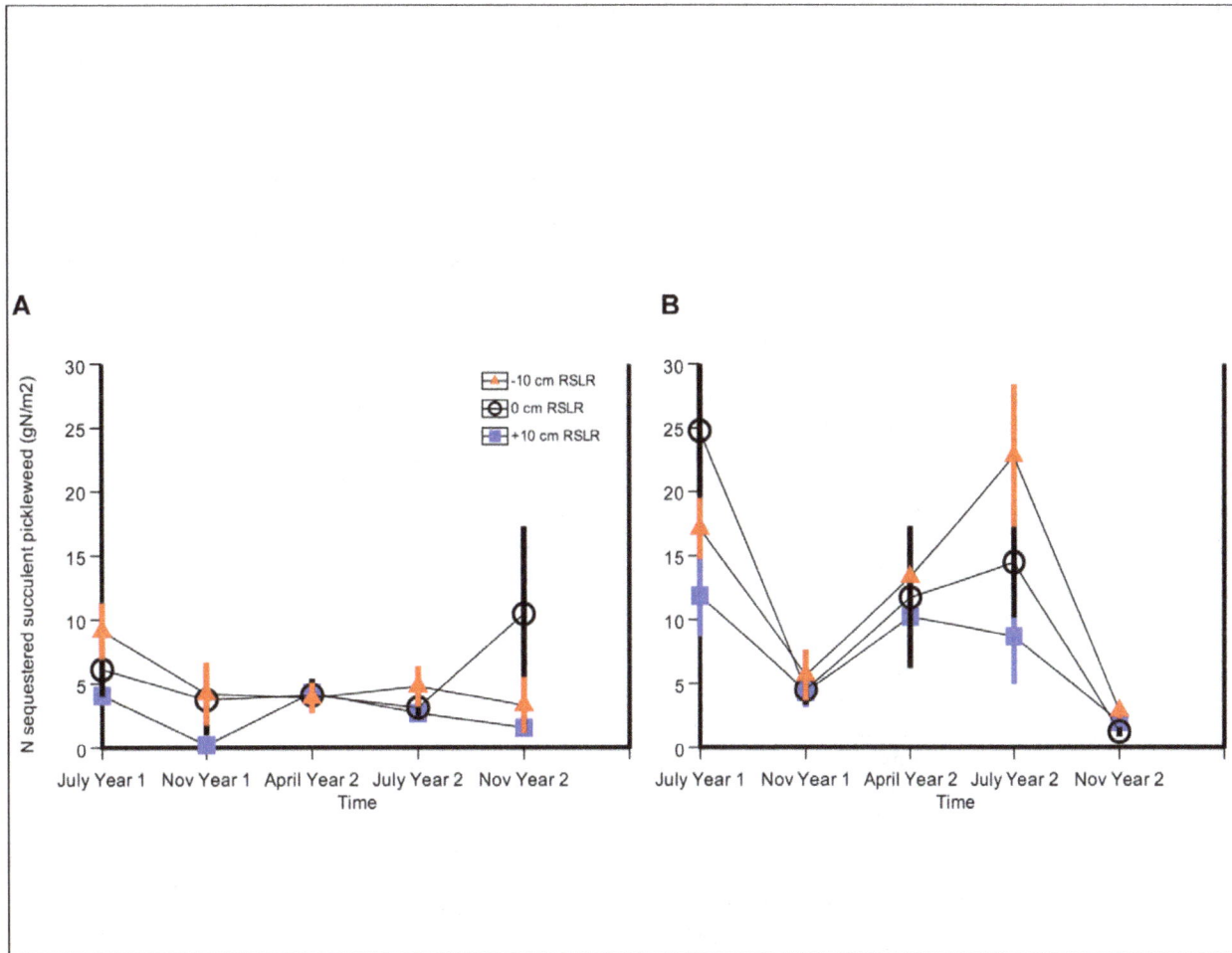

Figure 7. N sequestered in succulent pickleweed peaked in summer in added-N treatment. Total N sequestered in *S. pacifica* succulent tissue (gN m^{-2}) in a time-series depiction of a) control treatment (no N) and b) N-addition treatment (+N). Error bars depict standard error of the mean.

observed a much stronger growth response to N addition in aboveground biomass.

Salt marsh capacity to act as a coastal filter

As compared to other studies of salt marsh interception of land-derived N, Elkhorn Slough salt marsh appears to serve as a robust N trap. Notably high interception of externally added N has also been shown in the Great Sippewissett Salt Marsh, New England [6], even after 30 years of experimental fertilization treatments at low, high, and extra-high fertilization rates (0.9, 2.6 and 7.8 g N m−2 wk−1, respectively, in a N-P-K mix). However, in another New England study, salt marsh vegetation exposed to ∼70 μM NO$_3$$^-$ reached a saturation point for uptake, and became less effective at pollution control than the reference systems [50]. In a study in Portugal, the capacity of salt marshes to retain N depended on the age of the marsh, where the oldest marshes retained the most [51]. All of the above studies focused on low-marsh *Spartina spp.*, cordgrass, whereas there is no *Spartina* in Elkhorn Slough; species differences need to be taken into account. In terms of N application rates and loads, our study is closely matched to the extra-high fertilization treatment in Great Sippewissett Salt Marsh and exceeds the ∼70 μM NO$_3$$^-$ treatment by four orders of magnitude. Other studies of nutrient

enrichment in U.S. Pacific Coast, pickleweed-dominated salt marsh have shown that urea addition increases salt marsh productivity, alters community structure [37,52] and increases susceptibility to species invasions [53]. Organic forms of N, such as urea, depend on microbial mineralization for plant availability and are therefore considered "slow-release" applications [54]. In a greenhouse study of Elkhorn Slough pickleweed, a toxicity threshold was reached upon adding >7.0 g l^{-1} of urea-N [55], resulting in plant death. In comparison, our +N treatment exceeded that threshold amount – we applied 15 g l^{-1} of ammonium-nitrate-N biweekly – and was delivered in plant-available form (inorganic N), implying that "nitrogen burn" or lethal toxicity effects could have appeared more quickly than a slow-release form, yet did not over the course of two years. The difference in the greenhouse- and field study reinforces the importance of experiments that simulate real ecosystem conditions as closely as possible: plants in the field could intercept higher N loads than greenhouse cuttings. The difficulty in comparing Elkhorn Slough uptake to other marshes is that studies of interception of land-derived N were done primarily in Northeast U.S. *Spartina* marshes, and studies of Pacific Coast U.S. pickleweed marshes largely measured productivity or community structure. Taking into account species differences, organic or inorganic N application, and a range of response variables, Elkhorn Slough marsh appears to intercept a notably high amount of nitrogen.

Seasonality

The peak marsh growing season is summer in Elkhorn Slough marshes, but the highest concentrations of nutrients are delivered with winter rains (November – March). U.S. Pacific Coast marshes, in a Mediterranean-type climate, differ in this potential mismatch in timing from other North American marshes. For example, in the Mississippi River Delta, spring floods deliver N synchronous with peak growth of wetland grasses [56,57,58] (note, however, that N entering the Mississippi River Delta is the highest load in the US by an order of magnitude [30], so the synchronicity of delivery and plant uptake does not imply complete uptake.) Marsh plant uptake helps buffer nitrogen loading, as do terrestrial vegetative buffer strips in Elkhorn Slough (e.g., [59]), but they do not seem to be a comprehensive N trap given the timing mismatch [68]. Marsh plant uptake of N is not, therefore, a substitute for policies that reduce fertilizer N inputs and losses from land [22,60,61].

In the summer growing season, marsh plants grew more with N addition, relative to controls, than any other season. Nitrogen sequestration – our measure of the coastal filter – in succulent pickleweed (new-growth only), reached a maximum in the second summer. On the other hand, interestingly, our two-year experiment shows that a) marsh plants are taking up excess N in each season studied; and b) the total nitrogen sequestered, in all plant species and tissue types, reached a maximum in November of the second year, which demonstrates closer alignment in the timing of uptake and delivery of nitrogen in the estuary.

Nitrogen interactions with other global changes

Other studies exploring multiple-stressor interactions have shown varied responses to N-loading: dampened or amplified interactions, a switch in source/sink dynamics, and additive effects. In our experiment, when we found formal interactions (synergistic or additive), we found simple, additive interactions between the two changes of simulated sea-level rise and added nitrogen in the first year's summer growing season. In a Chesapeake Bay marsh, N additions led to a plant community shift towards C4 plants that diminished CO_2 uptake [62]. In a US Northeast *Spartina patens* marsh (Plum Island Sound, MA), short-term N additions created a source of the greenhouse gas nitrous oxide rather than a sink [63]; the role of salt marsh as a sink showed temporal variation, as did our study, relative to the growing season. In an earlier, ecosystem-scale study in the same estuary, (Plum Island Sound) water-column nitrate additions and predatory-fish reduction created synergistic amplified effects, increasing benthic microalgae biomass significantly in salt marsh creeks [64]. In a serpentine grassland in central California, four simulated, global changes – N deposition, elevated CO_2 concentration, warming and precipitation – did not interact synergistically in their effects on plant biodiversity; the treatments produced simple, additive combinations of single-factor effects [65]. These studies underscore the importance of assessing potential interactions between multiple human disturbances rather than extrapolating from single-factor experiments, in order to maintain ecosystem functions and services under global change.

Conclusions and implications for management

Our results have implications for management of both the elevation of marshes and agricultural best-management practices to limit nitrogen losses from land. For example, raising the level of the marsh platform – with halophyte species that play a role as

ecosystem engineers [66] or with added dredge sediments [46,67,68] – can increase vegetation productivity. Dredge sediment addition has raised questions about nutrient, metal, and pollution concentrations in those sediments. Given sources of sediment that have acceptably low levels of pollutants, sediment addition is an intervention that seems to support marsh survival and sustainability under conditions of relative sea-level rise [67,68,69]. In Elkhorn Slough, a potential management action is the adding of dredge materials to bare mudflat, lower in the estuarine intertidal than marsh, to prompt the growth of marsh vegetation [67]; restoration literature suggests that the dominant plant, pickleweed, can recruit from surrounding areas without re-planting efforts [70]. Our results suggest optimal heights for the marsh platform, within the MHW to MHHW range, to promote marsh productivity and N uptake.

Salt marsh distribution will change with sea-level rise – coastal wetlands could establish in areas where they may not have been documented currently and disappear from protected areas [71] – making tools for flexible land-use and conservation of greater importance. Exploring regulation and management strategies to mitigate greenhouse gas emissions and abate nutrient enrichment, at the same time, will be valuable to both conservation of coastal marshes and improvement of ocean water quality [60,61,72,73,74].

Materials and Methods

Study site

Elkhorn Slough (36°48′ N, 121°47′ W), located on the central coast of Monterey Bay, California, has one of the largest tracts of coastal salt marsh habitat in California, with 1,147 ha of marsh [75] (Fig. 8). The main channel of the Slough is part of the Monterey Bay National Marine Sanctuary and is surrounded by agricultural lands, with 24% of the slough watershed under production [76], primarily in heavily fertilized strawberries and vegetable row crops.

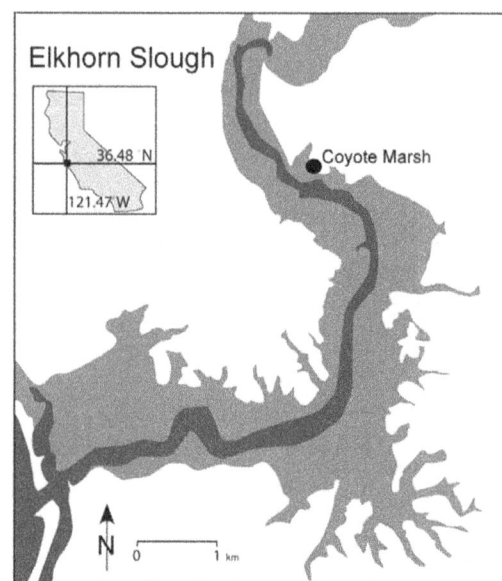

Figure 8. Location of Elkhorn Slough and experiment site. Moss Landing, California, on the coast of Monterey Bay. The experiment site, Coyote Marsh, is located in the Elkhorn Slough National Estuarine Research Reserve.

We established our experiment at Coyote Marsh, a high marsh in the Elkhorn Slough National Estuarine Research Reserve (ESNERR) (Fig. 8). Plant species at the site included *Sarcocornia pacifica* (pickleweed), which was the predominant cover, as well as *Jaumea carnosa* (fleshy jaumea), *Frankenia salina* (alkali heath), and *Distichlis spicata* (salt grass). All necessary permits were obtained for our field study through ESNERR, administered jointly by the National Oceanic and Atmospheric Administration and the California Department of Fish and Game.

Nitrogen delivery to the site. Nitrogen is delivered to marshes in surface runoff, groundwater, and inundation with estuarine-ocean water on flood tides. Therefore, although Coyote Marsh is subjected to flood tides no more than 4–6 times a month on average in winter, the marsh plants do get the 250–300 μM NO_3-N as an ambient dose. In Elkhorn Slough, 66% of nitrate in the main channel comes from terrestrial sources [77], as distinct from ocean upwelling, so even the flood tides are a majority of "land-derived" N.

Hydrodynamics and sediment characterization of the site. Elkhorn Slough has only one small, ephemeral river input (Carneros Creek) at the head. After major hydrologic changes to the Slough, it is considered starved of sediment delivery [48]. Finally, Elkhorn Slough is an ebb-dominated estuary [78], which tends to emphasize sediment loss with higher-velocity ebb waters. Paleoecological research indicates that a) the sediments are primarily inorganic (75% inorganic by weight); and b) sediment accumulation rate has been 2–5 mm/yr in the last 50 calendar years, and 1–2 mm/yr in the time period 200–50 years before the present [16]. Sediment size class is categorized as "fine silt": specifically, particle size distributions show a bimodal peak at 4 and 16 μm [16,17].

Experimental design

We crossed relative sea-level and nitrogen treatments in a fully factorial design to examine their potentially interacting effects on plant biomass and tissue nutrient concentrations. We used marsh elevation as a proxy for sea-level rise, and chose three elevations – with a fourth extreme sea-level rise simulation. Simulated sea levels were chosen to fall within the spectrum of IPCC (2001) scenarios (where +30 cm was the maximum predicted), or an ecologically significant amount of sedimentation addition [66]. Sediment addition incorporates predictions of more variability in precipitation and storm events with climate change [79]. The simulated sea levels were +10 cm; 0 cm (the ambient marsh platform); and −10 cm, simulating an increase in elevation of 10 cm, which might occur via a) sediment additions from more extreme storms; or b) management interventions to raise the marsh platform to promote marsh-plant survival. The fourth simulated sea level, which we refer to as "extreme sea level treatment" was +30 cm. Nitrogen additions simulated increased N in terrestrial surface runoff. The two levels of N treatment were 300 g N m^{-2} yr^{-1}, in the form of ammonium nitrate (NH_4NO_3), or no added nitrogen under ambient conditions. The added N is equal to an average five- to ten-fold addition of the conventional fertilizer used in the region on strawberry or vegetable fields [80]. We chose this level in accordance with other studies of marsh interception of land-derived N (e.g., a range of 10–90 times the recommended fertilizer for commercial oat crops [6,81]) [64], the potential for increased land conversion to agriculture in the watershed [75], and recent and forecasted exponential increases in agricultural synthetic fertilizer use [82,83]. We accounted for spatial variability across our site by establishing three blocks, each containing all treatment combinations (4 elevations and 2 N levels), for a total of 24 one-m^2 plots (Fig. 9). Control plots

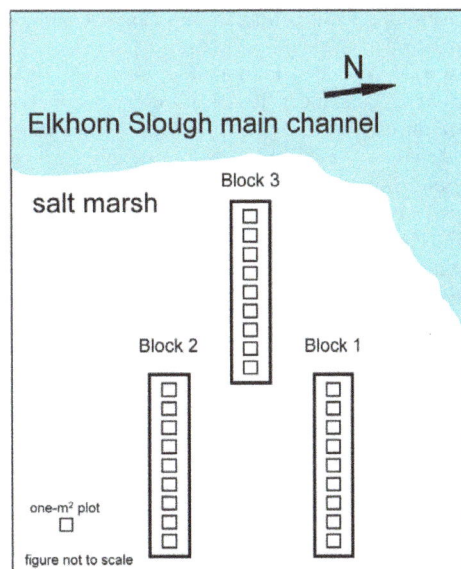

Figure 9. Diagram of block arrangement and experimental design. Coyote Marsh, Elkhorn Slough National Estuarine Research Reserve. Each block contained all treatment combinations (4 elevations and 2 N levels) plus one dig-control plot, for a total of 9 plots per block.

evaluated a possible digging effect by digging up and then replacing otherwise-unmanipulated marsh vegetation. There was one control plot in each block (Fig. 9): having determined that there was no significant digging effect for each analysis, we did not incorporate data from those plots.

Field methods

Elevation. We created the artificial sea-level rise treatment (adapted from [66]) by selecting a 1x1-m plot of marsh, removing vegetation with intact roots in a block of sediment, removing sediment beneath the vegetation layer (either 30 cm or 10 cm depth of sediment), and replacing the vegetation layer. A difference in marsh-plain elevation of 10 cm has been shown to have ecological effects [66,84]. The side-walls of plots were held in place with hardware cloth and landscape staples. Lowered plots did not have any drainage channels or other simulations of an ebb tide. Similarly, in raised plots, we removed vegetation with intact roots in sediment, in this case adding 10 cm of sediment underneath the root zone. Additional sediment was taken from the sediment-removal treatments in the same marsh: adding sediment beneath the root zone minimized any nutrient subsidies, corroborated by our findings that elevated plots (no N) show no difference in growth from the ambient-marsh-platform plots. The sediment did not compact any more than one cm (all plots were ≥9 cm elevated at the end of two years). The sediment in the raised plots could drain more readily out the sides of the hardware-cloth retaining structures than lowered plots, which could release marsh plants from waterlogging and salinity stress [18]. Each plot was at least 3 m away from any other plot.

Extreme sea-level rise simulation. Marsh plots lowered 30 cm simulated a sea-level rise that we estimate to be greater in magnitude than 30 cm, because the plots did not drain and had no system to simulate an ebb tide. Plants were inundated in water with a salinity of ~35 (practical salinity units), typical of the main channel Slough and the Pacific Ocean. Plots did drain occasionally, with no intervention, in a pattern that was not correlated with any variables we measured. We refer to this scenario as "extreme sea-level rise simulation" since it is a rapid and almost-continual inundation of marsh.

Nutrient addition. We added ammonium nitrate (NH_4NO_3) to designated plots in the amount of 15 gN m^{-2} every two weeks. We did not fertilize during July and August of each year, because summer nutrient levels in the Slough are lowest and fertilizer applications are low, becoming high again in October [43]. Therefore, we added a total of 300 gN m^{-2} yr^{-1} to fertilized plots. We dissolved NH_4NO_3 pellets in 1 L of main-channel Slough water and added them to treatment plots; we added 1 L of Slough water to each control plot.

Biomass harvest. We measured the impacts of sea-level change and nitrogen addition on plant biomass, above- and belowground, and plant physiological measures of tissue nitrogen concentration and resource allocation. We harvested a 10×50 cm swath of aboveground vegetation from each meter-square plot on the following dates: July and November 2008, and April, July and November 2009. The swath was taken from a randomly-chosen quarter of a plot with the following constraints: the 50-cm edge was always internal to the plot to avoid edge effects, we harvested a given 10×50 cm area only once in the two years, and we stopped harvesting when all plants in a plot were visibly dead. Once harvested, we sorted plants by species. We separated succulent (new) and woody (perennial) tissue for *Sarcocornia pacifica* only. All plant material was dried in a laboratory oven at 60°C for at least 48 hours; weighed; and a portion ground with a ball mill (Spex 8000, Spectrum Chemicals and Laboratory Products, CA and NJ, USA). We used a C:N analyzer (Elementar varioMAX, Elementar, Germany) in order to obtain tissue nitrogen concentration.

In the second year only (2009), we harvested root biomass with a 5-cm-diameter sediment corer, taking 20-cm-deep cores. We isolated plant material through root-washing by hand and categorized roots as fine or coarse. The approximate diameter cutoff between fine and coarse roots was 0.5 mm. We dried the roots in a laboratory oven at 60°C for at least 48 hours, weighed them, and ground all material in a ball mill. We analyzed %N in November root data only (due to limited sample size, C:N analyzed with a Costech ECS 4010, Costech, CA, USA).

Analytical methods

To assess treatment effects on plant aboveground biomass, we grouped all plant species (which includes succulent and woody tissue biomass of *Sarcocornia pacifica*) in each plot and used a General Linear Model with repeated measures in SYSTAT v12 (Systat Software. Inc., Chicago, IL, USA). We tested for a block effect, and where it was insignificant – in all analyses but one – removed it as a factor. Therefore, independent factors were N-level, relative sea-level (RSL) and their interaction. We log-transformed biomass data to conform to a normal distribution. The repeated measures analysis incorporated all 24 plots over 5 harvests (July, November, April, July November). Similarly, to assess treatment effects on plant tissue nitrogen concentration (mg N per gram of plant tissue), we used a General Linear Model with repeated measures analysis, where data were log-transformed. We ran a post-hoc comparison for repeated measures, with a Bonferroni correction for pairwise comparisons, to assess which seasons might be different than each other. To assess experimental effects on root biomass and shoot:root ratios, we used a factorial ANOVA on each of two harvests. In any analysis where there were significant interactions, we explored the data visually to interpret patterns. To test for main effects of one factor, N addition, we used a paired t-test between the groups: average of response variable at reference-level of N (no addition) and average of response variable with added N. We set the significance level for all analyses at $\alpha = 0.05$, *a priori*.

Supporting Information

Figure S1   Marsh plants are vulnerable to sea-level rise simulation. Simulation of +30 cm sea-level rise resulted in the death of all salt marsh plants before the summer of Year Two of the experiment (bar graph), where plant tissue N concentrations increased with N treatment (XY graph). Error bars depict standard error.

Figure S2   Nitrogen concentration in aboveground plant tissue increased strongly in plots with nitrogen addition. N concentration ([N]) (mgN g^{-1} plant tissue) in a) July 2008; b) Nov 2008; c) July 2009; and d) Nov 2009 harvests. Four out of five harvests are shown. Control treatment (no N) is shown in grey, and N-addition treatment (+N) in green. Error bars depict standard error of the mean.

Acknowledgments

We would like to thank the Elkhorn Slough National Estuarine Research Reserve; F.S. Chapin, III, M. FitzSimmons, K. Johnson, and K. Wasson for improving the manuscript; and P. Raimondi for improving the statistical analyses. We thank our lab group for constructive comments – especially K. Hulvey – and all field assistants: M. Kornfield, T. Burdick; E. Hodges; C. Waslohn; S. Wheatley; R. Laughman; P. Beattie; C. Morozumi; and T. Kemper, with particular thanks to Steve Legnard. J. Nelson thanks Leor, Aolani, Yair, and Karyn.

Author Contributions

Advised by ESZ: JLN. Advised by the academic committee (Drs. F. Stuart Chapin, III, K. Wasson, and M. FitzSimmons, who are listed in the Acknowledgments): JLN. Raimondi advised during Analyzed the Data: JLN. Performed the experiments: JLN. Analyzed the data: JLN ESZ. Contributed reagents/materials/analysis tools: JLN. Wrote the paper: JLN.

References

1. Chapin FS, Zavaleta ES, Eviner VT, Naylor RL, Vitousek PM, et al. (2000) Consequences of changing biodiversity. Nature 405: 234–242.
2. IPCC (2007) Climate Change 2007: The Physical Science Basis. Contribution of Working Group I to the Fourth Assessment Report of the Intergovernmental Panel on Climate Change. Cambridge, United Kingdom, and New York, NY, USA: Cambridge University Press. 996 p.
3. Vitousek PM, Mooney HA, Lubchenco J, Melillo JM (1997) Human domination of Earth's ecosystems. Science 277: 494–499.
4. Little C (2000) The biology of soft shores and estuaries; Crawley MJ, Little C, Southwood TRE, Ulfstrand S, editors. Oxford & New York: Oxford University Press. 252 p.
5. Teal JM (1962) Energy flow in the salt marsh ecosystem of Georgia. Ecology 43: 614–624.
6. Brin LD, Valiela I, Goehringer D, Howes B (2010) Nitrogen interception and export by experimental salt marsh plots exposed to chronic nutrient addition. Marine Ecology-Progress Series 400: 3–17.
7. Howes BL, Weiskel PK, Goehringer DD, Teal JM (1996) Interception of freshwater and nitrogen transport from uplands to coastal waters: the role of saltmarshes. In: Nordstrom KF, Roman CT, editors. Estuarine Shores: Evolution, Environments, and Human Alterations New York: Wiley. 287–310.
8. Kennedy VS, editor (1984) The Estuary as a Filter. Orlando, FL: Academic Press. 511 p.
9. Emmett R, Llanso R, Newton J, Thom R, Hornberger M, et al. (2000) Geographic signatures of North American West Coast estuaries. Estuaries 23: 765–792.
10. Zedler JB (1996) Coastal Mitigation in Southern California: The Need for a Regional Restoration Strategy. Ecological Applications 6: 84–93.

11. Day JW, Christian RR, Boesch DM, Yanez-Arancibia A, Morris J, et al. (2008) Consequences of climate change on the ecogeomorphology of coastal wetlands. Estuaries and Coasts 31: 477–491.

12. Stevenson JC, Kearney MS (2009) Impacts of global climate change and sea-level rise on tidal wetlands. In: Silliman BR, Grosholz ED, Bertness MD, editors. Human Impacts on Salt Marshes: A Global Perspective. Berkeley: University of California Press. 171–206.

13. Rahmstorf S (2007) A semi-empirical approach to projecting future sea-level rise. Science 315: 368–370.

14. Mahall BE, Park RB (1976) Ecotone between *Spartina foliosa* Trin and *Salicornia virginica* L in salt marshes of northern San Francisco Bay. 2. Soil water and salinity. Journal of Ecology 64: 793–809.

15. Mitsch WJ, Gosselink JG (2000) Wetlands. New York: Wiley.

16. Watson EB, Wasson K, Pasternack GB, Woolfolk A, Van Dyke E, et al. (2011) Applications from paleoecology to environmental management and restoration in a dynamic coastal environment. Restoration Ecology 19: 1–11.

17. Gillespie A, Schaffner A, Watson E, Callaway JC (2011) Morro Bay sediment loading update. Morro Bay, CA: Moor Bay National Estuary Program.

18. Schile LM, Callaway JC, Parker VT, Vasey MC (2011) Salinity and inundation influence productivity of the halophytic plant *Sarcocornia pacifica*. Wetlands 31: 1165–1174.

19. Cloern JE (2001) Our evolving conceptual model of the coastal eutrophication problem. Marine Ecology-Progress Series 210: 223–253.

20. Howarth RW, Marino R (2006) Nitrogen as the limiting nutrient for eutrophication in coastal marine ecosystems: Evolving views over three decades. Limnology and Oceanography 51: 364–376.

21. National_Research_Council (2000) Clean coastal waters: understanding and reducing the effects of nutrient pollution. Washington DC: National Academies Press. 405 p.

22. Boesch DF (2002) Challenges and opportunities for science in reducing nutrient over-enrichment of coastal ecosystems. Estuaries 25: 886–900.

23. Zedler JB, Kercher S (2005) Wetland resources: status, trends, ecosystem services, and restorability. Annual Review of Environment and Resources 30: 39–74.

24. Hopkinson CS, Giblin AE (2008) Nitrogen dynamics of coastal salt marshes. In: Capone DG, Bronk D, Mulholland MR, Carpenter EJ, editors. Nitrogen in the marine environment. Second ed. Burlington, MA: Academic Press. 991–1036.

25. Morris JT (1991) Effects of nitrogen loading on wetland ecosystems with particular reference to atmospheric deposition. Annual Review of Ecology and Systematics 22: 257–279.

26. Haines B, Dunn EL (1985) Coastal marshes. In: Chabot B, Mooney HA, editors. Physiological Ecology of North American Plant Communities. New York: Chapman and Hall. 323–346.

27. Diaz RJ, Rosenberg R (2008) Spreading dead zones and consequences for marine ecosystems. Science 321: 926–929.

28. Howarth R, Chan F, Conley DJ, Garnier J, Doney SC, et al. (2011) Coupled biogeochemical cycles: eutrophication and hypoxia in temperate estuaries and coastal marine systems. Frontiers in Ecology and the Environment 9: 18–26.

29. Rabalais NN, Turner RE, Diaz RJ, Justic D (2009) Global change and eutrophication of coastal waters. ICES Journal of Marine Science 66: 1528–1537.

30. Bricker SB, Longstaff B, Dennison W, Jones A, Boicourt K, et al. (2007) Effects of nutrient enrichment in the nation's estuaries: a decade of change. Silver Spring, MD: National Centers for Coastal Ocean Science. 328 p.

31. Bianchi TS (2007) Biogeochemistry of Estuaries. Oxford and New York: Oxford University Press. 706 p.

32. Seitzinger SP, Kroeze C (1998) Global distribution of nitrous oxide production and N inputs in freshwater and coastal marine ecosystems. Global Biogeochemical Cycles 12: 93–113.

33. McGlathery KJ, Sundback K, Anderson IC (2007) Eutrophication in shallow coastal bays and lagoons: the role of plants in the coastal filter. Marine Ecology-Progress Series 348: 1–18.

34. Short FT, Short CA (1984) The seagrass filter: purification of estuarine and coastal waters. In: Kennedy VS, editor. The Estuary as a Filter. Orlando, FL/London: Academic Press. 395–414.

35. Adam P (1990) Saltmarsh Ecology. Cambridge: Cambridge University Press.

36. Lowe PB (1999) Marsh loss in Elkhorn Slough, California: Patterns, mechanisms, and impact on shorebirds. San Jose, CA: San Jose State University. 65 p.

37. Boyer KE, Fong P, Vance RR, Ambrose RF (2001) *Salicornia virginica* in a Southern California salt marsh: Seasonal patterns and a nutrient-enrichment experiment. Wetlands 21: 315–326.

38. Covin JD, Zedler JB (1988) Nitrogen effects on *Spartina foliosa* and *Salicornia virginica* in the salt marsh at Tijuana Estuary, California. Wetlands 8: 51–65.

39. Simoes MP, Calado ML, Madeira M, Gazarini LC (2011) Decomposition and nutrient release in halophytes of a Mediterranean salt marsh. Aquatic Botany 94: 119–126.

40. Hopkinson CS, Schubauer JP (1984) Static and dynamic aspects of nitrogen cycling in the salt-marsh graminoid *Spartina alterniflora*. Ecology 65: 961–969.

41. Elsey-Quirk T, Seliskar DM, Gallagher JL (2011) Nitrogen pools of macrophyte species in a coastal lagoon salt marsh: Implications for seasonal storage and dispersal. Estuaries and Coasts 34: 470–482.

42. Jannasch HW, Coletti LJ, Johnson KS, Fitzwater SE, Needoba JA, et al. (2008) The Land/Ocean Biogeochemical Observatory: A robust networked mooring system for continuously monitoring complex biogeochemical cycles in estuaries. Limnology and Oceanography-Methods 6: 263–276.

43. Johnson KS (2004–2010) Land Ocean Biogeochemical Observatory.

44. Nixon SW (1995) Coastal marine eutrophication: a definition, social causes, and future concerns. Ophelia 41: 199–219.

45. Mendelssohn IA, Morris JT (2000) Eco-physiological controls on the productivity of *Spartina alterniflora* Loisel. In: Weinstein MP, Kreeger DA, editors. Concepts and Controversies in Tidal Marsh Ecology. Dordrecht, the Netherlands; Boston, MA, USA; and London, UK: Kluwer Academic Publishers. 59–80.

46. Day JW, Ko JY, Rybczyk J, Sabins D, Bean R, et al. (2004) The use of wetlands in the Mississippi Delta for wastewater assimilation: a review. Ocean & Coastal Management 47: 671–691.

47. Turner R (2010) Beneath the salt marsh canopy: loss of soil strength with increasing nutrient loads. Estuaries and Coasts: 1–10.

48. Turner RE, Howes BL, Teal JM, Milan CS, Swenson EM, et al. (2009) Salt marshes and eutrophication: An unsustainable outcome. Limnology and Oceanography 54: 1634–1642.

49. Nyman JA, Walters RJ, Delaune RD, Patrick WH (2006) Marsh vertical accretion via vegetative growth. Estuarine Coastal and Shelf Science 69: 370–380.

50. Drake DC, Peterson BJ, Galvan KA, Deegan LA, Hopkinson C, et al. (2009) Salt marsh ecosystem biogeochemical responses to nutrient enrichment: a paired N-15 tracer study. Ecology 90: 2535–2546.

51. Sousa AI, Lillebø AI, Caçador I, Pardal MA (2008) Contribution of *Spartina maritima* to the reduction of eutrophication in estuarine systems. Environmental Pollution 156: 628–635.

52. Boyer KE, Zedler JB (1999) Nitrogen addition could shift plant community composition in a restored California salt marsh. Restoration Ecology 7: 74–85.

53. Martone RG, Wasson K (2008) Impacts and interactions of multiple human perturbations in a California salt marsh. Oecologia 158: 151–163.

54. University_of_California_Cooperative_Extension (2012): University of California Agriculture and Natural Resources.

55. Griffith KA (2008) The ecology of a parasitic plant and its host plant in a central California salt marsh. Santa Cruz: University of California.

56. Mitsch WJ, Day JW (2006) Restoration of wetlands in the Mississippi-Ohio-Missouri (MOM) River Basin: Experience and needed research. Ecological Engineering 26: 55–69.

57. Mitsch WJ, Day JW, Gilliam JW, Groffman PM, Hey DL, et al. (2001) Reducing nitrogen loading to the Gulf of Mexico from the Mississippi River Basin: Strategies to counter a persistent ecological problem. Bioscience 51: 373–388.

58. Mitsch WJ, Day JW, Zhang L, Lane RR (2005) Nitrate-nitrogen retention in wetlands in the Mississippi river basin. Ecological Engineering 24: 267–278.

59. Los Huertos MW (1999) Nitrogen dynamics in vegetative buffer strips receiving nitrogen runoff in Elkhorn Slough Watershed, California [PhD dissertation]: University of California, Santa Cruz.

60. Faeth P, Greenhalgh S (2002) Policy synergies between nutrient over-enrichment and climate change. Estuaries 25: 869–877.

61. Foley JA, DeFries R, Asner GP, Barford C, Bonan G, et al. (2005) Global consequences of land use. Science 309: 570–574.

62. Langley JA, Megonigal JP (2010) Ecosystem response to elevated CO(2) levels limited by nitrogen-induced plant species shift. Nature 466: 96–99.

63. Moseman-Valtierra S, Gonzalez R, Kroeger KD, Tang JW, Chao WC, et al. (2011) Short-term nitrogen additions can shift a coastal wetland from a sink to a source of N(2)O. Atmospheric Environment 45: 4390–4397.

64. Deegan LA, Bowen JL, Drake D, Fleeger JW, Friedrichs CT, et al. (2007) Susceptibility of salt marshes to nutrient enrichment and predator removal. Ecological Applications 17: S42–S63.

65. Zavaleta ES, Shaw MR, Chiariello NR, Mooney HA, Field CB (2003) Additive effects of simulated climate changes, elevated CO2, and nitrogen deposition on grassland diversity. Proceedings of the National Academy of Sciences of the United States of America 100: 7650–7654.

66. Fogel BN, Crain CM, Bertness MD (2004) Community level engineering effects of *Triglochin maritima* (seaside arrowgrass) in a salt marsh in northern New England, USA. Journal of Ecology 92: 589–597.

67. Elkorn_Slough_National_Estuarine_Research_Reserve (2011) Tidal Wetlands Project. Moss Landing, CA.

68. Brew DS, Williams PB (2010) Predicting the impact of large-scale tidal wetland restoration on morphodynamics and habitat evolution in South San Francisco Bay, California. Journal of Coastal Research 26: 912–924.

69. Brand A (2012) North San Francisco Bay Salt Pond marsh restoration trajectory. Ecological Engineering *in press*.

70. Lindig-Cisneros R, Zedler JB (2002) Halophyte recruitment in a salt marsh restoration site. Estuaries 25: 1174–1183.

71. Craft C, Clough J, Ehman J, Joye S, Park R, et al. (2009) Forecasting the effects of accelerated sea-level rise on tidal marsh ecosystem services. Frontiers in Ecology and the Environment 7: 73–78.

72. Nicholls RJ, Marinova N, Lowe JA, Brown S, Vellinga P, et al. (2011) Sea-level rise and its possible impacts given a "beyond 4 degrees C world" in the twenty-first century. Philosophical Transactions of the Royal Society a-Mathematical Physical and Engineering Sciences 369: 161–181.

73. Ahrens TD, Beman JM, Harrison JA, Jewett PK, Matson PA (2008) A synthesis of nitrogen transformations and transfers from land to the sea in the Yaqui Valley agricultural region of Northwest Mexico. Water Resources Research 44.

74. Gruber N, Galloway JN (2008) An Earth-system perspective of the global nitrogen cycle. Nature 451: 293–296.

75. Caffrey J, Brown M, Tyler WB, Silberstein M, editors (2002) Changes in a California Estuary: A Profile of Elkhorn Slough. Moss Landing, CA: Elkhorn Slough Foundation. 280 p.

76. Phillips BM, Stephenson M, Jacobi J, Ichikawa G, Silberstein M, et al. (2002) Land use and contaminants. In: Caffrey JM, Brown M, Tyler WB, Silberstein M, editors. Changes in a California Estuary: A Profile of Elkhorn Slough. Moss Landing, CA: Elkhorn Slough Foundation. 237–253.

77. Plant JN, Needoba JA, Fitzwater SE, Coletti LJ, Jannasch HW, et al. (2009) Linking agriculture, nitrogen inputs and ecosystem metabolism in Elkhorn Slough, on time scales from hours to years. Coastal and Estuarine Research Federation: Estuaries and Coasts in a Changing World. Portland, OR.

78. Nidzieko NJ (2010) Tidal asymmetry in estuaries with mixed semidiurnal/diurnal tides. Journal of Geophysical Research-Oceans 115.

79. Field CB, Daily GC, Davis FW, Gaines S, Matson PA, et al. (1999) Confronting climate change in California: ecological impacts on the Golden State. The Union of Concerned Scientists and The Ecological Society of America.

80. Breschini SJ, Hartz TK (2002) Presidedress soil nitrate testing reduces nitrogen fertilizer use and nitrate leaching hazard in lettuce production. HortScience 37: 1061–1064.

81. Valiela I, Teal JM, Sass WECMS (1973) Nutrient retention in salt marsh plots experimentally fertilized with sewage sludge. Estuarine and Coastal Marine Science.

82. Tilman D, Fargione J, Wolff B, D'Antonio C, Dobson A, et al. (2001) Forecasting agriculturally-driven global environmental change. Science 292: 281–284.

83. Galloway JN, Townsend AR, Erisman JW, Bekunda M, Cai ZC, et al. (2008) Transformation of the nitrogen cycle: Recent trends, questions, and potential solutions. Science 320: 889–892.

84. Ward KM, Callaway JC, Zedler JB (2003) Episodic colonization of an intertidal mudflat by native cordgrass (Spartina foliosa) at Tijuana Estuary. Estuaries 26: 116–130.

15

Quantifying Uncertainties in N_2O Emission Due to N Fertilizer Application in Cultivated Areas

Aurore Philibert[1,2]*, Chantal Loyce[1,2], David Makowski[1,2]

1 INRA, UMR 211 Agronomie, Thiverval Grignon, France, **2** AgroParisTech, UMR 211 Agronomie, Thiverval Grignon, France

Abstract

Nitrous oxide (N_2O) is a greenhouse gas with a global warming potential approximately 298 times greater than that of CO_2. In 2006, the Intergovernmental Panel on Climate Change (IPCC) estimated N_2O emission due to synthetic and organic nitrogen (N) fertilization at 1% of applied N. We investigated the uncertainty on this estimated value, by fitting 13 different models to a published dataset including 985 N_2O measurements. These models were characterized by (i) the presence or absence of the explanatory variable "applied N", (ii) the function relating N_2O emission to applied N (exponential or linear function), (iii) fixed or random background (i.e. in the absence of N application) N_2O emission and (iv) fixed or random applied N effect. We calculated ranges of uncertainty on N_2O emissions from a subset of these models, and compared them with the uncertainty ranges currently used in the IPCC-Tier 1 method. The exponential models outperformed the linear models, and models including one or two random effects outperformed those including fixed effects only. The use of an exponential function rather than a linear function has an important practical consequence: the emission factor is not constant and increases as a function of applied N. Emission factors estimated using the exponential function were lower than 1% when the amount of N applied was below 160 kg N ha^{-1}. Our uncertainty analysis shows that the uncertainty range currently used by the IPCC-Tier 1 method could be reduced.

Editor: Carl J. Bernacchi, University of Illinois, United States of America

Funding: This work was funded by the French Research Agency (ANR project ORACLE: Opportunities and Risks of Agrosystems & forests in response to CLimate, socio-economic and policy changEs in France (and Europe)). The funders had no role in study design, data collection and analysis, decision to publish, or preparation of the manuscript.

Competing Interests: The authors have declared that no competing interests exist.

* E-mail: Aurore.Philibert@grignon.inra.fr

Introduction

Nitrous oxide (N_2O) is a greenhouse gas (GHG) with a global warming potential approximately 298 times greater than that of CO_2 [1]. N_2O emissions increased by almost 17% from 1990 to 2005 [2]. The nitrogen (N) cycle is complex and N_2O emissions are determined by many factors [3]. Natural and anthropogenic N_2O is emitted as a result of nitrification (oxidation of ammonia) and denitrification (nitrate reduction), and these processes are influenced by applications of mineral N fertilizer and manure to agricultural soils [4,5]. N applications are recognized as the major source of anthropogenic nitrous oxide emission [6,7]. N_2O emissions are also influenced by other management practices (e.g., tillage [8]), soil and climate characteristics (e.g. soil water content) [9,10,11,12,13].

For countries unable to provide local statistics, N_2O emission can be estimated by the IPCC-Tier 1 method. In this approach, direct N_2O emission from N inputs is calculated as Y_N inputs $= (F_{SN}+F_{ON}+F_{CR}+F_{SOM})*EF_1+ (F_{SN}+F_{ON}+F_{CR}+F_{SOM})$-$_{FR}*EF_{1FR}$, where F_{SN} is the annual amount of synthetic N fertilizer applied to soils, F_{ON} is the annual amount of organic N applied to soils, F_{CR} is the annual amount of N in crop residues, F_{SOM} is the annual amount of N in mineral soils, EF_1 is the emission factor for N_2O emissions from N inputs and FR indicates that the value concerned is for flooded rice [14]. For all crops other than flooded rice, the relationship between N_2O emission from N fertilizer and the dose of N applied can be expressed as

$Y=EF*X$, where Y represents N_2O emissions due solely to N fertilization, X is the amount of synthetic and organic N applied and EF (emission factor) is the amount of N_2O emitted per unit of applied N. In the United Nations Framework Convention on Climate Change [15], 56% of developed countries reported using the Tier 1 method of the IPCC to estimate N_2O emission from agricultural soils in 2006, and half the published N_2O emission inventories are based on this approach [16].

The EF value of 1.25%, set in 1999 [17], was calculated from the following linear regression: $Y=0.0125*X$, where Y is the emission rate (in kg N_2O-N ha^{-1} yr^{-1}) and X is the fertilizer application rate (in kg N ha^{-1} yr^{-1}), based on 20 experiments [18]. A background emission of 1 kg N_2O-N ha^{-1} yr^{-1} (i.e., emission for X=0) was obtained in five experiments. The new value of EF used by the IPCC after 2006 (1%; [14]) was estimated from a larger dataset, including N_2O emission measurements from studies on both crops and grassland [10].

Several recent studies have improved the estimation of N_2O emission further. Process-based models, such as the DNDC model [12] have been used to calculate the N_2O emission factor as a function of the organic carbon content of the soil, fertilizer type and weather conditions, and the DAYCENT model [19] has been used to calculate N_2O emissions as a function of soil class, daily weather, historical vegetation cover and land management practices, such as the type of crop grown, fertilizer additions and cultivation events. As these models describe the nitrogen cycle in detail, they may require long computation times and many input variables and are therefore difficult to implement [20]. Various statistical models have also recently been proposed for the estimation of N_2O emission from global datasets. For example, linear regression models have been used [20,21], and a nonlinear model based on an exponential function was proposed in another study [13].

The IPCC-Tier 1 method used for the estimation of N_2O emissions due to N fertilization includes three main sources of uncertainty on N_2O emission: (i) the uncertainty concerning the equation relating N_2O emission to applied N, (ii) the uncertainty concerning the equation parameters and (iii) the uncertainty about the amount of applied N (X).

In the IPCC-Tier 1 method, N_2O emission is assumed to be linearly related to applied N, but this assumption has been challenged; some authors [22,23] suggest that N_2O emission may instead increase exponentially as a function of applied N, and an exponential relationship between Y and X was also considered in the N_2O mitigation protocol proposed by Millar et al. [24]. A nonlinear relationship between Y and X was also considered by Stehfest and Bouwman [10]. There is currently no consensus concerning the most appropriate function for describing the relationship between N_2O emission and applied N at the global scale.

Uncertainty about the true value of the model parameter EF is another source of concern, for two reasons. First, N_2O emission measurements are known to be highly variable, both within a given site-year and between site-years. For a given site-year, N_2O emission varies principally due to climatic conditions, such as variations in the timing and intensity of rainfall, which modify microbial activity and the rates of gaseous emission [25]. For example, N_2O emissions must be measured after a period of rain to detect peaks in emission. Many factors may be responsible for variability between site-years, including differences in management practices (e.g. type of N fertilizer), soil characteristics and weather conditions between sites and years [20], by modifying chemical exchanges in agricultural soils. Duration of the experiment [10] and method used to measure emissions [11] may also affect N_2O emission measurements. Second, the emission factor can be estimated by several different statistical methods, some based on fixed-parameter models (i.e., classical regression) and others based on mixed-effect models or Bayesian methods. The sensitivity of EF to the statistical method used for its estimation has never been evaluated.

Finally, the amounts of N applied can be estimated from regional and national statistics and from interviews with farmers [10,26,27], but the actual amounts of N applied are not perfectly known and vary from year to year.

In this study, we focused on the first two of these sources of uncertainty: the equation of the model and the values of the model parameters. We fitted 13 different models to the dataset of Stehfest and Bouwman [10], and calculated uncertainty ranges on average N_2O emissions from a subset of these models, comparing our ranges with those currently used by the IPCC.

Materials and Methods

Database

The dataset is a global compilation of nitrous oxide (N_2O) and nitric oxide (NO) emissions extracted from peer-reviewed publications appearing between 1979 and 2004, established by Stehfest and Bouwman [10]. Readers should refer to the original paper by Stehfest and Bouwman for a more complete presentation of the data.

The dataset (available from http://www.pbl.nl/en/publications/2006/N2OAndNOEmissionFrom AgriculturalFieldsAndSoilsUnderNaturalVegetation) includes 1891 measurements of N_2O and NO emissions in natural and agricultural fields from 387 publications. As we focused on calculation of the emission factor associated with fertilizer applications in agricultural fields (EF), we excluded the following experiments from the initial dataset: (i) 418 experiments carried out in natural areas, (ii) 360 experiments including measurements of NO emission only, (iii) 57 experiments on organic soils (not concerned by EF), (iv) 25 experiments including the use of chemicals or additives considered to inhibit nitrification (also excluded by Stehfest and Bouwman [10]), (v) 8 experiments in grazing systems (also excluded by Stehfest and Bouwman [10]), (vi) 38 experiments in which the amounts of applied N exceeded 500 kg N ha^{-1} yr^{-1} (given that the maximum amounts of N applied to agricultural fields has been estimated at 400 kg N ha^{-1} [20,27,28]).

We finally worked with a dataset including 985 measurements of N_2O emission in agricultural fields extracted from 203 publications, corresponding to a set of experiments encompassing various soil and climatic characteristics and types of fertilization (Figs. 1 and 2).

The distribution of N_2O measurements and amounts of applied N are presented in Table 1 for the entire dataset and for each continent separately. The largest amount of data was available for the temperate-continental climate (460), followed by the temperate-oceanic climate (258) and the tropics-warm humid climate (104). Only 80, 44, 21, 12 and 6 data were collected for the subtropical-summer rains, subtropical-winter rains, tropic-seas dry, boreal and cool tropics climates, respectively.

Statistical Analysis

Statistical Models

Thirteen models relating N_2O emission to the amount of applied N were fitted to the data (Table 2). These models were characterized by (i) the presence or absence of the explanatory variable "applied N" (X), (ii) the function relating emission to applied N (an exponential or linear function), (iii) fixed or random background emission (i.e., emission for X = 0), and (iv) fixed or random N effect.

The first 11 models (with L, NL, N, 0, F, and R standing for linear, nonlinear, nitrogen effect, no nitrogen effect, fixed parameter and random parameter, respectively) can be expressed as:

Model NL-N-FF: (1) $Y_{ijk} = \exp(\mu_0 + \mu_1 X_{ij}) + \varepsilon_{ijk}$

with $\varepsilon_{ijk} \sim N(0, \tau^2)$

Model NL-0-R: (2) $Y_{ijk} = \exp(\alpha_{0i}) + \varepsilon_{ijk}$

with $\varepsilon_{ijk} \sim N(0, \tau^2)$ and $\alpha_{0i} \sim N(\mu_0, \sigma_0^2)$

Model NL-N-RF: (3) $Y_{ijk} = \exp(\alpha_{0i} + \mu_1 X_{ij}) + \varepsilon_{ijk}$

with $\varepsilon_{ijk} \sim N(0, \tau^2)$ and $\alpha_{0i} \sim N(\mu_0, \sigma_0^2)$

Model NL-N-FR: (4) $Y_{ijk} = \exp(\mu_0 + \alpha_{1i} X_{ij}) + \varepsilon_{ijk}$

with $\varepsilon_{ijk} \sim N(0, \tau^2)$ and $\alpha_{1i} \sim N(\mu_1, \sigma_1^2)$

Model NL-N-RR: (5) $Y_{ijk} = \exp(\alpha_{0i} + \alpha_{1i} X_{ij}) + \varepsilon_{ijk}$

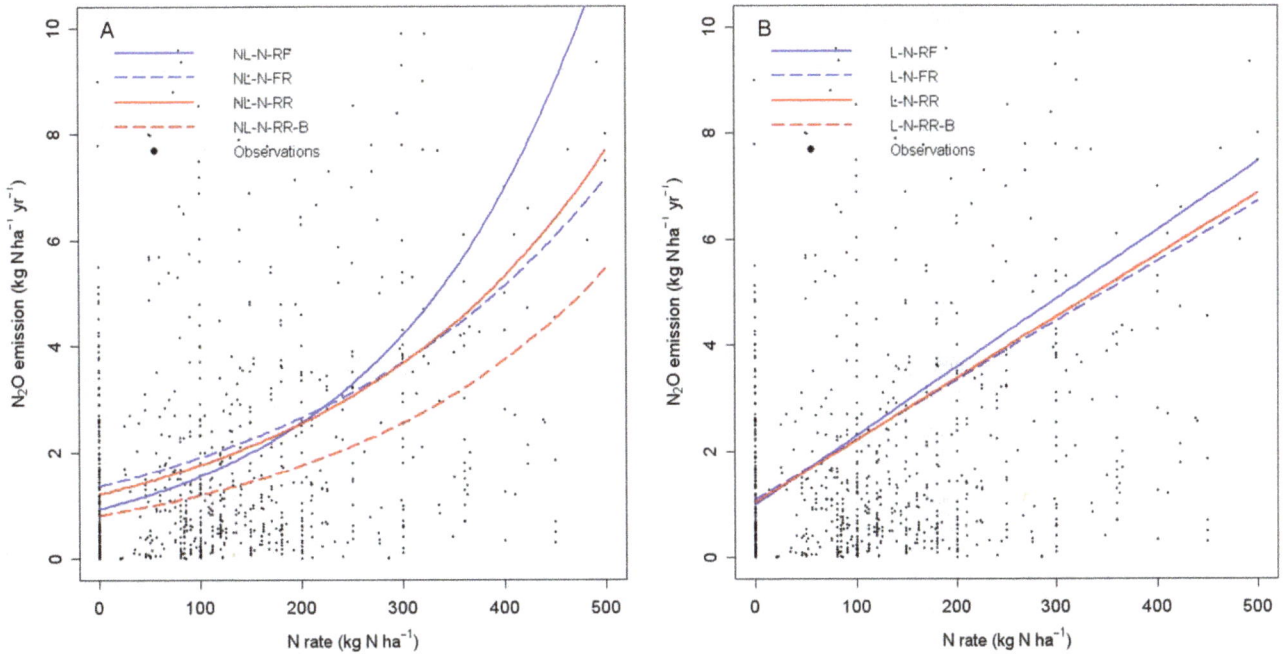

Figure 1. Fitted response curves obtained with the four selected nonlinear models (A) and the four selected linear models (B). Black points correspond to N$_2$O data (96.04% of available observations are displayed; the other data are too extreme for graphical presentation).

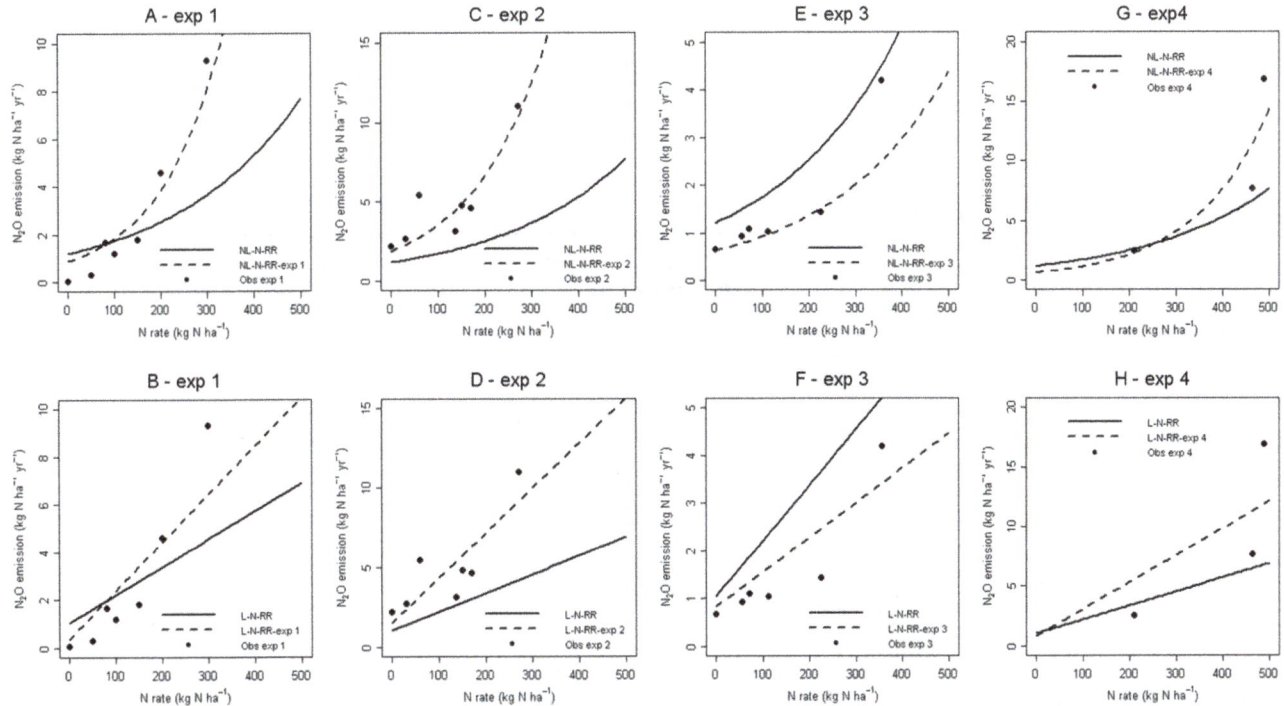

Figure 2. Fitted response curves for four experiments (exp 1: (A–B), exp 2: (C–D), exp 3: (E–F) and exp 4: (G–H)). For each published experiment, mean response (solid black line) and experiment-specific response (dotted black line) were calculated with model NL-N-RR (A, C, E, G) and model L-N-RR (B, D, F, H). Black points represent N$_2$O data averaged over replicates.

Table 1. Minimal, maximal, median and mean values of nitrous oxide (N_2O) and amount of applied N (N rate) for the world and for North America, South America, Asia, Europe and Oceania.

Variable	Continent/world	min	median	mean	max	Number of data
	World	0.003	1.07	2.4	46.44	985
	Asia	0.01	0.53	1.11	15.60	124
	Europe	0.004	1.25	2.53	31.73	453
N_2O (kg N ha^{-1} yr^{-1})	North America	0.004	0.93	2.16	26.9	306
	Oceania	0.016	1.39	2.45	15	26
	South America	0.003	1.56	4.67	46.44	76
	World	0	100	124	500	985
	Asia	0	120	139.8	423	124
	Europe	0	100	132	500	453
N rate (kg N ha^{-1})	North America	0	92	115.3	450	306
	Oceania	0	66	108.6	500	26
	South America	0	0	90.96	360	76

with $\varepsilon_{ijk} \sim N(0,\tau^2)$, $\alpha_{0i} \sim N(\mu_0,\sigma_0^2)$ and $\alpha_{1i} \sim N(\mu_1,\sigma_1^2)$

Model L-0-F: (6) $Y_{ijk} = \mu_0 + \varepsilon_{ijk}$
with $\varepsilon_{ijk} \sim N(0,\tau^2)$

Model L-N-FF: (7) $Y_{ijk} = \mu_0 + \mu_1 X_{ij} + \varepsilon_{ijk}$

with $\varepsilon_{ijk} \sim N(0,\tau^2)$

Model L-0-R: (8) $Y_{ijk} = \alpha_{0i} + \varepsilon_{ijk}$

with $\varepsilon_{ijk} \sim N(0,\tau^2)$ and $\alpha_{0i} \sim N(\mu_0,\sigma_0^2)$

Model L-N-RF: (9) $Y_{ijk} = \alpha_{0i} + \mu_1 X_{ij} + \varepsilon_{ijk}$

with $\varepsilon_{ijk} \sim N(0,\tau^2)$ and $\alpha_{0i} \sim N(\mu_0,\sigma_0^2)$

Model L-N-FR: (10) $Y_{ijk} = \mu_0 + \alpha_{1i} X_{ij} + \varepsilon_{ijk}$

with $\varepsilon_{ijk} \sim N(0,\tau^2)$ and $\alpha_{1i} \sim N(\mu_1,\sigma_1^2)$

Model L-N-RR: (11) $Y_{ijk} = \alpha_{0i} + \alpha_{1i} X_{ij} + \varepsilon_{ijk}$

with $\varepsilon_{ijk} \sim N(0,\tau^2)$, $\alpha_{0i} \sim N(\mu_0,\sigma_0^2)$ and $\alpha_{1i} \sim N(\mu_1,\sigma_1^2)$

where Y_{ijk} is the N_2O emission (kg N ha^{-1} yr^{-1}) measured in the

Table 2. Characteristics of the 13 statistical models for N_2O emission.

Model name	Linear	Amount of N applied	Intercept	Effect of the amount of N applied	AIC	% AIC	BIC	% BIC	DIC
NL-N-FF	No	Yes	Fixed	Fixed	5513.1	23.0	5527.8	22.6	–
NL-0-R	No	No	Random	–	5091.9	13.6	5106.5	13.3	–
NL-N-RF	No	Yes	Random	Fixed	4553.9	1.6	4573.5	1.5	–
NL-N-FR	No	Yes	Fixed	Random	4598.9	2.6	4618.5	2.5	–
NL-N-RR	No	Yes	Random	Random	4482.7	0	4507.1	0	–
NL-N-RR-B	No	Yes	Random	Random	–	–	–	–	4196.71
L-0-F	Yes	No	Fixed	–	5653.9	20.5	5663.7	20.1	–
L-N-FF	Yes	Yes	Fixed	Fixed	5512.1	17.4	5526.8	17.2	–
L-0-R	Yes	No	Random	–	5268.5	12.3	5283.2	12.0	–
L-N-RF	Yes	Yes	Random	Fixed	5117.4	9.0	5136.9	8.9	–
L-N-FR	Yes	Yes	Fixed	Random	4698.0	0.1	4717.5	0	–
L-N-RR	Yes	Yes	Random	Random	4693.2	0	4717.6	0.002	–
L-N-RR-B	Yes	Yes	Random	Random	–	–	–	–	4421.63

Models were characterized by their response function (linear or exponential), the use of the explanatory variable 'amount of applied N', the use of random effects for the intercept and/or the effect of the amount of N applied, values of the Akaïke and Schwartz criteria (AIC and BIC), and of the deviance information criterion (DIC) for Bayesian models. % AIC and % BIC indicate the percentage increase in AIC and BIC with respect to the best linear and nonlinear models.

i^{th} published experiment ($i = 1 \dots 203$), the jth applied N dose ($j = 1 \dots N_i$), and the kth replicate ($k = 1 \dots K_{ij}$), X_{ij} is the jth applied N dose (kg N ha^{-1}) in the ith published experiment, μ_0 is the mean background emission, α_{0i} is the published experiment-specific background emission (random), μ_1 is the mean applied N effect, α_{1i} is the published experiment-specific applied N effect (random), and ε_{ijk} is the residual error term. The random terms α_{0i}, α_{1i} and ε_{ijk} were assumed to be independent and normally distributed. Models including correlated α_{0i} and α_{1i} were also fitted to the data but, as their outputs were very similar to the outputs of the models with independent random parameters, they were not considered further. Note that, in nonlinear models (1–5), the N$_2$O response does not follow a normal distribution, even if its parameters α_{0i} and α_{1i} do, due to the use of an exponential function to relate emissions to model parameters.

In the linear models (6–11), the parameter μ_1 corresponds to the emission factor EF used by the IPCC. In the nonlinear models based on an exponential function (1–5), N$_2$O emission per unit of applied N is not constant; instead, it increases as a function of X if μ_1 is positive. In the models including one or two random parameters (2–5 and 8–11), the response of N$_2$O to the amount of applied N is assumed to follow the same function (linear or exponential) in all experiments, but the parameters of these models (background emission α_{0i}, effect of applied N α_{1i}, or both) were assumed to vary between experiments. Distributions of α_{0i} and α_{1i} describe the between-experiment variability of background emission and N fertilizer effect. An intercept was included in all statistical models to account for background anthropogenic N$_2$O emission [18]. The values of the μ_0, μ_1, σ_0, σ_1, and τ parameters of models 1–11 were estimated by an approximate maximum likelihood method, with the nlme R statistical package [29].

Two additional models, NL-N-RR-B and L-N-RR-B, were defined. These models were based on the equations of models NL-N-RR and L-N-RR, respectively, but their parameters were estimated by a Bayesian method implemented with a Markov chain Monte Carlo algorithm (MCMC). Normal and independent prior probability distributions were defined for μ_0 and μ_1; μ_0, $\mu_1 \sim N(0,1000)$. Uniform and independent prior probability distributions were defined for τ, σ_0, σ_1; τ, σ_0, $\sigma_1 \sim U(0,100)$. Under these assumptions, μ_0 and μ_1 had a prior mean of zero and a prior standard deviation of 32, which is quite large given the measured values, which ranged from 0.003 to 46.44 in our dataset. These distributions represent a broad *a priori* distribution with respect to the data obtained. For example, the 95% credibility interval derived from the prior distributions ranged from -6272.3 to 6331.8 N$_2$O kg N ha^{-1} yr^{-1} for X = 100 kg N ha^{-1}. Posterior distributions of the parameters of models NL-N-RR-B and L-N-RR-B were calculated with WinBUGS software [30], with three chains of 100,000 MCMC iterations. Convergence was checked with the Gelman-Rubin method [31].

Model Assessment and Uncertainty Analysis

The Akaike information criterion (AIC) and the Schwartz criterion (BIC) [32,33] were calculated for the first 11 models, and the deviance information criterion (DIC) [34] was calculated for the two Bayesian models. Lower values of AIC, BIC or DIC are considered to indicate better models. Note that the weighting of the experiments according to their lengths did not reduce AIC, BIC or DIC.

We calculated the 95% confidence intervals for each model by a bootstrap method [35,36]; data were sampled, with replacement, 500 times, and each model was fitted to each of the generated samples. For the two Bayesian models, 95% credibility intervals

for the predicted N$_2$O emissions were calculated from the parameter values generated by the MCMC algorithm.

The predictions generated by the three best non-Bayesian linear models, the three best non-Bayesian exponential models (selected with AIC and BIC criteria) and the two Bayesian models were compared with the N$_2$O emissions calculated by the IPCC-Tier1 method: Y = EF*X, where EF is taken as 0.01 [14]. The range of uncertainty on predicted N$_2$O emissions for the IPCC method was calculated from the minimum and maximum values of EF (0.003 and 0.03, respectively) reported by the IPCC [14]. The emissions due to applied N calculated with the IPCC method were compared with the predictions of the eight selected models minus the values predicted at X = 0.

This uncertainty range was then compared with each of the confidence intervals for the eight selected models. We also compared the lower limit of the IPCC uncertainty range with the lowest of the eight 2.5 percentiles calculated for the eight selected models, and the upper limit of the IPCC uncertainty range with the highest of the eight 97.5 percentiles of the eight selected models. The most extreme 2.5 and 97.5 percentiles obtained with the eight selected models can be interpreted as best-case and worst-case emission scenarios, respectively. They correspond to the lowest and highest limits of the confidence intervals calculated for the eight models.

The code used for statistical analysis is available, on request, from the corresponding author.

Results

Parameter Values

The estimated value of parameter μ_0 (mean background emission) ranged from -0.21 to 0.88 for nonlinear models and from 0.69 to 2.78 for linear models (Table 3). The between-model variability of the estimated values of μ_1 (mean applied N effect) was small: estimated values ranged from 0.0033 to 0.0050 for nonlinear models and from 0.0113 to 0.0138 for linear models. For both linear and nonlinear models, the estimated values of μ_1 were lower when the effect of applied N was considered a random

Table 3. Estimated values of the parameters of the 13 models.

Model name	μ_0	σ_0	μ_1	σ_1	τ
NL-N-FF	0.25 (0.096)	–	0.0042 (0.0003)	–	3.96
NL-0-R	0.87 (0.077)	0.84	–	–	2.84
NL-N-RF	−0.068 (0.092)	0.83	0.0050 (0.0003)	–	2.09
NL-N-FR	0.31 (0.068)	–	0.0033 (0.0005)	0.0043	2.13
NL-N-RR	0.19 (0.09)	0.72	0.0037 (0.0004)	0.0025	1.94
NL-N-RR-B	−0.21 (0.13)	0.92	0.0038 (0.0005)	0.0032	1.91
L-0-F	2.40 (0.14)	–	–	–	4.26
L-N-FF	0.69 (0.19)	–	0.0138 (0.0011)	–	3.96
L-0-R	2.78 (0.27)	3.44	–	–	2.91
L-N-RF	0.99 (0.28)	3.16	0.0130 (0.0010)	–	2.67
L-N-FR	1.09 (0.11)	–	0.0113 (0.0017)	0.0195	2.12
L-N-RR	1.04 (0.13)	0.70	0.0117 (0.0017)	0.0187	2.08
L-N-RR-B	1.04 (0.14)	0.76	0.0117 (0.0017)	0.0189	2.08

The standard deviations of the estimators of μ_0 and μ_1 are indicated in brackets.

effect (Table 3). For example, the estimated value of μ_1 was 0.005 for NL-N-RF, but only 0.0037 for NL-N-RR.

μ_0 was less accurately estimated than μ_1; the coefficient of variation (standard deviation/estimated value) was lower for μ_1 than for μ_0. For a given type of function (linear or exponential) estimates of σ_0 and σ_1 (between-experiment standard deviation of background emission and applied N effects, respectively) were similar between models. The estimated values of τ (standard deviation of model residuals) were lower for models with random parameters and for those containing the explanatory variable X.

Model Selection

The lowest AIC and BIC values were obtained for the nonlinear model including two random effects (NL-N-RR) (Table 2). Thus, models based on an exponential function outperformed models based on a linear function. This result was confirmed by the DIC values obtained for the two Bayesian models: DIC was lower with the exponential function.

AIC and BIC values were much higher in models in which applied N (X) was not included as an explanatory variable. The AIC and BIC values of the NL-0-R model were 5091.9 and 5106.5, respectively, whereas the AIC and BIC values of the NL-N-RF model were 4553.9 and 4573.5, respectively (Table 2).

Models including one or two random effects outperformed those including only fixed effects. The best linear model was L-N-RR on the basis of AIC, and L-N-FR, on the basis of BIC. The NL-N-FF (no random effect) model had an AIC of 5513.1 and a BIC of 5527.8, whereas both these values were much lower (AIC = 4482.7 and BIC = 4507.1) for the NL-N-RR (two random parameters) model. Models including one or two random effects had similar AIC and BIC values; the use of one random effect rather than two did not increase AIC and BIC by more than 2.6% and 9% for the nonlinear and linear models, respectively (Table 2).

The AIC and BIC values of models (1), (2), (6), (7) and (8) were more than 10% higher than those for the best nonlinear and linear models, and were therefore not considered further. We therefore considered only models (3), (4), (5), (9), (10) and (11) and the two Bayesian models in our subsequent estimations of N_2O emission.

Estimation of N_2O Emissions with the Selected Models

Figure 1 shows N_2O emissions estimated with the eight selected models. The rate of increase of N_2O emissions with the amount of N applied was greater with the NL-N-RF model (Fig. 1A) than with the other two nonlinear models (NL-N-FR and NL-N-RR). The predicted increase in the amount of N_2O emitted per unit increase in the amount of applied N was thus lower when the effect of applied N was defined as a random effect. Similar results were obtained for linear models, for which the highest rate of increase in N_2O emissions with the amount of N applied was obtained for the L-N-RF model, which had a fixed slope (Fig. 1B). These results are consistent with the estimated parameter values reported in Table 3.

The emissions predicted by the Bayesian model NL-N-RR-B were the lowest for all values of applied N (Fig. 1A), due to the low estimated value of the intercept for this model (Table 3). The amounts of emission predicted by the L-N-RR model and its Bayesian counterpart (L-N-RR-B) were very similar and were essentially undistinguishable.

Figure 2 shows the fitted response curves obtained with the best linear and nonlinear models, NL-N-RR and L-N-RR, for four experiments. Considering experiment-specific responses, the nonlinear model better fitted the emissions measured at high N doses in experiments 1, 2, and 4, and the emissions measured at low N doses in experiments 3 and 4. Between-experiment variability was high for N_2O emissions (Figure 2) and could be accounted for by

the experiment-effects included in the mixed-effect models. The residual standard error was lower with NL-N-RR than with L-N-RR (see values of τ in Table 3).

Comparison with the Emissions Estimated with the IPCC-Tier 1 Method

We determined the ranges of N_2O emissions (Fig. 3) covered by the eight models considered in Figure 1, either taking into account the uncertainty on the estimated parameter values (dark gray area) or not taking this uncertainty into account (light gray area). The final values predicted by the models were calculated by subtracting the predicted value at X = 0 (background emission) from the value actually predicted for a given amount of applied N. This graphical presentation made it possible to compare our models with the N_2O emissions predicted with the IPCC-Tier 1 method. The estimates of N_2O emission obtained with an emission factor of 1% (as used by the IPCC) were within the range of values covered by the eight selected models (Fig. 3B), but the range of uncertainty for emissions estimated with the IPCC-Tier 1 method was larger than that for the eight selected models. The upper limit of the uncertainty range for the IPCC method was much higher than that defined by the highest value of the eight 97.5 percentiles of the eight selected models, particularly for N applications below 300 kg ha^{-1}, as generally practiced in farmers' fields (Fig. 3). The lower limits of the uncertainty ranges for the IPCC method and for our models were more similar.

This result was confirmed (Fig. 4) by comparing the estimates of N_2O emissions due to applied N obtained with the eight models with those obtained by the IPCC-Tier 1 method for four different amounts of applied N. These amounts of applied N correspond to the average amounts applied in western, eastern and southern Africa, worldwide, Europe and eastern Asia [10]. The purpose of Figure 4 was to compare model predictions for contrasted applied N doses, not to calculate average emissions at the continental scale. The uncertainty ranges obtained with the IPCC method were indeed larger than those defined by the highest of the 97.5 percentiles and the lowest of the 2.5 percentiles for the eight selected models (Fig. 4). The upper limits of the IPCC uncertainty ranges were systematically higher than the highest 97.5 percentile obtained with our models. We also found that the emissions predicted by the IPCC method were very similar to those obtained with the linear models, but systematically higher than the emissions predicted by the nonlinear models (Fig. 3).

Discussion

Our analysis was carried out with the dataset of Stehfest and Bouwman [10] because this dataset includes a large number of data of N_2O emissions in agricultural soils. These data were collected under various conditions characterized by different measurement methods (e.g. from short to long periods of measurements), different soils, climates, and crops. The variability of these conditions and their effects on N_2O emission were taken into account in our analysis using random parameter models. In these models, the N_2O emission was related to applied N using linear or nonlinear functions including two random parameters (α_{0i} and α_{1i}). The probability distribution of these parameters describes the between-study variability of the parameter values due to i) the heterogeneity of the experimental protocols (i.e. length of the experiment and measurement frequency) and ii) the variability of soil, climate, and crop characteristics. This approach accounts for both the heterogeneity of the measurement protocols and the variability of the environments.

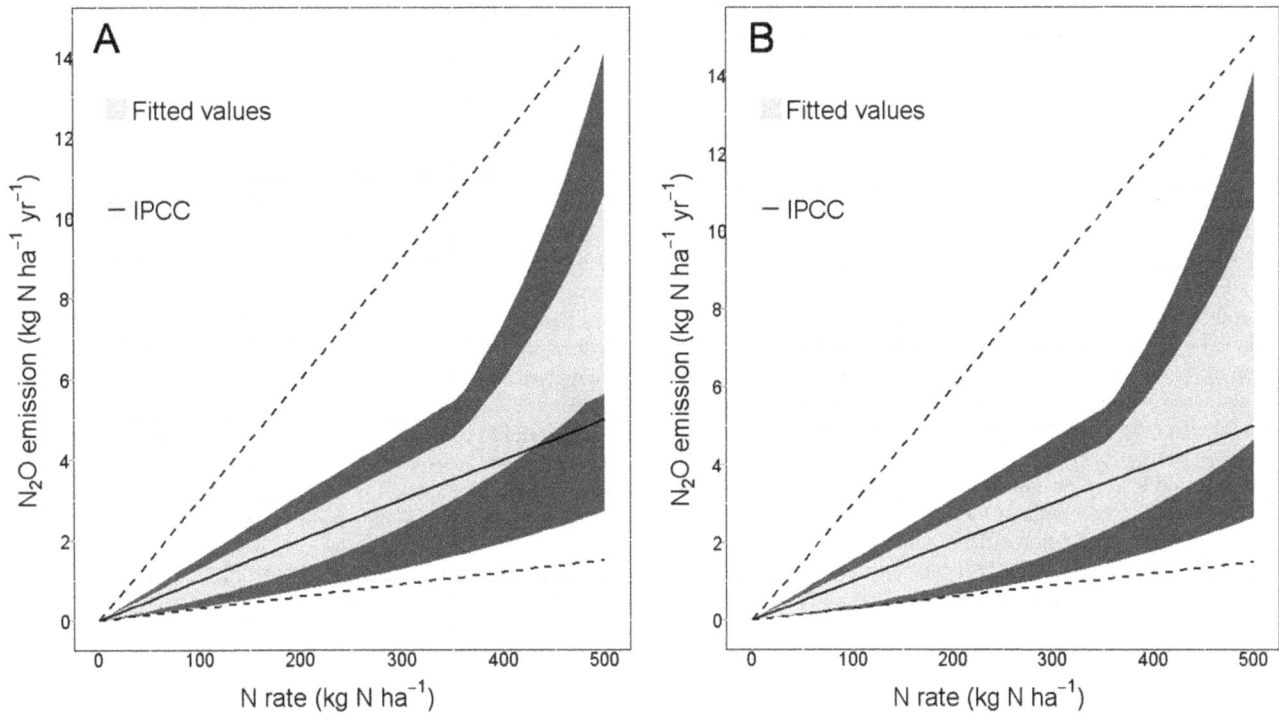

Figure 3. Predicted N$_2$O emissions and uncertainty ranges for our eight selected models and the IPCC-Tier 1 method. The light gray area represents the uncertainty in the model equations and includes the mean values predicted by six models (A) or by eight models (6 non-Bayesian models +2 Bayesian models) (B). The dark gray area represents the uncertainty in model equations and parameter values. The upper and lower limits of the dark gray area indicate the worst-case and best-case scenarios, respectively, defined from six models (A) or from eight models (6 non-Bayesian models +2 Bayesian models) (B). The solid black line and the dotted lines indicate the N$_2$O emissions predicted with an EF of 1% and the uncertainty range of the IPCC-Tier1 method, respectively.

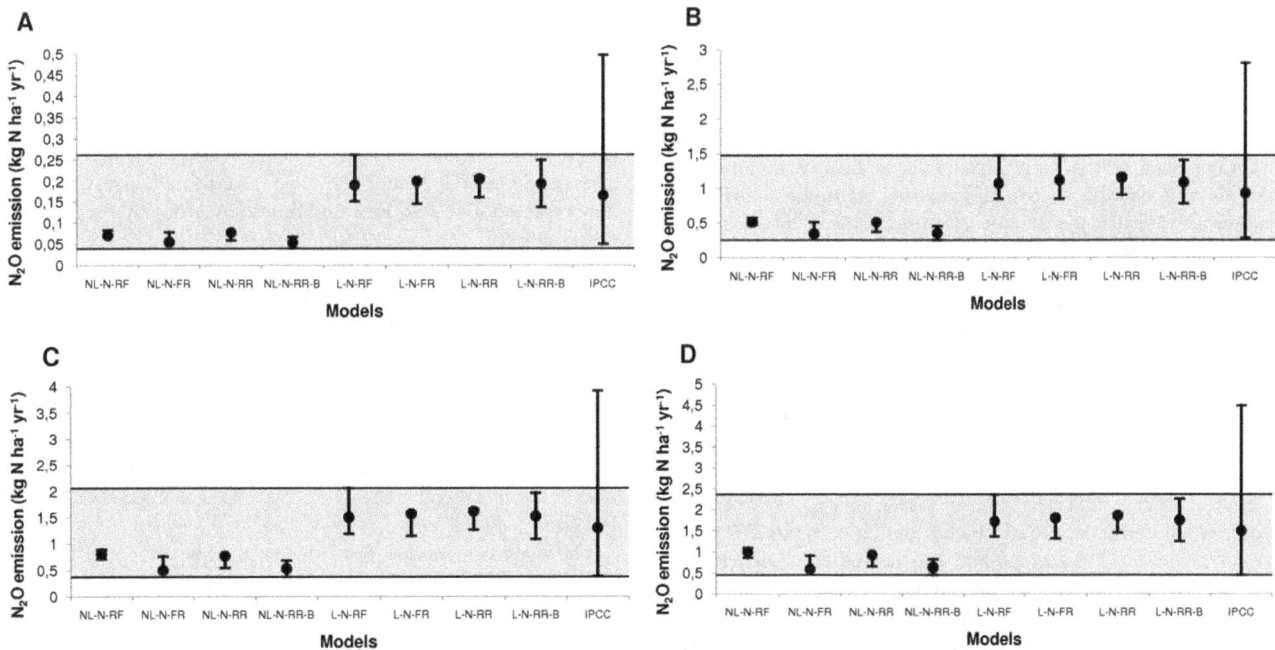

Figure 4. Predicted N$_2$O emissions due to N fertilization and 95% confidence intervals (CI) for each model, and predicted values and uncertainty ranges for the IPCC-Tier 1 method. The light gray area corresponds to the values covered by the 95% CI of our eight models. The amounts of N applied were A) 16.62 kg N ha^{-1}, B) 93.6 kg N ha^{-1}, C) 130.74 kg N ha^{-1}, D) 149.58 kg N ha^{-1} (average amounts of applied N for western, eastern and southern Africa, worldwide, Europe, and eastern Asia respectively). N$_2$O emissions were estimated by subtracting the value corresponding to the application of no N from the value for each amount of N applied.

Exponential models outperformed linear models for the three statistical criteria considered: AIC, BIC and DIC. However, the differences were small. The AIC of the best linear model was only 4.7% higher than the AIC of the best exponential model. The use of an exponential function rather than a linear model has an important practical consequence: EF is not constant and increases as a function of applied N.

Our results indicate that EF is lower than the estimated value used by the IPCC-Tier 1 (i.e., 1% of applied N) if the amount of N applied is below 160 kg N ha^{-1} for the NL-N-RF model, 240 kg N ha^{-1} for the NL-N-FR model and 220 kg N ha^{-1} for the NL-N-RR model. According to Spiertz [27], farmers may apply amounts of nitrogen fertilizer below these thresholds in ecological low-input cropping systems and in some technological high-input systems. Consequently, the use of an exponential model rather than a linear model is likely to decrease estimates of N_2O emissions in many cases. According to Hoben et al. [23], the current IPCC-Tier 1 method could lead to an underestimation of N_2O emission if the true response is exponential. Our results suggest that this is the case only for the application of large amounts of N fertilizer.

McSwiney and Robertson [22] suggested that the use of a nonlinear model instead of a linear model leads to a greater estimated reduction in N_2O emission for a moderate reduction in the amount of applied N, with little or no yield penalty. Our results do not entirely support this statement, because little difference was observed between the two types of models for doses of up to about 200 kg N ha^{-1} (Fig. 1). For example, if the amount of applied N is decreased from 150 kg N to 120 kg N (minus 20%), the resulting reduction of N_2O calculated with the NL-N-RR model is 0.22 kg N ha^{-1} yr^{-1}, slightly less than that calculated with the current IPCC emission factor (0.01*30 = 0.3 kg N ha^{-1} yr^{-1}). With the same model, the reduction induced by a decrease from 350 kg N ha^{-1} to 280 kg N ha^{-1} (minus 20%) is much larger, reaching 1 kg N ha^{-1} yr^{-1}; this value is higher than the reduction calculated with the IPCC emission factor (0.01*70 = 0.7 kg N ha^{-1} yr^{-1}). The estimated reduction of N_2O emission induced by a decrease in the amount of applied N is greater with the nonlinear model than with the linear model only for high N doses.

According to the AIC and BIC values obtained, models including one or two random effects outperformed models including fixed effects only. Mixed-effect models are commonly used in meta-analysis studies [37] and are recommended for the analysis of repeated measurements on the same individuals [38]. In the dataset of Stehfest and Bouwman [10], N_2O emissions were measured for several amounts of N applied in the same published-experiment. It was therefore appropriate to estimate N_2O emissions with mixed-effect models including one or two random effects in our study (Fig. 2). Models including one or two random effects performed similarly (less than 10% difference in AIC and BIC values), but the estimated effect of the amount of N applied on the amount of N_2O emitted tended to be lower when the amount of N applied was considered as a random effect.

Several models had very similar performances. We therefore used an ensemble approach based on eight models for the estimation of N_2O emissions and the definition of uncertainty ranges. The confidence intervals obtained with the models were used to define lower and upper limits, corresponding to the best-case and worst-case scenarios, respectively. These confidence intervals represent the uncertainty in average N_2O emissions over all experiments, but they do not describe the between-experiment variability of N_2O emission. The range of uncertainty defined here is relevant for the Tier 1 method and useful for explorations of the consequences of N applications for average N_2O emissions, taking into account the uncertainty due to model equations and

parameter estimations. The lower limit of our uncertainty range is close to that defined by the IPCC-Tier 1, although our lower limit is slightly higher than that of the IPCC for applications of large amounts of N. Our upper limit is much lower than the upper limit of the IPCC range, particularly for total N applications below 300 kg ha^{-1}, as commonly used in agriculture. Thus, the upper limit of the IPCC range gives an estimated N_2O emission of 9 kg N ha^{-1} yr^{-1} for a dose of 300 kg N ha^{-1}, whereas the upper limit of our uncertainty range (i.e., the highest upper limit of the confidence intervals of the eight models considered) gave an estimated emission value of only 4.7 kg N ha^{-1} yr^{-1}. This result is consistent with the findings of Leip et al. [12], suggesting that the uncertainty on estimates of N_2O emissions was overestimated when derived from experimental data variances, which largely compensate at large scales.

It is useful to compare our uncertainty ranges with other ranges calculated with process-based models [39], top-down methods [40] and hierarchical Bayesian models [16].

Our uncertainty range for the average N dose applied in North America – 0.49–1.88 kg N ha^{-1} yr^{-1} – is similar to the 95% confidence interval proposed by Del Grosso et al. [39] for the United States (133–304 Gg N yr^{-1} i.e. 0.99–2.27 kg N ha^{-1} yr^{-1} with the cropland area of North America reported by Stehfest and Bouwman [10]).

Our uncertainty range for the average N dose applied at the world scale (93.6 kg ha^{-1} of applied N, as reported by Stehfest and Bouwman [10]) (Fig. 4B) – 0.25–1.48 kg N ha^{-1} yr^{-1} – is lower and narrower than the interval proposed by Crutzen et al. [40] (2.8–4.68 kg N ha^{-1} yr^{-1}). However, it is difficult to compare these intervals, due to the use of a top-down method by Crutzen et al. Furthermore, these authors did not consider direct emission due to N fertilizer only, instead also taking into account indirect emissions from leaching and atmospheric deposition [13].

The 95% confidence interval calculated by Berdanier and Conant [16] with a hierarchical Bayesian linear model is similar to our uncertainty ranges for the four regions of the world presented in Figure 4. The two intervals overlap in all four regions, but our intervals tend to have lower upper and lower limits. For example, Berdanier and Conant [16] reported an interval of 0.05–0.46 for Africa, for a N fertilizer dose of 16.62 kg N ha^{-1} [10], whereas our interval was 0.04–0.26 kg N ha^{-1} yr^{-1} for the average N fertilizer dose reported for West, East and Southern Africa by Stehfest and Bouwman [10].

When between-experiment variability was taken into account, the experiment-specific N_2O estimated with our models covered a wider range of values. Thus, for applied N levels of 100 kg N ha^{-1} and with the NL-N-RR model, the 90% percentile for N_2O emission was 1.79 kg N ha^{-1} yr^{-1}, the 95% percentile was 2.52 kg N ha^{-1} yr^{-1} and the 99% percentile was 5.03 kg N ha^{-1} yr^{-1}, all these values being higher than the 1 kg N ha^{-1} yr^{-1} of the IPCC-Tier 1 method. Thus, N_2O emission has 1% chance to exceed 5 kg N ha^{-1} yr^{-1} for an N fertilizer dose of 100 kg ha^{-1}.

The nonlinear models presented in this paper should be used with caution for estimating average N_2O emissions at the country and continental scales. The average output value of a nonlinear model is not strictly equal to the output value obtained with the average input value. In order to calculate the average N_2O emission in a given country with a nonlinear model, the best approach is i) to determine the distribution of applied N fertilizer doses in this country, ii) to run the model for all doses, and iii) to take the average of all the model outputs. However, this approach requires the knowledge of the distribution of N fertilizer dose.

We focused on the Tier 1 approach of the IPCC, but the proposed exponential models could be extended to take several other environmental variables, such as climatic characteristics, soil types and fertilizer type, into account. This possibility has already been explored by Lesschen et al. [13], who took several variables into account (type of fertilizer, crop residues, atmospheric deposition, land use, soil type and precipitation) and by Leip et al. [12], who calculated the stratified emission factor as a function of soil organic carbon content, fertilizer type (mineral fertilizer or manure) and weather conditions. Such variables could be included in our models, for the estimation of region-specific N_2O emissions, taking local characteristics into account.

Author Contributions

Conceived and designed the experiments: AP CL DM. Performed the experiments: AP. Analyzed the data: AP DM. Contributed reagents/materials/analysis tools: AP DM. Wrote the paper: AP CL DM.

References

1. IPCC (2007) Climate Change 2007: The Physical Science Basis. Contribution of Working Group I to the Fourth Assessment Report of the IPCC. Cambridge: Cambridge University Press.
2. Smith PD, Martino Z, Cai D, Gwary H, Janzen P, et al. (2007) Agriculture. In: Climate Change 2007: Mitigation. Contribution of Working Group III to the Fourth Assessment Report of the Intergovernmental Panel on Climate Change. Cambridge: Cambridge University Press.
3. Galloway JN, Dentener FJ, Capone DG, Boyer EW, Howarth RW, et al. (2004) Nitrogen cycles: past, present and future. Biogeochemistry 70: 153–226.
4. IPCC (2001) Climate Change 2001: The Scientific Basis: Contribution of Working Group I to the Third Assessment Report of the IPCC. Cambridge: Cambridge University Press. 881p.
5. Mosier A, Kroeze C, Nevison C, Oenema O, Seitzinger S, et al. (1998) Closing the global atmospheric N2O budget: nitrous oxide emissions through the agricultural nitrogen cycle; OECD/IPCC/IEA Phase II Development of IPCC Guidelines for National Greenhouse Gas Inventories. Nutrient Cycling in Agroecosystems 52: 225–248.
6. Davidson EA (2009) The contribution of manure and fertilizer nitrogen to atmospheric nitrous oxide since 1860. Nature Geoscience 2: 659–662.
7. Snyder CS, Bruulsema TW, Jensen TL, Fixen PE (2009) Review of greenhouse gas emissions from crop production systems and fertilizer management effects. Agriculture, Ecosystems & Environment 133: 247–266.
8. Rochette P (2008) No-till only increases N2O emissions in poorly-aerated soils. Soil and Tillage Research 101: 97–100.
9. Rochette P, Tremblay N, Fallon E, Angers DA, Chantigny MH, et al. (2010) N2O emissions from an irrigated and non-irrigated organic soil in eastern Canada as influenced by N fertilizer addition. 61: 186–196.
10. Stehfest E, Bouwman L (2006) N2O and NO emission from agricultural fields and soils under natural vegetation: summarizing available measurement data and modeling of global annual emissions. Nutrient Cycling in Agroecosystems 74: 207–228.
11. Rochette P, Worth DE, Lemke RL, McConkey BG, Pennock DJ, et al. (2008) Estimation of N2O emissions from agricultural soils in Canada. I. Development of a country-specific methodology. Canadian Journal of Soil Science 88: 641–654.
12. Leip A, Busto M, Winiwarter W (2011) Developing spatially stratified N2O emission factors for Europe. Environmental Pollution 159: 3223–3232.
13. Lesschen JP, Velthof GL, de Vries W, Kros J (2011) Differentiation of nitrous oxide emission factors for agricultural soils. Environmental Pollution 159: 3215–3222.
14. IPCC (2006) Agriculture, Forestry and Other Land Use, Volume 4. In: 2006 IPCC Guidelines for National Greenhouse Gas Inventories. Japan: Institute for Global Environmental Strategies.
15. Lokupitiya E, Paustian K (2006) Agricultural soil greenhouse gas emissions. Journal of Environment Quality 35: 1413–1427.
16. Bernadier AB, Conant RT (2012) Regionally differentiated estimates of croplands N2O emission reduce uncertainty in global calculations. Global Change Biology 18: 928–935.
17. IPCC (1999) N2O: Direct Emissions from Agricultural Soils. In: Background Papers: IPCC Expert Meetings on Good Practice Guidance and Uncertainty Management in National Greenhouse Gas Inventories. 361–380.
18. Bouwman AF (1996) Direct emission of nitrous oxide from agricultural soils. Nutrient Cycling in Agroecosystems 46: 53–70.
19. Del Grosso SJ, Ojima DS, Parton WJ, Stehfest E, Heistemann M (2009) Global scale DAYCENT model analysis of greenhouse gas emissions and mitigation strategies for cropped soils. Global and Planetary Change 67: 44–50.
20. Roelandt C, Van Wesemael B, Rounsevell M (2005) Estimating annual N2O emissions from agricultural soils in temperate climates. Global Change Biology 11: 1701–1711.
21. Freibauer A, Kaltschmitt M (2003) Controls and models for estimating direct nitrous oxide emissions from temperate and sub-boreal agricultural mineral soils in Europe. Biogeochemistry 63: 93–115.
22. McSwiney CP, Robertson GP (2005) Nonlinear response of N2O flux to incremental fertilizer addition in a continuous maize (Zea mays L.) cropping system. Global Change Biology 11: 1712–1719.
23. Hoben JP, Gehl RJ, Millar N, Grace PR, Robertson GP (2011) Nonlinear nitrous oxide (N2O) response to nitrogen fertilizer in on-farm corn crops of the US Midwest. Global Change Biology 17: 1140–1152.
24. Millar N, Robertson GP, Grace PR, Gehl RJ, Hoben JP (2010) Nitrogen fertilizer management for nitrous oxide (N2O) mitigation in intensive corn (maize) production: an emissions reduction protocol for US Midwest agriculture. Mitigation and Adaptation Strategies for Global Change 15: 185–204.
25. Skiba U, Smith KA (2000) The control of nitrous oxide emissions from agricultural and natural soils. Chemosphere Global Change Science 2: 379–386.
26. Food and Agricultural Organisation (2011) FAO statistic database (FAOSTAT). Roma. Available: http://faostat.fao.org.
27. Spiertz JHJ (2010) Nitrogen, sustainable agriculture and food security. A review. Agronomy for Sustainable Development 30: 43–55.
28. Tilman D, Cassman KG, Matson PA, Naylor R, Polasky S (2002) Agricultural sustainability and intensive production practices. Nature 418: 671–677.
29. Pinheiro J, Bates D (2000) Mixed-effects Models in S and S-PLUS. 2nd ed. New York: Springer.
30. Lunn DJ, Thomas A, Best N, Spiegelhalter D (2000) WinBUGS - a Bayesian modelling framework: concepts, structure, and extensibility. Statistics and Computing 10: 325–337.
31. Brooks SP, Gelman A (1998) General methods for monitoring convergence of iterative simulations. Journal of Computational and Graphical Statistics 7: 434–455.
32. Akaike H (1974) A new look at the statistical model identification. IEEE Transactions on Automatic Control 19: 716–723.
33. Burnham KP, Anderson DR (2002) Model selection and multimodel inference: A practical Information-Theoretic Approach. New York: Springer. 2nd Ed.
34. Spiegelhalter DJ, Best NG, Carlin BP, van der Linde A (2002) Bayesian measures of model complexity and fit. Journal of the Royal Statistical Society: Series B 64: 583–639.
35. Efron B, Tibshirani R (1986) Bootstrap methods for standard errors, confidence intervals and other measures of statistical accuracy. Statistical Science 1: 54–75.
36. Efron B, Tibshirani R (1993) An Introduction to the Bootstrap (Chapman & Hall, London).
37. Philibert A, Loyce C, Makowski D (2012) Assessment of the quality of meta-analysis in agronomy. Agriculture, Ecosystems & Environment 148: 72–82.
38. Davidian M, Giltinan DM (1995) Non linear mixed effect models for repeated measurement data. Chapman & Hall. 359p.
39. Del Grosso SJ, Ogle SM, Parton WJ, Breidt FJ (2010) Estimating uncertainty in N2O emissions from US cropland soils. Global Biogeochemical Cycles 24: 12pp.
40. Crutzen PJ, Mosier AR, Smith KA, Winiwarter W (2008) N2O release from agro-biofuel production negates global warming reduction by replacing fossil fuels. Atmospheric Chemistry and Physics 8: 389–395.

Climate Change Disproportionately Increases Herbivore over Plant or Parasitoid Biomass

Claudio de Sassi*, Jason M. Tylianakis

School of Biological Sciences, University of Canterbury, Christchurch, New Zealand

Abstract

All living organisms are linked through trophic relationships with resources and consumers, the balance of which determines overall ecosystem stability and functioning. Ecological research has identified a multitude of mechanisms that contribute to this balance, but ecologists are now challenged with predicting responses to global environmental changes. Despite a wealth of studies highlighting likely outcomes for specific mechanisms and subsets of a system (e.g., plants, plant-herbivore or predator-prey interactions), studies comparing overall effects of changes at multiple trophic levels are rare. We used a combination of experiments in a grassland system to test how biomass at the plant, herbivore and natural enemy (parasitoid) levels responds to the interactive effects of two key global change drivers: warming and nitrogen deposition. We found that higher temperatures and elevated nitrogen generated a multitrophic community that was increasingly dominated by herbivores. Moreover, we found synergistic effects of the drivers on biomass, which differed across trophic levels. Both absolute and relative biomass of herbivores increased disproportionately to that of plants and, in particular, parasitoids, which did not show any significant response to the treatments. Reduced parasitism rates mirrored the profound biomass changes in the system. These findings carry important implications for the response of biota to environmental changes; reduced top-down regulation is likely to coincide with an increase in herbivory, which in turn is likely to cascade to other fundamental ecosystem processes. Our findings also provide multitrophic data to support the general concern of increasing herbivore pest outbreaks in a warmer world.

Editor: Justin Wright, Duke University, United States of America

Funding: CdS is supported by a University of Canterbury Doctoral Scholarship and a Hellaby Trust Fellowship. JMT is funded by a Rutherford Discovery Fellowship, administered by the Royal Society of New Zealand. This research was funded by the Marsden Fund (UOC-0705) and the Miss E.L. Hellaby Indigenous Grassland Research Trust. The funders had no role in study design, data collection and analysis, decision to publish, or preparation of the manuscript.

Competing Interests: The authors have declared that no competing interests exist.

* E-mail: cdesassi@gmail.com

Introduction

Global environmental changes affect all living organisms, with complex consequences for biodiversity, ecosystem structure and function [1,2]. Predicting generalities in the direction of such changes represents one of the major challenges in ecology. However, the complexity of this task is exacerbated by the great variability of responses observed, across biomes, space, time, and scales of biotic organization [3]. Climate has effects at all levels of organization, from population dynamics to community composition and species-specific responses [4,5], and it has strong impacts on ecosystems and their services [1,6,7]. A wealth of studies have shown that climate warming, provided it is not too extreme, generally increases plant net primary production [8]. However, warming has also been shown to have positive effects on herbivore population size and herbivory [9], which may counteract the increased plant growth. Furthermore, the net effect of climate on herbivores will result both from direct and plant-mediated effects and from top-down control by natural enemies, and this complexity may be partly responsible for the highly-variable responses of herbivores to different environmental change drivers [5].

The net ecosystem balance arising from the combination of these effects therefore depends on the relative response of individual trophic levels. A vast body of literature has addressed the effect of climate on plant-herbivore and prey-predator systems, but it disproportionately represents studies looking at pairs of interacting species, rather than larger modules or communities at once [3,5]. Despite the insights on specific mechanisms (e.g., phenological mismatches, shifts in competition, prey defense and palatability) gained from this approach, such studies do not allow generalizations to be made on the *relative* impact of climate or other change drivers at different trophic levels. In fact, only a handful of investigations have specifically considered overall responses at different trophic levels. For example, Voigt and colleagues focused on covariance in the response to multiple climatic factors of community composition, at different trophic levels [10] and functional groups [11], and concluded that sensitivity (i.e population fluctuations) to climate increases with trophic level. Focusing on a model system including a raptor bird, four passerine species and two caterpillar species, Both *et al.* [12] showed that the response of consumers is weaker than that of their resource. However, this result is contrasted by a recent study showing a climate-induced increase in synchrony between food demand and availability in a similar caterpillar-passerine system [13]. This latter result indicates that variability in species responses may not necessarily match on the overarching community-wide response. Nevertheless, these results imply that climate change is likely to prompt changes in the trophic structure of communi-

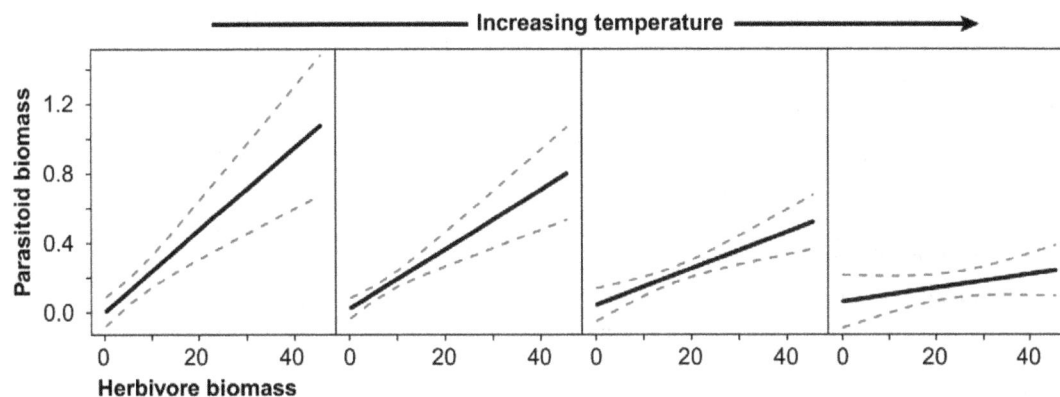

Figure 1. The correlation between herbivore biomass and parasitoid biomass along the temperature gradient. Panels are ordered from left to right (increasing temperature). Thus, the left plot represents the herbivore-parasitoid biomass relationship at the lowest temperature and the right plot is for the highest temperature. Black lines represent fitted values from out mixed effects model, dashed lines show the upper and lower range of the standard error. This plot highlights the decreasing slope of biomass correlation with increasing temperature.

ties, which could directly or indirectly affect ecosystem processes such as nutrient cycling, herbivory and predation [14,15].

Finally, in addition to indirect effects on species through changes at adjacent trophic levels, organismal responses to climate could be altered by co-occurring changes in the biotic and abiotic environment, such that recent literature has called for the integration of multiple drivers in global change research [5,16].

For example, biologically-available nitrogen deposition in non-agricultural systems has increased rapidly and become a major driver of biotic change [17]. As well as generally increasing net primary productivity (NPP), nitrogen has been shown to alter plant competitive interactions [5,18,19] and drive biodiversity losses [20,21]; effects that can percolate to higher trophic levels [22]. Despite the logical assumption that nitrogen will, in contrast to temperature, only affect herbivores via bottom-up effects [5], the interaction of the direct effect of temperature with changes in basal resource availability triggered by nitrogen, create a complex interplay that shows more context dependence than either effect in isolation [23,24]. Finally, the combined impact of warming and nitrogen on natural enemies is largely unknown, though N deposition tends to benefit predators [5], while climate warming can destabilize predator-prey interactions [25]. Thus, the interactive effects of temperature and N on plant growth [26] and herbivores [5], complicated by the general absence of data on their effect on natural enemies, suggest that these drivers may have complex, non-additive effects on trophic balance.

In this study, we examine how biomass at three trophic levels (plants, lepidopteran herbivores and their parasitoids) responds to co-occurring increases in temperature and nitrogen. We use seminatural grasslands as a model system, due to their global ubiquity [27] and importance for grazing agriculture. Furthermore, they are known to respond to N addition [20], and more strongly to warming than to other climate drivers such as CO_2 concentration and drought [28].

We use a field experiment along an altitudinal gradient, combined with an artificial warming experiment under controlled field conditions, and measure how total biomass of plants, herbivores and parasitoids, as well as parasitism rates, respond to elevated temperature and nitrogen treatments.

Materials and Methods

Study Site: Altitudinal Gradient Experiment

We established our experiment near Lewis Pass, North Canterbury, New Zealand (Appendix S1). The valley is located at the foothills of the Southern Alps, and ranges from 600 to 1,700 m elevation. The climate is cool and humid, with a mean annual rainfall of 1560 mm and a mean annual temperature of 9.1°C [29]. The wider experimental area is characterized by montane tussock grassland, dominated by native species in the genus *Festuca*, *Poa*, *Rytidosperma*, and *Chionochloa* at higher altitudes. These species are typical of semi-arid to humid, montane and subalpine zones in New Zealand [30]. The inter-tussock ground is generally dominated by stock-palatable Eurasian species (particularly *Agrostis capillaris*, *Anthoxanthum odoratum*, *Trifolium* spp.), which were over-sown after forest clearing in the late 1800s. At present, the area is farmed at very low intensity, with a stock density of less than 1 sheep per hectare, and no nitrogen fertilizer is applied.

Experimental Design and Sampling of Altitudinal Gradient

To generate a climatic gradient, we used an elevation gradient as a 'space for time substitution' [31,32]. We established five vertical transects of three plots, each at 150 m intervals of elevation, such that there was a total of 300 m difference in altitude between the lowest and the highest plot in each transect. Transects were at least 600 m apart (twice the vertical length of each individual transect, see Table 1 and Figure 1 in Appendix S1).

All plots had a similar incline and vegetation type, and faced north or north-west. Note, however, that analyses incorporated transect as a random (blocking) factor, so any environmental differences among transects would not confound treatment effects. To maintain similar characteristics, transects were not all positioned at exactly the same elevation, so plots ranged from 650 m at the lowest point to 1073 m a.s.l at the highest (423 m of total elevation span). This provided a total temperature gradient of 2.83°C across all plots (the average temperature in each plot over the entire period of data recording ranged from 3.89 to 6.72°C). This temperature gradient falls within the range of temperature increases predicted for the region within the next 100 years [33] and provided a better fit to the data than using changes in extreme temperatures (maxima or minima) as predictors (data not shown).

Table 1. Elevational gradient experiment: coefficient table for the combined effect of the drivers on A) plant biomass and the effect of the drivers and resource (plant or herbivore host respectively) biomass on total biomass of B) herbivores and C) parasitoids.

A) Plants

	Values	Std. Error	df	t-value	P-value	
(Intercept)	9830.12	10020.47	13	0.98	0.345	
Temperature	−232.27	1873.78	9	−0.12	0.904	
Nitrogen	10577.54	10907.60	13	0.97	0.350	
Temperature : nitrogen	−1408.00	2042.77	13	−0.69	0.503	
B) Herbivores						
(Intercept)	−43.10	10.53	12	−4.09	0.002	**
Plant biomass	0.0005	0.0002	12	2.37	0.036	*
Temperature	9.32	1.94	9	4.80	0.001	**
Nitrogen	−10.87	13.60	12	−0.80	0.440	
Temperature : nitrogen	2.43	2.53	12	0.96	0.355	
C) Parasitoids						
(Intercept)	0.07	0.18	11	0.38	0.713	
Herbivore biomass	0.051	0.01	11	4.41	0.001	**
Temperature	−0.003	0.03	9	−0.07	0.943	
Nitrogen	−0.29	0.20	11	−1.45	0.174	
Herbivore biomass : warming	−0.007	0.002	11	−3.72	0.003	**
Temperature : nitrogen	0.04	0.04	11	1.26	0.230	

Asterisks indicate level of significance (. $\leq$0.1, * $\leq$0.05, ** $\leq$0.01).

Similarly, we found a strong correlation ($R^2 > 0.91$) between yearly mean temperature and the mean temperature of the growing season (September to January). Mean growing season temperature did not increase the fit of the data and we therefore used the annual mean for consistency with our sampling regime.

Local topography may create significant microclimatic variation, which could modify temperature over short vertical distances that override the more general altitudinal trends [34]. This allowed us to test the effects of temperature *per se*, partially uncoupled from the effects of other environmental variables that co-vary with elevation (such as oxygen availability and radiation [32]. Temperature was recorded in each plot using Hobo series ProV2 data loggers, protected by a sun shield, logging temperature at 1 h intervals from February to December 2009. We used the overall mean site temperature for this period as a predictor variable in the analysis.

At each elevation, we established a 24×12 m sampling plot. We further subdivided each plot into two 12×12 m subplots, and randomly assigned one of these to a nitrogen addition treatment (addition or control with no added N). This resulted in a split-plot design, with temperature varying at the scale of plots (n = 15), blocked by transects (n = 5), and N treatments applied to subplots (n = 30) nested within plots. The N fertilization treatment comprised a total application of 50 Kg ha^{-1} yr^{-1}, which falls within the current range of globally-observed rates of atmospheric deposition [35]. Precise N deposition rates for the study region are not known, but expansion of dairy farming across New Zealand is driving rapid increases in N fertilizer application [36], which will likely impact adjacent semi-natural grasslands. Nitrogen fertilizer

was applied in the form of Calcium Ammonium Nitrate (CAN) granules (Ravensdown LTD, New Zealand). This form of fertilizer combines fast and slower release of biologically-available nitrogen, and has been used previously to simulate atmospheric deposition [21].

We began N addition in September 2008, by adding 40% of the total year budget (20 Kg ha^{-1} yr^{-1}, 1066 g CAN per subplot) and applying the remaining 60% in 4 pulses, evenly distributed over the next 12 months, by sprinkling the dry granules throughout the treated subplot. Fertiliser addition continued at a rate of 50 Kg ha^{-1}yr^{-1} until sampling was completed in December 2009.

Although initial sampling of insects began in October 2008, here we present data only from samples where biomass was measured, which were those collected from May to December 2009, i.e., approximately a year after starting the nitrogen fertilization treatment. To minimize disturbance and depletion of caterpillars in the experimental area, we subdivided each 12 x 12 m subplot into 4 strips of 3×12 m each, and sequentially sampled one strip only during each sampling round. This allowed monthly sampling, but ensured a time window of at least 4 months before re-sampling of the same section. This timeframe is substantially longer than the average larval life stage of Lepidoptera in our study area, and therefore prevented bias in the abundance of any sample caused by depletion from previous sampling rounds.

We searched all the tussocks within the 3×12 m strip at each sampling round. Plant searches involved thorough teasing apart of denser vegetation to locate any hidden larvae.

Study Site: Artificial Warming Experiment

We set up an artificial warming experiment adjacent to the University of Canterbury field station at Cass in the Waimakariri River catchment, South Island of New Zealand (Figure 2 in Appendix S2). The Cass field station lays at 640 m a.s.l., a mean annual rainfall of 1300 mm (1918-1965) is uniformly distributed throughout the year, and typical monthly mean air temperatures range from 1.6°C (July) to 15.7°C (February). Snow lies for some days each winter (June-September). The climate of the area is described in detail by Greenland [37].

The study area is embedded in a montane short-tussock grassland with very similar characteristics to the environment of the altitudinal gradient experiment, although the two experimental locations are over 60 Km apart and belong to different catchments. The intertussock area is dominated by Eurasian grasses oversown for pastoral purposes. The area surrounding the Field Station shows a strong dominance of *Agrostis capillaris* and *Anthoxanthum odoratum* [38].

Experimental Design and Sampling of Warming Experiment

The experiment comprised a 2×2 factorial design, with warming and nitrogen as treatments with two levels each (control vs. elevated) and five true replicates per treatment combination, totaling 20 plots. We dug a 24 m by 19 m experimental area in October 2008, to a depth of 20 cm to establish the 20 3.5×3.5 m plots (12.25 m^2), each separated by a 1 m corridor. We then leveled the ground and installed custom-made electric heating cables (Argus Heating Ltd, Christchurch, New Zealand: coiled copper wire on fiberglass core and silicon coating) in half of the plots, and dummy cables in the remaining (unheated) plots. Each plot was fitted with two coils of 45 m meters each, resulting in a spacing of 14 cm between cable lines. Heating power totals 940 Watts per plot or a power density of 76W/m^2. Similar power

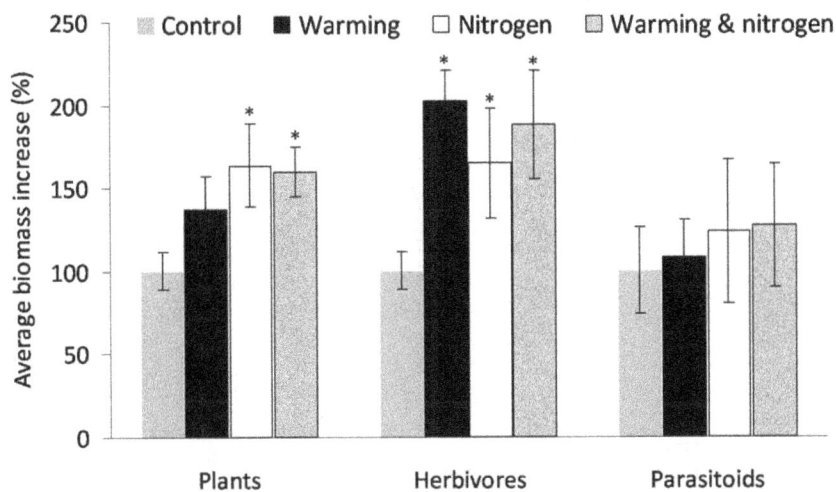

Figure 2. The effect of the global change drivers (warming, nitrogen and their combination) on the percentage increase in biomass relative to the control treatment for plants, herbivores and parasitoids. As the percentage increase is scaled within trophic levels, this graph allows a direct comparison of the effect of the drivers within and across trophic levels (plant, herbivores and parasitoids). Asterisks depict significant differences relative to the group control.

output has been recommended [39] and successfully used in previous underground heating experiments [40].

In each warmed plot, we installed three thermocouples (Type E, Chromel-Constantan, Campbell Scientific, USA) at 10 cm depth and standardized position relative to the heating cables (1 directly above the cable, 1 between two heating cables, 1 between the other two thermocouples), to capture any potential temperature differences within the plot driven by distance from the heating cables. In each control (unheated) plot, we installed 1 thermocouple at the same depth. The thermocouple in the control plot provided a baseline measure of ambient temperature so that the warmed plots could be kept at a constant temperature above ambient.

We homogenized all the extracted soil by mixing it with a digger to remove any confounding nutrient or bio-geochemical gradient before re-installing it in the experimental area and leveling the ground to ensure constant ground depth relative to the cables We planted well established (at least 3 month old) individuals of four species of tussock grasses, which were common to the general area and also present in the altitudinal gradient experiment (50× *Poa cita*, 50× *Festuca novae-zelandiae*, 12× *Chionochloa rigida* and 12× *Chionochloa flavecens* per plot), in a consistent composition and layout for each plot). This resulted in each plot being planted with 144 individual plants, amounting to 2880 tussocks in total. We completed the set up and planting in January 2009 (see Figure 3 in Appendix S2). To minimize water stress to the recently-planted tussocks in the height of the first summer, we installed an automated watering system, which ran for half an hour at dawn and after sunset until May 2009. We first activated the warming treatment in April 2009, after the plants had established for over three months. However, this required adjustment and tuning of temperature differences, and the experiment was fully operational by late June 2009. We paired each warming plot with its spatially-closest control plot to keep the warmed treatments at 3°C above ambient, logging the temperature of all thermocouples every minute using two Campbell CR1000 (Campbell Scientific, USA) data loggers. The average temperature of the thermocouples in the warming plots is used against the control plot to switch the power on and off as required (See Figures 4 and 5 in Appendix S2 for details on the temperature control). The three degrees of warming

achieved in this experiment is in line with the temperature gradient we found in the field experiment and with the predictions of global (and New Zealand) warming scenarios for the next 100 years [33].

The nitrogen treatment application, using the same fertilizer as the gradient experiment, began shortly after planting, by adding the equivalent of 25 Kg ha^{-1} yr^{-1} in late January 2009. Applications reached a total of 50 Kg ha^{-1} yr^{-1} with three evenly-distributed applications during the rest of the year. Fertilization treatments continued in 2010 with five applications of 10 Kg ha^{-1} yr^{-1}, one every two months except the winter months of July and August, where the plots were often covered in snow. The decision to use five applications arose from a tradeoff between maximizing frequency of applications (to resemble natural deposition), yet applying enough to practically allow even application across the entire treated plot.

We began sampling insects in January 2010, that is, a full year after plot establishment and planting. Sampling continued at monthly intervals until June 2010 (i.e. mid winter, when snow cover made sampling impractical), and resumed at monthly intervals from September to December 2010, totaling 11 sampling rounds. To minimize disturbance and depletion of caterpillars in the experimental area, we sampled half of each plot during each round, alternating between the two halves. This ensured a time window of at least 8 weeks before re-sampling of the same section. Sampling entailed visually searching for caterpillars on tussock plants, teasing apart the dense vegetation to find any hidden larvae.

Both the artificial and the gradient experiment present a number of caveats in their design: using natural-gradient studies has limitations in the ability to explain the response of communities to temperature changes, as populations may already have adapted to the different conditions [32]. Additionally, changes to mean temperatures following global warming may be strongly influenced by changes to frequency and magnitude of extreme temperature events [33], which remain unaccounted for in our study. Similarly, artificial warming experiments such as the one presented in this study can be criticized for the necessarily small scale, and limitations of any heating method used in simulating global change [41]. However, most experiments to date have used one of these methods. In this study, we used both a large-scale field experiment combined with a manipulative controlled field experiment, finding largely consistent results that provide a good degree of confidence that the patterns found were due to the generalities of communities' response to simulated global-change drivers, rather than spurious effects of any particular experimental approach.

Insect Rearing

For both experiments, we identified each individual larva to morphospecies. To allow collection of parasitoids, we individually reared all larvae to maturity (emergence of the adult moth or parasitoid) in a climate-controlled room, with a constant temperature of 16 degrees, relative humidity of 60% and a light cycle of 16L:8D. The feeding protocols varied according to the species requirements. All parasitoids were identified to species level where possible, and to morphospecies for organisms lacking a recognized classification. We sought the expertise of two taxonomists to help with the identification: John S. Dugdale developed a larval key for Lepidoptera and confirmed the identity of all the tachinid flies, and Jo Berry validated hymenopteran morphospecies and formally identified all known species. The individual rearing of every herbivore larva allowed us to estimate the rate of parasitism (proportion of larvae from which a parasitoid emerged).

Larvae that died during rearing (22% in the altitudinal gradient and 42% in the warming experiment) were excluded from all analyses. A total of 4224 caterpillars (39 species) and 860 parasitoids (41 species) were included for the altitudinal gradient experiment, whilst the artificial warming experiment comprised 893 caterpillars (26 species) and 331 parasitoids (21 species).

Biomass Measurements

To estimate effects of temperature and N on larval biomass, we weighed the caterpillars (Mettler Toledo analytical balance accurate to 0.0001 g) directly after collection for all samples. We estimated total herbivore biomass as the sum of the larval weight of all individuals in each plot. As we could not always observe parasitoids as soon as they emerged, there is a risk that individual parasitoid weight could be biased by the time between emergence and being discovered. Furthermore, unlike herbivore mass, which was measured directly after collection, parasitoid body mass can only be measured at emergence, and could therefore be strongly determined by the age at which the host larva was brought into the laboratory for rearing, and the food provided to the growing larva. Therefore, to avoid the possibility that these effects could generate spurious differences across treatments, we calculated the total parasitoid biomass for a plot by multiplying the total counts of each species by the average weight of that species. We obtained each species average by weighing 20 adult individuals of each species, or all individuals for the rarer species (less than 20 individuals).

To estimate plant biomass without disruptive sampling of the plots, we estimated the total tussock volume in each plot. To obtain the total tussock volume, we first calculated the mean tussock volume per plot by measuring a subset of randomly-selected tussocks (20 in the warming experiment, 30 in the gradient experiment. We measured basal circumference and height from the ground to the highest leaf, and then calculated the cylinder volume [42]. After obtaining the average tussock volume for each plot, we multiplied it by the total count of tussock individuals. To convert plant volume to biomass, we measured the volume of 10 tussock plants from our glasshouse cultures following the same procedure as above. We then clipped them to ground level and dried the leaf material at 60°C for 48 hours. We used

a linear regression to test how well volume approximated dry weight, and found a significant relationship ($F_{1,8} = 20.68$, $P = 0.001$, $R^2 = 0.72$).

Data Analysis

We carried out all analyses using R version 2.12.0 (2010 The R Foundation for Statistical Computing). To account for our split-plot design in the gradient experiment, we used general linear mixed effects models [43], within the nlme package [44]. We used total (summed) plant, herbivore or parasitoid biomass as the response variable, with a Gaussian error distribution. We included nitrogen treatment as a fixed factor and temperature as a (fixed) variate, with plots nested in transects as random effects. Biomass of consumer trophic levels is likely to be highly correlated with the biomass of the trophic level below (its resource). Therefore, we included the biomass of plants as a variate in the model predicting herbivore biomass, and herbivore biomass in the model for parasitoids.

We initially included all possible interactions, then simplified the model by removing non-significant interaction sequentially, each time assessing changes in Akaike Information Criterion (AIC) scores before any further simplification. This allowed us to determine if the effects of the drivers on consumer biomass persisted after accounting for variation explained by resource biomass, i.e. if there was any direct effect of the drivers beyond the bottom-up, resource-biomass-mediated effects. To highlight the differential response between trophic groups, we calculated an herbivore to plant biomass ratio and a parasitoid to herbivore biomass ratio. We used a logit transformation for these ratios to meet the assumptions of normality, then tested them each as a response variable in a mixed effects model with the two global change drivers as a predictor and a Gaussian error distribution.

In addition to biomass changes, the activity of natural enemies may respond to the treatments (e.g., higher activity due to higher metabolic rates with increasing temperature, or altered attack rates as host quality changes under elevated N). Because such a response may not have been captured by looking solely at changes in biomass, we tested the response of overall parasitism rates to the drivers. We modeled parasitism rates using a generalized linear mixed model with a binomial error distribution, carried out in the lme4 package [45]. The proportion of all herbivores that were parasitised was the response variable, and the drivers temperature and nitrogen as predictors.

To test for changes in biomass at each trophic level in the artificial warming experiment, given the full factorial design, we used general linear models (the lm function in the base package of R). We used total (summed) plant, herbivore or parasitoid biomass as the response variable, with temperature and nitrogen as fixed factors. We followed the same procedure as in the altitudinal gradient experiment by including resource biomass (biomass of plants as a variate in the model predicting herbivore biomass, and herbivore biomass in the model for parasitoids) alongside the drivers, including all interactions and subsequently simplifying the model as above. To highlight changes in total biomass within each trophic level, we also tested the relative percentage increase in biomass (arcsine square root transformed to meet the assumptions of normality and homoscedasticity) compared with the control treatment for each trophic level. We did not carry out this analysis in the gradient experiment because the use of temperature as a variate rather than the categorical (warming vs control) used here did not allow an equally effective comparison. We tested biomass ratio using the same procedure as in the altitudinal gradient experiment. To test the response of parasitism rates to the driver treatments, we used binomial errors and a logit link function in the glm function of the base package in R.

Results

Altitudinal Gradient Experiment

In the gradient experiment, we found no effect of the drivers on plant biomass (Table 1A). Herbivore biomass was positively correlated with plant biomass, but a strong effect of temperature on herbivores remained even after controlling for resource biomass (Table 1B). In contrast, a trend towards a positive effect of nitrogen on herbivore biomass ($t=1.82$, d.f $=9$, $P=0.090$) disappeared when plant biomass was included in the model. Parasitoid biomass was positively correlated with host resource biomass (Table 1C), but did not respond directly to the treatments. Interestingly, we found a negative interaction between herbivore biomass and temperature, such that the positive relationship between herbivore biomass and parasitoid biomass was significantly weaker at higher temperatures (Table 1C, Figure 1).

We also found that increasing temperature led to an overall increase in the biomass ratio of herbivores to plants ($t=3.66$, d.f. $=9$, $P=0.005$) and a tendency for a decrease in the ratio of parasitoid biomass to herbivore biomass ($t=-1.91$, d.f. $=9$, $P=0.084$, see Table 1 in Appendix S3). Concordantly, we found a negative effect of both drivers on parasitism rates, with a significant interaction such that the drivers acted sub-additively (Temperature: $Z=-2.15$, $P=0.031$, Nitrogen: $Z=-2.11$, $P=0.034$, interaction: $Z=1.98$, $P=0.047$).

Warming Experiment

In the warming experiment, relative biomass responses to each driver differed across the different trophic levels (Table 2, Figure 2). There was no significant relative change in plant biomass at high temperature, but there was a significant increase in the nitrogen treatment (both in the absolute biomass, Table 2A, and in the mean ($\pm$ SE) percent change relative to control $= +63.8\% \pm24.9$, $P=0.016$), which remained when temperature and nitrogen were combined ($+59.9\% \pm$ SE 15.2, non-significant warming x nitrogen interaction: Table 2A). In contrast, herbivore biomass on average doubled in response to temperature (relative change of $+102\%$ ±18.6, $P=0.006$, for absolute change in total biomass see Table 2B) and was marginally higher in the nitrogen treatment ($+64.7\pm$ SE 32.9, $P=0.062$), with combined treatments showing a weakly sub-additive effect ($+88.1\%$ $\pm$ SE 33.1, $P=0.095$). Herbivore total biomass was positively correlated with plant biomass but, nevertheless, retained a positive effect of temperature (Table 2B), consistent with the altitudinal gradient experiment. In contrast, the marginally-significant ($P=0.062$) main effect of nitrogen on herbivores disappeared after plants were included in the model, providing some evidence that nitrogen effects on herbivores were indeed bottom-up.

Finally, parasitoid relative biomass did not differ from control under any treatment combination ($P>0.1$ in all cases). After including herbivore biomass alongside the treatments predicting total parasitoid biomass, we found a positive correlation between resource and consumer biomass and only a trend ($P<0.1$) towards a negative effect of temperature on parasitoid biomass after controlling for the effect of herbivore biomass (Table 2C).

Overall, the observed changes in biomass at higher temperatures led to an increase in the biomass ratio between herbivores and plants ($t=2.50$, $P=0.023$) and a tendency towards a decrease in the ratio of parasitoid to herbivore biomass ($t=-1.78$, $P=0.093$, see Table 1 in Appendix S3). In contrast, we found

Table 2. Artificial warming experiment: coefficient table for the combined effect of the drivers on A) plant biomass, and the effect of the drivers and resource (plant or herbivore host respectively) biomass on total biomass of B) herbivores and C) parasitoids.

A) Plants	Value	Std.Error	df	t-value	P-value	
(Intercept)	2036.54	377.49	12	5.3949	<0.001	**
Warming	751.10	480.56	12	1.5630	0.144	
Nitrogen	1299.46	480.56	12	2.7040	0.019	*
Warming : nitrogen	−828.69	679.62	12	−1.2193	0.246	
B) Herbivores						
(Intercept)	−0.18	1.37	15	0.13	0.898	
Plant biomass	0.002	0.0005	15	3.68	0.002	**
Warming	2.84	1.22	15	2.33	0.034	*
Nitrogen	0.15	1.34	15	0.16	0.910	
Warming : nitrogen	−1.70	1.69	15	−1.01	0.330	
C) Parasitoids						
(Intercept)	0.03	0.04	15	0.62	0.548	
Herbivore biomass	0.02	0.007	15	2.84	0.012	*
Warming	−0.08	0.05	15	−1.81	0.098	.
Nitrogen	−0.03	0.04	15	−0.60	0.556	
Warming : nitrogen	0.06	0.06	15	0.95	0.358	

Asterisks indicate level of significance (. ≤0.1, * ≤0.05, ** ≤0.01, *** ≤0.001).

no effect of the drivers on parasitism rates in the warming experiment.

Discussion

We found distinct responses of biomass at different trophic levels under elevated temperature and nitrogen and, overall, these results were consistent between two experiments that strongly differed in spatial scale and design. In particular, herbivore biomass increased significantly more than plant or parasitoid biomass at higher temperature, and this generated an increased ratio of herbivore to plant biomass with warming. Our findings of greatly increased herbivore biomass at higher temperature support hypotheses of increased herbivory based on data from agricultural systems (reviewed by Rustad et al. [8], Bale et al. [9] and Throop and Lerdau [46]) and paleological records [47,48].

In contrast to warming, the strength of the nitrogen direct effect on relative biomass change decreased from plants (strong positive response) to herbivores (marginally-significant positive response) to parasitoids (no response), suggesting that bottom-up effects, or increases in resource availability, had decreasing strength or efficiency moving up the food chain. The importance of bottom-up effects was emphasized by the significant effect of plant on herbivore, and herbivore on parasitoid biomass in both experiments, and after controlling for these effects, nitrogen had no significant effect on herbivore or parasitoid biomass. In light of these findings, the role of natural enemies in controlling herbivore populations is likely to be strongly impaired by both drivers (an hypothesis supported by the significant reduction in parasitism rates under elevated temperature or nitrogen).

In addition to biomass, rates of herbivory are also predicted to increase at higher levels of nitrogen availability, which could in turn support larger herbivore populations [49]. Although there may have been a top-down reduction in plant biomass due to elevated herbivory, this was not sufficient to outweigh the effect of nitrogen on plant growth, a finding congruent with that of Throop [46], who showed that positive impacts of N on shoot biomass were typically not significantly suppressed by herbivory. In our experiments, we found that both plants and herbivores substantially gained biomass under elevated nitrogen, indicating a generally more productive system at the plant and herbivore level, but not at the parasitoid level.

Interestingly, under elevated levels of both temperature and nitrogen, we observed a higher increase in biomass of herbivores than parasitoids, whilst the increase of herbivores and plants was qualitatively similar. In other words, the presence of nitrogen as a second driver mitigated the strong difference in response between plants and herbivores at higher temperatures. This result highlights the importance of considering the co-occurrence of global change drivers; under a scenario of global warming with no increase in nitrogen deposition, herbivores show a clearly stronger response than plants and parasitoids. However, under a realistic scenario of co-occurring drivers [3,5,16], the difference in response, particularly between plants and herbivores, may be less than expected when considering each driver in isolation.

Contrastingly, we found strong evidence in both field experiments that natural enemies were not able to respond as positively to increased herbivore (host) resource availability under a changing environment. Parasitoid biomass did not significantly increase under any treatment and, importantly, showed a significantly lower response than herbivores at higher temperature. Moreover, both experiments qualitatively showed a net negative effect of temperature on parasitoids. Once we had accounted for the predictable correlation between herbivore biomass and parasitoid biomass, we found a trend for a negative effect of temperature on parasitoid biomass in the artificial warming experiment. Similarly, in the elevation gradient experiment, we found that the biomass correlation between parasitoids and herbivores was weaker at higher temperatures, and led to a decreasing parasitoid-herbivore biomass ratio. These results suggest that parasitoids were not able to counteract the strong response of herbivores (perhaps because parasitoid population responses were too slow), and this effectively generated a situation of predator-release under elevated temperature. This view is also supported by the significantly lower rates of parasitism found in the gradient experiment. Even though parasitoids attacked significantly more hosts under elevated temperature (results not shown), this increase was not proportionate to the increase in host abundance, which generated a lower proportion of hosts parasitised. It must also be noted that parasitism rates did not differ significantly across treatments in the artificial warming experiment. However, due to the small distance between plots, it is plausible that parasitoids could display behavioral choices to attack hosts across different treatments depending on their availability at a given time. Therefore, the parasitism results in the artificial warming experiment should be taken cautiously and should not undermine the validity of the results we obtained under natural field conditions in the elevational gradient study.

Our results contrast with those of Andrew and Hughes [50], who found no evidence for increased ratios of herbivores to parasitoids and other natural enemies along a latitudinal gradient. However, the scope and methodology of their study shows substantial differences from ours. Andrew and Hughes sampled all arthropods by knocking them down from the host plant using a pyrethrum/water solution. They thus obtained data on abundance and biomass of the major insect taxa sorted into feeding groups, but would have also included 'tourist' species, which may not have been feeding on the plants or herbivores. In contrast, we reared parasitoids from living hosts, which incorporates host-selection effects on parasitoids, and provides a measure of biomass that directly relates to the ecosystem function of parasitism. Thus, our estimate of biomass is obtained from parasitoids that are not merely present, but also able to interact successfully with their host. Nevertheless, the results of Andrew and Hughes imply that sampling of free-living adult parasitoids could lead to different results.

Strengthened top-down control by generalist predators observed under warming in terrestrial systems [51] suggests that more specialized natural enemies such as parasitoids may be less responsive than generalists. Temperature is known to increase metabolic rates of mobile predators such as spiders [25] whilst, in contrast, parasitoid development is dependent on their host, which may constrain (e.g., through changes in host phenology or quality) their ability to adapt to change.

Previous studies on tri-trophic food chains concluded that parasitoids are unlikely to effectively counteract the response of herbivores to climate change [52], and specifically suggested that bottom-up forces may be more important than top-down control by the parasitoids [53]. Our findings are congruent with this suggestion, and show a severe limitation in the ability of parasitoids to effectively control herbivore populations. Our findings have concerning implications for biological control of herbivore pests, and suggest that herbivores will be the most likely to benefit and thrive in a changing environment.

Supporting Information

Appendix S1 Altitudinal gradient location details. GPS coordinates, altitude and mean temperature of each sampling plot, and photographic map of their location.

Appendix S2 Artificial warming experiment: study location and experimental temperature control. Location of the two experiments in the landscape of the South Island of New Zealand, photographic timeline of the artificial warming experiment set up and graphic details of the temperature control system.

Appendix S3 Biomass ratio analyses. Coefficient tables for the analyses of parasitoid-herbivore and herbivore-plant biomass ratios

Acknowledgments

We thank Owen T. Lewis for providing useful comments on previous versions of this manuscript. Mike and Cliff Cox kindly provided access on private land and accommodation in the field. Rebecca Jackson, Michael Bartlett and Kirsty Trotter provided help in the field and in the lab. Jenny Ladley, Andrew Winther and John Hunt helped with the set up of the artificial warming experiment.

Author Contributions

Conceived and designed the experiments: CdS JMT. Performed the experiments: CdS. Analyzed the data: CdS. Contributed reagents/ materials/analysis tools: CdS JMT. Wrote the paper: CdS.

References

1. Chapin FS, Zavaleta ES, Eviner VT, Naylor RL, Vitousek PM, et al. (2000) Consequences of changing biodiversity. Nature 405: 234–242.
2. Thomas CD, Cameron A, Green RE, Bakkenes M, Beaumont LJ, et al. (2004) Extinction risk from climate change. Nature 427: 145–148.
3. Gilman SE, Urban MC, Tewksbury J, Gilchrist GW, Holt RD (2010) A framework for community interactions under climate change. Trends in Ecology & Evolution 25: 325–331.
4. Parmesan C (2006) Ecological and evolutionary responses to recent climate change. Annual Review of Ecology Evolution and Systematics 37: 637–669.
5. Tylianakis JM, Didham RK, Bascompte J, Wardle DA (2008) Global change and species interactions in terrestrial ecosystems. Ecology Letters 11: 1351–1363.
6. Parmesan C, Yohe G (2003) A globally coherent fingerprint of climate change impacts across natural systems. Nature 421: 37–42.
7. Tscharntke T, Tylianakis J (2010) Conserving complexity: Global change and community-scale interactions. Biological Conservation 143: 2249–2250.
8. Rustad LE, Campbell JL, Marion GM, Norby RJ, Mitchell MJ, et al. (2001) A meta analysis of the response of soil respiration, net nitrogen mineralization, and aboveground plant growth to experimental ecosystem warming. Oecologia 126: 543–562.
9. Bale JS, Masters GJ, Hodkinson ID, Awmack C, Bezemer TM, et al. (2002) Herbivory in global climate change research: direct effects of rising temperature on insect herbivores. Global Change Biology 8: 1–16.
10. Voigt W, Perner J, Davis AJ, Eggers T, Schumacher J, et al. (2003) Trophic levels are differentially sensitive to climate. Ecology 84: 2444–2453.
11. Voigt W, Perner J, Jones TH (2007) Using functional groups to investigate community response to environmental changes: two grassland case studies. Global Change Biology 13: 1710–1721.
12. Both C, van Asch M, Bijlsma RG, van den Burg AB, Visser ME (2009) Climate change and unequal phenological changes across four trophic levels: constraints or adaptations? Journal of Animal Ecology 78: 73–83.
13. Vatka E, Orell M, Rytkonen S (2011) Warming climate advances breeding and improves synchrony of food demand and food availability in a boreal passerine. Global Change Biology 17: 3002–3009.
14. Kishi D, Murakami M, Nakano S, Maekawa K (2005) Water temperature determines strength of top-down control in a stream food web. Freshwater Biology 50: 1315–1322.
15. Petchey OL, McPhearson PT, Casey TM, Morin PJ (1999) Environmental warming alters food-web structure and ecosystem function. Nature 402: 69–72.
16. Didham RK, Tylianakis JM, Gemmell NJ, Rand TA, Ewers RM (2007) Interactive effects of habitat modification and species invasion on native species decline. Trends in Ecology and Evolution 22: 489–496.
17. Vitousek PM, Aber JD, Howarth RW, Likens GE, Matson PA, et al. (1997) Human alteration of the global nitrogen cycle: Sources and consequences. Ecological Applications 7: 737–750.
18. Brooker RW (2006) Plant-plant interactions and environmental change. New Phytologist 171: 271–284.
19. Zavaleta ES, Shaw MR, Chiariello NR, Thomas BD, Cleland EE, et al. (2003) Grassland responses to three years of elevated temperature, CO2, precipitation, and N deposition. Ecological Monographs 73: 585–604.
20. Stevens CJ, Dise NB, Mountford JO, Gowing DJ (2004) Impact of nitrogen deposition on the species richness of grasslands. Science 303: 1876–1879.
21. Clark C, Tilman D (2008) Loss of plant species after chronic low-level nitrogen deposition to prairie grasslands. Nature 451: 712–715.
22. Richardson SJ, Press MC, Parsons AN, Hartley SE (2002) How do nutrients and warming impact on plant communities and their insect herbivores? A 9-year study from a sub-Arctic heath. Journal of Ecology 90: 544–556.
23. Thompson PL, St-Jacques MC, Vinebrooke RD (2008) Impacts of climate warming and nitrogen deposition on alpine plankton in lake and pond habitat: an in vitro Experiment. Arctic, Antarctic, and Alpine Research 40: 192–198.
24. Wallisdevries MF, Van Swaay CAM (2006) Global warming and excess nitrogen may induce butterfly decline by microclimatic cooling. Global Change Biology 12: 1620–1626.
25. Rall BC, Vucic-Pestic O, Ehnes RB, Emmerson M, Brose U (2011) Temperature, predator-prey interaction strength and population stability. Global Change Biology 16: 2145–2157.
26. Reich PB, Hobbie SE, Lee T, Ellsworth DS, West JB, et al. (2006) Nitrogen limitation constrains sustainability of ecosystem response to CO2. Nature 440: 922–925.
27. Hooper DU, Chapin FS, Ewel JJ, Hector A, Inchausti P, et al. (2005) Effects of biodiversity on ecosystem functioning: A consensus of current knowledge. Ecological Monographs 75: 3–35.
28. Bloor JMG, Pichon P, Falcimagne R, Leadley P, Soussana J-F (2010) Effects of Warming, Summer Drought, and CO2 Enrichment on Aboveground Biomass Production, Flowering Phenology, and Community Structure in an Upland Grassland Ecosystem. Ecosystems 13: 888–900.
29. Williams PA, Courtney SP (1995) Site Characteristics and Population Structures of the Endangered Shrub Olearia-Polifa (Wilson Et Garnock-Jones), Nelson, New-Zealand. New Zealand Journal of Botany 33: 237–241.
30. Rose AB, Suisted PA, Frampton CM (2004) Recovery, invasion, and decline over 37 years in a Marlborough short-tussock grassland, New Zealand. New Zealand Journal of Botany 42: 77–87.
31. Pickett STA (1989) Space-for-time substitution as an alternative to long-term studies.; Likens GE, editor. Heidelberg: Springer Verlag. 110–135 p.
32. Hodkinson ID (2005) Terrestrial insects along elevation gradients: species and community responses to altitude. Biological Reviews 80: 489–513.
33. IPCC (2007) (Intergovernmental Panel on Climate Change) The physical science basis. Geneva. 21 p.
34. Weiss SB, Murphy DD, White RR (1988) Sun, Slope, and Butterflies - Topographic Determinants of Habitat Quality for Euphydryas-Editha. Ecology 69: 1486–1496.
35. M.E.A (2005) (Millenium Ecosystem Assesment) Ecosystems and human wellbeing: scenarios.: Island Press, Washington D.C., USA.
36. Austin D, Cao K, Rys G (2007) Modeling Nitrogen Fertilizer Demand in New Zealand. Wellington: Ministry of Agriculture and Forestry.
37. Greenland DE (1977) Weather and climate at Cass; Burrows CJ, editor. Christchurch, New Zealand: Department of Botany, University of Canterbury. 418 pp p.
38. Barratt B, Ferguson C, Logan R, Barton D, Bell N, et al. (2005) Biodiversity of indigenous tussock grassland sites in Otago Canterbury and the central North Island of New Zealand I. The macro-invertebrate fauna. Journal of the Royal Society of New Zealand 35: 287–301.
39. Peterjohn WT, Melillo JM, Bowles FP, Steudler PA (1993) Soil Warming and Trace Gas Fluxes - Experimental-Design and Preliminary Flux Results. Oecologia 93: 18–24.
40. Melillo JM, Steudler PA, Aber JD, Newkirk K, Lux H, et al. (2002) Soil warming and carbon-cycle feedbacks to the climate system. Science 298: 2173–2176.
41. Kimball BA CM, Wang S, Xingwu L, Luo C, Morgan J ans Smith D (2008) Infrared heater arrays for warming ecosystem field plots. Global Change Biology 14: 309–320.
42. Laliberte E, Norton DA, Tylianakis JM, Scott D (2010) Comparison of Two Sampling Methods for Quantifying Changes in Vegetation Composition Under Rangeland Development. Rangeland Ecology & Management 63: 537–545.
43. Bolker BM, Brooks ME, Clark CJ, Geange SW, Poulsen JR, et al. (2009) Generalized linear mixed models: a practical guide for ecology and evolution. Trends in Ecology & Evolution 24: 127–135.
44. Pinheiro J, Bates D, DebRoy S, Sarkar D, Team RDC (2011) nlme: Linear and Nonlinear Mixed Effects Models.
45. Bates D, Maechler M (2010) lme4: Linear mixed-effects models. R package version 0.999375–37/r1127. ed: http://R-Forge.R-project.org/projects/lme4/.
46. Throop HL (2005) Nitrogen deposition and herbivory affect biomass production and allocation in an annual plant. Oikos 111: 91–100.
47. Wilf P, Labandeira CC (1999) Response of plant-insect associations to Paleocene-Eocene warming. Science 284: 2153–2156.
48. Currano ED, Wilf P, Wing SL, Labandeira CC, Lovelock EC, et al. (2008) Sharply increased insect herbivory during the Paleocene-Eocene thermal maximum. Proceedings of the National Academy of Sciences of the United States of America 105: 1960–1964.
49. Throop HL, Lerdau MT (2004) Effects of nitrogen deposition on insect herbivory: Implications for community and ecosystem processes. Ecosystems 7: 109–133.
50. Andrew NR, Hughes L (2005) Arthropod community structure along a latitudinal gradient: Implications for future impacts of climate change. Austral Ecology 30: 281–297.
51. Barton BT, Schmitz OJ (2009) Experimental warming transforms multiple predator effects in a grassland food web. Ecology Letters 12: 1317–1325.
52. Hoover JK, Newman JA (2004) Tritrophic interactions in the context of climate change: a model of grasses, cereal aphids and their parasitoids. Global Change Biology 10: 1197–1208.
53. Tuda M, Matsumoto T, Itioka T, Ishida N, Takanashi M, et al. (2006) Climatic and intertrophic effects detected in 10-year population dynamics of biological control of the arrowhead scale by two parasitoids in southwestern Japan. Population Ecology 48: 59–70.

17

Greenhouse Gas Flux and Crop Productivity after 10 Years of Reduced and No Tillage in a Wheat-Maize Cropping System

Shenzhong Tian[1], Yu Wang[2], Tangyuan Ning[1]*, Hongxiang Zhao[1], Bingwen Wang[1], Na Li[1], Zengjia Li[1], Shuyun Chi[3]*

1 State Key Laboratory of Crop Biology, Shandong Key Laboratory of Crop Biology, National Engineering Laboratory for Efficient Utilization of Soil and Fertilizer Resources, Shandong Agricultural University, Taian, Shandong, China, 2 Shandong Rice Research Institute, Jinan, Shandong, China, 3 College of Mechanical and Electronic Engineering, Shandong Agricultural University, Taian, Shandong, China

Abstract

Appropriate tillage plays an important role in mitigating the emissions of greenhouse gases (GHG) in regions with higher crop yields, but the emission situations of some reduced tillage systems such as subsoiling, harrow tillage and rotary tillage are not comprehensively studied. The objective of this study was to evaluate the emission characteristics of GHG (CH_4 and N_2O) under four reduced tillage systems from October 2007 to August 2009 based on a 10-yr tillage experiment in the North China Plain, which included no-tillage (NT) and three reduced tillage systems of subsoil tillage (ST), harrow tillage (HT) and rotary tillage (RT), with the conventional tillage (CT) as the control. The soil under the five tillage systems was an absorption sink for CH_4 and an emission source for N_2O. The soil temperature positive impacted on the CH_4 absorption by the soils of different tillage systems, while a significant negative correlation was observed between the absorption and soil moisture. The main driving factor for increased N_2O emission was not the soil temperature but the soil moisture and the content of nitrate. In the two rotation cycle of wheat-maize system (10/2007–10/2008 and 10/2008–10/2009), averaged cumulative uptake fluxes of CH_4 under CT, ST, HT, RT and NT systems were approximately 1.67, 1.72, 1.63, 1.77 and 1.17 t ha^{-1} $year^{-1}$, respectively, and meanwhile, approximately 4.43, 4.38, 4.47, 4.30 and 4.61 t ha^{-1} $year^{-1}$ of N_2O were emitted from soil of these systems, respectively. Moreover, they also gained 33.73, 34.63, 32.62, 34.56 and 27.54 t ha^{-1} yields during two crop-rotation periods, respectively. Based on these comparisons, the rotary tillage and subsoiling mitigated the emissions of CH_4 and N_2O as well as improving crop productivity of a wheat-maize cropping system.

Editor: Gil Bohrer, The Ohio State University, United States of America

Funding: This work was financial supported by Financial supported by the National Science and Technology Research Projects of China (2012BAD14B07), the Special Research Funding for Public Benefit Industries (Agriculture) of China (201103001), and the Nature Science Fund of China (30900876 and 31101127). The funders had no role in study design, data collection and analysis, decision to publish, or preparation of the manuscript.

Competing Interests: The authors have declared that no competing interests exist.

* E-mail: ningty@163.com (TN); chishujun1955@163.com (SC)

Introduction

Methane (CH_4) and nitrous oxide (N_2O) play a key role in global climate change [1]. The global warming potential of these gases are respectively 25 and 298 times that of carbon dioxide (CO_2) [2]; thus, the release of these gases is a crucial contributory factor to increasing loads of greenhouse gases (GHG). According to estimations of the IPCC [3], the fluxes of CH_4 and N_2O from agricultural sources account for 50% and 80% of the total emission of these gases, respectively. There have been many studies on CO_2 emission in different ecosystems [4–6]; however, emissions of CH_4 and N_2O emission have been researched incompletely, especially in agricultural ecosystems [7,8].

In general, appropriate soil tilling may reduce GHG emissions because the emissions from soil are strongly affected by tilling, results that have been found by many studies [8–10], most of which, have reported the emissions of CH_4 and N_2O under conventional tillage (CT) and no-tillage (NT) systems in different sites [5,9]. However, the both generally revealed two extremes in maintenance the soil organic carbon stock and crop productivity,

agricultural environment protection, the results in these aspects showed the regional character and sometimes they did not suit for agricultural sustainable development [8,9,11]. In which case, some reduced tillage systems such as subsoiling (ST), harrow tillage (HT) and rotary tillage (RT) have been introduced [11], and sometimes they as more important tillage practices combination with no tillage were used to in rotation-tillage systems, which changed some soil environment factors and crop yield [12]. Although these reduced systems were frequently used and developed rapidly due to they are not only advantageous to improve crop yield but also increase the utilization efficiency of soil, water and fertilizer [13,14], the emissions of CH_4 and N_2O under these systems were remain unclear.

The production, consumption and transport of CH_4 and N_2O in soil are strongly influenced by some soil factors. Many studies demonstrated that CH_4 uptake by soil is correlated with soil temperature [15,16] and the N_2O emission. The conditions of soil moisture and N concentration were also shown as two major driving factors of the emission of N_2O: the emission generally peaked during N fertilizer application and irrigation [17,18]. However, the effects of those factors on the emissions of CH_4 and N_2O under different tillage systems in the North China Plain are still not fully understood among available results.

Therefore, the aim of the present study was to quantify the emissions of CH_4 and N_2O under no tillage and three reduced tillage systems in the wheat-maize cropping system and to analyze the correlations between the two gas emissions and the soil temperature, moisture and nitrate content. The crop productivity of the wheat-maize cropping system during two crop-rotation periods was also analyzed.

Materials and Methods

Ethics Statement

This experiment was established in a long-term tillage and residue-management experiment site of Shandong Agricultural University. The farming operations of this experiment were similar to the rural farmers' operations and did not involve endangered or protected species; the operations were approved by the State Key Laboratory of Crop Biology, Shandong Key Laboratory of Crop Biology and National Engineering Laboratory for Efficient Utilization of Soil and Fertilizer Resources, Shandong Agricultural University.

Experimental Site

The study site was located at Tai'an (Northern China, 36°09′N, 117°09′E), which has the typical characteristics of the North China Plain. The average annual precipitation is 697 mm, and the average annual temperature is 13.0°C, with the minimum (−2.6°C) and maximum (26.4°C) monthly temperatures in January and July, respectively. The annual frost-free period is approximately 170–196 d in duration, and the annual sunlight time is 2627.1 hours. The soil is a loam with 40% sand, 44% silt, 16% clay. The characteristics of the surface soil (0–20 cm) were measured as follows: pH 6.8; soil bulk density 1.43g cm^{-3}; soil organic matter 1.36%; soil total nitrogen 0.13%; and soil total phosphorous 0.13%. The meteorological data during the experiment is shown in Figure 1.

Experimental Design

The study based on a 10-year tillage experiment, began in 2002, included no tillage (NT) and three reduced tillage systems involving subsoiling (ST), harrow tillage (HT), rotary tillage (RT), with the conventional tillage (CT) as the control. The treatments were arranged in a randomized block design with three replications. Each plot was 35 m long and 4 m wide.

The experimental site was cropped with a rotation of winter wheat (*Triticum aestivum* Linn.) and maize (*Zea mays L.*). Winter wheat (Jimai-22) was sown at a rate of 90 kg ha^{-1}, on 12 October 2007 and 15 October 2008, and was harvested 6 June 2008 and 10 June 2009. The basal fertilizer was added before sowing and contained with 225 kg N ha^{-1}, 150 kg P_2O_5 ha^{-1} and 105 kg K_2O ha^{-1}, and 100 kg N ha^{-1} was used as topdressing at the jointing stage with 160 mm of irrigation water. The maize was sown on 20 June 2008 and 22 June 2009, which included 66600 plants ha^{-1} and was harvested 8 October 2008 and 10 October 2009. For the maize, 120 kg N ha^{-1}, 120 kg P_2O_5 ha^{-1} and 100 kg K_2O ha^{-1} were used as a basal fertilizer, and 120 kg N ha^{-1} was used as topdressing at the jointing stage.

When the wheat and maize were harvested, the amount of residue retuned to each plot at an equal level according to the crop biomass and water content of residue, in order to ensure their amount and C content of the residue among treatment had no significant difference. The residue equal quality returned to the field, then pulverized using a residue chopper and mixed with the soil in the tillage operations. The operations of tillage and residue-management are shown in Table 1.

Gas Sampling and Analysis

The emission measurements of CH_4 and N_2O for each treatment were conducted using the static-chamber method [19]. According to some studies, there is optimal gas-sample collection duration in a day during which the sample can show the mean gas flux of the day [20–22]. Our previous study have showed that the ratios of the CH_4 flux between 9 a.m. and 10 a.m. and N_2O flux between 9 a.m. and 12 p.m. to the daily mean flux, respectively, verged 1 by the correction coefficient and regression analysis [23]. So, the CH_4 and N_2O were collected between 9 a.m. and 10 a.m. and between 9 a.m. and 12 p.m. respectively from October 2007 to August 2009 at approximately 1-month intervals [12,23]. Both

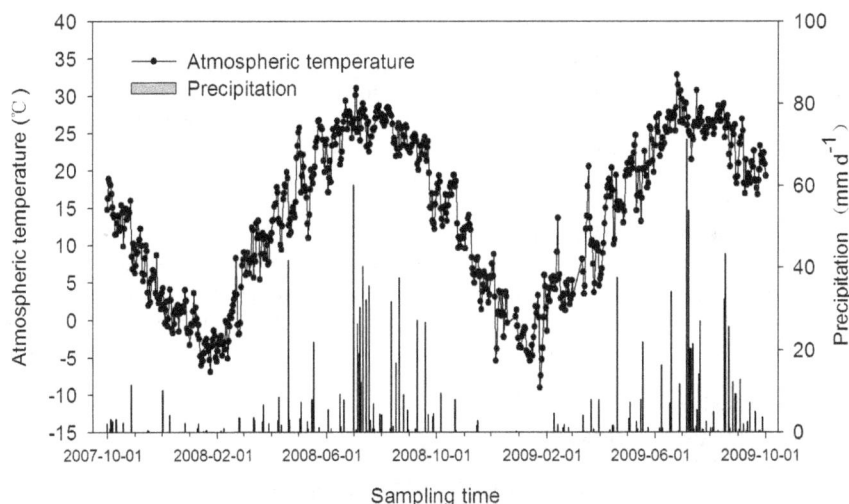

Figure 1. The atmospheric temperature and precipitation at the experimental site. The data were collected by the agricultural meteorological station approximately 500 m from the experiment field.

Table 1. The residue-management and tillage systems in the experimental plots.

Item	Tillage systems				
	CT	ST	HT	RT	NT
Wheat-residue					
Retention rate (t ha^{-1})	11.02	11.02	11.02	11.02	11.02
Residue-C (g kg^{-1})	51.27	51.27	51.27	51.27	51.27
Maize-residue					
Retention rate (t ha^{-1})	10.05	10.05	10.05	10.05	10.05
Residue-C (g kg^{-1})	52.15	52.15	52.15	52.15	52.15
Tillage					
Depth (cm)	25~35	40~45	12~15	10~15	–
Machine	moldboard	vibrating subsoil shovel	disc harrow	rototiller	–

CH_4 and N_2O were sampled at 5 minutes, 20 minutes and 35 minutes using a needle tube. The diurnal variations of CH_4 and N_2O in this study were collected from 2nd to 4th of May at 2-hour interval in 2009. All gases samples were collected at same time in order to avoid the order difference of sampling time. The atmospheric temperature, the temperature in the static chamber, the soil surface temperature and the soil temperature at a depth of 5 cm were determined simultaneously.

The samples were measured using a Shimadzu GC-2010 gas chromatograph. CH_4 was measured using a flame ionization detector with a stainless-steel chromatography column packed with a 5A molecular sieve (2 m long); the carrier gas was N_2. The temperatures of the column, injector and detector were 80°C, 100°C and 200°C, respectively. The total flow of the carrier gas was 30 ml min^{-1}, the H_2 flow was 40 ml min^{-1}, and the airflow was 400 ml min^{-1}. N_2O was measured using an electron-capture detector with a Porapak-Q chromatography column (4 m long); the carrier gas was also N_2. The temperatures of the column, injector and detector were 45°C, 100°C and 300°C, respectively. The total flow of the carrier gas was 40 ml min^{-1}, and the tail-blowing flow was 40 ml min^{-1}. The gas fluctuations were calculated based on the gas-concentration change over time per unit area.

The emission fluxes of CH_4 and N_2O were calculated using the following formula [19]:

$$F = \frac{60HMP}{8.314(273+T)} \frac{dc}{dt}$$

where F is the gas emission flux or uptake flux ($\mu g\ m^{-2}\ hour^{-1}$), 60 is the conversion coefficient of minutes and hours, H is the height of the static chamber (m), M is the molar mass of gas (g mol^{-1}), P is the atmospheric pressure (Pa), 8.314 is the ideal gas constant (J mol^{-1} K^{-1}), T is the average temperature in the static chamber (°C), and dc/dt is the slope of the line of the gas-concentration change over time.

The cumulative fluxes of CH_4 and N_2O were calculated by summing the products of the daily mean flux of two neighboring observations multiplying the days [6,24], the calculation formula as follows:

$$F_{cumulative} = \sum \left[\frac{(F_i + F_{i+1})}{2} \times d \right]$$

$$F_i = F_{avg} \times 24$$

where, $F_{cumulative}$ is the cumulative flux of CH_4 or N_2O (t ha^{-1} year^{-1}); F_i is the daily flux of observation i (t ha^{-1} d^{-1}); F_{i+1} is the daily flux of observation $i+1$ (t ha^{-1} d^{-1}); d is the days; F_{avg} is the mean flux of CH_4 or N_2O, which able to represent daily mean flux of CH_4 or N_2O according to the correction coefficient and regression analysis between the gas flux in observation duration and the daily total fluxes in diurnal variation.

Soil Sampling and Analysis

The meteorological data during the experiment (10/2007–10/2009) were by the agricultural meteorological station approximately 500 m from the experiment field. We measured soil temperature at a depth of 5 cm and the soil moisture in the 0–20 cm soil layers using a WET Sensor (WET brand, made in the UK). Soil samples (0–10 cm, 10–20 cm and 20–30 cm) were collected at five random positions of each plot and air dried to a constant weight, triturated and passed through a 2 mm sieve after thorough incorporation; they were then used to determine NO_3^-N using the UV colorimetric method [25].

Crop Yield

Winter wheat and maize were harvested at maturity, and the both harvest area was 9 m^2 in the central area of each plot to exclude edge effects. After air-drying, the grains were separated from the plants and oven-dried at 65°C for 48 h, and the dry weight was determined.

Statistical Analyses

The data were mapped using Sigma Plot 10.0, and all of the statistical analyses were performed using SPSS statistical software (SPSS Inc., Chicago, IL). The standard deviation (S.D.) and least significant difference (LSD) were calculated to compare the treatment means.

Results

Seasonal Variation of CH_4 Uptake

The soil acted an absorption sink for CH_4 in all tillage systems, which significantly varied by different soil tillage (Figure 2). The CH_4 flux during the sampling period (10/2007~10/2009) ranged from 6.31 to 41.36 $\mu g\ m^{-2}\ h^{-1}$ under CT, from 4.03 to 33.45 μg

$m^{-2} h^{-1}$ under ST, from 5.30 to 35.21 $\mu g\ m^{-2}\ h^{-1}$ under HT, from 6.78 to 32.54 $\mu g\ m^{-2}\ h^{-1}$ under RT, and from 1.10 to 26.21 $\mu g\ m^{-2}\ h^{-1}$ under NT. Moreover, the fluxes tended to respond to the change of atmospheric temperature during the sampling time (Figure 2), with higher uptake fluxes in the summer and lower in the winter in the different tillage systems.

Seasonal Variation of N₂O Emission

The emitting of N_2O from the soil was observed under the different treatments (Figure 3), but the differences were small among the treatments in the same sampling time. However, the peak N_2O emission coincided with the irrigation and fertilization periods, which had the highest emission fluxes of all of the periods ($P<0.01$), and the flux did not accord with the trend of atmospheric temperature. Meanwhile, the flux in all of the samples of N_2O ranged from 14.07 to 130.39 $\mu g\ m^{-2}\ h^{-1}$ under CT, from 14.20 to 126.43 $\mu g\ m^{-2}\ h^{-1}$ under ST, from 12.68 to 134.93 $\mu g\ m^{-2}\ h^{-1}$ under HT, from 10.81 to 126.42 $\mu g\ m^{-2}\ h^{-1}$ under RT, and from 13.04 to 128.29 $\mu g\ m^{-2}\ h^{-1}$ under NT.

Diurnal Variations of CH₄ and N₂O Under CT and NT

The diurnal flux variations of CH_4 uptake and N_2O emission significantly differed between the CT and NT treatments (Figure 4). In both of the treatments, the CH_4 uptake exhibited the lowest and highest fluxes at 6 a.m. and 2 p.m., respectively, but the lowest and highest values of CH_4 absorption were 12.99 and 23.77 $\mu g\ m^{-2}\ h^{-1}$, respectively, under CT and 12.23 and 23.03 $\mu g\ m^{-2}\ h^{-1}$, respectively, under NT (Figure 4a). The emission troughs and peaks of N_2O under CT and NT were observed at 4 a.m. and 4 p.m., respectively, with flux values of

16.90 and 41.89 $\mu g\ m^{-2}\ h^{-1}$ under CT and 13.57 and 34.38 $\mu g\ m^{-2}\ h^{-1}$ under NT, respectively (Figure 4b).

Cumulative Emissions of CH₄ and N₂O

The cumulative fluxes of CH_4 and N_2O under the different tillage systems in the two rotation cycle of wheat-maize systems were shown in Table 2. The highest cumulative uptake flux of CH_4 presented in the RT treatment with 1.85 t $ha^{-1}\ year^{-1}$ in the first rotation cycle of wheat-maize system (10/2007–10/2008), which was higher 8.3%, 13.7%, 22.5% and 60.9% than CT, ST, HT and NT treatments, respectively. But in the second rotation cycle (10/2008–10/2009), the highest one was the ST treatment. The order of total cumulative uptake fluxes of CH_4 under five tillage systems in two years was RT>ST>CT>HT>NT. The cumulative emission flux of N_2O ranged from 4.18 to 4.63 t $ha^{-1}\ year^{-1}$ in the first year and from 4.15 to 4.67 t $ha^{-1}\ year^{-1}$ in the second year, the total cumulative emission flux of N_2O ordered of NT>HT>CT>ST>RT, the flux under NT was only higher 7.2% than that of under RT.

Seasonal Variations of the Soil Temperature, Moisture and NO₃⁻N

A significant difference of the soil temperature at 5 cm depth was measured in the different periods. The changes under the different tillage systems were related to the atmospheric temperature (Figure 5a). The averaged soil temperature in all of the periods under RT was higher than under the other tillage methods. The soil moisture of the 0–20 cm layer varied among the different treatments and was related to precipitation or irrigation (Figure 5b). The averaged moisture level of the 0–20 cm layer was highest in the NT treatment. Similarly, higher nitrate contents

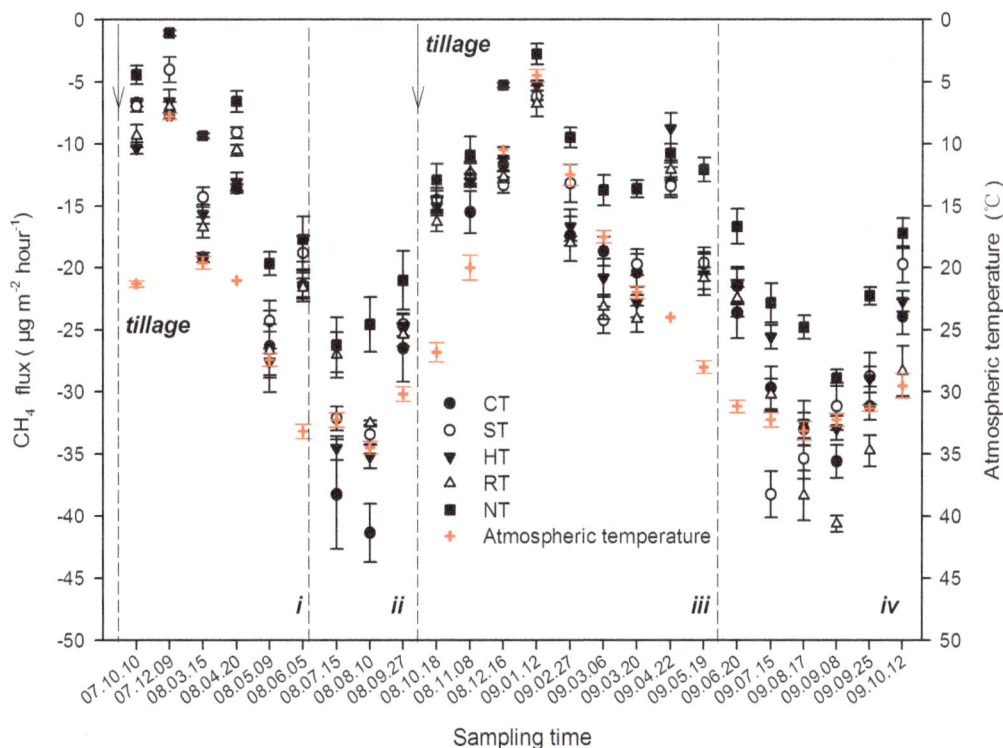

Figure 2. The seasonal characteristics of CH₄ flux under the different tillage systems. i and iii were the periods of wheat growth in 2007–2008 and 2008–2009, respectively; ii and iv were the periods of maize growth in 2008 and 2009, respectively. The arrows indicate the time of tilling, and the dotted lines distinguish the different periods of crop growth. The data are means ± SD (n = 3).

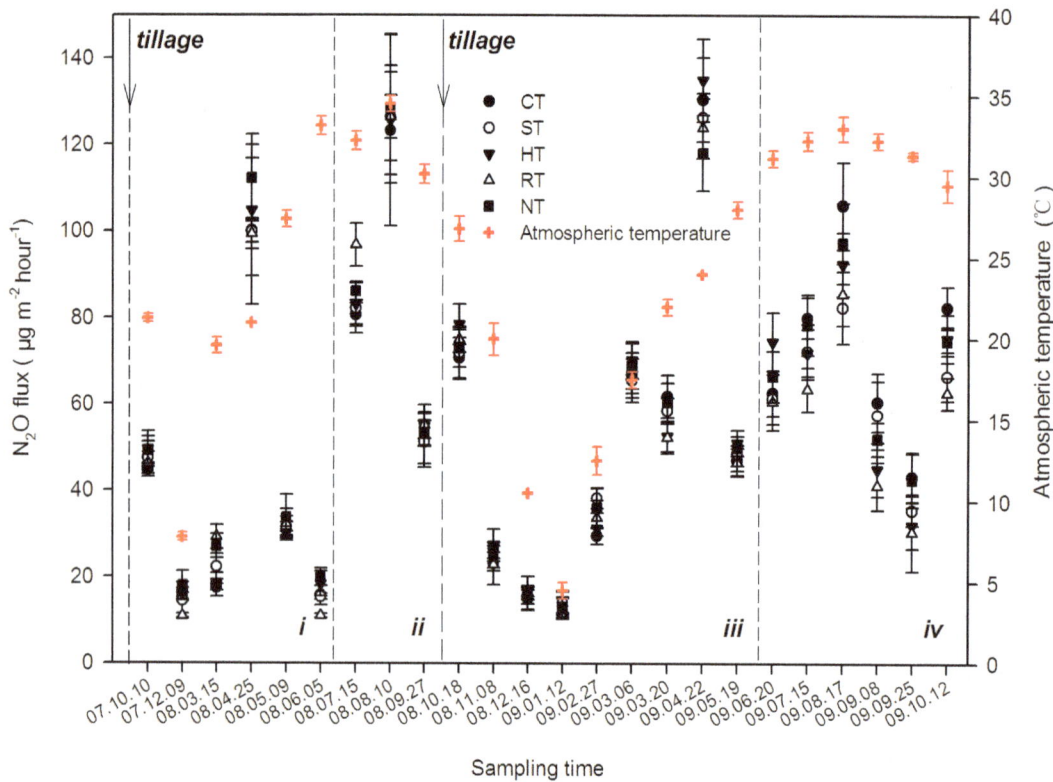

Figure 3. The seasonal characteristics of N$_2$O flux under the different tillage systems. i and iii were the periods of wheat growth in 2007–2008 and 2008–2009, respectively; ii and iv were the periods of maize growth in 2008 and 2009, respectively. The arrows indicate the time of tilling, and the dotted lines distinguish the different periods of crop growth. The data are means ± SD (n = 3).

were measured under the HT, NT, RT and ST treatments (Figure 5c); the levels in the NT, RT and ST treatments were higher than in the CT treatment by 4.21%, 2.42% and 1.40%, respectively.

Regression Analysis between CH$_4$, N$_2$O and Soil Factors

The absorption of CH$_4$ by the soil in different tillage systems was strongly affected by the soil temperature and soil moisture, and the uptake flux showed a positive correlation with the soil temperature ($R^2 = 0.44$, $P < 0.01$; Figure 6a), and a negative correlation was observed with the soil moisture ($R^2 = 0.36$, $P < 0.01$; Figure 6b).

The N$_2$O emission flux and soil temperature were not significantly correlated in this study. However, the N$_2$O emission flux was significantly related to the soil moisture ($R^2 = 0.63$, $P < 0.01$; Figure 7a) and the content of nitrate ($R^2 = 0.50$, $P < 0.01$; Figure 7b), which promoted the N$_2$O emission.

Crop Yields

The crop productivity of the wheat-maize rotation in the two years differed among the tillage systems (Table 3). The highest total productivity in two crop-rotation periods were measured in the ST treatment, with 34.63 t ha^{-1}, which was higher 2%, 6.2%, 0.2% and 25.7% than in the CT, HT, RT and NT treatment. The NT system showed the lowest productivity, only producing 27.54 t ha^{-1} yield in two crop-rotation periods.

Discussion

Effects of Soil Factors on CH$_4$ Uptake and N$_2$O Emission

In general, the emissions of CH$_4$ and N$_2$O in different seasons are affected by the soil temperature [26,27]. In this study, the CH$_4$ uptake and N$_2$O emission were lower in winter and higher in summer (Figures 2 and 3) and significantly related with the change of the soil temperature (Figures 2 and 6a). Similar results have been indicated in previous studies [28,29]. However, the CH$_4$ uptake flux usually decreased with the irrigation (Figures 3 and 6b). Many studies reported that soil moisture was a limiting factor for CH$_4$ absorption by the soil [16,30], leading to reduced rates of gas and O$_2$ diffusion [17].

Sometimes, the seasonal characteristics of N$_2$O emission exhibit varied trends in different sites, generally affected by temperature, N fertilization or irrigation in a significant linear relationship [27,31], and higher NO$_3$-N concentrations in wet soils promote the activity of nitrification and denitrification [18]. The higher moisture of the soil in this study promoted N$_2$O emission (Figures 5b and 7a), and the emission also increased with N fertilizer application (Figures 5c and 7b). In addition, there was no significant correlation between the N$_2$O emission and soil temperature in the seasonal variation in this study, similar results also found by other studies [32].

Tillage Effect on CH$_4$ Uptake and N$_2$O Emission

The fluxes of CH$_4$ uptake and N$_2$O emission were related to some soil factors (Figures 6 and 7), and these related soil factors general varied by the different tillage systems (Figure 5), because the changed soil structure and microflora by tillage would further

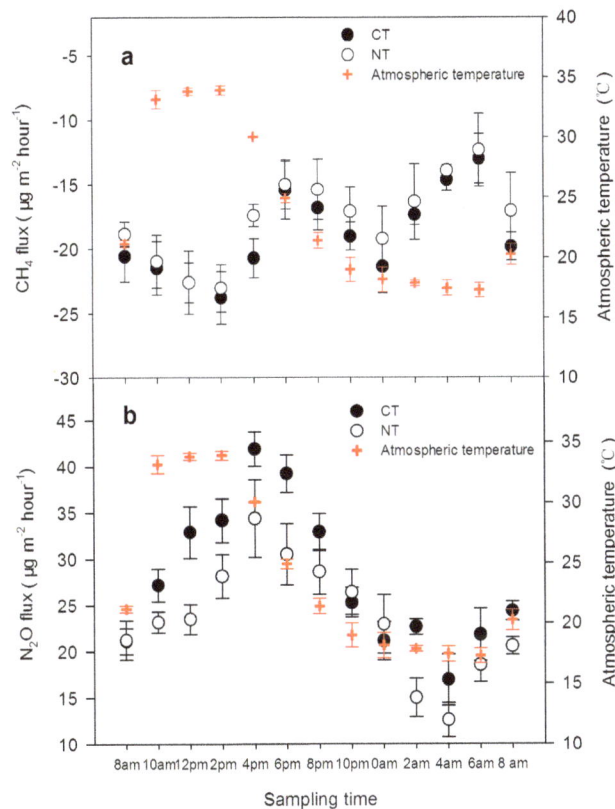

Figure 4. The diurnal flux variations of CH₄ uptake (a) and N₂O emission (b) in the CT and NT treatments. The data were collected from 2ⁿᵈ to 4ᵗʰ of May at 2-hour intervals in 2009. The data are means ± SD (n = 3).

CH_4 uptake was observed under NT (Figure 2), which was consistent with many previous studies [13,14]. The highest moisture and the lowest temperature conditions were also found in the soil of the NT treatment (Figures 5a and 5b). Some previous studies have indicated that the excessive wet condition of soil generally led to compaction of the soil surface without tillage, which blocked CH_4 from entering into soil for oxidation in dryland farming systems [32–35]. However, others opposite results also indicated that NT could increase CH_4 oxidation because the reduced disturbance of soil could increase the activity of methane-oxidizing bacteria [36,37], while disturbance may negatively affect CH_4 uptake by the soil [38], but sometimes this effect from disturbance is small and can largely be ignored [9,38]. Because the variations of these factors are important driving factors for soil microflora and also responded sensitive by the changed soil structure by tillage.

Similarly, the HT treatment, which had the highest emission of N_2O from the soil because the content of NO_3-N was the highest compared with the other treatments (Figures 5c and 7b). The emission flux of N_2O usually peaked after the stage of N-fertilizer application or irrigation [10], and the highest flux was observed in reduced systems (HT treatment). In some cases, there was a risk of increasing the emission of N_2O under NT or reduced tillage [9]. These methods mostly produced decreased N_2O emission relative to that reported by many previous studies [39,40] because little N_2O was generally produced with a better mixture of soil and residue, and the predominant form of nitrogen was NO_3-N or NH_4-N [14].

Tillage Effect on Crop Yield

Little information was reported on reduced tillage systems effect on crop yields. Sometimes, subsoiling was regarded an effective method to increase wheat production [41–43]. In this study, the highest crop productivity of two rotation periods was shown under the ST and RT systems (Table 3). However, the NT system in this study had the lowest crop total yield, and a similar trends were reported by other studies [5,10]. However, the yields under NT showed dissimilar results in different sites, in which generally reported with the NT increase crop yields compared to conventional tillage [44,45]. Furthermore, relative to other tillage systems, no-tillage grain yields and profits are often dramatically lower during the first few years of adoption. In fact, no tillage have not been widely applied in the world or at least China, because there are actual or perceived problems in crop grown using the

drive emissions of the N_2O and CH_4 from soil [8,15]. For example, a soil with better permeability was a larger absorption sink of CH_4 [5]. In this study, the highest uptake flux of CH_4 was observed in the RT treatment (Table 2), which also contained the highest averaged soil temperature (Figure 5a). The regression analysis showed that there was a significant positive correlation between CH_4 uptake and soil temperature (Figure 6a). In contrast, under the five tillage systems used in this study, the lowest flux of

Table 2. The cumulative emissions of CH_4 and N_2O under the different tillage systems.

Item	Cumulative flux (t ha⁻¹ year⁻¹)				
	CT	ST	HT	RT	NT
The first rotation cycle of wheat-maize system (10/2007–10/2008)					
Cumulative uptake flux of CH₄	1.71b	1.63c	1.51d	1.85a	1.15e
Cumulative emission flux of N₂O	4.18d	4.30c	4.37c	4.45b	4.64a
The second rotation cycle of wheat-maize system (10/2008–10/2009)					
Cumulative uptake flux of CH₄	1.63d	1.81a	1.74b	1.67c	1.20e
Cumulative emission flux of N₂O	4.67a	4.46c	4.57b	4.15d	4.57b
The total cumulative flux in the two years					
CH₄	3.34	3.43	3.25	3.53	2.34
N₂O	8.85	8.76	8.94	8.60	9.22

Different small letters in the same line indicate $P<0.05$. n = 3.

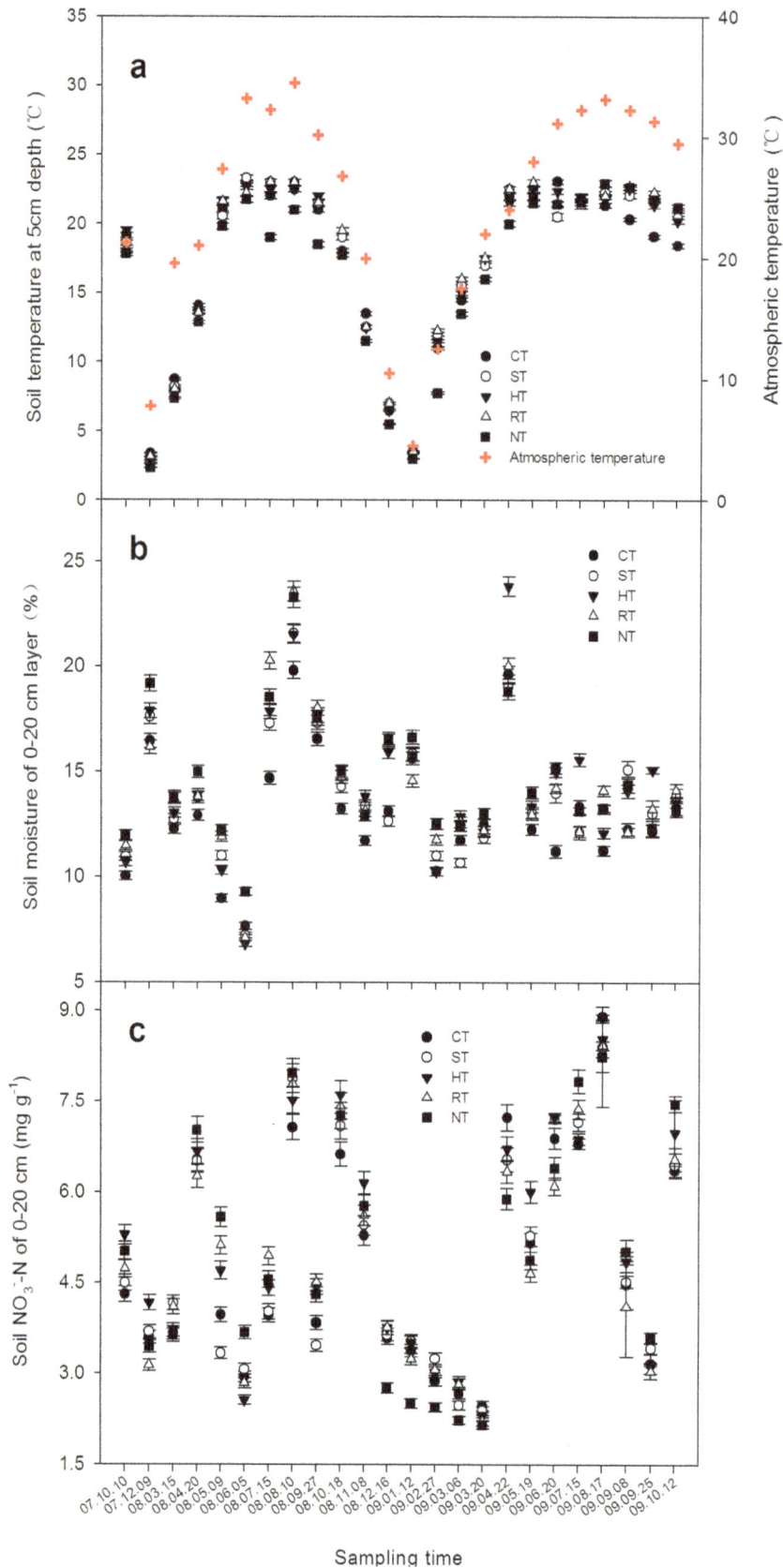

Figure 5. The seasonal variations of the soil temperature at a depth of 5 cm (a), soil moisture at 0–20 cm (b), and soil NO$_3^-$N content at 0–20 cm (c) under the different tillage systems. The data are means $\pm$ SD (n = 3).

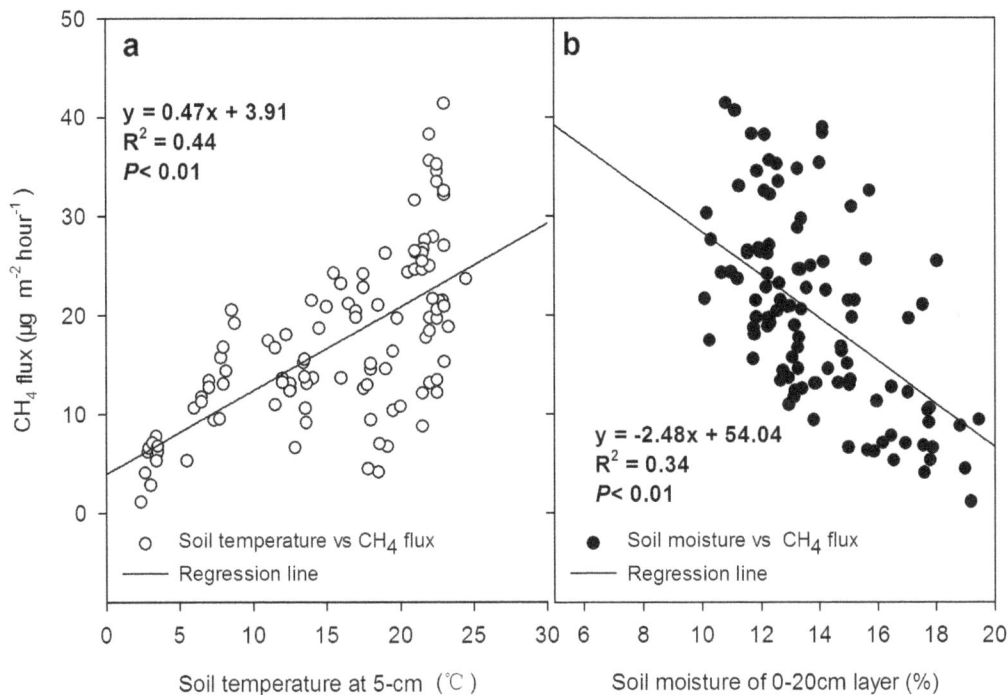

Figure 6. The regression analysis of the CH$_4$ flux and soil temperature at 5-cm depth (a, n = 105) and soil moisture of the 0–20 cm layer (b, n = 105).

NT, it may have limited adoption by growers. Some specific problems include lower early-season soil temperatures, reduced seed germination and emergence, below-optimal plant populations, poorer weed control, delayed plant development and maturity, increased grain moisture content, and lower grain yield potential [46–51]. Yet no tillage system effects on yield are highly dependent upon soil type, drainage, climate/latitude, and crop rotation [52].

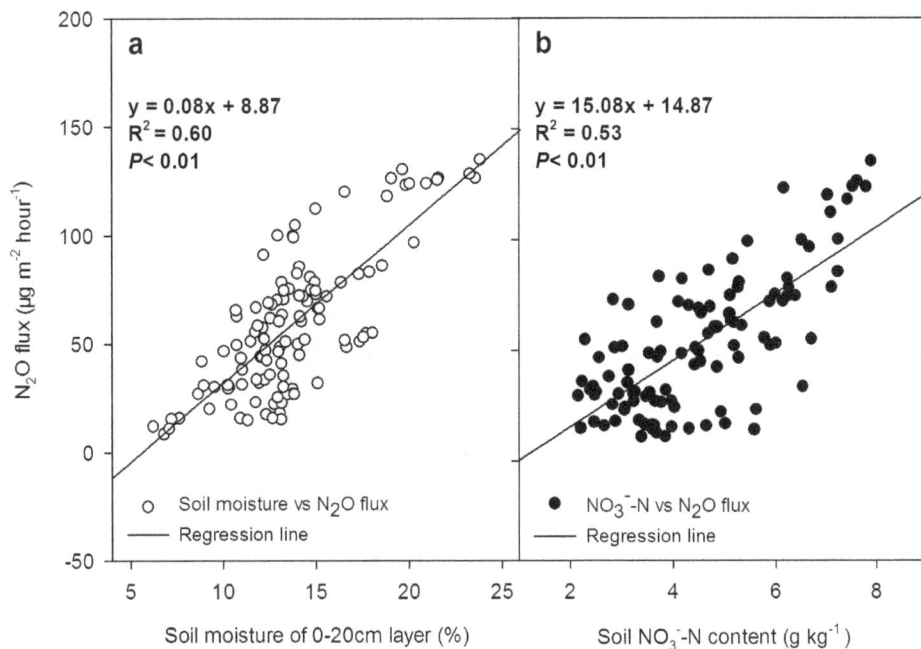

Figure 7. The regression analysis of the N$_2$O flux and soil moisture of the 0–20 cm layer (a, n = 105) and the soil NO$_3^-$N of the 0–20 cm layer (b, n = 105).

Table 3. The crop productivity in the different tillage systems.

Wheat-maize cropping-rotation system	Crop yield (t ha^{-1})				
	CT	ST	HT	RT	NT
The first rotation cycle					
Wheat (10/2007~06/2008)	5.90 d	6.18 c	6.58 a	6.35 b	4.65 e
Maize (06/2008~10/2008)	12.19 a	10.98 b	10.10 c	9.92 d	8.83 e
The second rotation cycle					
Wheat (10/2008~06/2009)	5.85 c	6.90 a	6.17 b	6.06 b	4.95 d
Maize yield (06/2009~10/2009)	10.03 c	10.58 b	9.76 d	12.23 a	9.11 e
Total productivity	33.95 b	34.63 a	32.62 c	34.56 a	27.54 d

Different small letters in the same line indicate $P<0.05$. n = 3.

Conclusions

The fluxes of CH_4 and N_2O were significantly affected by no tillage and reduced tillage systems, which also impacted soil factors, such as the soil temperature, moisture and nitrate content. These factors were related with the changes of CH_4 and N_2O fluxes. No-tillage did not reveal a better result for mitigating emissions of CH_4 and N_2O, as well as maintaining a high level of crop yield in this cropping rotation system. Comparison to the conventional tillage and no tillage, under the reduced tillage systems such as subsoiling and rotary tillage, may mitigate emissions of CH_4 and N_2O, and also gain a high level of crop productivity in the wheat-maize cropping rotation system. Sometimes, the both also as an important rotation tillage systems was applied in this region, in particular the subsoiling. The results also provide information on optional tillage in rotation tillage systems for mitigating GHG emissions and improving crop yield.

Author Contributions

Conceived and designed the experiments: ST TN ZL SC. Performed the experiments: ST YW HZ BW NL. Analyzed the data: ST TN. Contributed reagents/materials/analysis tools: ST. Wrote the paper: ST TN.

References

1. Forster P, Ramaswamy V, Artaxo P, Berntsen T, Betts R, et al. (2007) Changes in atmospheric constituents and in radiative forcing. In: Solomon S, Qin D, Manning M, Chen Z, Marquis M, et al., ed. Climate Change 2007: The Physical science basis. Contribution of working group I to the fourth assessment report of the intergovernmental panel on climate change. Cambridge University Press, Cambridge, United Kingdom and New York, USA.
2. Bouwman AF (1990) Exchange of greenhouse gases between terrestrial ecosystems and the atmosphere. In: Bouwman AF, ed. Soils and the Greenhouse Effect. Wiley, Chichester, 61–127.
3. IPCC (2007) Climate change 2007: The physical science basis. Contribution of working group I to the fourth assessment report of the intergovernmental panel on climate change. Cambridge University Press, Cambridge, United Kingdom and New York, USA.
4. Paustian K, Andren O, Janzen HH, Lal R, Smith P, et al. (1997) Agricultural soil as a C sink to offset CO_2 emissions. Soil Use Manage 13: 230–244.
5. Dong YH, Ouyang Z (2005) Effects of organic manures on CO_2 and CH_4 fluxes of farmland. Chin J Appl Ecol 16: 1303–1307.
6. Chatskikh D, Olesen JE (2007) Soil tillage enhanced CO_2 and N_2O emissions from loamy sand soil under spring barley. Soil Till Res 97: 5–18.
7. Willison TW, Webster CP, Goulding KWT, Powlson DS (1995) Methane oxidation in temperate soils: effective of land use and the chemical form of nitrogen fertilizer. Chemosphere 30: 539–546.
8. Lee J, Six J, King AP, Van Kessel C, Rolston DE (2006) Tillage and field scale controls on greenhouse gas emissions. J Environ Qual 35: 714–725.
9. Six J, Ogle SM, Breidt FJ, Conant RT, Mosier AR, et al. (2004) The potential to mitigate global warming with no-tillage management is only realized when practised in the long term. Glob Change Biol 10: 155–160.
10. Bhatia A, Sasmal S, Jain N, Pathak H, Kumar R, et al. (2010) Mitigating nitrous oxide emission from soil under conventional and no-tillage in wheat using nitrification inhibitors. Agr Ecosyst Environ136: 247–253.
11. Zhang HL, Gao WS, Chen F, Zhu WS (2005) Prospects and present situation of conservation tillage. J Chin Agr Uni 10: 16–20.
12. Tian SZ, Ning TY, Zhao HX, Wang BW, Li N, et al. (2012) Response of CH_4 and N_2O Emissions and Wheat Yields to Tillage Method Changes in the North China Plain. PLoS one 7(12): e51206.
13. Elder JW, Lal R (2008) Tillage effects on gaseous emissions from an intensively farmed organic soil in North Central Ohio. Soil Till Res 98: 45–55.
14. Bai XL, Zhang HL, Chen F, Sun GF, Hu Q, et al. (2010) Tillage effects on CH_4 and N_2O emission from double cropping paddy field. Transac CSAE 26: 282–289.
15. Dunfield P, Knowles R, Dumont R, Moore TR (1993) Methane production and consumption in temperate and subarctic peat soils. Soil Biolog Biochem25: 321–326.
16. Zhang XS, Shen SH, Li J (2006) Soil CH_4 uptake in winter wheat field in the North China Plain. J Nanjing Inst Meteorol 29: 181–188.
17. Le Mer J, Roger P (2001) Production, oxidation, emission and consumption of methane by soils: A review. Eur J Soil Biol 37: 25–50.
18. Gregorich EG, Rochette P, Vandenbygart AJ, Angers DA (2005) Greenhouse gas contributions of agricultural soils and potential mitigation practices in Eastern Canada. Soil Till Res 83: 53–72.
19. Robertson G (1993) Fluxes of nitrous oxide and other nitrogen trace gases from intensively managed landscapes: a global perspective. In: Harpwr LA, Mosier AR, Duxbury JM, Rolston DE, ed. Agricultural ecosystem effects on trace gases and global climate change. ASA Special Publication No. 55. ASA, CSSA, SSSA, Madison, wi 95–108.
20. Ma J, Xu H, Cai ZC, Yagi K (2007) Diurnal variation of CH_4 emission from rice field as affected by rice plant. Soils 39(6): 859–862.
21. Zheng XH, Wang MX, Wang YS, Shen RX, Li J, et al. (1998) Comparison of manual and automatic methods for measurement of methane emission from rice paddy fields. Adv Atmos Sci 15(4): 569–579.
22. Wan YF, Lin ED, Li YE, Gao QZ, Qin XB (2005) Studies on closing time in measuring greenhouse gas emission from dry cropland by static chamber method. Chin J Agrometeorol 27(2): 122–124.
23. Tian SZ, Ning TY, Chi SY, Wang Y, Wang BW, et al. (2012) Diurnal variations of the greenhouse gasses emission and their optimal observation duration under different tillage systems. Acta Ecologica Sinica 32: 879–888.
24. Liang W, Shi Y, Zhang H, Yue J, Huang GH (2007) Greenhouse gas emissions from northeast China rice fields in fallow season. Pedosphere 17: 630–638.
25. Bao SD (2000) Soil and Agricultural Chemistry Analysis. China Agriculture Press, Beijing.
26. Hassink J (1995) Density fractions of soil macroorganic matter and microbial biomass as predictors of C and C mineralization. Soil Biolog Biocheml 27: 1099–1108.
27. Groffman PM, Hardy JP, Driscoll CT, Fahey TJ (2006) Snow depth, soil freezing, and fluxes of carbon dioxide, nitrous oxide and methane in a northern hardwood forest. Glob Change Biol 12: 1748–1760.
28. Qi YC, Dong YS, Zhang S (2002) Methane fluxes of typical agricultural soil in the north china plain. Rural Ecol Environ 18: 56–60.
29. Dijkstra FA, Morgan JA, von Fischer JC, Follett RF (2011) Elevated CO_2 and warming effects on CH_4 uptake in a semiarid grassland below optimum soil moisture. J Geoph Res 116: 1–9.

30. Von Fischer JC, Butters G, Duchateau PC, Thelwell RJ, Siller R (2009) In situ measures of methanotroph activity in upland soils: A reaction diffusion model and field observation of water stress, J Geoph Res 114:G01015.

31. Cui RY, Chen ZZ, Chen SQ (2001) Progress in research on soil respiration of grasslands. Acta Ecologica Sinica 21: 315–325.

32. Koponen HT, Flojt L, Martikainen PJ (2004) Nitrous oxide emissions from agricultural soils at low temperatures: a laboratory microcosm study. Soil Biolog Biochem 36: 757–766.

33. Omonode RA, Vyn TJ, Smith DR, Hegymegi P, Ga'l A (2007) Soil carbon dioxide and methane fluxes from long-term tillage systems in continuous corn and corn–soybean rotations. Soil Till Res 95: 182–195.

34. Ahmad S, Li C, Dai G, Zhan M, Wang J, et al. (2009) Greenhouse gas emission from direct seeding paddy field under different rice tillage systems in central China. Soil Till Res 106: 54–61.

35. Li DM, Liu MQ, Cheng YH, Wang D, Qin JT, et al. (2011) Methane emissions from double-rice cropping system under conventional and no tillage in southeast China. Soil Till Res 113: 77–81.

36. Smith P, Goulding KWT, Smith KA, Powlson DS, Smith JU, et al. (2001) Enhancing the carbon sink in European agricultural soils: including trace gas fluxes in estimates of carbon mitigation potential. Nut cycle Agroecosyst 60: 237–252.

37. Hütsch BW (1998) Tillage and land use effects on methane oxidation rates and their vertical profiles in soil. Biol Fert Soils 27: 284–292.

38. Robertson GP, Paul EA, Harwood RR (2000) Greenhouse gases in intensive agriculture: Contributions of individual gases to the radiative forcing of the atmosphere. Science 289: 1922–1925.

39. Kessavalou A, Mosise AR, Doran JW (1998) Fluxes of carbon dioxide, nitrous oxide, and methane in grass sod and winter wheat-fallow tillage management. J Environ Qual 27: 1094–1104.

40. Lal R (2004) Soil carbon sequestration impacts on global climate change and food security. Science 304: 1623–1627.

41. He J, Li HW, Gao HW (2006) Subsoiling effect and economic benefit under conservation tillage mode in Northern China. Transact CSAE 22: 62–67.

42. Gong XJ, Qian CR, Yu Y, Zhao Y, Jiang YB, et al. (2009) Effects of Subsoiling and No-tillage on Soil Physical Characters and Corn Yield. J Maize Sci 17: 134–137.

43. Huang M, Wu JZ, Li YJ, Yao YQ, Zhang CJ, et al. (2009) Effects of different tillage management on production and yield of winter wheat in dryland. Transact CSAE 25: 50–54.

44. He J, Li HW, Rasaily RG, Wang QJ, Cai GH, et al. (2011) Soil properties and crop yields after 11 years of no tillage farming in wheat–maize cropping system in North China Plain. Soil Till Res 113: 48–54.

45. He J, Wang QJ, Li HW, Liu LJ, Gao HW (2009b) Effect of alternative tillage and residue cover on yield and water use efficiency in annual double cropping system in North China Plain. Soil Till Res 104: 198–205.

46. Mock JJ, Erbach DC (1977) Influence of conservation-tillage environments on growth and productivity of corn. Agron J 69: 337–340.

47. Kovar JL, Barber SA, Kladivko EJ, Griffith DR (1992) Characterization of soil temperature, water content, and maize root distribution in two tillage systems. Soil Till Res 24: 11–27.

48. Lund MG, Carter PR, Oplinger ES (1993) Tillage and crop rotation affect corn, soybean, and winter wheat yields. J Prod Agr 6: 207–213.

49. Swan JB, Higgs RL, Bailey TB, Wollenhaupt NC, Paulson WH, et al (1994) Surface residue and in-row treatment effects on long-term no-tillage continuous corn. Agron J 86: 711–718.

50. Fortin MC, Pierce FJ (1990) Developmental and growth effects of crop residues on corn. Agron J 82: 710–715.

51. Fortin MC (1993) Soil temperature, soil water, and no-till corn development following in-row residue removal. Agron J 85, 571–576.

52. Boomsma CR, Santini JB, West TD, Brewer JC, McIntyre LM, et al. (2010) Maize grain yield responses to plant height variability resulting from crop rotation and tillage system in a long-term experiment. Soil Till Res 106: 227–240.

18

Variation in Yield Gap Induced by Nitrogen, Phosphorus and Potassium Fertilizer in North China Plain

Xiaoqin Dai*, Zhu Ouyang, Yunsheng Li*, Huimin Wang

Key Laboratory of Ecosystem Network Observation and Modeling, Institute of Geographic Sciences and Natural Resources Research, Chinese Academy of Sciences, Beijing, China

Abstract

A field experiment was conducted under a wheat-maize rotation system from 1990 to 2006 in North China Plain (NCP) to determine the effects of N, P and K on yield and yield gap. There were five treatments: NPK, PK, NK, NP and a control. Average wheat and maize yields were the highest in the NPK treatment, followed by those in the NP plots among all treatments. For wheat and maize yield, a significant increasing trend over time was found in the NPK-treated plots and a decreasing trend in the NK-treated plots. In the absence of N or P, wheat and maize yields were significantly lower than those in the NPK treatment. For both crops, the increasing rate of the yield gap was the highest in the P omission plots, i.e., 189.1 kg ha^{-1} yr^{-1} for wheat and 560.6 kg ha^{-1} yr^{-1} for maize. The cumulative omission of P fertilizer induced a deficit in the soil available N and extractable P concentrations for maize. The P fertilizer was more pivotal in long-term wheat and maize growth and soil fertility conservation in NCP, although the N fertilizer input was important for both crops growth. The crop response to K fertilizers was much lower than that to N or P fertilizers, but for maize, the cumulative omission of K fertilizer decreased the yield by 26% and increased the yield gap at a rate of 322.7 kg ha^{-1} yr^{-1}. The soil indigenous K supply was not sufficiently high to meet maize K requirement over a long period. The proper application of K fertilizers is necessary for maize production in the region. Thus, the appropriate application of N and P fertilizers for the growth of both crops, while regularly combining K fertilizers for maize growth, is absolutely necessary for sustainable crop production in the NCP.

Editor: Carl J. Bernacchi, University of Illinois, United States of America

Funding: This study was financially supported by the Strategic Science Plan of Institute of Geographic Sciences and Natural Resources Research (2012ZD004)and National Science and Technology Support Projects of Ministry of Science and Technology of China (2013BAD05B03). The funders had no role in study design, data collection and analysis, decision to publish, or preparation of the manuscript.

Competing Interests: The authors have declared that no competing interests exist.

* E-mail: daixq@igsnrr.ac.cn (XD); liys@igsnrr.ac.cn (YL)

Introduction

The North China Plain (NCP) is the largest and the most important agricultural production region in China. It covers approximately 18.3% of total national farm lands (18 million hectares) and produces 21.6% of the total grain yield of edible crops in the country [1]. The main crops grown are wheat and maize, periodically rotated. The NCP now supplies more than 50% of the nation's wheat and 33% of its maize [2]. Grain production and the maintenance of soil fertility in the NCP are very important for the food security and agricultural sustainability of China.

Nutrient availability is the most yield-limiting factor. To produce higher yields, the over-application of chemical fertilizers has been a common practice in wheat-maize rotation systems and has led to severe environmental problems [3]. Improved nutrient management practices are urgently needed to maximize crop yields and maintain soil fertility while minimizing environmental impacts. The ability to better identify crop response to the application of fertilizers, soil indigenous nutrient supply capability, and the maintenance of soil fertility over time are crucial to the development of improved nutrient management practices. Various long-term experiments have been conducted to test the effects of fertilization on yield or soil fertility throughout the world [4–14],

but over long time scales, crop response to the application of N, P and K fertilizers and soil indigenous nutrient supply capability has seldom been clearly understood.

The yield and soil fertility gap between a full NPK fertilizer plot and a fertilizer omission plot was used as a good diagnostic tool to assess the extent of macronutrient limitations. Mussgnug et al (2006) analyzed the yield gap resulting from nutrient limitation on a degraded soil in the Red River Delta (RRD) in Northern Vietnam. They found that in the absence of K application, the yield gap for rice and maize respectively averaged 1.7 Mg ha^{-1} and 3.4 Mg ha^{-1}, while when N or P was omitted the yield gap was less. Potassium was the most yield-limiting macronutrient in RRD of Vietnam [15]. The factors that primarily limit increasing crop yields varied by crop and region [16]. The objectives of the study were to examine the effects of the continuous use of inorganic fertilizers on wheat and maize yield and soil fertility gap to elucidate crop response to fertilizer inputs and the long-term N, P and K supply capability in the NCP.

Materials and Methods

Experimental Site

A long-term field experiment was conducted from 1990 to 2006 at the Yucheng Comprehensive Experimental Station (36°57′ N,

Table 1. Physical and chemical properties of soil at different layers at Yucheng, Shandong Province of China in 1990.

Soil parameters	0–20 cm	20–40 cm
Soil texture	Silt loam	Sandy loam
Clay (%)	17.3	13.7
Silt (%)	61.4	11.4
Sand (%)	21.3	74.9
Bulk density (g cm^{-3})	1.31	1.48
Field capacity (%)	23.3	22.7
Organic C (g kg^{-1})	5.4	4.8
Total N (g kg^{-1})	0.41	0.36
Total P (g kg^{-1})	1.11	1.13
Total K (g kg^{-1})	10.7	14.0
Available N (mg kg^{-1})	46.1	41.4
Available P (mg kg^{-1})	8.3	5.4
Available K (mg kg^{-1})	149.0	125.4

116°36′ E, 28 m a.m.s.l.) of the Chinese Academy of Sciences (CAS) located in Shandong Province, which lies in the NCP. This area represents the moderate- to high-yielding region of the NCP, where the dominant cropping system is double-cropped wheat-maize. The annual mean precipitation is 575 mm, with approximately 70% of the total precipitation falling between June and September. The soil is classified as fluvo-aquic loam soil. Before the initiation of the experiment, soil samples were collected from depths of 0–20 cm and 20–40 cm by stepwise soil augers in the experimental field in 1990. All samples from the same soil layers were mixed thoroughly to create one homogenous sample, and a representative sample was drawn to determine the soil texture, bulk density, organic carbon, total N, P and K, and available N, P and K. The physical and chemical properties of the soil in 1990 are listed in Table 1.

Experimental Design and Treatments

The study consisted of five treatments: no fertilization (control); N, P, and K applied (NPK); N and P applied (NP; K omission); N and K applied (NK; P omission); P and K applied (PK; N omission), with four replications performed for each condition in a randomized complete block design (Table 2). The application rates

Table 2. Treatment and fertilizer nutrient rates (kg ha^{-1}) applied to winter wheat and maize at Yucheng, Shandong Province of China.

Treatments	Winter wheat			Maize		
	N	P$_2$O$_5$	K$_2$O	N	P$_2$O$_5$	K$_2$O
NPK	253	90	171	262	52	298
PK	0	90	171	0	52	298
NK	253	0	171	262	0	298
NP	253	90	0	262	52	0
Control	0	0	0	0	0	0

of N, P and K, if they were applied, were the same in all of the treatments: 253 kg N ha^{-1} as urea (46% N), 90 kg P$_2$O$_5$ ha^{-1} as single superphosphate (16% P$_2$O$_5$), and 171 kg K$_2$O ha^{-1} as potassium sulfate (50% K$_2$O) for winter wheat and 262 kg N ha^{-1} as urea (46% N), 52 kg P$_2$O$_5$ ha^{-1} as single superphosphate (16% P$_2$O$_5$), and 298 kg K$_2$O ha^{-1} as potassium sulfate (50% K$_2$O) for maize (Table 2). Phosphate was applied before sowing and incorporated during land preparation. Winter wheat received 40% of the total N as a basal dressing before sowing, whereas the remainder was top-dressed in two splits (40% at green-up and 20% at flowering). Approximately 50% of the K was applied before sowing and 50% top-dressed at green-up (the first leaf growing in the early of spring growed to 1–2 cm from soil surface and the phenomena was seen on a half of wheat seedling in the field). During the maize season, all of the P and K fertilizers and 35% of the N fertilizer were applied as a basal fertilizer, and 65% of the N was applied at the elongation stage. Each plot area measured 6 m×5 m, and all plots were isolated from one another by a concrete wall to a soil depth of 1 m.

Winter wheat (*Triticum aestivum*) was double-cropped with maize (*Zea mays*), and the crops were grown along the border of the four blocks. Winter wheat was sown in mid-October and harvested in early June of the next year. Maize was sown in mid-June and harvested in early October of the same year. The winter wheat was seeded at a rate of 225 kg ha^{-1}, and the maize was planted at a rate of 75000 seeds ha^{-1}. Surface irrigations were conventionally conducted according to the conditions of crop growth and soil moisture. Wheat and maize were harvested to the level of the soil surface; thus, the stubble left in the field was negligible. However, the roots were left in the soil. All straws were removed from the field. All plots were kept free of weeds by the application of an herbicide. Pesticides were applied in accordance with good practices for crop protection.

Sampling and Chemical Analyses

At harvestable maturity, the grain yields were determined over the whole plot area. Soils samples were collected at a depth of 20 cm at six randomly selected points in every plot after wheat and maize were harvested each year. The soil samples were mixed thoroughly from all cores to obtain a representative soil sample for each plot. The soil samples were air-dried and sieved through a 1-mm mesh to perform available N, Olsen-P and available K measurements and through a 0.149-mm mesh to estimate the organic carbon and total N content. Soil available N was determined using the alkali-hydrolytic diffusion method [17]. Olsen-P was measured by ascorbic acid-molybdate blue colorimetry [18]. Available K was extracted with an ammonium acetate solution (NH$_4$OAc, 1 mol/L) and then determined with a flame photometer [17]. Soil organic carbon was determined using the wet oxidation method of Walkley and Black, and total N was measured following Kjeldahl digestion and distillation [17].

Data Analyses

The results of yield and soil fertility in the first three years of the experiment were not used because in the autumn of 1993 soybean was sown. Therefore, the results reported are from 1994 to 2006. Crop yield losses caused by N, P, K or NPK omission were calculated from the differences between an NPK treatment and the PK, NK, NP or control plot during the same years. The yield differences were called the "yield gap" induced by N (GYG$_N$), P (GYG$_P$), K (GYG$_K$) or NPK (GYG$_{NPK}$) fertilizer omission to assess the extent of macronutrient limitations in the study. The differences in soil available N, Olsen-P and available K between the NPK treatment and the PK, NK, NP or control plots were

A

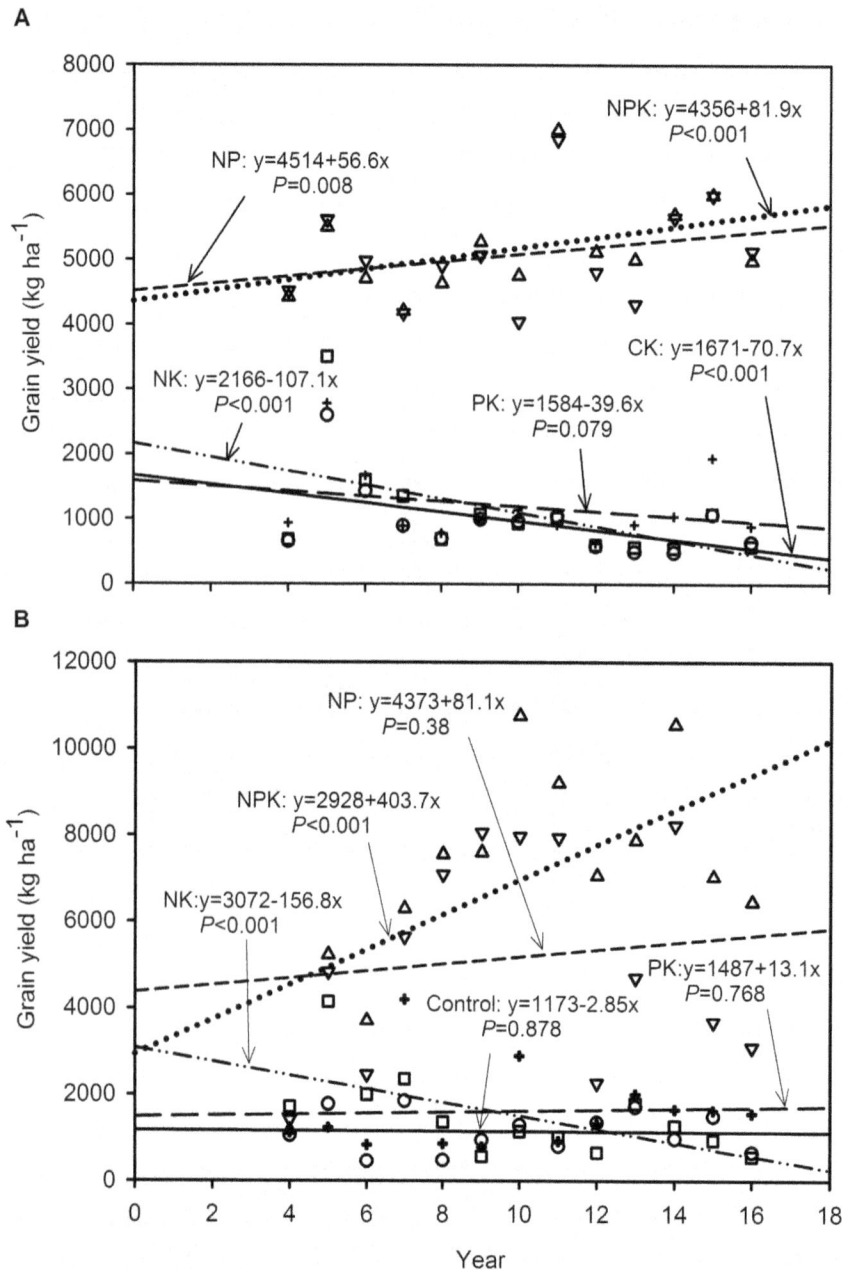

B

Figure 1. Grain yield of wheat (a) and maize (b) of different fertilizer treatments (Control, circle; NPK, triangle up; NP, triangle down; NK, square; PK, plus) under wheat-maize rotation system from 1994 to 2006 at Yucheng, Shandong Province of China.

calculated to evaluate the variation in soil fertility when nutrients were limited. The soil nutrient differences were called the "soil nutrient gap (SNG)" induced by N (SNG_N), P (SNG_P), K (SNG_P), or NPK (SNG_{NPK}) fertilizer omission.

All statistical analyses including the analysis of variance and regression were conducted using the SPSS package (version 16.0). Considering the data over time was lack of independence, the repeated measures ANOVA between differences in mean through years among treatments for various parameters (grain yield, soil organic C, total N, available N, Olsen-P, available K) were analyzed using a Fisher's protected least significant difference (LSD) test at $P = 0.05$. A mixed models with random block and

block×treatment were done to assess trends (slopes) of grain yield, yield gap and various soil nutrients gap over the years. The P-values of the slopes were used to test whether the observed changes were significantly different from 0.

Results

Grain Yield and Yield Gap

The fertilizer treatments significantly affected the grain yields (Table 3). The wheat and maize yields were highest in the NPK treatment, followed by those in the NP treatment. In the PK or NK treatments, the yields of wheat and maize were significantly

Table 3. Multi-year average grain yield of wheat and maize, as affected by various treatments with fertilizer in a wheat-maize rotation system at Yucheng, Shandong Province of China (mean±standard error).

Treatment	Wheat (kg ha^{-1})	Maize (kg ha^{-1})
Control	963±155 b	1145±133 c
NPK	5174±201 a	6965±724 a
NP	5080±216 a	5184±666 b
NK	1094±218 b	1504±270 c
PK	1188±165 b	1618±269 c

Different letters within a column indicate a significant difference at P<0.05 between fertilizer treatments.

lower than those in the NP and NPK treatments. Compared to those in the NPK treatment, the yields of wheat and maize decreased, respectively, by 77% with the N omission, 79% and 78% with the P omission, 2% and 26% with the K omission, and 81% and 84% with no fertilizer. The grain yield losses were much greater in either N or P omission than in plots fertilized combined N and P. Over time wheat yield significantly increased at a rate of 81.9 kg ha^{-1} yr^{-1} in NPK plots ($P=0.008$), and maize yield also significantly increased at a rate of 403.7 kg ha^{-1} yr^{-1} in NPK plots ($P<0.001$, Fig. 1). The yields of wheat and maize in NK treatment decreased significantly at a rate of 107.1 kg ha^{-1} yr^{-1} ($P<0.001$) and 156.8 kg ha^{-1} yr^{-1} ($P<0.001$) over time during the study period, respectively (Fig. 1). The long-term omission of K fertilizer also significantly reduced the maize yield compared to the maize yield of the NPK-treated plot (Table 3). But both crops yields had no significant upward or downward trends when K fertilizer was omitted (Fig. 1).

The variation in the yield gap of wheat and maize between the NPK and PK, NK, NP and control plots is illustrated in Figure 2. For wheat, the yield gap significantly increased at a rate of 152.6 kg ha^{-1} yr^{-1} for the control plots ($P<0.001$), a rate of 189.1 kg ha^{-1} yr^{-1} for the NK plots ($P<0.001$) and a rate of 121.5 kg ha^{-1} yr^{-1} for the PK plots ($P<0.001$). In the NP plots, the yield gap also gradually increased with the cumulative effect of K omission, but the relationship was not significant ($P=0.246$). However, for maize, when the N, P K or NPK fertilizer was omitted, the yield gap significantly increased with the continuous omission of nutrients. The significant increase in the yield gap was 390.6 kg ha^{-1} yr^{-1} for the PK treatments ($P<0.001$), 560.6 kg ha^{-1} yr^{-1} for the NK treatments ($P<0.001$), 322.7 kg ha^{-1} yr^{-1} for the NP treatments ($P<0.001$), and 406.6 kg ha^{-1} yr^{-1} for the control treatments ($P<0.001$). For both crops, the increasing rate of the yield gap was highest in the P omission plots, which indicated that the cumulative effect of P omission was the most significant for wheat and maize in the region. The slopes of the yield gap were the second highest in the N omission plots for both crops, which indicated that N fertilizer inputs were important for wheat and maize growth, especially the latter. It is worth noting that the slopes of the maize yield gap increased significantly with continuous K omission. This result indicates that the soil indigenous K supply was not sufficiently high to meet the maize K requirement over a long period.

Soil Nutrient Status

Long-term continuous treatment with different fertilizers significantly affected the soil organic carbon, available N, Olsen-

P and available K concentrations (Table 4). The multi-year average soil organic carbon content was significantly lower in the control, NK and PK plots than in the NPK and NP plots for wheat and maize. The decrease was 6.7% to 15.6% for wheat and 7.4% to 15.8% for maize. The different treatments except control had no significant influences on soil total N. The available N, Olsen-P and available K concentrations were higher in the plots treated with the corresponding N, P or K fertilizer inputs (Table 4). For wheat, the available N concentration increased significantly in the acquired N fertilizer inputs plots compared to that in the PK plots. The Olsen-P concentration in plots with P fertilizer inputs increased significantly by 7.2 to 12.3 times the concentration in the control plots. Meanwhile, the Olsen-P concentration also increased by 9% and 62% in the NP and PK plots, respectively, compared with that in the NPK plots, indicating that the P uptake of the crops was lower in the NP and PK plots. Similar to the Olsen-P concentration, the available K concentration increased significantly in the plots with K fertilizer inputs, and the increase was as high as 1.2 to 2.3 times the K concentration in the control plots. In the NK and PK plots, the available K concentration increased by 48% and 55%, respectively, compared to that in the NPK plots.

The differences in soil nutrients between the NPK and PK, NK, NP or NPK omission plots are shown in Figure 3&4. During the wheat growth season, the continuous omission of nutrients had no significant effects on the difference in the soil available N concentration (Fig. 3a). The continuous omission of P fertilizer significantly enhanced the differences in the soil Olsen-P concentration in the NPK plots ($P<0.001$), indicating that the long-term absence of P fertilizer resulted in a great deficit in the soil available P concentration. Similarly to the P omission plots, the control plots also possessed a significantly enhanced soil Olsen-P gap ($P<0.001$). However, a surplus concentration of Olsen-P was observed when the continuous N was omitted ($P=0.006$, Fig. 3b). Also a surplus concentration of available K was observed when the continuous P and N inputs were omitted ($P<0.001$) due to a large amount of K fertilizer inputs and lower crop uptake. When the K fertilizer was omitted, the soil available K concentration gap significantly increased ($P=0.001$ for NP and $P=0.002$ for Control), indicating a deficit in soil available K in the NP and Control plots.

During the maize growth season, the soil available N concentration gradually became deficient when the nutrients were omitted, but only when P fertilizer was omitted, the effects were significant (Fig. 4a). Like wheat, P fertilizer omission induced a severe deficit in soil Olsen-P ($P<0.001$). Although the PK plots feature a larger amount of P fertilizer inputs and lower crop uptake, no significant surplus in the soil Olsen-P concentration was observed (Fig. 4b). In the plots fed K fertilizer inputs, a surplus in the soil available K concentration was observed, but the effects were not significant (Fig. 4c). It is worthy to note that the NP plots induced a deficit in soil available K (Table 4). Although the effects were not significant, maize growth was significantly affected due to the shortage of soil available K.

Discussion

The response of wheat and maize growth of combined N and P fertilization (i.e. NP) was much greater than that of K fertilizer application combined with N or P (i.e. NK and PK treatments) (Table 3). The continuous 16-yr omission of N or P fertilizer significantly reduced the wheat and maize yield (Table 3). The grain yield in the N omission plots exhibited no significant decreasing trends over time (Fig. 1a). On the one hand, the

A

B

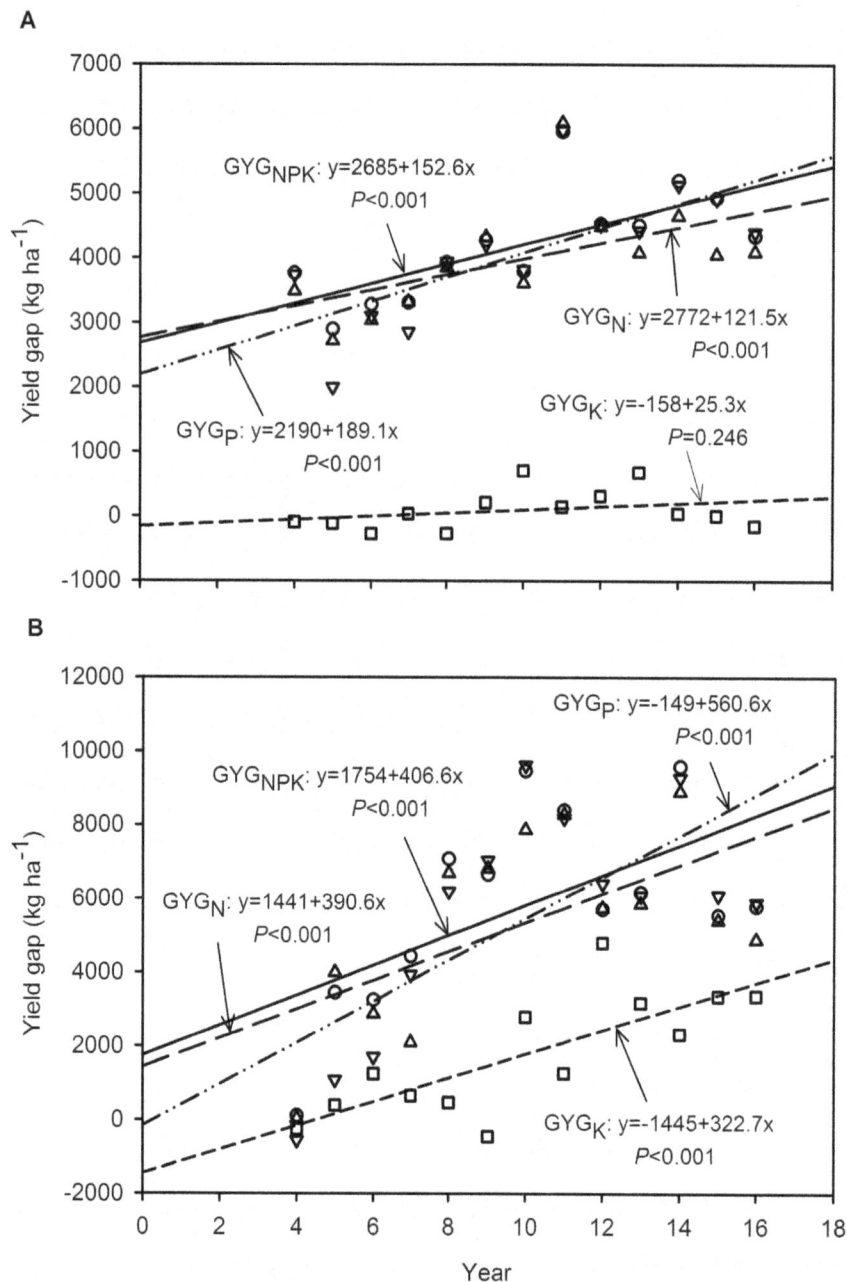

Figure 2. Yield gap variation of wheat (a) and maize (b) by N (GYG$_N$, the yield difference between NPK and PK plots in each year, triangle up), P (GYG$_P$, the difference between NPK and NK plots, triangle down), K (GYG$_K$, the difference between NPK and NP plots, square) or NPK (GYG$_{NPK}$, the difference between NPK and Control plots, circle) fertilizer omission from 1994 to 2006 at Yucheng, Shandong Province of China.

continuous N omission significantly reduced the soil available N concentration, but no significant effects were observed on the total N concentration (Table 4). The differences in soil available N between the NPK and PK plots also exhibited no significant increase over time (Fig. 3a&4a). It might be possible that atmospheric nitrogen deposition in the NCP soil, N mineralization and lower nitrogen depletion may have attributed to the trend in available N over the years in the nitrogen omitted plots. Liu et al. (2006) and Zhang et al. (2008) reported that the bulk deposition of inorganic N in the NCP is approximately 30 kg/ha per year,

which has a significant effect on agricultural systems [19,20]. But yield gap significantly increased over time when the N was omitted (Fig. 2), indicating that the soil indigenous N supply combined with the deposition of N could not meet the N requirement for wheat and maize growth in the NCP.

Both of the wheat and maize yields had significant downward trends in P omission treatment (Fig. 1) and the yield gap of both crops significantly increased with the continuous omission of P fertilizer (Fig. 2), indicating that the cumulative absence of P fertilizers significantly inhibited wheat and maize growth. Tang

Table 4. Multi-year average soil nutrient status at wheat and maize harvest, as affected by various treatments with fertilizer in a wheat-maize rotation system at Yucheng, Shandong Province of China (mean±standard error).

Crops	Treatments	Organic C (g kg^{-1})	Total N (g kg^{-1})	Available N (mg kg^{-1})	Olsen-P (mg kg^{-1})	Available K (mg kg^{-1})
Wheat	Control	5.5±0.19 c	0.56±0.04b	52.5±5.15b	3.1±0.28c	104.5±5.50c
	NPK	6.5±0.21a	0.66±0.05a	63.3±4.35a	25.2±1.95b	226.1±16.0b
	NP	6.3±0.15ab	0.67±0.05a	62.7±4.07a	27.5±2.04b	91.6±8.72c
	NK	5.9±0.17abc	0.64±0.05a	62.5±4.17a	3.2±0.31c	333.4±40.3a
	PK	5.8±0.19bc	0.56±0.04ab	49.1±4.20b	40.8±3.98a	349.3±35.3a
Maize	Control	5.7±0.13c	0.66±0.02ab	56.9±6.11b	2.7±0.32c	106.5±5.22c
	NPK	6.7±0.15a	0.71±0.05a	65.8±6.45a	22.1±2.47b	283.7±24.6b
	NP	6.5±0.15ab	0.69±0.04ab	66.6±6.05a	21.7±1.78b	90.9±4.46c
	NK	6.0±0.13bc	0.66±0.04ab	68.4±6.76a	3.6±0.54c	423.8±43.7a
	PK	5.8±0.14bc	0.62±0.03b	52.1±5.90b	33.5±2.32a	438.8±39.8a

For each crop, different letters within a column indicate a significant difference at $P<0.05$ between fertilizer treatments.

et al. (2008) also reported that wheat and maize yields without P fertilization significantly decreased over time in several other sites in China [21]. The results are different from those reported under European conditions, most likely due to differences in the P supplying abilities of the different soils [22]. It is noted that the continuous omission of P fertilizer significantly reduced the soil Olsen-P concentration for both wheat and maize (Table 4). Meanwhile, the differences in the Olsen-P concentration between the NPK and NK plots significantly increased with continuous P omission (Fig. 3b&4b). The results indicate that the long-term absence of P fertilizer resulted in a great deficit in soil available P. It is interesting that the continuous omission of P fertilizer increased the available N concentration gap between the NPK and NK plots, and the effects were significant for maize (Fig. 4a), which indicates that a large amount of available N was depleted or lost, although the plots were fed N fertilizer inputs. The cumulative absence of P fertilizer intensified the shortage of N for crop growth. Because the treatments fertilized without P had a lower the number of cultivable microorganisms, microbial biomass and community functional diversity than in the treatments with P fertilization [23], which possibly induced the lower N and P mineralization in treatments fertilized without P.

The yield response to K fertilizer was much lower than that to N or P fertilizer (Table 3), most likely due to the high inherent soil K levels, which were in excess of the crop K demands. Most soils of the alluvial floodplain in Asia are high in K, and K is a rare limiting factor [24]. Shen et al. (2004) conducted a 14-yr field trial in Hebei Province of northern China and indicated that the application of N and P enhanced rice yields, while K had no yield-increasing effect due to the large resource of available soil K [25]. In our study, wheat and maize yield in the plots of combined NP and K fertilizer significantly improved, and the yield gap of maize significantly increased with continuous K omission; however, that of wheat yield did not (Fig. 1&2). Compared to that in the NPK plots, the soil available K concentration significantly decreased in the NP plots (Table 4). The differences in soil available K between NPK and NP also improved with the continuous omission of K, and the effects were significant during the wheat season (Fig. 3c). Although soil available K was gradually depleted, the soil indigenous K supply was sufficiently high to meet the requirement

for normal wheat growth over the 16-yr period of the study. If the K fertilizer was continuously omitted for a longer period, the soil available K deficit would possibly inhibit wheat growth, which must be further verified. In addition, the soil available K gap between the NPK and NP plots during maize growth season showed an increasing trend with the omission of K, but the differences were not significant (Fig. 4c). This behavior is because the long-term omission of K fertilizer significantly reduced the maize yield, which resulted in lower K uptake. The results show that long-term maize production is much more sensitive to the absence of K than wheat production in the NCP. The results are supported by those of Tan et al. (2007), who concluded that the effect of K fertilizer on maize was higher than that on wheat under the wheat-maize rotation system of Hebei [26].

In our study, the combination of N and P fertilizers mostly sustained soil organic carbon, total N and available N, P and K levels over time (Table 4). This result is similar to that of Shen et al. (2004), who reported that the soil organic carbon and total N concentrations remained stable over time [25]. Manna et al. (2005) also showed that the recommended NPK plots are adequate for maintaining a constant SOC content under the sub-humid and semi-arid tropical conditions of India over a long period [27]. When N fertilizer was omitted, soil available P and K significantly improved in the plots with P and K fertilizer (PK). Meanwhile, when P fertilizer was omitted, soil available K was significantly accumulated in the plots with N and K fertilizer (NK, Table 4). Because the deficiency of other nutrients inhibited crop production, both crops exhibited lower P or K uptake. The application of inorganic fertilizer allowed the crops to exceed production needs and thus resulted in a substantial build-up of available P or available K, which inevitably induced resource waste and ecological pollution.

Conclusions

The P fertilizer was more pivotal in the long-term growth of wheat and maize and the conservation of soil fertility in the NCP, although the N fertilizer input was important for the growth of both crops as well. Although the crop yield response to K fertilizer was much lower than to N or P fertilizer, the proper application of K fertilizer is also necessary, especially for maize production in the

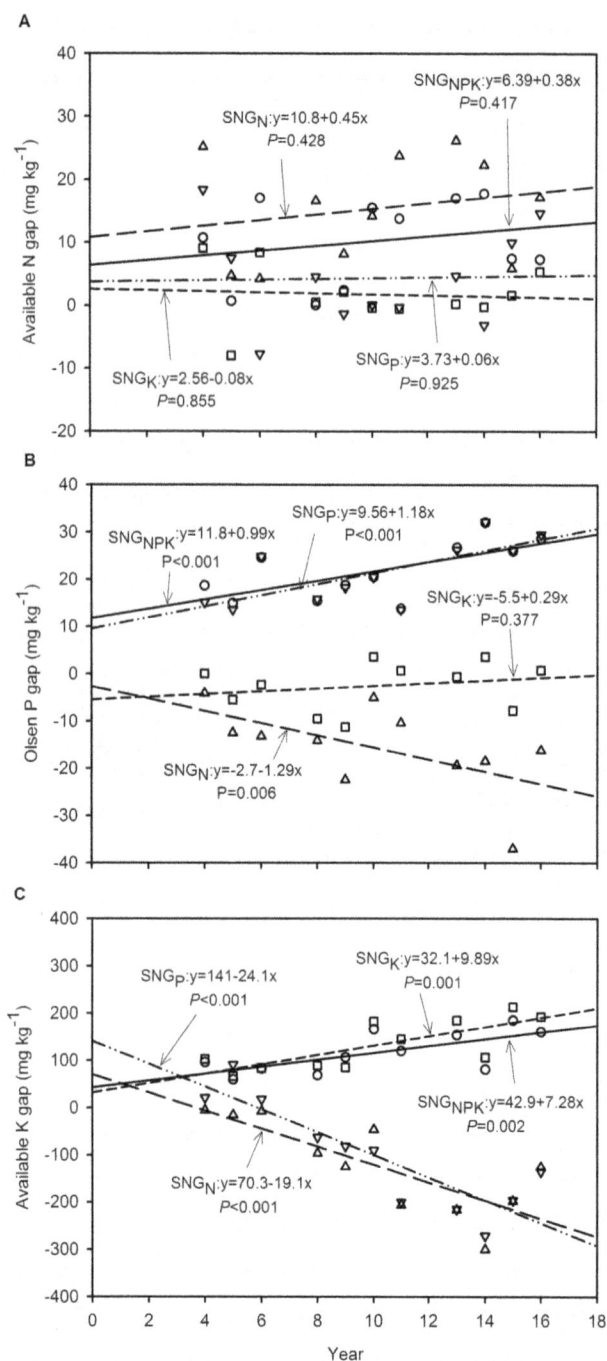

Figure 3. Soil available N (a), Olsen-P (b) and available K (c) gap variation by N (SNG$_N$, soil available nutrient difference between NPK and PK plots, triangle up), P (SNG$_P$, the difference between NPK and NK plots, triangle down), K (SNG$_K$, the difference between NPK and NP plots, square) or NPK (SNG$_{NPK}$, the difference between NPK and Control plots, circle) fertilizer omission from 1994 to 2006 during wheat season at Yucheng, Shandong Province of China.

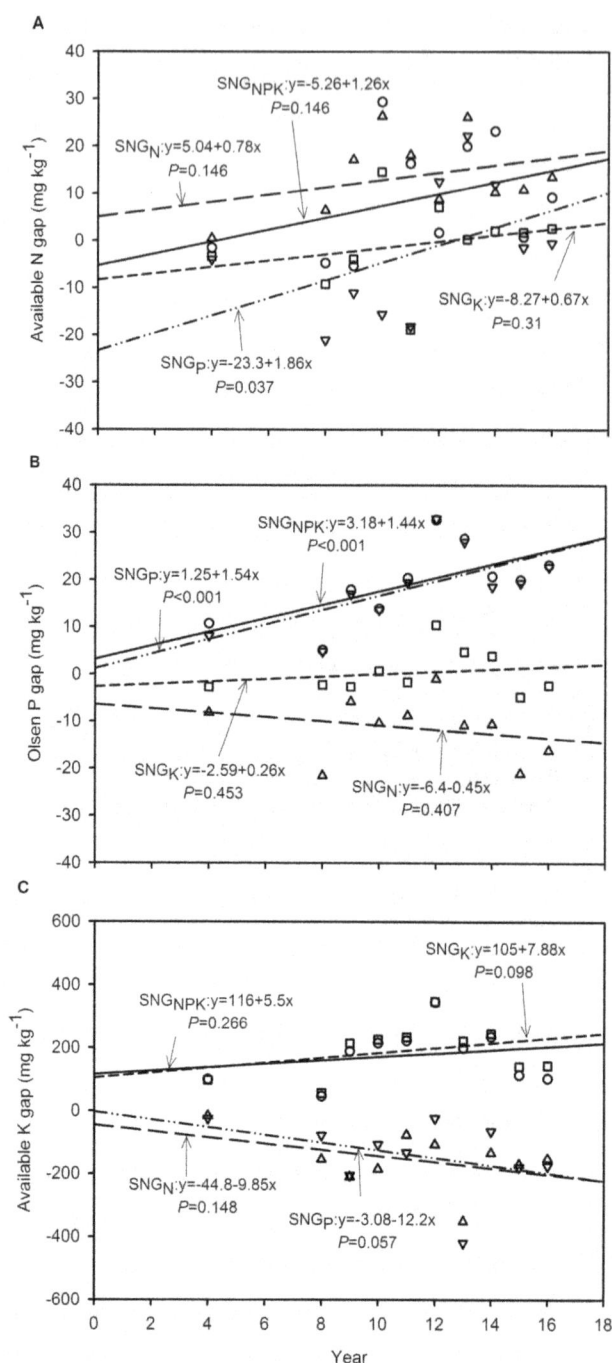

Figure 4. Soil available N (a), Olsen-P (b) and available K (c) gap variation by N (SNG$_N$, soil available nutrient difference between NPK and PK plots, triangle up), P (SNG$_P$, the difference between NPK and NK plots, triangle down), K (SNG$_K$, the difference between NPK and NP plots, square) or NPK (SNG$_{NPK}$, the difference between NPK and Control plots, circle) fertilizer omission from 1994 to 2006 during maize season at Yucheng, Shandong Province of China.

region. Thus, the appropriate application of N and P fertilizers for both crops, in combination with regular K fertilizers for maize, is absolutely necessary in terms of sustainable crop production in the NCP. However, a longer-range study is required to verify whether the soil indigenous K supply could continue to meet the requirement for wheat growth over many years.

Acknowledgments

The authors are grateful to Zhenrong Tian in the Yucheng Comprehensive Experimental Station of the Chinese Academy of Sciences (CAS) for many sampling work and Lynn M. Johnson in Cornell University for statistical analysis. Authors also thank the academic editor and anonymous reviewers for their constructive comments, which helped in improving the manuscript.

References

1. Yu Q, Saseendran SA, Ma L, Flerchinger GN, Green TR, et al. (2006) Modeling a wheat-maize double cropping system in China using two plant growth modules in RZWQM. Agr Syst 89: 457–477.
2. China Statistics Bureau (2001) China Statistics Bureau, Annual Agricultural Statistics of Henan Province. Beijing, China Statistics Press. 112–115.
3. Vitousek PM, Naylor R, Crews T, David MB, Drinkwater LE, et al. (2009) Nutrient imbalances in agricultural development. Science 324: 1519–1520.
4. Berzsenyi Z, Győrffy B, Lap D (2000) Effect of crop rotation and fertilization on maize and wheat yields and yield stability in a long-term experiment. Eur J Agron 13: 225–244.
5. Bi LD, Zhang B, Liu GR, Li ZZ, Liu YR, et al. (2009) Long-term effects of organic amendments on the rice yields for double rice cropping systems in subtropical China. Agr Ecosyst Environ 129: 534–541.
6. Cai ZC, Qin SW (2006) Dynamics of crop yields and soil organic carbon in a long-term fertilization experiment in the Huang-Huai-Hai Plain of China. Geoderma 136: 708–715.
7. Fan T, Stewart BA, Wang Y, Luo J, Zhou G.(2005) Long-term fertilization effects on grain yield, water-use efficiency and soil fertility in the dryland of Loess Plateau in China. Agr Ecosyst Environ 106: 313–329.
8. Glendining MJ, Powlson DS, Poulton PR, Bradbury NJ, Palazzo D, et al. (1996) The effects of long-term applications of inorganic nitrogen fertilizer on soil nitrogen in the Broadbalk wheat experiment. J Agr Sci 127: 347–363.
9. Gong W, Yan XY, Wang JY, Hu TX, Gong YB (2009) Long-term manuring and fertilization effects on soil organic carbon pools under a wheat-maize cropping system in North China Plain. Plant Soil 314: 67–76.
10. Jiang D, Hengsdijk H, Dai T, de Boer W, Jing Q, et al. (2006) Long-term effects of manure and inorganic fertilizers on yield and soil fertility for a winter wheat-maize system in Jiangsu, China. Pedosphere16: 25–32.
11. Lu R, Shi Z (1998) Effect of long-term fertilization on soil properties. In: Lu R, Xie J, Cai G, Zhu Q (2010) editors. Soil-plant nutrients principles and fertilizer. Beijing: Chemical Industry Press. 102–110.
12. Wang YC, Wang E, Wang DL, Huang SM, Ma YB, et al. (2010) Crop productivity and nutrient use efficiency as affected by long-term fertilization in North China Plain. Nutr Cycl Agroecosys 86: 105–119.
13. Yan XY, Gong W (2010) The role of chemical and organic fertilizers on yield, yield variability and carbon sequestration – results of a 19-year experiment. Plant Soil 331: 471–480.
14. Zhang H, Xu M, Zhang F (2009) Long-term effects of manure application on grain yield under different cropping systems and ecological conditions in China. J Agr Sci 147: 31–42.
15. Mussgnug F, Becker M, Son TT, Buresh RJ, Vlek PLG (2006) Yield gaps and nutrient balances in intensive, rice-based cropping systems on degraded soils in the Red River Delta of Vietnam. Field Crop Res 98: 127–140.
16. Mueller ND, Gerber JS, Johnston M, Ray DK, Ramankutty N, et al. (2012) Closing yield gaps through nutrient and water management. Nature 490: 254–257.
17. Page AL, Miller RH, Keeney DR (1982) Methods of soil analysis. Part 2. Madison, WI: American Society of Agronomy. 539–871.
18. Olsen SR, Cole CV, Watanabe FS, Dean LA (1954) Estimation of available phosphorus in soils by extraction with sodium bicarbonate. USDA Circular 939.
19. Liu X, Ju X, Zhang Y, He C, Kopsch J, et al. (2006) Nitrogen deposition in agroecosystems in the Beijing area. Agr Ecosyst Environ 113: 370–377.
20. Zhang Y, Liu XJ, Fangmeier A, Goulding KTW, Zhang FS (2008) Nitrogen inputs and isotopes in precipitation in the North China Plain. Atmos Environ 42: 1436–1448.
21. Tang X, Li J, Ma Y, Hao X, Li X (2008) Phosphorus efficiency in long-term (15-years) wheat-maize cropping systems with various soil and climate conditions. Field Crop Res 108: 231–237.
22. Blake L, Mercik S, Koerschens M, Moskal S, Poulton PR, et al. (2000) Phosphorus content in soil, uptake by plants and balance in three European long-term field experiments. Nutr Cycl Agroecosys 56: 263–275.
23. Zhong W, Cai Z (2007) Long-term effects of inorganic fertilizers on microbial biomass and community functional diversity in a paddy soil derived from quaternary red clay. Appl Soil Ecol 36: 84–91.
24. Bajwa MI (1994) Soil K status, K fertilizer usage and recommendation in Pakistan. Potash Review No. 3/1994. Subject 1, 20th Suite. Basel: International Potash Institute. 67.
25. Shen J, Li R, Zhang F, Fan J, Tang C, et al. (2004) Crop yields, soil fertility and phosphorus fractions in response to long-term fertilization under the rice monoculture system on a calcareous soil. Field Crop Res 86: 225–238.
26. Tan D, Jin J, Huang S, Li S, He P (2007) Effect of long-term application of K fertilizer and wheat straw to soil on crop yield and soil K under different planting systems. Agr Sci China 6: 200–207.
27. Manna MC, Swarup A, Wanjari RH, Ravankar HN, Mishra B, et al. (2005) Long-term effect of fertilizer and manure application on soil organic carbon storage, soil quality and yield sustainability under sub-humid and semi-arid tropical India. Field Crop Res 93: 264–280.

Author Contributions

Conceived and designed the experiments: ZO YL. Performed the experiments: XD YL. Analyzed the data: XD HW. Wrote the paper: XD.

19

Evaluating Status Change of Soil Potassium from Path Model

Wenming He[1,2]**, Fang Chen**[1,3]*****

1 Key Laboratory of Aquatic Botany and Watershed Ecology, Wuhan Botanical Garden, Chinese Academy of Sciences, Moshan, Wuchang, Wuhan, Hubei Province, China, **2** Graduate University of Chinese Academy of Sciences, Beijing, China, **3** International Plant Nutrition Institute, Wuhan, China

Abstract

The purpose of this study is to determine critical environmental parameters of soil K availability and to quantify those contributors by using a proposed path model. In this study, plot experiments were designed into different treatments, and soil samples were collected and further analyzed in laboratory to investigate soil properties influence on soil potassium forms (water soluble K, exchangeable K, non-exchangeable K). Furthermore, path analysis based on proposed path model was carried out to evaluate the relationship between potassium forms and soil properties. Research findings were achieved as followings. Firstly, key direct factors were soil S, ratio of sodium-potassium (Na/K), the chemical index of alteration (CIA), Soil Organic Matter in soil solution (SOM), Na and total nitrogen in soil solution (TN), and key indirect factors were Carbonate (CO_3), Mg, pH, Na, S, and SOM. Secondly, path model can effectively determine direction and quantities of potassium status changes between Exchangeable potassium (eK), Non-exchangeable potassium (neK) and water-soluble potassium (wsK) under influences of specific environmental parameters. In reversible equilibrium state of $wsK \underset{\beta}{\overset{\alpha}{\rightleftarrows}} neK \underset{\chi}{\overset{\gamma}{\rightleftarrows}} eK$, K balance state was inclined to be moved into β and χ directions in treatments of potassium shortage. However in reversible equilibrium of $wsK \underset{\theta}{\overset{\varepsilon}{\rightleftarrows}} eK \underset{\omega}{\overset{\lambda}{\rightleftarrows}} neK$, K balance state was inclined to be moved into θ and λ directions in treatments of water shortage. Results showed that the proposed path model was able to quantitatively disclose moving direction of K status and quantify its equilibrium threshold. It provided a theoretical and practical basis for scientific and effective fertilization in agricultural plants growth.

Editor: Raffaella Balestrini, Institute for Plant Protection (IPP), CNR, Italy

Funding: The study was supported by the National Natural Science Foundation of China (41171243), the National Key Technology R&D Program (2012BAD15B01) and the Cooperated Program with the International Plant Nutrition Institute (IPNI-HB-33). The funders had no role in study design, data collection and analysis, decision to publish, or preparation of the manuscript.

Competing Interests: The authors have declared that no competing interests exist.

* E-mail: fchen@ipni.ac.cn

Introduction

The status and transformation of potassium in the soil is of great significance to crop growth [1,2]. Investigations on potassium status gained a numerous achievements[3,4]. The effectiveness of potassium in soil is controlled by four forms, e.g., mineral potassium, non-exchangeable potassium (neK), exchangeable potassium (eK) and water-soluble potassium (wsK) which can be transformed into each other [5–8]. Absorption capacity of crops and application of farming fertilizer have influence on the potassium forms, thus impact on the release and fixation of potassium in the soil [9]. Within these different forms of soil potassium, there exists a complex and dynamic chemical balance, and it greatly depends on the situation of each form and its environment conditions [10–13]. However, there are few experimental studies on ternary systems involving K^+ in spite of the plant nutritional importance of K^+ in soils [3,5,14]. Research indicated that the amount of K ions retained on the solid phase of soil was much larger than that dissolved in the soil solution. It suggested that a root segment become easily permeable to Ca^{2+} followed by Na^+ and K^+ with aging [15,16]. Thus, it is indispensable to take into account the cation exchange processes

in modeling the response of the soil solution to fertilizer application and transport of cationic solutes in soils.

The transform of soil non-exchangeable potassium to soil exchangeable potassium and soil soluble potassium is a slowly process. It is hard to be determined by using the routine method of ion exchange, although it shows that sodium tetraphenylborate method is more accurate than conventional methods to reflect changes in soil potassium through long-term field fixed experiments of potash fertilization [17,18]. Path analysis is a statistical technique that distinguishes coefficient and causation by partitioning coefficients into direct and indirect effects. Researchers used path analysis to analysis phosphorus retention capacity in allophanic and non-allophanic andisols and soil organic matter effects on phosphorus sorption is successful [19,20]. The method of path coefficients, which was proposed by Wright (1921), was effective in disclosing relationships between variables by diagrams [21–25]. However, path coefficients approach has not yet applied to the study on soil potassium. The potassium absorb efficiency of plant root is affected by multiple factors [26], so it is not suitable to make estimation by use a single parameter [27–29]. The above mentioned studies focused on qualitatively descriptive analysis of single factor or multifactor, and failed to quantitatively disclose the dynamic process of potassium status [30].

In this study, we conducted soil K ions exchange experiments for soil samples and further calculated their selectivity coefficients, and explored potassium status on basis of path model to determine dynamic changes of potassium status and specific environmental parameters. The model assumptions were related to the adsorption mechanism at molecular level. The first step, we attempted to examine different indicators of potassium in the soils, and then to analyze direct and indirect correlations between water-soluble potassium (wsK), non-exchangeable potassium (neK) and exchangeable potassium (eK), and further to calculate path coefficients of different potassium status, finally to construct path model for describing the changes of potassium status. The model was calibrated and validated by using 48 cotton rhizosphere/nonrhizosphere soils samples to test its accuracy. We quantitatively investigated on interaction process of plant-soil-microorganisms and relationship between environmental parameters and potassium release through path coefficient model, and the model was used to predict the multi-component ion exchange equilibrium in soil.

Materials and Methods

Hereby, I, along with coauthor, confirm that no specific permissions were required for our experiment locations/activities since this experiment field belongs to our institute and for scientific research only. And the field studies did not involve endangered or protected species.

In this study, meadow soil was selected as experimental agents in order to investigate potassium nutrient supply and physiological mechanism (Table 1). Rhizosphere soil samples were collected from 90 days cotton plant. When collecting soil samples, we firstly loosed root zone soil to collect rhizosphere soil, and uprooted whole plant cotton, and then gently shook off root zone soil, and finally get root surface adhesion soil. The non-rhizosphere soil samples were collected 10–15 cm depth from surface.

For purpose of this study, we proposed a detail research scheme to evaluate status changes of soil potassium from path model (Fig. 1). We conducted soil K ions exchange experiments for soil samples, and further calculated their selectivity coefficients, and explored potassium status on basis of path model to determine dynamic changes of potassium status and specific environmental parameters. Soil samples were **N**on-**R**hizosphere soil in **OPT**imum of **K** of **H**igh efficiency genotype cotton (NROPTH); **N**on-**R**hizosphere soil in **OPT**imum of **K** of **L**ow efficiency genotype

cotton (NROPTL); **N**on-**R**hizosphere soil in **S**hortage of **K** of **H**igh efficiency genotype cotton (NRSKH); **N**on-**R**hizosphere soil in **S**hortage of **K** of **L**ow efficiency genotype cotton (NRSKL); **N**on-**R**hizosphere soil in **S**hortage of **W**ater of **H**igh efficiency genotype cotton (NRSWH); **N**on-**R**hizosphere soil in **S**hortage of **W**ater of **L**ow efficiency genotype cotton (NRSWL); **N**on-**R**hizosphere soil in **S**hortage of **W**ater and **K** of **H**igh efficiency genotype cotton (NRSWKH); **N**on-**R**hizosphere soil in **S**hortage of **W**ater and **K** of **L**ow efficiency genotype cotton (NRSWKL); **R**hizosphere soil in **OPT**imum of **K** of **H**igh efficiency genotype cotton (ROPTH); **R**hizosphere soil in **OPT**imum of **K** of **L**ow efficiency genotype cotton (ROPTL); **R**hizosphere soil in **S**hortage of **K** of **H**igh efficiency genotype cotton (RSKH); **R**hizosphere soil in **S**hortage of **K** of **L**ow efficiency genotype cotton (RSKL); **R**hizosphere soil in **S**hortage of **W**ater of **H**igh efficiency genotype cotton (RSWH); **R**hizosphere soil in **S**hortage of **W**ater of **L**ow efficiency genotype cotton (RSWL); **R**hizosphere soil in **S**hortage of **W**ater and **K** of **H**igh efficiency genotype cotton (RSWKH); **R**hizosphere soil in **S**hortage of **W**ater and **K** of **L**ow efficiency genotype cotton (RSWKL). Firstly, we divided experiment soil into 8 treatments, and then implemented 256 soil samples test (Fig. 2). We measured 33 parameters of the different attributes of 256 soil samples. Those parameters were selected based on standards of Hashimoto and Kang research methods of phosphorus[19,31]. Humic acid (NHA) in soil which can be extracted by NaOH solution (0.1 mol·L^{-1}, pH = 3), humic acid (PHA) in soil which can be extracted by $Na_2P_2O_4$ solution (0.05 mol·L^{-1}, pH = 9.2), humus linked to iron (HMi) and humic linked to clay (HMc). Key experimental sampling test datasets were listed in Table 2–Table 6. Secondly, statistical analysis was further implemented to investigate relationship between environmental parameters and potassium status. Datasets were normalized, followed by correlation analysis (Table 7), 8 out of 16 parameters were selected thus to determine the most important parameters. Path coefficients of absolute value of represented the size effect on potassium morphology change. The size of the "+" meant the same as the arrow direction with arrow, "−" represents in contrast to the arrow direction with arrow. Finally, path model of potassium status was constructed on basis of direct and indirect correlated parameters. In this model, "e1" represents unknown variable and its impact factor, the straight line with arrows stands for direct impact factor, and double arc arrow is direction of interaction between parameters.

Table 1. Properties of mead soil for experiments.

Soil properties		Average value
Mean compact density (g·cm^{-3})		1.54
Particle composition	Clay (<0.002 mm, %)	16.8
	Sand (2-0.05 mm, %)	61.7
	Silt (0.05-0.002 mm, %)	21.4
Mineral composition	Smectite (%)	4.0
	Vermienlite (%)	25.0
	Intergrade mineral hydromica (1.4 nm, %)	49.0
	Kaolinite (%)	19.0
Cations composition	exchangeable Ca (mmol·kg^{-1})	35.4
	exchangeable Mg (mmol·kg^{-1})	18.1
	exchangeable Na (mmol·kg^{-1})	2.0
	FeO (noncrystalline iron extracted with Tamm's solution, g·kg^{-1})	49.0

Laboratory Measurement of Different Parameters of Rhizosphere Soil of Cotton

Measurement of total contents of soil elements (by X-ray fluorescence, XRF): Tab. *2* K, Mg, Na, S, Al, Si, Fe, Ca, P, S, Cl, CO_3.

Measurement of different forms of potassium: (6) Tab. *3* water soluble potassium (wsK), exchangeable potassium (eK), non-exchangeable potassium (neK).
Step 1: Soil solution preparation: water soluble potassium leaching agent is secondary distilled water, neK of leaching agent is 1.0 mol/L NH4OAc solution, eK of leaching agent is 1.0 mol/L HNO3 solution.
Step 2: To measure wsK, eK and neK of the above soil solution at normal temperature.
Step 3: To measure wsK, eK and neK after preprocessing as followings,
a) Firstly, soil is heated up to 600 ^{0}C for 6 hours,
b) Secondly, heated soil is cooled down till normal temperature
c) Thirdly, soil is mixed with 20% humic acid and laid down for 45 days.
d) Fourthly, after 45 days, make preparation for soil solution following step 1.
e) Finally, to measure wsK, eK and neK of the above soil solution.

Measurement of silicate dissolving potassium bacteria: Tab *4*
1 To isolate the silicate-dissolving bacteria of the soil,
2 Being incubated in the culture medium with crystal of silicon dioxide.
3 To measure the concentration of active silicon ions in the culture medium after different culturing time, to select the bacterial strains of the strongest silicate-dissolving ability with silicon molybdenum blue spectrophotometry.
4 Strong silicate-dissolving potassium bacteria are gained and can be measured.

Soil organic matter (SOM): to be measured by potassium dichromate volumetric method. Tab. *5*

Measurement of Total Carbon (TC), Total nitrogen (TN): (2) Tab. *5* (1) When testing TC and TN in soil solution, deionized water and soil sample (screened by 0.047mm<R<0.5mm sieve) are mixed and lixiviated according to the ratio 20ml: 14g, (2) And then (after 36h), it was filtered by filter membrane (Whatman GF/C-1822-047-0.45μm). The filtrate is measured at 800^{0}C using XTA module of Multi N/C 2100.

Measurement of humic acid: (4) Tab. *5* humic acid from NaOH (NHA), humic acid-iron complex (HMi), humic acid from sodium pyrophosphate (PHA), humic linked to clay (HMc). Humic acid extraction is based on Pallo's method.

Indices analysis: Tab. *6* Na/K, index of compositional variability (ICV), weathering and leaching coefficients (BA), Silica sesquioxide ratio (SAF), Oxidation Reduction Potential (ORP, water, 25^{0}C), pH (water, 25^{0}C), CIA, chemical index of alteration.

Statistical analysis of experiment data: normalization, correlation analysis, coefficient analysis

Path Model simulation of exchange of different forms of potassium

Equilibrium shift analysis of different forms of potassium

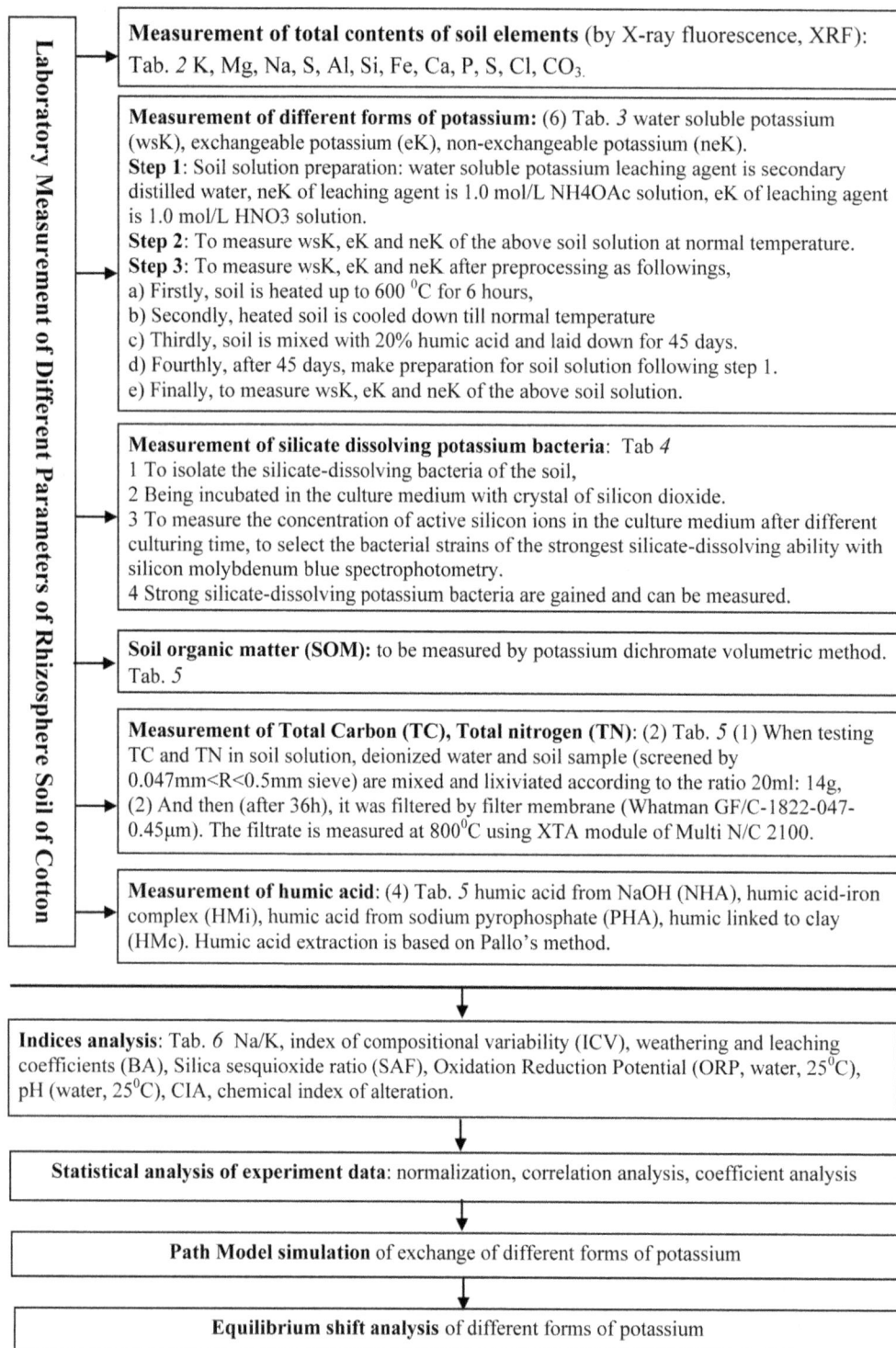

Figure 1. Workflow of evaluating status changes of soil potassium from Path Model.

Results

Path analysis of water-soluble potassium (wsK)

For purpose of path analysis of water-soluble potassium, direct and indirect impact factors of the path analysis were derived from multiple linear regressions coefficients of wsK, and correlation coefficients between soil properties. Direct coefficient variables of wsK include neK, eK, PHA, SOM, S, Na and Na/K; while indirect coefficient variables of wsK include CIA, Na, Mg, CO_3 and pH. Impact factor of path size of water-soluble potassium was determined by pH, CIA, Oxidation Reduction Potential (ORP), S, Na (Fig. 3 and Table 8). In the path model of wsK, direct effects of soil properties on the normalized value of wsK (ZwsK) were represented by single-headed arrows, while coefficients between soil properties were represented by double-headed arrows. Direct and indirect effects were indicated by value and marked with "±".

	OPT	SW	SK	SWK
HE&HP	Soil moisture (35%) KCl (6.83g)	Soil moisture (25%) KCl (6.83 g)	Soil moisture (35%) KCl (0g)	Soil moisture (25%) KCl (0g)
	Soil moisture (35%) KCl (6.83g)	Soil moisture (25%) KCl (6.83 g)	Soil moisture (35%) KCl (0g)	Soil moisture (25%) KCl (0g)
	Soil moisture (35%) KCl (6.83a)	Soil moisture (25%) KCl (6.83 g)	Soil moisture (35%) KCl (0g)	Soil moisture (25%) KCl (0g)

	OPT	SW	SK	SWK
LE&LP	Soil moisture (35%) KCl (6.83g)	Soil moisture (25%) KCl (6.83 g)	Soil moisture (35%) KCl (0g)	Soil moisture (25%) KCl (0g)
	Soil moisture (35%) KCl (6.83g)	Soil moisture (25%) KCl (6.83 g)	Soil moisture (35%) KCl (0g)	Soil moisture (25%) KCl (0g)
	Soil moisture (35%) KCl (6.83a)	Soil moisture (25%) KCl (6.83 g)	Soil moisture (35%) KCl (0g)	Soil moisture (25%) KCl (0g)

Figure 2. Diagram of experimental design to evaluate status changes of soil potassium.

Table 2. Total contents of soil elements of Laboratory measurement for different parameters of 256 soil samples (R, rhizosphere soil; NR, non-rhizosphere).

Sample	Na%	Mg%	Al%	Si%	Fe%	Mg%	Ca%	Pppm	CO$_3$%	Sppm	Clppm
NR1-16	0.44	0.39	6.49	36.86	3.35	1.49	0.37	491.01	49.74	219.34	119.8
R17-32	0.45	0.37	6.53	37.16	3.34	1.49	0.39	507.57	49.38	243.65	228.2
NR33-48	0.44	0.39	6.5	36.92	3.33	1.49	0.38	494.36	49.72	218.56	122.07
R49-64	0.44	0.39	6.51	36.94	3.34	1.49	0.39	483.04	49.64	251.93	217.78
N65-80	0.44	0.39	6.46	36.93	3.32	1.47	0.36	488.97	49.77	220.97	68.15
R81-96	0.47	0.42	6.3	36.38	3.29	1.46	0.4	515.89	50.45	248.77	80.32
NR97-112	0.44	0.38	6.49	37.04	3.35	1.48	0.37	506.34	49.62	218.38	82.48
R113-128	0.44	0.38	6.53	37.09	3.35	1.48	0.4	518.35	49.48	243.94	89.5
NR129-144	0.44	0.39	6.38	36.84	3.29	1.48	0.36	481.3	49.95	224.13	152.4
R145-160	0.45	0.39	6.39	36.94	3.27	1.47	0.38	512.73	49.84	237.7	201.2
NR161-176	0.43	0.38	6.49	37.09	3.32	1.49	0.37	484.59	49.58	225.39	142.88
R177-192	0.45	0.38	6.51	37.01	3.33	1.48	0.39	496.2	49.57	236.79	189.7
NR193-208	0.44	0.39	6.43	36.92	3.31	1.47	0.36	502.75	49.83	223.08	71.6
R209-224	0.45	0.38	6.43	37.08	3.32	1.46	0.39	491.22	49.64	245.74	81.48
NR225-240	0.44	0.38	6.46	37	3.31	1.47	0.38	509.11	49.7	228.9	90.5
R241-256	0.44	0.37	6.46	37.28	3.34	1.47	0.39	498.72	49.4	239.9	87.4

Table 3. Different forms of potassium of Laboratory measurement for different parameters of 256 soil samples (R, rhizosphere soil; NR, non-rhizosphere).

Sample	K(g/Kg)	wsK(ug/g)	neK(ug/g)	eK(ug/g)	wsK(ug/g, (weathered)	neK(ug/g) (weathered)	eK(ug/g) weathered
NR1-16	14.93	27.1	134.9	58.64	67.57	174.48	62.9
R17-32	14.9	12.03	54.21	14.32	59.36	155.15	66.59
NR33-48	14.9	20.01	134.13	50.84	76.97	177.27	55.87
R49-64	14.87	9.22	58.27	12.21	123.72	170.67	71.24
N65-80	14.7	24.8	121.55	40.27	53.02	190.22	59.93
R81-96	14.57	5.65	39.17	6.28	61.31	153.27	57.35
NR97-112	14.75	41.28	101.36	27.46	56.24	167.15	69.44
R113-128	14.8	6.05	42.83	5.23	58.41	149.91	52.24
NR129-144	14.77	38.55	158.29	56.41	68.09	190.55	65.03
R145-160	14.73	15.92	78.62	23.18	60.89	167.13	52.79
NR161-176	14.87	88.78	162.24	61.11	70.46	181.02	55.25
R177-192	14.83	18.61	70.07	19.27	64.82	185.07	59.89
NR193-208	14.73	13.26	99.74	30.21	63.65	171.35	62.95
R209-224	14.63	6.41	38.74	7.63	61.96	162.21	52.09
NR225-240	14.73	25.27	105.8	29.54	64.59	182.24	59.77
R241-256	14.73	6.56	44.8	5.31	72.96	154.08	60.7

The direct effects of soil properties on the ZwsK were termed path coefficients and were standardized partial regression coefficients for each of the soil properties in the multiple linear regression against the ZwsK. Indirect effects of soil properties on the ZwsK were determined from the product of the simple coefficient between soil properties and the path coefficient (i.e., one two-headed arrow and one single-headed arrow). The coefficient between the ZwsK and soil property was the sum of the entire path connecting two variables, as described by

$$r_{neK,wsK} = P_{wsK,neK} + r_{neK,PHA} \cdot P_{wsK,PHA} + r_{neK,S} \cdot P_{wsK,S} + r_{neK,Na/K} \cdot P_{wsK,Na/K} + r_{neK,CIA} \cdot P_{wsK,CIA} + r_{neK,Na} \cdot P_{wsK,Na} + r_{neK,SOM} \cdot P_{wsK,SOM} + r_{neK,eK} \cdot P_{wsK,eK} \quad (1)$$

Table 4. Silicate dissolving potassium bacteria of Laboratory measurement for different parameters of 256 soil samples (R, rhizosphere soil; NR, non-rhizosphere).

Sample	Bacteria of seedling stage ($\times 10^4$ CFUg^{-1})	Bacteria of budding stage ($\times 10^4$ CFUg^{-1})	Bacteria of wadding stage ($\times 10^4$ CFUg^{-1})
NR1-16	0.8	2.1	1.6
R17-32	1.5	5.2	1.4
NR33-48	0.7	4.3	1.7
R49-64	1.1	3.4	1.9
N65-80	0.9	17	14
R81-96	1.0	87	30
NR97-112	1.2	26	19
R113-128	1.9	39	24
NR129-144	0.1	1.2	0.5
R145-160	0.4	1.3	0.7
NR161-176	0.9	4.3	2.2
R177-192	0.3	1.7	0.6
NR193-208	0.8	1.9	1.6
R209-224	0.7	7.1	4.0
NR225-240	0.1	2.3	0.9
R241-256	1.2	6.3	2.8

Table 5. Soil organic matter, humic acid, Total Carbon (TC), Total nitrogen (TN) of Laboratory measurement for different parameters of 256 soil samples (R, rhizosphere soil; NR, non-rhizosphere).

Sample	TC%	TN%	SOM(%)	PHA%	NHA%	Hmi%	HMc%
NR1-16	4.51E-03	1.71E-03	7.88	4.53	2.59	1.47	0.92
R17-32	4.41E-03	5.74E-03	5.67	4.84	3.52	0.33	0.33
NR33-48	5.34E-03	2.02E-03	7.32	3.48	3.67	0.42	0.37
R49-64	3.74E-03	1.04E-02	5.54	4.11	2.55	0.45	0.32
N65-80	4.50E-03	1.64E-03	8.01	5.33	3.89	1.35	1.24
R81-96	3.64E-03	4.24E-03	5.79	5.26	4.58	0.42	0.23
NR97-112	3.01E-03	3.93E-03	8.53	3.62	3.28	0.66	0.41
R113-128	2.80E-03	1.31E-02	5.70	4.41	3.68	0.46	0.27
NR129-144	3.52E-03	3.46E-03	8.30	4.46	3.70	0.35	0.28
R145-160	3.22E-03	2.02E-02	5.38	5.00	3.82	0.30	0.30
NR161-176	3.18E-03	3.97E-03	7.84	4.90	1.75	0.88	0.45
R177-192	3.12E-03	2.31E-02	6.79	6.89	3.95	0.34	0.26
NR193-208	3.57E-03	3.24E-03	8.45	4.57	3.82	0.25	0.19
R209-224	3.79E-03	2.67E-02	5.79	5.43	4.22	0.44	0.36
NR225-240	4.30E-03	3.83E-03	8.01	4.94	4.61	0.52	0.32
R241-256	4.37E-03	2.75E-02	6.23	4.70	4.08	0.36	0.18

Table 6. Weathering indices of soil minerals of Laboratory measurement for different parameters of 256 soil samples (R, rhizosphere soil; NR, non-rhizosphere).

Sample	ORP(eV)	pH	ba	Na/K	ICA	saf	CIA
NR1-16	17.30	7.27	0.45	0.29	1.03	8.73	76.08
R17-32	18.65	6.84	0.44	0.30	1.03	8.77	75.82
NR33-48	13.80	7.20	0.45	0.29	1.03	8.75	76.02
R49-64	30.24	6.5	0.45	0.30	1.04	8.74	75.95
N65-80	22.66	6.79	0.45	0.30	1.03	8.8	76.18
R81-96	22.68	6.73	0.48	0.32	1.08	8.86	75.08
NR97-112	10.22	6.94	0.44	0.30	1.03	8.78	76.21
R113-128	65.80	6.09	0.45	0.30	1.03	8.74	76
NR129-144	18.50	6.93	0.45	0.30	1.04	8.88	75.88
R145-160	80.84	5.79	0.46	0.31	1.04	8.90	75.67
NR161-176	31.46	6.48	0.44	0.29	1.02	8.80	76.16
R177-192	62.82	5.85	0.45	0.30	1.03	8.75	75.92
NR193-208	19.33	6.87	0.45	0.30	1.03	8.83	76.06
R209-224	52.48	6.19	0.45	0.31	1.04	8.86	75.76
NR225-240	13.72	6.98	0.45	0.30	1.03	8.83	75.99
R241-256	20.54	6.86	0.45	0.30	1.04	8.87	75.84

where, r_{ij} is the simple correlation between the ZwsK and a parameter of soil property, P_{ij} is the path coefficient between the ZwsK and a parameter of soil property, and $r_{ij}P_{ij}$ is the indirect effect of a parameter of on the ZwsK. An uncorrelated residue (e_{wsk}) that represented the unexplained part of an observed variable in the path model was

$$e_{wsk} = (1 - R_{wsk}^2)^{0.5} = 0.2 \qquad (2)$$

where, R_{wsk}^2 is the coefficient of determination of the multiple regression equation between the ZwsK and the eight soil properties. Those parameters of soil properties were neK, PHA, S, Na/K, CIA, Na, SOM, eK. Backward regression analysis was performed with the ZwsK as the dependent variable and the eight soil properties were regarded as independent variables. Backward regression was a multiple regression procedure in which all the independent variables were entered into the regression equation at the beginning, and variables that did not contribute significantly to the fit of regression model were successively eliminated until only statistically significant variables remain. Fig. 4 and Fig. 5 were similar expressions.

In the path of water-soluble potassium, path contribution coefficients were quite different. Coefficient values were 2.10, −1.52 and 0.20 for wsK to neK, neK to wsK, and other variables to wsK respectively. It disclosed that most part of wsK was changed into neK (with values of 2.10). In the path of neK to wsK, coefficient was negative (with value of −1.52); it indicated that there was an inverse change from wsK to neK, a lot of wsK was changed into neK. Those dynamic changes indicated that wsK and eK reached equilibrium in soil solution, and this equilibrium represented its quick reaction process.

These changes impacted greatly on nutrients dynamic equilibrium of rhizosphere soil. This changed nutrients recycling process of the soil environment since continuous-release of humic acid had

direct effects on absorption and uptakes of those variables including neK, eK and PHA, SOM, S, Na and Na/K. It was clear that rhizosphere effects were likely to result in changes in root exudates, pH and ORP (eV). Consequently, it led to changes in pH values and population, quality and activity of microorganisms while pH change was caused by root exuded organic acids. Under conditions of different pH values, humic acid significantly correlated to adsorption and desorption of potassium. Humic acid of potassium adsorption and desorption presented an upward trend when it increased in initial concentration (pH was 4.0–8.0), however, desorption rate declined. The migration distance also showed very significant linear relationship within ranges of migrations between water-soluble potassium and exchangeable potassium. Organic matter nutrients and chemical efficiency of humic acid altered different variables such as physical, chemical, and biotic in root-soil interface.

Path analysis of non-exchangeable potassium (neK)

For path analysis of non-exchangeable potassium, we further investigated its direct and indirect impact factors. Results showed that direct coefficient variables of neK include eK, wsK, SOM, S, Na, Na/K, TN, CIA; and indirect coefficient variables of neK include Mg, CO_3. Impact factor of path size of neK was determined by eK, CO3, pH (Fig. 4 Table 9). Table 9 showed there was significant correlation between eK and neK (with coefficient of 0.81). In the Path of non-exchangeable potassium, contribution coefficient was 0.01 for wsK to neK, although wsK was determined by general effects of N, Na, S and CIA. The SOM was also an important contributor to eK and wsK of rhizosphere soil. Coefficients were 0.69 and 0.55 between SOM and eK, SOM and wsK, respectively.

The correlation between the ZneK and a soil property was the sum of the entire path connecting two variables, as described by

Table 7. Correlation analysis of environmental parameters of soil potassium.

	Z(SOM)	Z(CIA)	Z(Na/K)	Z(TC)	Z(TN)	Z(wsK)	Z(neK)	Z(eK)	Z(bacteria)	Z(ORP)	Z(pH)	Z(PHA)	Z(HMi)
Z(CIA)	0.61a												
Z(Na/K)	−0.52a	0.829b											
Z(TN)	−0.61a												
Z(wsK)	0.58a		−0.52a										
Z(neK)	0.80b	0.56a	−0.66b		−0.64b	0.77b							
Z(ek)	0.74b	0.51a	−0.67b		−0.62a	0.73b	0.98b						
Z(bacteria)	0.55a						0.53a						
Z(ORP)	−0.59a			−0.56a	0.66b								
Z(pH)	0.58a			0.66b	−0.68b					−0.95b			
Z(NHA%)											−0.57a		
Z(PHA%)		−0.51a	0.64			−0.65b	−0.51a						
Z(HMi%)							0.54a						
Z(HMc%)													0.93b
Z(CO₃)		−0.62b	0.56						−0.50a				
Z(K)		0.53a	−0.79b									−0.77b	

(Note: statistically, probabilities of a and b: a, $p<0.01$; b, $p<0.05$)

$$r_{\mathrm{neK,eK}} = P_{\mathrm{neK,eK}} + r_{\mathrm{neK,SOM}} \cdot P_{\mathrm{neK,SOM}} + \\ r_{\mathrm{neK,S}} \cdot P_{\mathrm{neK,S}} + r_{\mathrm{neK,Na/K}} \cdot P_{\mathrm{neK,Na/K}} + \\ r_{\mathrm{neK,TN}} \cdot P_{\mathrm{neK,TN}} + r_{\mathrm{neK,Na}} \cdot P_{\mathrm{neK,Na}} + \\ r_{\mathrm{neK,SOM}} \cdot P_{\mathrm{neK,SOM}} + r_{\mathrm{neK,wsK}} \cdot P_{\mathrm{neK,wsK}} \tag{3}$$

where, r_{ij} is the simple correlation coefficient between the ZneK and a soil property, P_{ij} is the path coefficient between the ZneK and a soil property, and $r_{ij}P_{ij}$ is the indirect effect of a soil property on the ZneK. An uncorrelated residue (e) that represents the unexplained part of an observed variable in the path model was

$$e_{nek} = (1 - R_{nek}^2)^{0.5} = 0.01 \tag{4}$$

where R_{nek}^2 is the coefficient of determination for the multiple regression equation between the ZneK and those soil properties. Those soil properties were wsK, SOM, S, Na/K, TN, Na, SOM, eK.

It disclosed that plant nutrition was indirectly influenced by nutrients release of microbial decomposition and organic matter. It was particularly obvious in nutrition release of low molecular weight organics in rhizosphere soil. High contents of associated ions (Na^+) were limit factors that control crop yields and soil nutrients use. Ammonium nitrogen was likely to reduce fixation of potassium ions of soil, and increase risk of potassium leaching. NH_4^+ and K^+ were competitive ions for absorption of ORP (eV) of eK and neK in rhizosphere soil, and different status of nitrogen effects on efficient use of potassium. Therefore, suitable proportion of N and K is important for plant. Soil nitrogen availability determines potassium absorption of root surface and cell membrane. Root membrane could directly affected by NH_4^+, NO_3^-, K^+ when it absorbs potassium ions, and indirect affected by potential balance from assimilation of NH_4^+. Strong affinity of

K^+, high efficiency and fast speed of absorption of potassium are likely to improve efficient absorption of potassium.

Path analysis of exchangeable potassium (eK)

As to exchangeable potassium, it was quite different situation in comparison with those of water soluble and non-exchangeable potassium. It showed that direct coefficient variables of eK include neK, wsK, SOM, S, Na, Na/K, TN, CIA, HMi, PHA; and indirect coefficient variables of eK include Mg, CO_3. Impact factor of path size of eK was determined by neK, CO_3, pH, ORP, Na (Table 10).

From Fig. 5, it showed that neK was closely related to wsK (with correlation coefficients of 0.72), and there was an inverse proportion between ZneK and ZwsK (the normalized value of neK and wsK) in condition of changes of micro-bioorganic and geochemical environmental factors in rhizosphere soil. Path contribution coefficients were 1.10 and 0.04 for neK to eK, and wsK to eK. It suggests that neK is more likely to be changed into eK while the possibility of wsK to eK is relatively low.

The correlation between the normalized value of eK (ZeK) and a soil property is the sum of the entire path connecting two variables, as described by

$$r_{\mathrm{neK,eK}} = P_{\mathrm{eK,neK}} + r_{\mathrm{neK,SOM}} \cdot P_{\mathrm{eK,SOM}} + \\ r_{\mathrm{neK,Na/K}} \cdot P_{\mathrm{eK,Na/K}} + r_{\mathrm{neK,Na}} \cdot P_{\mathrm{eK,Na}} + \\ r_{\mathrm{neK,HMi}} \cdot P_{\mathrm{eK,HMi}} + r_{\mathrm{neK,TN}} \cdot P_{\mathrm{eK,TN}} + \\ r_{\mathrm{neK,wsK}} \cdot P_{\mathrm{eK,wsK}} + r_{\mathrm{neK,S}} \cdot P_{\mathrm{eK,S}} \tag{5}$$

where r_{ij} is the simple correlation coefficient between the ZeK and a soil property, P_{ij} is the path coefficient between the ZeK and a soil property, and $r_{ij}P_{ij}$ is the indirect effect of a soil property on the ZeK. An uncorrelated residue (e) that represents the unexplained part of an observed variable in the path model was

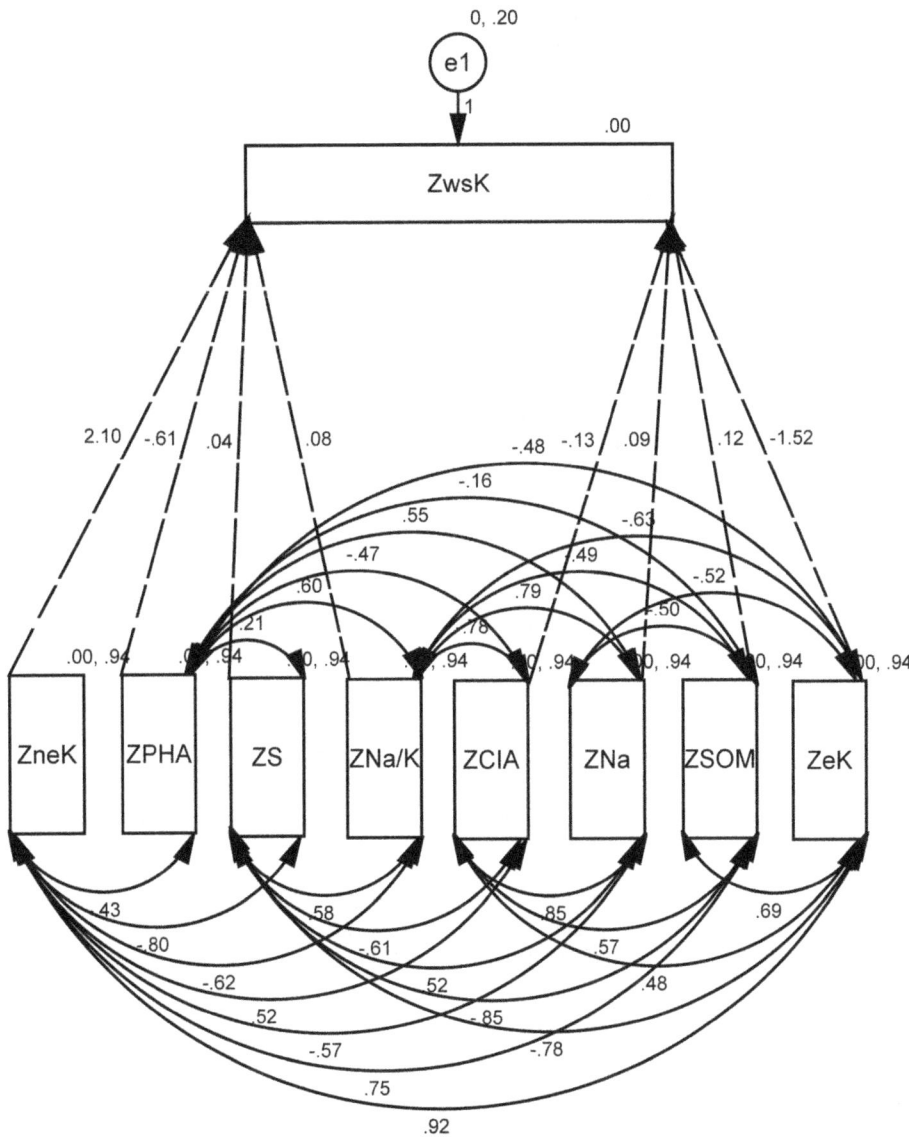

Figure 3. Path model of water-soluble potassium (wsK, solid line for double arrow, dashed-line for single arrow, and single arrow flows into ZwsK).

$$e_{ek} = \left(1 - R_{ek}^2\right)^{0.5} = 0.01 \qquad (6)$$

Where, R_{ek}^2 is the coefficient of determination of the multiple regression equation between the ZwsK and the eight soil properties. Those soil properties are neK, SOM, Na/K, Na, HMi, TN, wsK, S. Backward regression analysis was performed with the ZeK as the dependent variable and the eight soil properties as independent variables. Backward regression is a multiple regression procedure in which all the independent variables are entered into the regression equation at the beginning, and variables that do not contribute significantly to the fit of the regression model are successively eliminated until only statistically significant variables remain.

Cation exchange capacity (CEC) or the number of exchangeable cation depends on the type and number of clay and organic matter in the soil. To a certain extent, TN, SOM, HMi and the PHA and the CIA are able to well reflect type and number of clay

and organic matte in rhizosphere soil. Proper pH value of solution, hot water pressure, concentration increases of magnesium carbonate and ion are useful to exchangeable potassium (neK) when it is absorbed by crops. A plenty of CO_2 could be produced by root respiration and soil microbe respiration, and is fixed in form of CO_3^{2-}. For stable pH values, it is beneficial for organic acid to reach in equilibrium of absorption in soil solution. The exchangeable potassium and fixed potassium which comes from layers of secondary clay mineral are released when there occurs hydrolysis and cation substitution resulted by chemical element H^+. Those variables, such soil organic matter (ZSOM), iron-binding Humic (ZHMi), and water-soluble nitrogen (ZTN), not only provide nutrition for cotton root, but also changed soil chemical compositions and absorption characteristics of potassium in rhizosphere soil. Additionally, concomitant phenomenon (ZS) appears between S (the lost chemical element in soil) and potassium. Antagonism occurred between Na and K, and it

Table 8. Direct and indirect correlation coefficients of water-soluble potassium (wsK).

Variables	Direct correlation coefficients	Indirect correlation coefficients	Path contrition coefficients
Z(PHA)	−0.65	−0.32	−0.38
Z(S)	−0.56	0.17	−0.8
Z(Na)	−0.55	−1.15	−0.78
Z(Na/K)	−0.52	−0.01	−0.01
Z(TN)	−0.41	0.23	−0.38
Z(ORP)	−0.22	0.02	−1.28
Z(Mg)	−0.14	−0.79	−0.37
Z(CO₃)	−0.07	−0.53	−0.24
Z(pH)	0.14	−0.35	−1.87
Z(HMi)	0.39	0.2	0.29
Z(K)	0.41	0.02	−0.01
Z(CIA)	0.47	2.02	−1.61
Z(SOM)	0.58	0.06	0.25
Z(eK)	0.73	0.05	0.14
Z(neK)	0.77		

implied that sodium salt was able to have significant effect on plant (ZNa/K, ZNa).

From above analysis, we conclude that key direct factors of wsK, neK, and eK were S, Na/K, the CIA, SOM, Na, and TN, and key indirect factors were CO_3, Mg, pH, Na, S, and SOM. Whereas, direct impact factors of water soluble potassium (wsK) were PHA, S, Na, Na/K, K, the CIA, SOM, eK and neK. Each of its impact value was −0.65, −0.56, −0.55, −0.52, 0.41, 0.47, 0.58, 0.73, and 0.77 respectively. Indirect impact factors were Na, Mg, CO_3, pH, PHA, S, HMi, TN and CIA. Each of its impact value was −1.15, −0.79, −0.53, −0.35, −0.32, 0.17, 0.2, 0.23, and 2.02, respectively. Non exchangeable potassium direct impact factors were: S, Na/K, Na, TN, the CIA, wsK, SOM and eK. Each of its impact value was −0.86, −0.66, −0.64, −0.61, 0.56, 0.77, 0.8, and 0.98 respectively. Indirect impact factors were CO_3, eK, S, SOM, K, pH, Na and Mg. Each of its impact value was 1.18, 0.76, 0.19, −0.11, −0.14, −0.16, −0.34, and −0.67 respectively. Exchangeable potassium direct impact factors were S, Na/K, Na, TN, HMi, wsK, SOM and neK. Each of its impact value was −0.83, −0.67, −0.62, −0.55, 0.54, 0.73, 0.74, and 0.98 respectively. Indirect impact factors were CO_3, S, SOM, K, pH, Na, Mg and neK. Each of its impact value was −1.7, −0.28, 0.18, 0.21, 0.23, 0.46, 0.95, and 1.49 respectively. Absolute values of path coefficients represent the change ability of the potassium morphology, and the bigger value means the stronger ability to change potassium morphology. And "+"with and the arrow direction represent change direction of potassium morphology. In this way, we can quantitatively describe the K elements morphological changes.

Path model of soil potassium status changes of cotton

It is of great importance to further investigate on the equilibrium movement in soil. The following part will discuss the shift of dynamic balance among wsK, eK and neK, and their flow path. To test the reliability of the models, we collect another 48 cotton soil samples in the same experiment, using stepwise regression method, observation of potassium mobility and morphological change.

We carried out a plot experiment of potash application with high K-efficiency genotype (HEG) and low K-efficiency genotype

(LEG) cotton. The soil in each treatment weights 8.5 kg. Detail fertilizer schemes are listed in Table 11. LEG and HEG cottons' planting effects is shown in Fig. 6.

(1) Equilibrium shift in the path for non-exchangeable potassium

Supposed that $V_{wsK \to nek}$ is path for wsK to neK, $V_{ek \to nek}$ is path for eK to neK, therefore, equilibrium shift in path for non-exchangeable potassium can be expressed by contribution coefficient functions as below,

$$\begin{cases} y_{wsk} = 0.60 \cdot x_{neK} - 0.37 \cdot x_{PHA} + 5.87 \times 10^{-7} (R^2 = 0.87) \\ y_{neK} = 1.67 \cdot x_{wsk} + 0.62 \cdot x_{PHA} - 9.87 \times 10^{-7} \\ y_{eK} = 0.98 \cdot x_{neK} + 1.23 \times 10^{-6} (R^2 = 0.92) \\ y_{neK} = 1.02 \cdot x_{eK} - 1.25 \times 10^{-6} \end{cases} \quad (7)$$

Equilibrium occurs when X_{wsK}, X_{PHA}, X_{eK} satisfy with Function 8,

$$\begin{cases} 1.67 \cdot x_{wsk} + 0.62 \cdot x_{PHA} - 9.87 \times 10^{-7} = 1.02 \cdot x_{eK} - 1.25 \times 10^{-6} \\ x_{wsk} + 0.37 \cdot x_{PHA} = 0.61 \cdot x_{eK} - 1.57 \times 10^{-7} \end{cases} \quad (8)$$

Here, y_{nek} is the equilibrium threshold. And the value of x_{wsK}, x_{PHA}, x_{eK} is determined by soil characteristics. Calculated results from Equation 7 were listed in Table 12. In the table, column 1 and 2 were calculated from Equations 7. We used the wsK, eK, neK to construct dynamic equilibrium equations ($wsK \underset{\beta}{\overset{\alpha}{\rightleftharpoons}} neK \underset{\gamma}{\overset{\chi}{\rightleftharpoons}} eK$). The column 3 was normalized values of neK and the column 4 was the direction of balance movement of neK. These three column datasets were presented in Fig. 6.

In the reversible equilibrium of $wsK \underset{\beta}{\overset{\alpha}{\rightleftharpoons}} neK \underset{\chi}{\overset{\gamma}{\rightleftharpoons}} eK$, soil samples RSWKL, RSWH, RSWL, NRSWKL, NRSWL, K balance cycle were moved in α, χ direction movement; soil

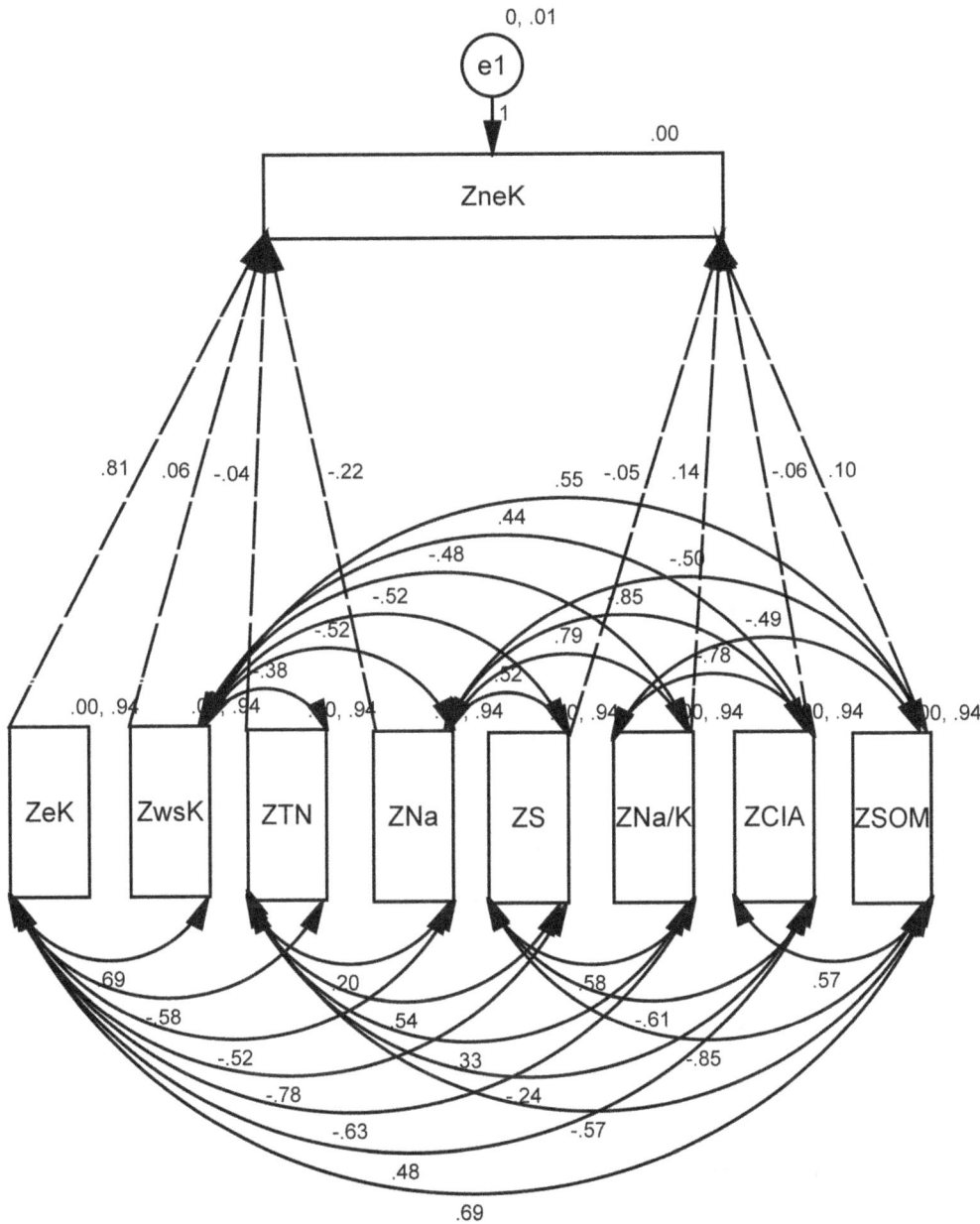

Figure 4. Path model of non-exchangeable potassium (neK, solid line for double arrow, dashed-line for single arrow, and single arrow flows into ZwsK).

samples RSWKH, RSKH, RSKL, ROPTH, ROPTL, NRSWKH, NRSWH, NROPTH, NROPTL, K balance cycle were moved in β and γ direction movement; soil samples NRSKH, NRSKL, K balance cycle were moved in β, χ direction movement.

In Fig. 6, neK reaches its equilibrium in following soil types which includes LEG in water stress treatment (SW), HEG after potassium stress treatment (SK), LEG after optimum fertilization treatment (OPT), and LEG after water stress treatment (SW). However, all the value in those soils with HEG in SW, LEG in SK, HEG in SW and SK, the values were all greater than the equilibrium, wsK reached its saturation. For other treatment soils, when wsK was lower than the equilibrium and PHA and eK was greater than equilibrium, neK changed into wsK. It proves that

eK equales to total K in matured crops. When eK is below a certain level in soil, plants can no longer get eK. When eK was at low levels, it was absorbed by other ions with stronger absorption. This reduces chance of potassium to enter into the soil solution. In this condition, neK begins to release through low concentrations of wsK and eK. However, the release capacity of neK depends on situations of crop types

(2) Equilibrium shift in the path for exchangeable potassium

Supposed that $V_{wsK \to ek}$ is path for wsK to eK, $V_{nek \to ek}$ is path for neK to eK, therefore, equilibrium shift in path for exchangeable potassium can be expressed by contribution coefficient functions as below,

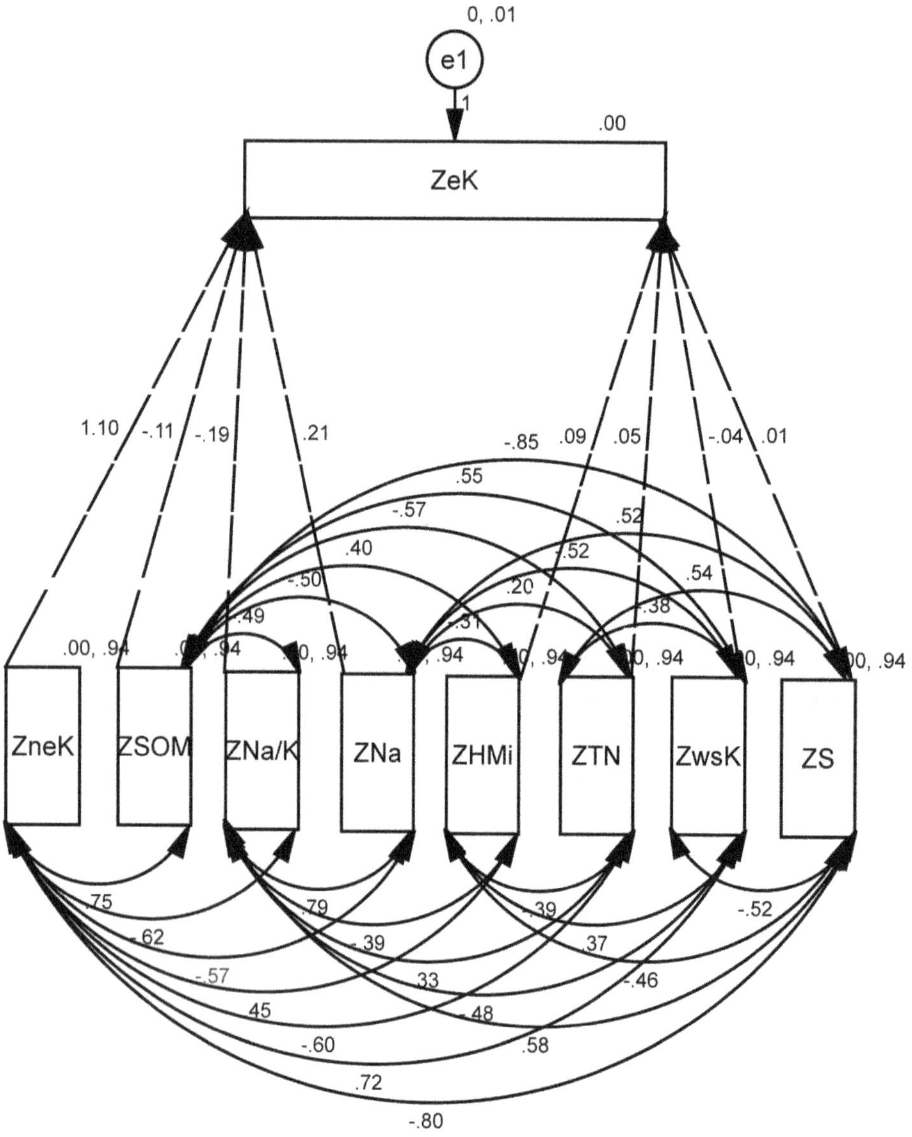

Figure 5. Path model of exchangeable potassium (eK, solid line for double arrow, dashed-line for single arrow, and single arrow flows into ZwsK).

$$\begin{cases} y_{wsk} = 0.98 \cdot x_{eK} + 3.328 \times 10^{-16} \quad (R^2 = 0.96) \\ y_{eK} = 1.02 \cdot x_{wsk} - 3.39 \times 10^{-16} \\ y_{neK} = 0.60 \cdot x_{eK} - 0.37 \cdot x_{PHA} + 7.89 \times 10^{-17} \quad (R^2 = 0.70) \\ y_{eK} = 1.67 \cdot x_{neK} + 0.62 \cdot x_{PHA} - 1.32 \times 10^{-16} \end{cases} \quad (9)$$

Equilibrium occurs when x_{wsK}, x_{PHA}, x_{neK} satisfy with Functions 11,

$$x_{wsk} = 1.64 \cdot x_{neK} + 0.61 \cdot x_{PHA} + 2.07 \times 10^{-16} \quad (10)$$

Here, y_{ek} is the equilibrium threshold. And the value of x_{wsK}, x_{PHA}, x_{neK} is determined by soil characteristics. Calculated results from Equations 9 and 10 are listed in Table 12. In this table, column 5 and 6 are calculated from Equations 9 and 10. We use

the wsK, eK, neK constructs the dynamic equilibrium equations ($wsK \overset{\varepsilon}{\underset{\theta}{\rightleftarrows}} eK \overset{\lambda}{\underset{\omega}{\rightleftarrows}} neK$). The column 7 is normalized values of neK and the column 8 is the direction of movement of balance of eK. These three column datasets are described in Fig. 7.

In the reversible equilibrium of $wsK \overset{\varepsilon}{\underset{\theta}{\rightleftarrows}} eK \overset{\lambda}{\underset{\omega}{\rightleftarrows}} neK$, soil samples SWH, NRSWL, K balance cycle were moved in ε, λ direction movement; soil samples RSWKL, RSWH, RSWL, K balance cycle were moved in θ, λ direction movement; soil samples RSWKH RSKH, RSKL ROPTH, ROPTL, NRSWKH, NRSKH, NRSKL, NROPTH, NROPTL K balance in soil samples, K balance cycle were moved in θ, ω direction movement respectively.

As is shown in Fig. 7, neK reaches its equilibrium in following soil plots, including LEG (SW), HEG (SK), LEG (OPT), and LEG (SW). And the exchangeable K reaches its equilibrium in soil types of LEG (SW). For other treatment soils, such as LEG (SW and

Table 9. Direct and indirect correlation coefficients of non-exchangeable potassium (neK).

Variables	Direct correlation coefficients	Indirect correlation coefficients	Path contrition coefficient
Z(S)	−0.86	0.19	−0.25
Z(Na/K)	−0.66	0.02	0.03
Z(TN)	−0.64	0.03	−0.03
Z(Na)	−0.61	−0.34	−0.37
Z(PHA)	−0.46	0.03	0.16
Z(ORP)	−0.41	0.10	−0.42
Z(Mg)	0.04	−0.67	−0.34
Z(CO3)	0.10	1.18	0.56
Z(K)	0.44	−0.14	0.2
Z(pH)	0.46	−0.16	−0.50
Z(HMi)	0.48	0.07	0.06
Z(CIA)	0.56	0.05	−0.07
Z(wsK)	0.77	0.01	0.01
Z(SOM)	0.8	−0.11	−0.14
Z(eK)	0.98	0.76	0.69

SK), LEG (SW), LEG (SK), HEG (SW and SK), HEG (OPT), neK changes to eK more or less. In soil of HEG (SW and SK), eK changes to wsK when wsK was lower than equilibrium. In soil of LEG (SW and SK), neK changes to eK when wsK of soil was greater than that of crops. Soils of both genotypes cotton (HEG and LEG) can efficiently absorb exchangeable potassium.

During a lot of vermiculite in the soil, those negative charges caused by vermiculite isomorphous substitution are near p-site, the electrostatic attraction of potassium ion became bigger, the adsorption capacity much more than other type (2:1) minerals, therefore, wsK and neK are fixed and the soil changed into deficiency. In general, drying accelerated the fixation of wsK

adsorption, but neK still moved in the direction of eK in the reversible equilibrium in HEG rhizosphere soil because rhizosphere soil exchangeable potassium content was low to meet the needs of the cotton growth in the SW and SWK plant. When moisture was adequate, redox reaction of soil became strengthened and oxidation reduction potential was reduced. For example, Fe^{2+} and Mn^{2+} ions were increased rapidly, Fe^{2+} and Mn^{2+} replaced p-site potassium of soil colloids and part of i-site potassium, and this improved the content of rapidly-available potassium and biological effective parts. In addition, the dissolved iron and manganese mineral alteration and also increased the release of mineral K.

Table 10. Direct and indirect correlation coefficients in exchangable potassium (eK).

Variables	Direct correlation coefficients	Indirect correlation coefficients	Path contrition coefficient
Z(S)	−0.83	−0.28	0.36
Z(Na/K)	−0.67	−0.03	−0.04
Z(TN)	−0.62	−0.04	0.04
Z(Na)	−0.55	0.46	0.54
Z(PHA)	−0.51	−0.06	−0.24
Z(ORP)	−0.38	−0.16	0.6
Z(Mg)	0.07	0.95	0.49
Z(CO3)	0.11	−1.7	−0.81
Z(pH)	0.47	0.23	0.71
Z(K)	0.48	0.21	−0.3
Z(CIA)	0.51	−0.07	0.1
Z(HMi)	0.54	−0.09	−0.09
Z(wsK)	0.73	−0.01	−0.02
Z(SOM)	0.74	0.18	0.2
Z(neK)	0.98	1.49	1.44

Table 11. Fertilizing rates of different treatments (unit: g).

Treatments	OPT	SW	SK	SWK
Urea	6.65	6.65	6.65	6.65
$Na_2PO_4.2H_2O$	11.72	11.72	11.72	11.72
KCl	6.83	6.83	0	0
$CaCO_3$	2.13	2.13	2.13	2.13
H_3BO_3	1.00	1.00	1.00	1.00
$ZnSO_4.7H2O$	1.87	1.87	1.87	1.87
$MgSO_4.7H2O$	2.61	2.61	2.61	2.61

OPT, optimum fertilization treatment; *SW*, water limited; *SK*, potassium limited; *SWK*, water & potassium limited.

Discussion and Conclusions

In this study, changes of potassium status in soil were investigated through our proposed path model on basis of laboratory experimental datasets. Through this investigation, following findings were achieved.

Firstly, changes of potassium status in the rhizosphere soil were controlled by different environmental variables. wsK, neK and eK were controlled by some common factors and have their specular characteristics respectively. Those parameters, such as soil organic matter (SOM), Na and Na/K and S, were significant direct coefficient factors which control dynamic process of wsK, neK and eK. CO_3 and Mg are significant indirect coefficient factors which control dynamic of wsK, neK and eK. The pH is the primary factor which controls path of the wsK, neK and eK. Organic acids of HEG root system secretion increased soil acidity; the amount of fixed potassium was reduced. PHA surface is porous structure, PHA increased the adsorption of wsK in the experiment, and HMi could promote the moving to wsK, eK in the K form exchange. Both of them were negative correlation significantly negative correlation. First, under acid condition, hydroxyl aluminum and aluminum ions can occupy the potassium fixation point stronger. At the same time, the larger diameter of hydroxy aluminium ions

entered into the interlayer of mineral, formed as the "island" and played a supporting role, provided the potassium ion diffusion of mobile channel in the interlayer. These reduced the soil potassium fixation. Second, water protonated in acidic condition, the radius of H_3O^+ 12–13.3 nm were similar to K^+, H_3O^+ can produced competitive adsorption in soil where fixed on K^+ parts, so potassium reduced fixation. Those dynamic processes, such as eK to wsK, low concentration wsK and eK to neK, are likely to occur, in condition that wsK is lower than equilibrium value and PHA and eK are greater than equilibrium values. eK is changed neK when wsK is greater than equilibrium value. In soil of HEG (SW and SK), eK is changed into wsK when plant wsK is lower than soil wsK. In soil of LEG (SW and SK), neK is changed into eK when plant wsK is higher than soil wsK. Both genotype cottons can effectively absorb eK in treatment of stress of potassium (SK).

Secondly, it discloses that SOM, S, TN, HMi and PHA in rhizosphere soil determined dynamic balance of potassium status which includes adsorption and desorption, precipitation and dissolution, complexion and chelating. Special characteristics of humus, e.g., acidic and hydrophilic, cationic exchange, complexing capacity and high absorption capacity, could improves the sustained-release effect of potassium fertilizer. Many variables, pH, ORP, etc., are closely related to transformation and utilization of potassium fertilizer. The pH change in soil is mainly caused by coupling effects of nutrient uptake by plant roots and the secretion of organic acid and absorption imbalance of positive ion and negative ion leads to rhizosphere pH change, however, different factors affect imbalance of absorption of positive ion and negative ion. Dynamic equilibrium of potassium is affected by equilibrium constants, temperature, and products of this equilibrium system (wsK). Therefore, it is useful for cotton planting to improve to appropriate temperature, to alter SOM, TN and pH in soil, and to reduce the equilibrium constant to positive reaction. Meanwhile, it is good to choose HEG cotton for better use soil wsK. It is important to understand ways to slow down the fast response and accelerate speed of slow response in potassium balance system.

Finally, path model can effectively determine direction and quantities of potassium status changes between eK, neK and wsK under influences of specific environmental parameters. The proposed path model was able to quantitatively disclose moving

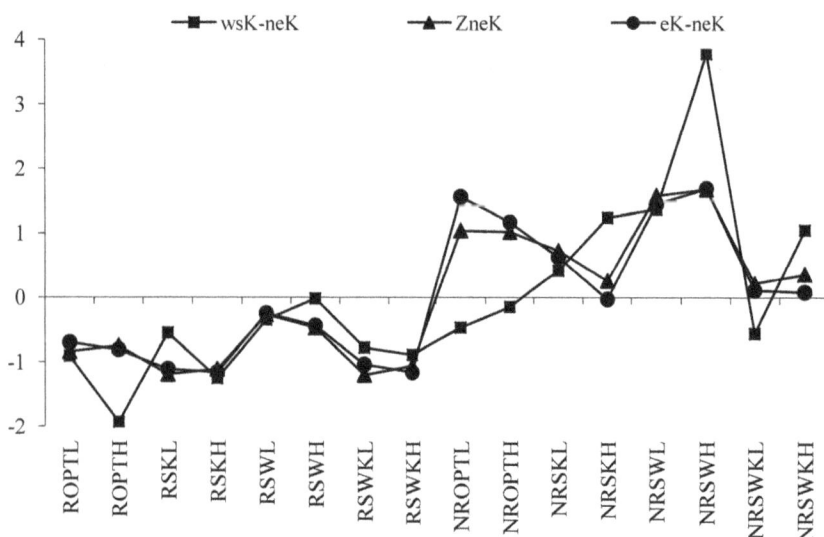

Figure 6. Dynamic equilibrium of non-exchangeable potassium in rhizosphere/non-rhizosphere soils.

Table 12. Dynamic equilibrium of exchangeable/non-exchangeable potassium (*neK/eK*) in rhizosphere/non-rhizosphere soils.

Variables	$wsK \underset{\beta}{\overset{\alpha}{\rightleftharpoons}} neK \underset{\chi}{\overset{\gamma}{\rightleftharpoons}} eK$			Movement direction at balance state	$wsK \underset{\theta}{\overset{\varepsilon}{\rightleftharpoons}} eK \underset{\omega}{\overset{\lambda}{\rightleftharpoons}} neK$			Movement direction at balance state
	$V_{wsk \to neK}$	$V_{ek \to neK}$	ZneK		$V_{wsk \to eK}$	$V_{nek \to eK}$	ZeK	
RSWKH	−0.89	−1.16	−1.06	β, χ	−0.78	−1.39	−1.14	θ, ω
RSWKL	−0.78	−1.04	−1.2	α, χ	−0.78	−1.51	−1.02	θ, λ
RSWH	−0.02	−0.44	−0.47	α, χ	−0.19	−0.5	−0.44	θ, λ
RSWL	−0.35	−0.25	−0.27	α, χ	−0.32	−0.28	−0.24	θ, λ
RSKH	−1.25	−1.16	−1.11	β, χ	−0.8	−1.79	−1.14	θ, ω
RSKL	−0.55	−1.11	−1.19	β, γ	−0.82	−1.19	−1.09	θ, λ
ROPTH	−1.93	−0.81	−0.75	β, χ	−0.65	−2.11	−0.79	θ, ω
ROPTL	−0.91	−0.70	−0.84	β, γ	−0.51	−1.48	−0.69	θ, λ
NRSWKH	1.05	0.08	0.36	β, χ	0.14	1.43	0.08	ε, λ
NRSWKL	−0.56	0.11	0.22	α, χ	−0.45	0.54	0.11	θ, ω
NRSWH	3.78	1.69	1.68	β, χ	3.24	1.28	1.66	ε, λ
NRSWL	1.37	1.45	1.58	α, χ	0.79	2.73	1.42	θ, λ
NRSKH	1.24	−0.03	0.26	β, χ	0.92	0.17	−0.03	ε, λ
NRSKL	0.42	0.63	0.73	β, χ	0.11	1.45	0.61	θ, ω
NROPTH	−0.15	1.17	1.02	β, γ	−0.12	1.76	1.14	θ, ω
NROPTL	−0.47	1.56	1.04	β, γ	0.23	0.90	1.53	θ, ω

direction of K status and quantify its equilibrium threshold. It provided a theoretical and practical basis for scientific and effective fertilization in agricultural plants growth. The significance of this study is that we are able to implement investigation on the equilibrium movement of potassium status through path model analysis. It discloses the gradual dynamic process of potassium status, and to decouple the sophisticated interactions between different variables. It is useful to guide the use of potassium fertilizer for cotton crops in practical. The model provides a biogeochemical theory basis of controlling moisture in cotton plant, applying potash fertilizer and organic fertilizer, and appropriately reducing nitrogen fertilizer.

However, environmental parameters such as soil moisture and temperature are some of the most important factors that influenced crop growth. Observation of dynamic change of soil moisture and temperature in every other time-step is quite sophisticated for us to design the control experiments while it involves in more than 4 times treatments, which means that we have more than 500 pots. By limitation, we had to set the soil moisture into two states, water sufficient (>35%) and deficient (<25%), and temperature were set to indoor temperature.

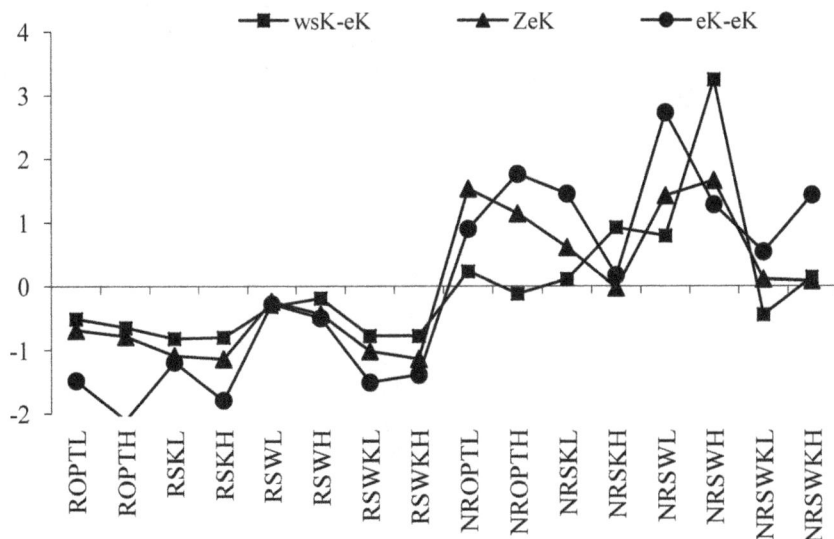

Figure 7. Dynamic equilibrium of exchangeable potassium in rhizosphere/non-rhizosphere soils.

Anyway, we were very aware of the importance of soil moisture and temperature impacts on control experiment potassium status. The next step of our work is to further investigate efficient utilization of potassium and the influence of soil moisture and temperature on exchange of potassium forms. We hope this study is able to gain knowledge of promoting high efficient use of potassium for crops.

Supporting Information

Appendix S1   Explanation of symbols.

References

1. Mahmood-Ul-Hassan M, Rashid M, Rafique E (2011) Nutrients transport through variably structured soils. Soil Science and Plant Nutrition 57: 331–340.
2. Osaki M, Matsumoto M, Shinano T, Tadano T (1996) A root-shoot interaction hypothesis for high productivity of root crops. Soil science and plant nutrition 42: 289–301.
3. Wada S-I, Seki H (1994) Ca-K-Na exchange equilibria on a smectitic soil: Modeling the variation of selectivity coefficient. Soil Science and Plant Nutrition 40: 629–636.
4. Puente M, Li C, Bashan Y (2004) Microbial Populations and Activities in the Rhizoplane of Rock-Weathering Desert Plants. II. Growth Promotion of Cactus Seedlings. Plant Biology 6: 643–650.
5. Wada S-I, Masuda K (1995) Control of salt concentration of soil solution by the addition of synthetic hydrotalcite. Soil Science and Plant Nutrition 41: 377–381.
6. Kobayashi H, Masaoka Y, Takahashi Y, Ide Y, Sato S (2007) Ability of salt glands in Rhodes grass (Chloris gayanaKunth) to secrete Na+and K+. Soil Science and Plant Nutrition 53: 764–771.
7. Han MY, Zhang LX, Fan CH, Liu LH, Zhang LS, et al. (2011) Release of nitrogen, phosphorus, and potassium during the decomposition of apple (Malus domestica) leaf litter under different fertilization regimes in Loess Plateau, China. Soil Science and Plant Nutrition 57: 549–557.
8. He H, Zhou J, Wu Y, Zhang W, Xie X (2008) Modeling the interaction of urbanization and surface water quality environment. Environmental Forensics 9: 215–225.
9. Eick MJ, Sparks DL, Bar-Tal A, Feigenbaum S (1990) Analyses of adsorption kinetics using a stirred-flow chamber: II. Potassium-calcium exchange on clay minerals. Soil Science Society of America Journal 54: 1278–1282.
10. McCune B, Caldwell BA (2009) A single phosphorus treatment doubles growth of cyanobacterial lichen transplants. Ecology 90: 567–570.
11. Schneider A-K, Schröder B (2012) Perspectives in modelling earthworm dynamics and their feedbacks with abiotic soil properties. Applied Soil Ecology 58: 29–36.
12. He H, Jim C (2012) Coupling model of energy consumption with changes in environmental utility. Energy Policy 43: 235–243.
13. He H, Jim C (2010) Simulation of thermodynamic transmission in green roof ecosystem. Ecological Modelling 221: 2949–2958.
14. Brouder S, Cassman K (1994) Evaluation of a mechanistic model of potassium uptake by cotton in vermiculitic soil. Soil Science Society of America Journal 58: 1174–1183.
15. Nakahara O, Wada S-I (1995) Surface Complexation Model of Cation Adsorption on Humic Soils. Soil Science and Plant Nutrition 41: 671–679.
16. Uroz S, Calvaruso C, Turpault M-P, Frey-Klett P (2009) Mineral weathering by bacteria: ecology, actors and mechanisms. Trends in microbiology 17: 378–387.
17. Puente M, Bashan Y, Li C, Lebsky V (2004) Microbial populations and activities in the rhizoplane of rock-weathering desert plants. I. Root colonization and weathering of igneous rocks. Plant Biology 6: 629–642.
18. Hinsinger P, Fernandes Barros ON, Benedetti MF, Noack Y, Callot G (2001) Plant-induced weathering of a basaltic rock: experimental evidence. Geochimica et Cosmochimica Acta 65: 137–152.
19. Hashimoto Y, Kang J, Matsuyama N, Saigusa M (2012) Path Analysis of Phosphorus Retention Capacity in Allophanic and Non-allophanic Andisols. Soil Science Society of America Journal 76: 441–448.
20. Kang J, Hesterberg D, Osmond DL (2009) Soil Organic Matter Effects on Phosphorus Sorption: A Path Analysis. Soil Science Society of America Journal 73: 360.
21. Lambers H, Mougel C, Jaillard B, Hinsinger P (2009) Plant-microbe-soil interactions in the rhizosphere: an evolutionary perspective. Plant and Soil 321: 83–115.
22. Wright S (1921) Correlation and causation. Journal of agricultural research 20: 557–585.
23. Wright S (1920) The relative importance of heredity and environment in determining the piebald pattern of guinea-pigs. Proceedings of the National Academy of Sciences of the United States of America 6: 320.
24. Wright S (1921) Systems of mating. I. The biometric relations between parent and offspring. Genetics 6: 111.
25. Wright S (1921) Systems of mating. II. The effects of inbreeding on the genetic composition of a population. Genetics 6: 124.
26. Kulahci F (2011) A risk analysis model for radioactive wastes. J Hazard Mater 191: 349–355.
27. Berthrong ST, Jobbágy EG, Jackson RB (2009) A global meta-analysis of soil exchangeable cations, pH, carbon, and nitrogen with afforestation. Ecological Applications 19: 2228–2241.
28. Kocev D, Naumoski A, Mitreski K, Krstić S, Džeroski S (2010) Learning habitat models for the diatom community in Lake Prespa. Ecological Modelling 221: 330–337.
29. He C, Cui K, Duan A, Zeng Y, Zhang J (2012) Genome-wide and molecular evolution analysis of the Poplar KT/HAK/KUP potassium transporter gene family. Ecol Evol 2: 1996–2004.
30. Kamewada K (1996) Application of "Four-Plane Model" to the Adsorption of K+, NO3-, and SO42-from a Mixed Solution of KNO3and K2SO4on Andisols. Soil Science and Plant Nutrition 42: 801–808.
31. Kang J, Hesterberg D, Osmond DL (2009) Soil organic matter effects on phosphorus sorption: A path analysis. Soil Science Society of America Journal 73: 360–366.

Acknowledgments

We are grateful to invaluable review comments of the editor and two anonymous reviewers to make the paper perfect. We would like to express our special thanks to Dr. Feng WU of the Institute of Earth Environment, Chinese Academy of Sciences for his kind technical support.

Author Contributions

Conceived and designed the experiments: WH FC. Performed the experiments: WH. Analyzed the data: WH. Contributed reagents/materials/analysis tools: FC WH. Wrote the paper: WH FC.

20

Organic vs. Conventional Grassland Management: Do ^{15}N and ^{13}C Isotopic Signatures of Hay and Soil Samples Differ?

Valentin H. Klaus[1]*, **Norbert Hölzel**[1], **Daniel Prati**[2], **Barbara Schmitt**[2], **Ingo Schöning**[3], **Marion Schrumpf**[3], **Markus Fischer**[2], **Till Kleinebecker**[1]

1 University of Münster, Institute of Landscape Ecology, Münster, Germany, **2** University of Bern, Institute of Plant Sciences, Bern, Switzerland, **3** Max-Planck-Institute for Biogeochemistry, Jena, Germany

Abstract

Distinguishing organic and conventional products is a major issue of food security and authenticity. Previous studies successfully used stable isotopes to separate organic and conventional products, but up to now, this approach was not tested for organic grassland hay and soil. Moreover, isotopic abundances could be a powerful tool to elucidate differences in ecosystem functioning and driving mechanisms of element cycling in organic and conventional management systems. Here, we studied the δ^{15}N and δ^{13}C isotopic composition of soil and hay samples of 21 organic and 34 conventional grasslands in two German regions. We also used $\Delta\delta^{15}$N (δ^{15}N plant - δ^{15}N soil) to characterize nitrogen dynamics. In order to detect temporal trends, isotopic abundances in organic grasslands were related to the time since certification. Furthermore, discriminant analysis was used to test whether the respective management type can be deduced from observed isotopic abundances. Isotopic analyses revealed no significant differences in δ^{13}C in hay and δ^{15}N in both soil and hay between management types, but showed that δ^{13}C abundances were significantly lower in soil of organic compared to conventional grasslands. $\Delta\delta^{15}$N values implied that management types did not substantially differ in nitrogen cycling. Only δ^{13}C in soil and hay showed significant negative relationships with the time since certification. Thus, our result suggest that organic grasslands suffered less from drought stress compared to conventional grasslands most likely due to a benefit of higher plant species richness, as previously shown by manipulative biodiversity experiments. Finally, it was possible to correctly classify about two third of the samples according to their management using isotopic abundances in soil and hay. However, as more than half of the organic samples were incorrectly classified, we infer that more research is needed to improve this approach before it can be efficiently used in practice.

Editor: Shuijin Hu, North Carolina State University, United States of America

Funding: The work has been funded by the DFG Priority Program 1374 "Infrastructure-Biodiversity-Exploratories" (FI 1246/6-1, FI1246/9-1, HO 3830/2-2, SCHR 1181/2-1). The funders had no role in study design, data collection and analysis, decision to publish, or preparation of the manuscript.

Competing Interests: The authors have declared that no competing interests exist.

* E-mail: v.klaus@uni-muenster.de

Introduction

Distinguishing organic and conventional products is a major issue of food security and authenticity and much research on method development has been conducted to tackle this issue [1]. Stable isotope analysis was proven to give important insight in ecosystem functioning and was successfully used to detect differences between organic and conventional agriculture [2,3,4]. Since Nakano et al. [5] proposed the use of natural abundances of stable isotopes to separate organic and conventional products, several studies tested this approach successfully for fruits, vegetables and other plant products [1,6,7,8,9,10] as well as for beef [11,12] and milk [13], but not for grassland hay or soil samples. Differences among organic and conventional plant products were mostly attributed to differences in δ^{15}N isotopic signatures of applied fertilizers [14], because organic farming abandons the use of synthetic mineral fertilizers. While such conventional (synthetic) N sources exhibit δ^{15}N values close to 0‰, organic N sources such as cattle dung or slurry are strongly enriched in δ^{15}N [15]. Consequently, organic farming products are mostly enriched in δ^{15}N compared to conventional ones due to the replacement of synthetic N sources by organic fertilizers. However, in nature δ^{15}N abundances in plants are affected by a multiplicity of factors such as type and degree of mycorrhization, the chemical type of N-compounds taken up or further soil characteristics, which can be barely separated from each other [16,17].

Similarly, δ^{13}C in plant and soil are also of broad ecological interest [18]. In C3 plants, which represent Central European grassland vegetation, δ^{13}C abundances in biomass are first of all affected by water availability and drought stress, but show also significant interactions with nutrient availability and fertilization [2,19]. Additionally, δ^{13}C values are related to a different contribution of CO_2 from soil respiration to plant photosynthesis y [6,20] and thus contain valuable ecological information related to agricultural management. Furthermore, was shown to be related to functional aspects of plant communities [18].

Although grasslands play a central role in the production of organic meat and dairy products [21], and proportions of organic grasslands have increased significantly during the last decade [22], stable isotope analysis was so far not used to distinguish between soils and yield (hay) of organically and conventionally managed grasslands. As organic fertilizers can even in grasslands lead to higher $\delta^{15}N$ values in soil and vegetation [23], this might give the ability to classify organic and conventional plant products using isotopic abundances [7]. Moreover, isotopic abundances are related to important ecosystem processes affecting nutrient cycling and balances [24] and thus bear the potential to elucidate possible differences in ecosystem functioning of organic vs. conventional grasslands, which are otherwise difficult to detect.

Here, we studied 21 organic and 34 conventional grasslands in two German regions and analyzed $\delta^{13}C$ and $\delta^{15}N$ of soil and hay (plant biomass). Furthermore, $\Delta\delta^{15}N$ values ($\delta^{15}N$ plant - $\delta^{15}N$ soil) were calculated to estimate differences in nitrogen dynamics. We also assessed the time since organic certification to test for temporal trends. In detail, we analyzed whether (a) differences in isotopic abundances among organic and conventional grasslands exist and whether there are (b) significant trends in isotopic composition with time since conversion to organic management. Additionally (c), we used discriminant analysis to deduce the respective management type from the isotopic composition of hay and/or soil samples.

Methods

Ethics statement

Field work permits were given by the responsible state environmental offices of Baden-Württemberg, Thüringen, and Brandenburg (according to § 72 BbgNatSchG).

Study design

We studied agriculturally used permanent grasslands in two regions in Germany which belong to the *Biodiversity Exploratories* project [25]: (I) *Hainich-Dün* in Thuringia in central Germany situated in and around the National Park Hainich and (II) the Biosphere Reserve *Schwäbische Alb* in Baden-Württemberg in south-west Germany. In grasslands of both regions Cambisols occur, while in the Schwäbische Alb Leptosols and in Hainich-Dün Stagnosols and Vertisols can also be found. Grassland types could be categorized as pastures, meadows and mown pastures [25]. To get information on land use for each grassland, farmers and land owners were annually questioned about the amount and type of fertilizer (kg N ha^{-1}) from 2006 to 2010 [26]. We chose organic and conventional grasslands from a randomly selected dataset of 50 plots in each region. Organic management of grasslands abandons pesticides and synthetic fertilizers, restricts livestock density and the use of organic fertilizers from animal husbandry to a maximum of 170 kg N*ha^{-1}*a^{-1} (European Union, 2008). Accordingly, study plots can be distinguished in two sub-sets: organic plots which are managed according to an official organic farming certificate [27] and uncertified (conventional) plots, where management goes against certification criteria. Please note that all unfertilized but not certified grasslands were excluded from the analysis. Finally, we used 17 conventional plots per study region as well as 17 organic plots at Hanich-Dün and 4 organic plots at the Schwäbische Alb for comparison. Duration of organic management of the grasslands differs from 3 up to 20 years. Although we have no detailed data on the management prior conversion to organic farming, it seems to be likely that at least some of the organic grasslands were already previously managed at a low to medium intensity. The proportion of legumes varied widely among study plots (from 0.0 to 60.5%) but not among organic and conventional management (data not shown). C4 plants are generally no regular component of Central European grassland vegetation.

Field work and chemical analyses

Soil sampling was conducted in early May 2011. On each plot mixed samples of 14 soil cores from 0 to 10 cm depth were collected using a split tube sampler with a diameter of 5 cm. Cores were taken along two 20 m transects at each plot. Roots were removed from the samples in the field and soil samples were air-dried, sieved to <2 mm and ground. For $\delta13C$ analyses, soil samples were weighted into tin capsules and treated with sulphurous acid inside the capsules to remove carbonates. Samples were dried again at 70°C before combustion in an oxygen stream using an elemental analyser (NA 1110, CE Instruments, Milan, Italy). Evolved $CO2$ was analysed using an isotope ratio mass spectrometer (IRMS; Delta C or DELTA+XL, Thermo Finnigan MAT, Bremen, Germany). For biomass sampling, we harvested aboveground community biomass in four quadrates of 0.25 m^2 from mid-May to mid-June 2011 in both regions simultaneously. Temporary fences ensured that no mowing or grazing took place before yield was sampled. Plant material was dried for 48 h at 80°C and ground to fine powder for lab analyses. Both isotopic abundances of biomass samples and $\delta^{15}N$ of soil samples were determined by mass spectrometry (Finnigan MAT DeltaPlus with Carlo Erba Elementar Analysator with ConFlo II Interface). Isotope ratios are given in per mille (‰), whereby $\delta^{13}C$ is relative to the international reference standard v-PDB using NBS19 and $\delta^{15}N$ relative to AIR-N$_2$ [28].

The nearby climate station in *Hainich-Dün*, located in Schönstedt (193 m A.S.L.), revealed that the precipitation prior to sampling (April and May 2011) was 70% lower than the mean of the same month from 2003–2010 (33.8 mm instead of 116.2 mm) [29]. At the *Schwäbische Alb*, the next climate station is situated in Münsingen-Rietheim (732 m A.S.L.). Compared to the same month in the period 2005–2010, precipitation in April and May 2011 was 20% lower (153.8 mm instead of 191.5 mm) [30].

Data analysis

Multiple analysis of variance (ANOVA) models were used to examine differences among management types (organic vs. conventional) while also accounting for farm or land owner ($n = 7$), soil type ($n = 4$), study region ($n = 2$) and grasslands type (meadow, pasture, mown pasture) including all two-way interactions with management type. Therefore, the lm() and step() functions in R for stepwise reduction of explanatory variables were used. Obtaining model results by using the anova() output assured that the order of variables entering the model had no effect on significance gained for the respective variable, because all other variables were taken into account previously as co-variables. Model assumptions were checked using diagnostic plots function. Similarly, also applying the lm() function analysis of co-variance (ANCOVA) models were calculated to relate the isotopic composition to the time since certification in all organic plots including farm, soil type and study region. Soil texture (proportions of clay, silt and sand) was also incorporated in lm() models but removed for final analysis as they did not add to explained variance. Additionally, we employed the qda() function from the MASS package [31] to perform Quadratic Discriminant Analysis for the classification of organic and conventional grasslands based on ^{15}N and ^{13}C isotopic values in soil and biomass samples. To ensure normal distribution of variables log10 transformation was applied prior to all analyses where necessary. All statistical tests were performed with R [32].

Results

Farmer's questionnaires revealed that organic grasslands received average fertilizer applications of 21.6 ($\pm$37.0) kg N a^{-1}, while conventional grasslands received more than the threefold amount: 72.3 ($\pm$41.5) kg N a^{-1}. However, both management systems show wide variation in fertilization intensity at the plot level. Organic plots received only organic fertilizers, as prescribed by organic management guidelines, whereas conventional plots received mostly mineral fertilizers (75% of total fertilizer N).

Grasslands of both management types overlap widely in ^{13}C and ^{15}N isotopic abundances in soil and hay (Figure 1). Nevertheless, organic grasslands were characterized by significantly lower δ^{13}C abundances in soil (Figure 2a). Meanwhile, δ^{13}C in hay, δ^{15}N in hay and soil and $\Delta\delta^{15}$N did not differ between organic and conventional grasslands but only among farms, study regions and/or grassland types (Table 1). In none of the models the soil type was a significant predictor of ^{13}C and ^{15}N isotopic signals. Although this does not mean that isotopic signals are independent of further soil characteristics, it, nevertheless, underlines the comparability of the selected plots. Generally, analyses explained only 6 to 34% of the variance (Table 1).

When related to the time since certification, ^{13}C in hay and soil showed significant negative relationships, while the explained variance of the models increased considerably compared to previous models (Table 2). This was most significant for δ^{13}C in soil (Figure 2b). Neither δ^{15}N in soil or hay nor $\Delta\delta^{15}$N showed significant relationships with the time since certification (Table 2).

Due to significant effects of study region on some of the isotopic abundances (Table 3), quadratic discriminant analysis (QDA) was carried out using data which was centered (standardized) according to the regional mean. Using QDA, 60% of the samples could be correctly classified as organic or conventional using δ^{13}C and δ^{15}N isotopic information of soils, while 70% could be correctly classified using the isotopic information of the hay, assumingly due to higher variation in isotopic values of hay

compared to soil samples. Combining both, increased correct classifications only slightly up to 73%. In all cases analyses mismatched organic samples to a higher degree than conventional samples and classified significantly more organic samples wrongly as conventional ones (Table 3).

Discussion

Isotopic abundances of ^{15}N and ^{13}C can give insight in ecosystem functioning and were often shown to be a useful tool to separate between organic and conventional products [1,6]. In the case of organic management in grasslands, this is only partly true for δ^{15}N and δ^{13}C in soil and hay.

^{15}N in organic grasslands

Organic grasslands received only organic fertilizers and were characterized by significantly lower fertilization intensity, in line with Klaus et al. [33]. Nevertheless, this difference in fertilization regime did not imprint in the δ^{15}N signal of soil and hay samples. While δ^{15}N abundances of conventional synthetic fertilizers vary between -2 and 2‰, organic fertilizers such as cattle dung or slurry are strongly enriched in δ^{15}N (5 to 35‰) [15]. While experimental studies revealed an effect of organic compared to synthetic fertilization on the δ^{15}N isotopic composition of grassland soils and vegetation [23], it was likewise shown that both organic but also mineral fertilizers can lead to increasing δ^{15}N vales in soil and vegetation probably due to increased microbial activity and subsequent losses of ^{15}N depleted N from the system [17]. Furthermore, conventional management includes both the use of organic and mineral fertilizers and farmers may considerably change respective proportions among years [26]. In our study, the naturally diverse abiotic conditions such as strong variation in soil properties and land-use history among plots might have additionally impeded clear patterns of δ^{15}N in organic and conventional grasslands. This is supported by findings from Wrage et al. [34] giving poor to missing associations among N balances and δ^{15}N in soil and vegetation due to spatially heterogeneous conditions and short-term changes in stocking densities in pastures. Moreover, as only a certain proportion of the N stored

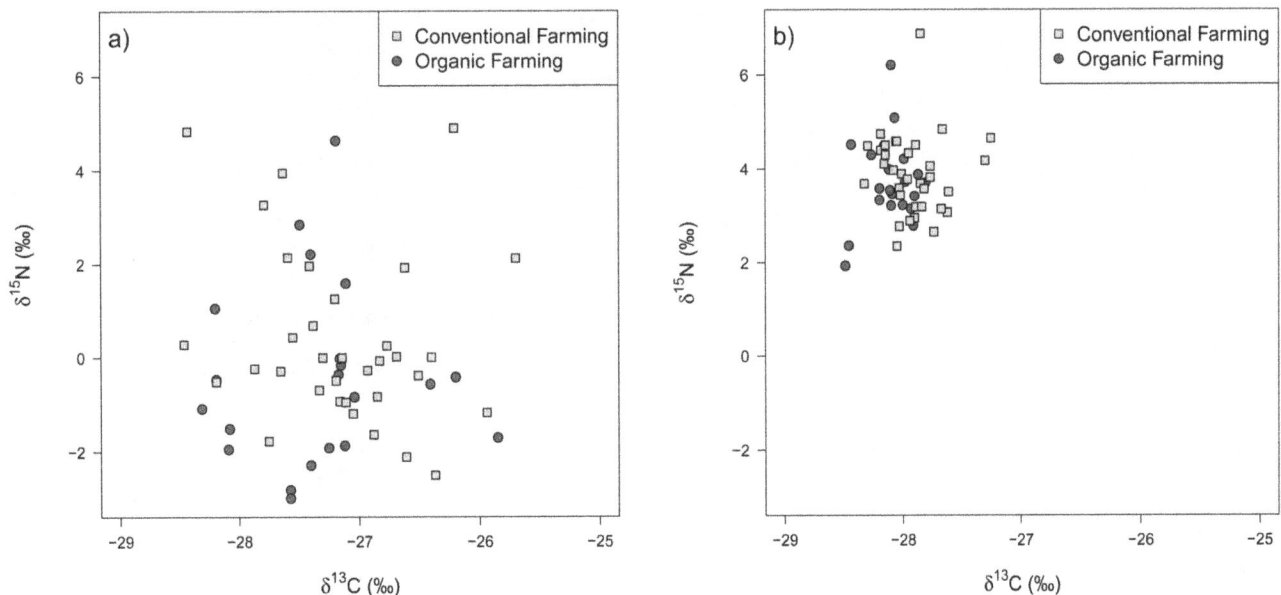

Figure 1. δ^{15}N and δ^{13}C composition of a) hay and b) soil samples of organic and conventional grasslands.

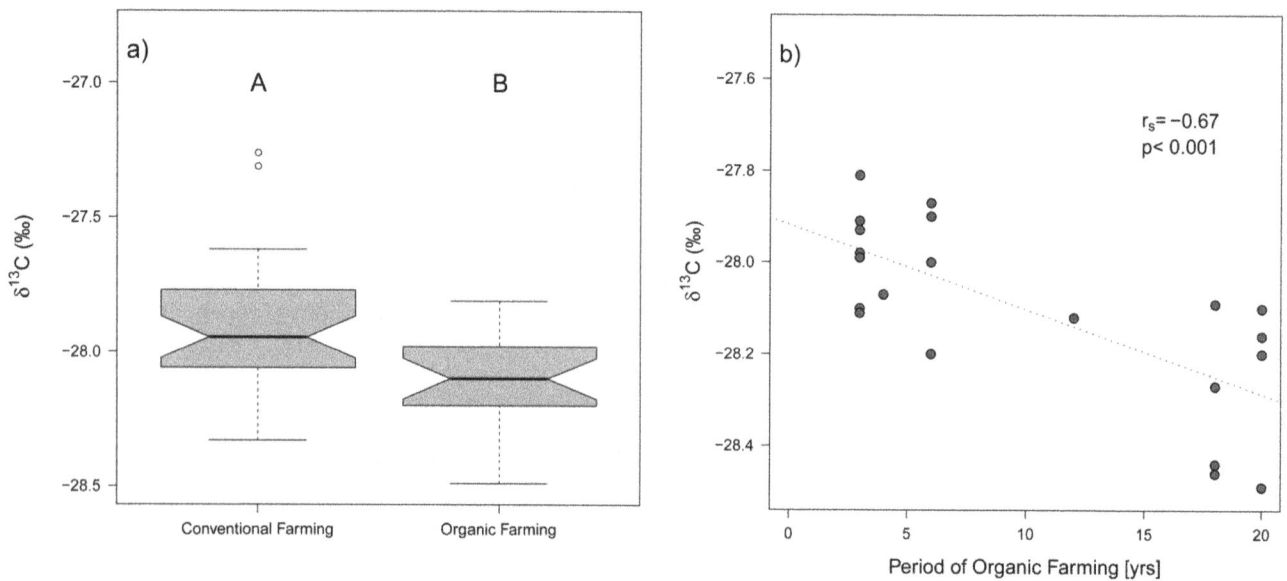

Figure 2. $\delta^{13}C$ **abundances in soil of a) organic vs. conventional grasslands and b) organic grasslands in relation to the time since certification (** $r_s = -0.70$**;** $p<0.001$**).** Letters indicate significant group differences according to ANOVA analyses (for details see Table 1 & 2).

in biomass and soil originates from fertilization, biologically fixed N (e.g. through legumes) have additionally diluted isotopic signals of fertilizer applications [35].

Missing differences in $\Delta\delta^{15}N$ between organic and conventional plots suggest that the studied grasslands do not substantially differ in nitrogen cycling [24]. Instead, the respective farm (or land owner) was responsible for most of the variation explained in $\Delta\delta^{15}N$ indicating that nitrogen cycling is strongly driven by individual decisions and practice at the farm level. Because plant species richness was shown decease $\Delta\delta^{15}N$ in plant and soil assumedly via nitrogen partitioning [35], missing or poor differences in plant diversity between organic and conventional grasslands [e.g. 33,36,37] could be another reason why $\Delta\delta^{15}N$ did not respond to organic grassland management. Contrary, in our study organic grasslands had higher mean plant species richness compared to conventional ones (t-test; $p<0.01$; data not shown), apparently not leading to significant differences in $\Delta\delta^{15}N$.

Haas et al. [21] suggest that lower fertilization intensity under organic management is associated with fewer losses of N to neighbouring habitats or to the ground water. However in our case, even after up to 20 years of organic farming ^{15}N isotopic abundances revealed no significant indication for fewer losses of N from the studied organic grasslands.

^{13}C in organic grasslands

We found significant differences in $\delta^{13}C$ abundances between organic and conventional soils and trends in decreasing $\delta^{13}C$ abundances with time since certification in both organic soil and hay samples. Missing statistical significances of soil types in the analyses suggested that this is not only caused by differences in water availability among study plots. However, clear differences in $\delta^{13}C$ in hay might only get apparent in dry years, because $\delta^{13}C$ isotopic patterns can differ considerably among seasons [38]. Thus, $\delta^{13}C$ abundances in soils are probably a more reliable indicator for water supply for plants, because they integrate $\delta^{13}C$ signals over a longer period of time, while $\delta^{13}C$ in hay strongly depends on the photosynthetic activity and stomatal conductance influenced by current weather conditions [2,38]. However, it is questionable whether differences in $\delta^{13}C$ in hay can get more pronounced than in the studied year as prior to sampling the weather was particularly dry.

It was shown that vegetation $\delta^{13}C$ decrease with plant diversity in grassland experiments due to facilitation and/or complementarity among plant species [39,40]. Assumedly, we also detected such a pattern of a positive feedback of higher plant diversity under organic management causing less water stress in organic grasslands. However, as mentioned above, organic and conventional grasslands

Table 1. Summary of multiple ANOVA models of isotopic abundances (no interaction with management was significant) ($n = 55$). "Grassland type" = pasture, meadow or mown pasture.[a]

	Adj. R^2	Effect of organic management	Farm	Region	Grassland type	Soil type
δ13C soil	0.13**	negative**	ns	ns	ns	ns
δ13C hay	0.34***	ns	*	***	ns	ns
δ15N soil	0.06*	ns	ns	*	ns	ns
δ15N hay	0.24*	ns	*	ns	*	ns
Δδ15N	0.20*	ns	**	ns	*	ns

[a]Significance levels: *** = $p<0.001$; ** = $0.001<p<0.01$; * = $0.01<p<0.05$.

Table 2. Summary of ANCOVA models of isotopic abundances testing for relationships with the time since certification (only organic plots, $n = 21$).

	Adj. R^2	Effect of time since certification	Farm	Region	Grassland type	Soil type
δ13C soil	0.49***	negative***	ns	ns	ns	ns
δ13C hay	0.73**	negative**	*	ns	ns	ns
δ15N soil	ns	ns	ns	ns	ns	ns
δ15N hay	ns	ns	ns	ns	ns	ns
Δδ15N	ns	ns	ns	ns	ns	ns

[a]Significance levels: *** $= p < 0.001$; ** $= 0.001 < p < 0.01$; * $= 0.01 < p < 0.05$.
"Grassland type" = pasture, meadow or mown pasture.[a]

do not always differ in plant diversity and more research is needed to give full evidence for reduced water stress under higher plant diversity in established permanent grasslands. Furthermore, changes in $\delta^{13}C$ with time since certification cannot be traced back to changes in plant diversity, because diversity was stable over time (data not shown).

There are several other mechanisms, which could have also occurred, but cannot be easily separated from the whole complex of factors influencing isotopic abundances. Georgi et al. [6] found organic biomass (vegetables) to be depleted in $\delta^{13}C$. They suggest higher soil respiration rates in organic soils due to higher microbial activity might cause a lowering of the $\delta^{13}C$ value of the soil CO_2 pool available to plants. As soil density fractions can have different ^{13}C signals [41], differences in the distribution of soil carbon among density fractions could also have also led to isotopic patterns.

Other reasons for lower $\delta^{13}C$ values in organic grasslands could be direct or indirect effects of (former) fertilization. As organic farms use less often corn (a C4 plant) to feed their livestock, conventional slurry is thus enriched in ^{13}C and may have led to higher $\delta^{13}C$ values in soil organic matter in conventional grasslands [42]. As the "conventional" organic matter deceases

after time since certification, this effect is likely to have caused decreasing $\delta^{13}C$ in soil with time since certification. Similarly, it was shown that high N availability can lead to higher $\delta^{13}C$ in plant biomass mostly due to structural changes in plant tissue further tightening drought conditions [43]. However, differences in $\delta^{13}C$ values of vegetables observed by Georgi et al. [6] were independent of optimal or reduced N supply. Finally, we can't rule out that further factors such as the local rainfall distribution also influenced observed $\delta^{13}C$ abundances.

Classification of organic and conventional samples

Although Rapisarda et al. [6] showed that approximately 90% of fruit samples could be correctly categorized as organic or conventional products using isotopic abundances, classification of grassland samples revealed rather weak results. Only one third of organic and conventional samples were correctly classified and a high proportion especially of organic samples were mismatched. This can be largely explained by missing differences in especially $\delta^{15}N$ abundances. These were significantly higher in organic fruits used in the study by Rapisarda et al. [6]. Thus, we have to conclude that separating organic and conventional soil and hay samples is barely possible using only $\delta^{15}N$ and $\delta^{13}C$ abundances.

Table 3. Results of quadratic discriminant analysis (QDA) of management types (organic vs. conventional) deduced from regionally standardized $\delta^{15}N$ and $\delta^{13}C$ isotopic abundances of soil and/or hay samples of grasslands.

Soil samples		Classified			
		organic	conventional	total	correct
Origin	organic	7	14	21	33%
	conventional	8	26	34	76%
				total	60%
Hay samples		Classified			
		organic	conventional	total	correct
Origin	organic	9	12	21	43%
	conventional	4	30	34	88%
				total	71%
Soil & hay samples		Classified			
		organic	conventional	total	correct
Origin	organic	10	11	21	48%
	conventional	4	30	34	88%
				total	73%

This is especially true, if we would also include uncertified but unfertilized grasslands, which occur frequently [33,44]. However, including further stable isotopes of S, O and H or other indicative substances might offer additional possibilities to improve the classification of organic vs. conventional samples from permanent grasslands [11,12].

Acknowledgments

We thank Harald Strauβ and Artur Fugman for friendly isotopic analyses and Svenja Agethen and Max Moenikes for help during laboratory work. Willi A. Brand, Heike Geilmann, Jessica Schäfer and Theresa Klötzing are acknowledged for ^{13}C analyses of soil samples. Furthermore, our thank goes to the managers of the exploratories, Swen Renner, Sonja Gockel annd Kerstin Wiesner for their work in maintaining the plot and project infrastructure; Simone Pfeiffer and Christiane Fischer giving support through the central office, Michael Owonibi for managing the central data base, and Markus Fischer, Eduard Linsenmair, Dominik Hessenmöller, Jens Nieschulze, Daniel Prati, Ingo Schöning, François Buscot, Ernst-Detlef Schulze, Wolfgang W. Weisser and the late Elisabeth Kalko for their role in setting up the Biodiversity Exploratories project.

Author Contributions

Conceived and designed the experiments: NH DP TK IS MS MF. Performed the experiments: VHK BS IS MS. Analyzed the data: VHK. Contributed reagents/materials/analysis tools: IS TK. Wrote the paper: VHK TK NH DP MF BS MS IS.

References

1. Camin F, Perini M, Bontempo L, Fabroni S, Faedi W, et al. (2011) Potential isotopic and chemical markers for characterizing organic fruits. Food Chemistry 125: 1072–1082.
2. Adams MA, Grierson PF (2001) Stable isotopes at natural abundance in terrestrial plant ecology and ecophysiology: An update. Plant Biology 3: 299–310.
3. Franke BM, Gremaud G, Hadorn R, Kreuzer M (2005) Geographic origin of meat - elements of an analytical approacho its authentication. European Food Research and Technology 221: 493–503.
4. Schwertl M, Auerswald K, Schäufele R, Schnyder H (2005) Carbon and nitrogen stable isotope composition of cattle hair: ecological fingerprints of production systems? Agriculture, Ecosystems and Environment 109: 153–165.
5. Nakano A, Uehara Y, Yamauchi A, (2003) Effect of organic and inorganic fertigation on yields, δ15N values, and δ13C values of tomato (Lycopersicon esculentum Mill. cv. Saturn). Plant and Soil 255: 343–349.
6. Georgi M, Voerkelius S, Rossmann A, Grassmann J, Schnitzler WH (2005) Multielement isotope ratios of vegetables from integrated and organic production. Plant and Soil 275: 93–100.
7. Rapisarda P, Calabretta ML, Romano G, Intrigliolo F (2005) Nitrogen metabolism components as a tool to discriminate between organic and conventional citrus fruits. Journal of Agricultural and Food Chemistry 53: 2664–2669.
8. Bateman AS, Kelly SD, Woolfe M (2007) Nitrogen Isotope Composition of Organically and Conventionally Grown Crops. Journal of Agricultural and Food Chemistry 55: 2664–2670.
9. del Amor FM, Navarro J, Aparicio PM (2008) Isotopic Discrimination as a Tool for Organic Farming Certification in Sweet Pepper. Journal of Envionmental Quality 37: 182–185.
10. Camin F, Moschella A, Miselli F, Parisi B, Versini G, et al. (2007) Evaluation of markers for the traceability of potato tubers grown in an organic versus conventional regime. Journal of the Science of Food and Agriculture 87: 1330–1336.
11. Boner M, Förstel H (2004) Stable isotope variation as a tool to trace the authenticity of beef. Analytical & Bioanalytical Chemistry 378: 301–310.
12. Schmidt O, Quilter JM, Bahar B, Moloney AP, Scrimgeour CM, et al. (2005) Inferring the origin and dietary history of beef from C, N and S stable isotope ratio analysis. Food Chemistry 91: 545–549.
13. Molkentin J (2009) Authentication of Organic Milk Using δ13C and the r-Linolenic Acid Content of Milk Fat. Journal of Agricultural and Food Chemistry 2009 57: 785–790.
14. Bateman AS, Kelly SD, Jickells TD (2005) Nitrogen Isotope Relationships between Crops and Fertilizer:Implications for Using Nitrogen Isotope Analysis as an Indicator of Agricultural Regime. Journal of Agricultural and Food Chemistry 53, 5760–5765.
15. Bateman AS, Kelly SD (2007) Fertilizer nitrogen isotope signatures. Isotopes in Environmental and Health Studies 43: 237–247.
16. Hobbie EA, Högberg P (2012) Nitrogen isotopes link mycorrhizal fungi and plants to nitrogen dynamics. New Phytologist 196: 367–382.
17. Kleinebecker T, et al. (submitted) 15N natural abundances reveal plant diversity effects on nitrogen dynamics in permanent grassland ecosystems. Submitted
18. de Bello F, Buchmann N, Casals P, Leps J, Sebastia MT (2009) Relating plant species and functional diversity to community δ13C in NE Spain pastures. Agriculture, Ecosystems and Environment 131: 303–307.
19. Högberg P, Johannisson C, Hallgren J-E (1993) Studies of 13Cin the foliage reveal interactions between nutrients and water in forest fertilization experiments. Plant and Soil 152: 207–214.
20. Šantrůčková H, Bird MI, Lloyd J (2000) Microbial processes and carbon-isotope fractionation in tropical and temperate grassland soils. Functional Ecology 14: 108–114.
21. Haas G, Wetterich F, Köpke U (2001) Comparing intensive, extensified and organic grassland farming in southern Germany by process life cycle assessment. Agriculture, Ecosystems and Environment 83: 43–53.
22. Schaack D, Iller S, Würtenberger E (2010) AMI-Marktbilanz Öko-Landbau 2010, Bonn: Agrarmarkt Informationsgesellschaft mbH.
23. Watzka M, Buchgraber K, Wanek W (2006) Natural N-15 abundance of plants and soils under different management practices in a montane grassland. Soil Biology & Biochemistry 38: 1564–1576.
24. Kahmen A, Wanek W, Buchmann N (2008) Foliar delta ^{15}N values characterize soil N cycling and reflect nitrate or ammonium preference of plants along a temperate grassland gradient. Oecologia 156: 861–870.
25. Fischer M, Bossdorf O, Gockel S, Hansel F, Hemp A, et al. (2010) Implementing large-scale and long-term functional biodiversity research: The Biodiversity Exploratories. Basic and Applied Ecology 11: 473–485.
26. Bluethgen N, Dormann CF, Prati D, Klaus VH, Kleinebecker T, et al. (2012) A quantitative index of land-use intensity in grasslands: Integrating mowing, grazing and fertilization. Basic and Applied Ecology 13: 207–220.
27. European Union (2008) Commission Regulations (EC) No 889/2008, Brussels.
28. Werner RA, Brand WA (2001) Referencing strategies and techniques in stable isotope ratio analysis. Rapid Communications in Mass Spectrometry 15: 501–519.
29. LUFTGEIST website. Available: http://www.luftgeist.de. Accessed 2013 Jun 3.
30. Wetterstation Riethein Lichse website. Available: http://www.wetterstation-rietheim-lichse.de/html/archiv.html. Accessed 2013 Jun 3.
31. Venables WN, Ripley BD (2002) Modern Applied Statistics with S. Fourth Edition. New York: Springer.
32. R Development Core Team (2011) R: A language and environment for statistical computing. R Foundation for Statistical Computing, Vienna.
33. Klaus VH, Kleinebecker T, Prati D, Fischer M, Alt F, et al. (2013) Does organic grassland farming benefit plant and arthropod diversity at the expense of yield and soil fertility? Agriculture, Ecosystems and Environment 177: 1–9.
34. Wrage N, Küchenmeister F, Isselstein J (2011) Isotopic composition of soil, plant or cattle hair no suitable indicator of nitrogen balances in permanent pasture. Nutrient Cycling in Agroecosystems 90: 189–199.
35. Gubsch M, Roscher C, Gleixner G, Habekost M, Lipowsky A, et al. (2011) Foliar and soil delta 15N values reveal increased nitrogen partitioning among species in diverse grassland communities. Plant Cell and Environment 34: 895–908.
36. Hole DG, Perkins AJ, Wilson JD, Alexander IH, Grice PV, et al. (2005) Does organic farming benefit biodiversity? Biological Conservation 122: 113–130.
37. Batáry P, Báldi A, Sárospataki M, Kohler F, Verhulst J, et al. (2010) Effect of conservation management on bees and insect-pollinated grassland plant communities in three European countries. Agriculture, Ecosystems and Environment 136: 35–39.
38. Neilson R, Hamilton D, Wishart J, Marriott CA, Boag B, et al. (1998) Stable isotope natural abundances of soil, plants and soil invertebrates in an upland pasture. Soil Biology and Biochemistry 30: 1773–1782.
39. Caldeira MC, Ryel RJ, Lawton JH, Pereira JS (2001) Mechanisms of positive biodiversity-production relationships: insights provided by δ13C analysis in experimental Mediterranean grassland plots. Ecology Letters 4: 439–443.
40. Jumpponen A, Mulder CPH, Huss-Danell K, Högberg P (2005) Winners and losers in herbaceous plant communities: insights from foliar carbon isotope composition in monocultures and mixtures. Journal of Ecology 93: 1136–1147.
41. Baisden WT, Amundson R, Cook AC, Brenner DL (2002) Turnover and storage of C and N in five density fractions from California annual grassland surface soils. Global Biogeochemical Cycles 16: 64–71.
42. Bol R, Moering J, Preedy N, Glaser B (2004) Short-term sequestration of slurry-derived carbon into particle size fractions of a temperate grassland soil. Isotopes in Environmental and Health Studies 40: 81–87.
43. Shangguan ZP, Shao MA, Dyckmans J (2000) Nitrogen nutrition and water stress effects on leaf photosynthetic gas exchange and water use efficiency in winter wheat. Environmental and Experimental Botany 44: 141–149.
44. Socher S, Prati D, Müller J, Klaus VH, Hölzel N, et al. (2012) Direct and productivity-mediated indirect effects of fertilization, mowing and grazing intensities on grassland plant species richness. Journal of Ecology 100: 1391–1399.

21

Yield and Economic Performance of Organic and Conventional Cotton-Based Farming Systems

Dionys Forster[1], **Christian Andres**[1]*, **Rajeev Verma**[2], **Christine Zundel**[1,3], **Monika M. Messmer**[4], **Paul Mäder**[4]

1 International Division, Research Institute of Organic Agriculture (FiBL), Frick, Switzerland, 2 Research Division, bioRe Association, Kasrawad, Madhya Pradesh, India, 3 Ecology Group, Federal Office for Agriculture (FOAG), Bern, Switzerland, 4 Soil Sciences Division, Research Institute of Organic Agriculture (FiBL), Frick, Switzerland

Abstract

The debate on the relative benefits of conventional and organic farming systems has in recent time gained significant interest. So far, global agricultural development has focused on increased productivity rather than on a holistic natural resource management for food security. Thus, developing more sustainable farming practices on a large scale is of utmost importance. However, information concerning the performance of farming systems under organic and conventional management in tropical and subtropical regions is scarce. This study presents agronomic and economic data from the conversion phase (2007–2010) of a farming systems comparison trial on a Vertisol soil in Madhya Pradesh, central India. A cotton-soybean-wheat crop rotation under biodynamic, organic and conventional (with and without Bt cotton) management was investigated. We observed a significant yield gap between organic and conventional farming systems in the 1st crop cycle (cycle 1: 2007–2008) for cotton (−29%) and wheat (−27%), whereas in the 2nd crop cycle (cycle 2: 2009–2010) cotton and wheat yields were similar in all farming systems due to lower yields in the conventional systems. In contrast, organic soybean (a nitrogen fixing leguminous plant) yields were marginally lower than conventional yields (−1% in cycle 1, −11% in cycle 2). Averaged across all crops, conventional farming systems achieved significantly higher gross margins in cycle 1 (+29%), whereas in cycle 2 gross margins in organic farming systems were significantly higher (+25%) due to lower variable production costs but similar yields. Soybean gross margin was significantly higher in the organic system (+11%) across the four harvest years compared to the conventional systems. Our results suggest that organic soybean production is a viable option for smallholder farmers under the prevailing semi-arid conditions in India. Future research needs to elucidate the long-term productivity and profitability, particularly of cotton and wheat, and the ecological impact of the different farming systems.

Editor: Jean-Marc Lacape, CIRAD, France

Funding: Biovision Foundation for Ecological Development, http://www.biovision.ch/; Coop Sustainability Fund, http://www.coop.ch/pb/site/nachhaltigkeit/node/64228018/Len/index.html; Liechtenstein Development Service (LED), http://www.led.li/en/home.html; Swiss Agency for Development and Cooperation (SDC), http://www.sdc.admin.ch/. The funders had no role in study design, data collection and analysis, decision to publish, or preparation of the manuscript.

Competing Interests: The authors have declared that no competing interests exist.

* E-mail: christian.andres@fibl.org

Introduction

The green revolution has brought about a series of technological achievements in agricultural production, particularly in Asia. Worldwide cereal harvests tripled between 1950 and 2000, making it possible to provide enough dietary calories for a world population of six billion by the end of the 20th century [1]. So far, global agricultural development has rather focused on increased productivity than on a more holistic natural resource management for food security and sovereignty. The increase in food production has been accompanied by a multitude of challenges and problems such as the exploitation and deterioration of natural resources, i.e. loss of soil fertility, strong decline of agro-biodiversity, pollution of water [2,3], and health problems associated with the use of synthetic plant protection products [4]. At present, more comprehensive system-oriented approaches are gaining momentum and are expected to better address the difficult issues associated with the complexity of farming systems in different locations and cultures [5].

The concept of organic agriculture builds on the idea of the efficient use of locally available resources as well as the usage of adapted technologies (e.g. soil fertility management, closing of nutrient cycles as far as possible, control of pests and diseases through management and natural antagonists). It is based on a system-oriented approach and can be a promising option for sustainable agricultural intensification in the tropics, as it may offer several potential benefits [6–11] such as: (i) A greater yield stability, especially in risk-prone tropical ecosystems, (ii) higher yields and incomes in traditional farming systems, once they are improved and the adapted technologies are introduced, (iii) an improved soil fertility and long-term sustainability of farming systems, (iv) a reduced dependence of farmers on external inputs, (v) the restoration of degraded or abandoned land, (vi) the access to attractive markets through certified products, and (vii) new partnerships within the whole value chain, as well as a strengthened self-confidence and autonomy of farmers. Critics contend that organic agriculture is associated with low labor productivity and high production risks [1,12–14], as well as high certification costs for smallholders [15]. However, the main criticism reflected in the scientific literature is the claim that organic agriculture is not able to meet the world's growing food demand, as yields are on average 20% to 25% lower than in conventional agriculture [16,17]. It should however be taken into account, that yield deviations among different crops and regions can be substantial depending on system and site characteristics [16,17]. In a meta-analysis by Seufert *et al.* [17] it is shown that yields in organic farming systems with good management practices can nearly match conventional yields, whereas under less favorable conditions they cannot. However, Reganold [18] pointed out that productivity is not the only goal that must be met in order for agriculture to be considered sustainable: The maintenance or enhancement of soil fertility and biodiversity, while minimizing detrimental effects on the environment and the contribution to the well-being of farmers and their communities are equally important as the above mentioned productivity goals. Farming systems comparison trials should thus - besides agronomic determinants - also consider ecological and economic factors over a longer period. These trials are inherently difficult due to the many elements the farming systems are comprised of, thus necessitating holistic research approaches in order to make comparisons possible [19].

Results from various farming systems comparison trials between organic and conventional management have shown, that even though yields may be slightly lower, organic farming systems exhibit several ecological and economic advantages, particularly long-term improvement of soil fertility [20–25]. However, most of the data has been obtained from trials in the temperate zones [20–26]. The little data available under tropical and subtropical conditions [9,27–29] calls for more long-term farming systems comparison trials to provide a better basis for decision making in these regions [17]. To address this issue, the Research Institute of Organic Agriculture (FiBL) has set up three farming systems comparison trials in Kenya, India and Bolivia, thereby encompassing different cropping systems and ethnologies. The main objective of these trials is to collect solid agronomic and socio-economic data on major organic and conventional agricultural production systems in the selected project regions. These trials will contribute to close the existing knowledge gap regarding the estimation of profitability of organic agriculture in developing countries (http://www.systems-comparison.fibl.org/). This paper presents results from cotton-based farming systems in India.

India is the second largest producer (after China) of cotton lint worldwide [30]. Cotton is a very important cash crop for smallholder farmers, but also one of the most exigent crops in terms of agrochemical inputs which are responsible for adverse effects on human health and the environment [27]. Genetically modified (GM) cotton hybrids carrying a gene of *Bacillus thuringiensis* (Bt) for protection against bollworm (*Helicoverpa* spp.) attack, have spread rapidly after their official introduction to India in 2002 [31,32]. By 2012, 7 million farmers cultivating 93% of India's total cotton area had adopted Bt cotton technology [32,33]. This high adoption rate might be attributed to the high pressure caused by cotton bollworms, and associated reductions in pesticide use upon the introduction of Bt cotton technology in India [34,35]. However, the discussion about the impacts of Bt cotton adoption remains highly controversial [36,37]. Giving focus to yields, advocates of Bt cotton claim that the technology has led to an increase in productivity of up to 60% [38–40] and in some cases even "near 100%" [41]. Opponents of Bt cotton on the other

hand attributed the yield gains, compared to the pre-Bt period, to other factors. These include (i) the increase of the area under cotton cultivation, (ii) the shift from traditional diploid cotton (*G. arboreum*, *G. herbaceum* which accounted for 28% of total cotton area in 2000) to tetraploid *G. hirsutum* species [42] and the widespread adoption of hybrid seeds, (iii) the increased use of irrigation facilities, (iv) the introduction of new pesticides with novel action (e.g. Imidacloprid seed treatment), and (v) the increased use of fertilizers in Bt cotton cultivation [43,44]. Critics of Bt crops also stress uncertainties concerning the impact of the technology on human health [45] and on non-target organisms [46], as well as the higher costs of Bt seeds [34,47].

While some argue that GM crops in general can contribute significantly to sustainable development at the global level [33,48], others state that there is no scientific support for this claim [49]. Considering economic benefits of Bt cotton, the same controversy prevails: Advocates claim sustainable socio-economic benefits and associated social development [33,36], while opponents claim Bt cotton to be responsible for farmer debt [50], thereby contributing to India's notoriety for farmers' suicides [37,51], a linkage which has been criticized as reductionist and invalid [52]. However, comparisons are mainly drawn between Bt and non-Bt cotton under conventional management in high-input farming systems. Organic cotton production systems - holding a minor percentage of the cotton growing area in India - are often neglected, and little information exists on the productivity and profitability of organic farming in India [53]. However, organic cotton production is slowly gaining momentum in the global cotton market [27]. GM cultivars are not compatible with the guidelines of organic agriculture [54]. Therefore, organic cotton producers have to refrain from Bt cotton hybrids. In addition, organic producers and processors have to take all possible measures to avoid contamination with Bt cotton in order not to lose organic certification.

While organic farming systems have attracted considerable interest of the scientific community [16,17,21,26], biodynamic farming systems are less common and little investigated. The biodynamic agricultural movement started in the early 1920s in Europe [55] and developed the international certification organization and label DEMETER. In India, the biodynamic movement started in the early 1990s (www.biodynamics.in). Preparations made from manure, minerals and herbs are used in very small quantities to activate and harmonize soil processes, to strengthen plant health and to stimulate processes of organic matter decomposition. Most biodynamic farms encompass ecological, social and economic sustainability and many of them work in cooperatives. One of the first initiatives in India was bioRe India Ltd. in Madhya Pradesh state (formerly called Maikaal cotton project), where several thousand farmers (2007–2010 between 4'700 and 8'800) produce organic cotton mainly for the European market (www.bioreindia.com). Although the farmers in this cooperative are trained in biodynamic farming, and follow the taught practices to a certain extent, the system is not certified as biodynamic. Nonetheless the products are declared as organic. The farming systems comparison trial presented here was set up in 2007 in Madhya Pradesh state, central India, and is embedded at the training and education center of bioRe Association (www.bioreassociation.org/bioresearch.html). The main aim of the trial is to assess the agronomic, economic and ecological performance of cotton-based farming systems under organic, biodynamic and conventional management including non-Bt and Bt cotton. In this paper, we present the yield and gross margin of cotton, soybean and wheat of the four different farming systems within the first four years after inception of the trial (considered as conversion period in this paper).

Materials and Methods

1 Site description and socioeconomic context

The trial site is located in the plains of the Narmada river belt in the Nimar Valley, Khargone district, Madhya Pradesh state, India ($22°8'30.3''$N, $75°4'49.0''$E), at an altitude of 250 meters above sea level. The climate is subtropical (semi-arid), with an average annual precipitation of 800 mm, which occurs in a single peak monsoon season usually lasting from mid-June to September. Temperatures range from $15°C$ to $49°C$ with a yearly average of $25°C$, and are highest in May/June and lowest in December/ January. Climatic data from 2007–2010 obtained near the trial are shown in Figure 1. The trial is located on a fertile Vertisol soil characterized by an average clay content of 600 g kg^{-1} soil, pH (H_2O) of 8.7, organic C content of 5.0 g kg^{-1} soil, and available P content (Olsen) of 7.0 mg kg^{-1} soil at the start of the trial. Vertisols have shrink-swell characteristics; they cover about 73 million ha of the subtropical (semi-arid) regions of India and are the predominant soil type in Madhya Pradesh [56].

Agriculture is the main livelihood activity in the project area. Farm sizes range from less than 1 ha to more than 10 ha, and soil fertility as well as access to irrigation water vary greatly throughout the region. The major crops in the region are cotton, soybean and wheat. Since 2002, Bt cotton has become very popular and is currently grown on more than 90% of the total area under cotton cultivation in Madhya Pradesh [57,58]. About 50% of India's organic cotton is produced in Madhya Pradesh [59]. The year consists of three seasons with distinctly different climatic characteristics: The Kharif (monsoon) season is characterized by the monsoon and lasts from June to October. Crops which require humid and warm condition are grown, for example cotton, or soybean. The Rabi (winter) season is characterized by lower temperatures and less rainfall; it lasts from November to March. Crops which require cool temperatures for vegetative growth are grown, for example wheat or chick pea. Finally, the Zaid (summer) season is characterized by hot temperatures and an extensive dry spell; it lasts from March to June. Only farmers with access to irrigation facilities or near river banks grow crops such as melons, gourds or cucumbers in this season. Longer duration crops such as cotton are cultivated during both Kharif and Rabi seasons.

2 Trial description

The farming systems comparison trial was established in 2007, and is expected to run for a period of 20 years. Before trial setup, the site was under conventional management by a local farmer. The homogeneity of the terrain was assessed before the implementation of the different farming systems with a test crop of unfertilized wheat (HA (0)) grown from December 2006 to

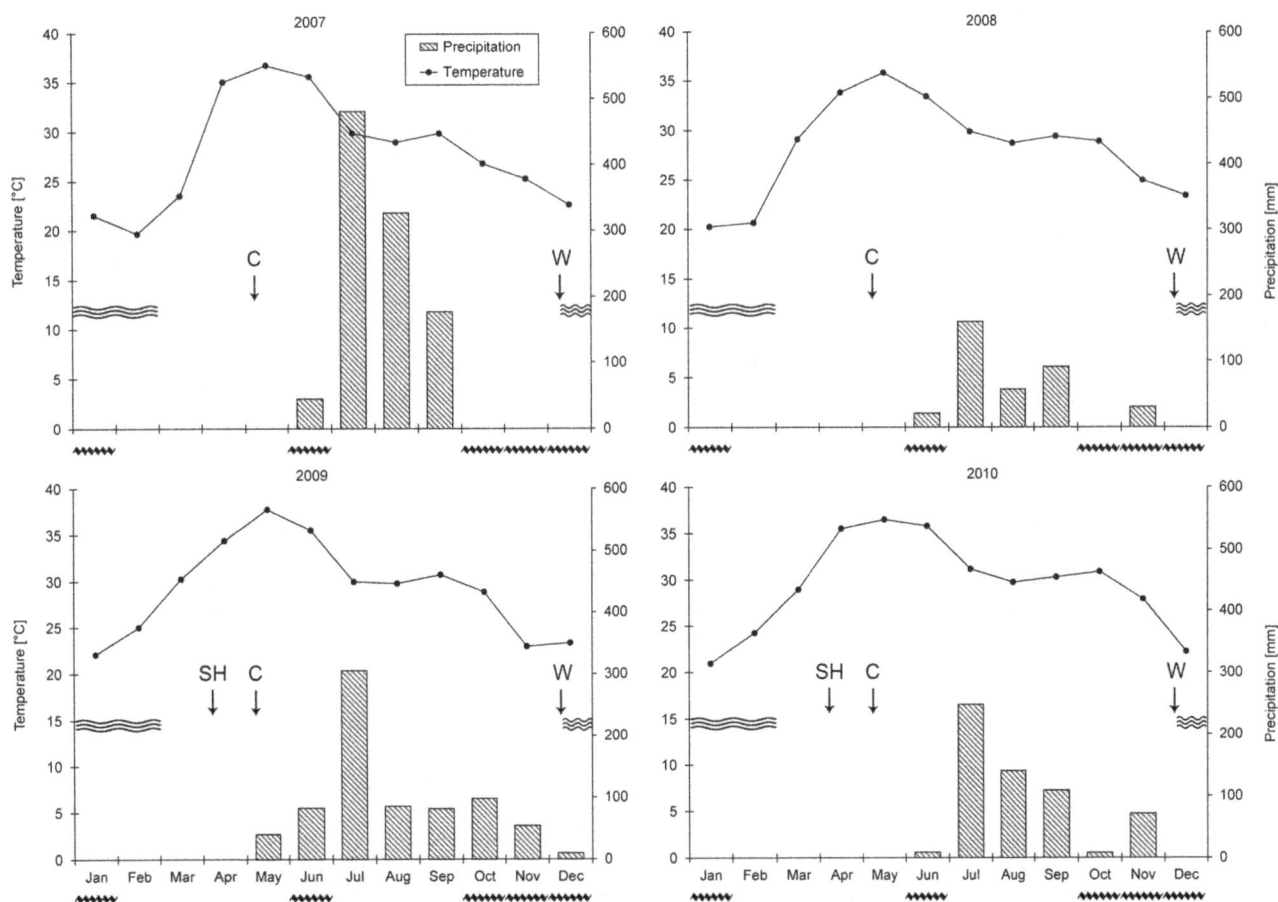

Figure 1. Temperature and precipitation recorded near the trial, Madhya Pradesh, India, 2007–2010, and irrigation practices in the farming systems comparison trial. Vertical arrows (↓) indicate flood irrigation prior to sowing of cotton (C), wheat (W) and sunn hemp (SH). Sunn hemp (green manure) was only grown in 2009 and 2010 on BIODYN and BIOORG plots before cotton sowing. Single closed undulating lines indicate period of drip and flood irrigation in cotton, multiple open undulating lines indicate period of flood irrigation in wheat (wheat received four to five flood irrigations).

Farming system	CYCLE 1								CYCLE 2								
	Year																
	2007				2008				2009				2010				2011	
	Quarter																	
	1	2	3	4	1	2	3	4	1	2	3	4	1	2	3	4	1	2	3	4
	Season																	
		Zaid	Kharif		Rabi	Zaid	Kharif		Rabi	Zaid	Kharif		Rabi	Zaid	Kharif		Rabi	Zaid	Kharif	

Strip 1 (plots 1 - 16)

BIODYN	HA (0)	Cotton		Soybean		Wheat	SH	Cotton		Soybean	Wheat	
BIOORG	HA (0)	Cotton		Soybean		Wheat	SH	Cotton		Soybean	Wheat	
CON	HA (0)	Cotton		Soybean		Wheat		Cotton		Soybean	Wheat	
CONBtC	HA (0)	Cotton		Soybean		Wheat		Cotton	Wheat 2	Soybean	Wheat	

Strip 2 (plots 17 - 32)

BIODYN	HA (0)	Soybean	Wheat	Cotton		Soybean	Wheat	SH	Cotton	
BIOORG	HA (0)	Soybean	Wheat	Cotton		Soybean	Wheat	SH	Cotton	
CON	HA (0)	Soybean	Wheat	Cotton		Soybean	Wheat		Cotton	
CONBtC	HA (0)	Soybean	Wheat	Cotton		Soybean	Wheat		Cotton	Wheat 2

Figure 2. Sequence of crops in different farming systems of the farming systems comparison trial 2007–2010. Seasons: Zaid (summer): March to June, Kharif (monsoon): June to October, Rabi (winter): November to March. HA (0) indicates the homogeneity assessment performed with unfertilized wheat before the implementation of the different farming systems. In 2009 and 2010 Bt cotton was uprooted 2 months earlier to grow a second wheat crop (wheat 2) to reflect common practice of local Bt cotton farmers.

March 2007 (Figure 2). The test crop was harvested using a 5×5 m grid. Data of wheat grain yield, organic C and pH of the soil were used for allocation of strips, blocks and plots (Figure S1).

The trial comprises two organic farming systems (biodynamic (BIODYN), organic (BIOORG)) and two conventional farming systems (conventional (CON), conventional including Bt cotton (CONBtC)). Details of the farming systems are shown in Tables 1 and S1. Organic and biodynamic farming were carried out according to the standards defined by the International Federation of Organic Agriculture Movements (IFOAM) [60] and DEMETER-International [61], respectively. Conventional farming systems followed the recommendations of the Indian Council of Agricultural Research (ICAR) [62] with a slight adjustment to represent local conventional farming systems: farmyard manure (FYM) was applied to account for the integrated nutrient management of local conventional farmers. BIODYN represented the predominant local organic practices, as farmers associated to bioRe India Ltd. (see above) are provided with the respective inputs and trained in biodynamic farming as practiced in the field trial. BIOORG represented general organic practices as practiced in various regions of India where organic cotton is grown (mainly Madhya Pradesh, Maharashtra and Gujarat [59]). CON represented the local conventional practices in Madhya Pradesh before the introduction of Bt cotton in 2002, and CONBtC represented the current local conventional practices.

The four farming systems mainly differed in the following aspects: Genetic material (cotton only), type and amounts of fertilizer inputs, green manures, plant protection, the use of biodynamic preparations (Table 1, Table S1), and crop sequence (Figure 2). Farming systems are extremely complex, whereby individual management practices are closely linked and interdependent. For instance, it is well known that chemical plant protection is in most cases only economically feasible under conditions of optimal fertilization. That means that we mirror to a certain extent the complexity of a system rather than analyzing effects of single factors, and we intended to mimic common regional practices for the respective farming systems with respect to all management practices, as specified above. This approach is quite common in farming systems research and reflects effects of the system as a whole [20,21], but does not allow to trace potential differences to individual practices. As a basis for the design of the organic and conventional farming systems served a farm survey of Eyhorn et al. [9] in the same region.

The two-year crop rotation consisted of cotton (*Gossypium hirsutum* L.), soybean (*Glycine max* (L.) Merr.) and wheat (*Triticum aestivum* L.) (Figure 2). While in organic farming systems green gram (*Vigna radiata*) was grown between cotton rows in all four years and sunn hemp (*Crotalaria juncea*) was used as a preceding green manure crop for cotton in 2009 and 2010 (crop cycle 2), none of these practices were followed in conventional farming systems. Both green manure crops were cut at flowering and incorporated to the soil. In order to compare the CON and CONBtC farming systems as a whole (rather than the effect of the Bt gene), both the fertilizer dose and crop sequence was adapted (Figure 2).

Fertilizer inputs relied mainly on synthetic products in conventional farming systems (depending on crop between 68 and 96% of total nitrogen (N_{total}) applied (Table S1)), whereas organic farming systems received nutrients from organic sources only (Table 1). Organic fertilizers were compost, castor cake, and FYM. Compost was prepared using crop residues, weeds, FYM, and slurry from biogas plants (fed with fresh FYM) as raw materials. FYM was also applied in both conventional farming systems. The relatively high levels of organic fertilizer inputs (Table 1) reflect practices of local smallholder farmers who usually apply some 18.5 t ha^{-1} fresh matter of compost to cotton. On average, compost and FYM contained 0.8-0.6-1.5% and 0.8-0.6-1.6% of N_{total}-P_2O_5-K_2O, respectively, whereas castor cake contained 3.3-0.9-0.9% of N_{total}-P_2O_5-K_2O. Compost and FYM were broadcasted on the field after land preparation and subsequently incorporated to the soil by bullock-drawn harrows in all farming systems; However, in the organic farming systems in cotton, only 50% of the compost was applied as basal fertilizer input, and the remaining 50% were applied in two equal split applications as top dressings, at square formation and flowering, respectively. Castor cake was applied plant to plant. Nutrient inputs by N-fixing green manure crops were not considered, but will be assessed in future studies (Table S1). Synthetic fertilizers applied in both conventional farming systems were Diammonium phosphate (DAP), Muriate of Potash (MOP), Single Super Phosphate (SSP) and Urea. MOP, SSP and Urea/DAP were applied as basal fertilizer input at sowing time, except in cotton where only 50% of Urea/DAP was applied as basal fertilizer input, and the remaining 50% as a single top dressing at flowering. Across all crops and years, input of N_{total} was 65 kg ha^{-1} in organic farming systems (BIODYN, BIOORG), 105 kg ha^{-1} in CON and 113 kg ha^{-1} in CONBtC (Table S1). The lower inputs of N_{total} in organic compared to conventional farming systems represent local organic practice. The difference in inputs of N_{total} between CON and CONBtC arises from adhering to

Table 1. Management of the different farming systems compared in a two-year rotation in central India (2007–2010).

Practices	Organic farming systems[1]		Conventional farming systems[2]	
	BIODYN (biodynamic)	**BIOORG (Organic)**	**CON (conventional)**	**CONBtC (conventional including Bt cotton)**
Genetic material (difference in cotton only)				
	Non-Bt cotton	Non-Bt cotton	Non-Bt cotton	Bt cotton
Fertilizer input				
Type and level (for nutrient inputs see Table S1)	aerobically composted crop residues, weeds, farmyard manure (FYM), and slurry; 19.5-7.7-12.0 t ha^{-1} to cotton-soybean-wheat	aerobically composted crop residues, weeds, farmyard manure (FYM), and slurry; 19.5-7.7-12.0 t ha^{-1} to cotton-soybean-wheat	mineral fertilizers (MOP, SSP, Urea, DAP (wheat only))	mineral fertilizers (MOP, SSP, Urea, DAP (wheat only))
	stacked FYM; 2.8-1.6-2.2 t ha^{-1} to cotton-soybean-wheat	stacked FYM; 2.8-1.6-2.2 t ha^{-1} to cotton-soybean-wheat	stacked FYM; 8.1-3.9-1.6 t ha^{-1} to cotton-soybean-wheat	stacked FYM; 8.1-3.9-1.6 t ha^{-1} to cotton-soybean-wheat
	castor cake; 0.1 t ha^{-1} to cotton (2007 & 2008 only)	castor cake; 0.1 t ha^{-1} to cotton (2007 & 2008 only)		
Green manure				
Type and timing of green manure	broadcasted sunn hemp (*Crotalaria juncea*) before cotton in 2009 and 2010 only	broadcasted sunn hemp (*Crotalaria juncea*) before cotton in 2009 and 2010 only	None	None
	hand sown green gram (*Vigna radiata*, 9'070 plants ha^{-1}) between cotton rows in all years	hand sown green gram (*Vigna radiata*, 9'070 plants ha^{-1}) between cotton rows in all years	None	None
Plant protection				
Weed control	bullock-drawn blade or tine harrows	bullock-drawn blade or tine harrows	bullock-drawn blade or tine harrows	bullock-drawn blade or tine harrows
	Hand weeding in cotton	Hand weeding in cotton	Hand weeding in cotton	Hand weeding in cotton
	None	None	Herbicide (2009 and 2010 in soybean and wheat only)	Herbicide (2009 and 2010 in soybean and wheat only)
Insect control and average number of applications per crop rotation (detailed product list, see Table S1)	organic (natural) pesticides 12.5	organic (natural) pesticides 12.25	synthetic pesticides 11.5	synthetic pesticides 11.0
Disease control	None	None	None	None
Special treatments	biodynamic preparations[3]	None	None	None

[1] in the text, BIODYN and BIOORG are referred to consistently as organic farming systems, [2] in the text, CON and CONBtC are referred to consistently as conventional farming systems, average dry matter content of organic fertilizers: 70%, DAP: Diammonium phosphate, MOP: muriate of potash, SSP: single super phosphate,
[3] biodynamic preparations entailed cow dung (BD-500) and silica powder (BD-501) both stored for six months, and a mixture of cow dung, chicken egg shell powder, basalt rock powder, and plant materials (yarrow, chamomile, stinging nettle, oak bark, dandelion, valerian) stored for 6 months in an open pit (cow pat pit = CPP).

recommendations by ICAR [62] who advocate systems with Bt cotton to be managed more intensively than systems with non-Bt cotton.

Pest management - including seed treatment - was done with organic (natural) pesticides in organic farming systems, while in conventional farming systems synthetic pesticides were used (Table 1). The type and number of pesticide applications in CON and CONBtC was the same to reflect local farmers' practices [63]. This practice was also confirmed in the survey of Kathage and Qaim [36] comparing conventional Bt and non-Bt cotton in the period 2006–2008 conducted in the four states Maharashtra, Karnataka, Andhra Pradesh, and Tamil Nadu.

The BIODYN system received small amounts of biodynamic preparations (Table 1) consisting of organic ingredients (cow manure, medicinal plants), and mineral compounds (quartz, basalt) which are intended to activate the soil and increase plant health [64]. No significant amounts of nutrients were added by these applications. For further details of biodynamic practices see Carpenter-Boggs *et al.* [64].

With cotton, soybean and wheat the trial represents a cash crop-based farming system in a two-year crop rotation, which is typical for the Nimar Valley in the plains of the Narmada river belt, were the trial is located. Cotton was grown from May to February, except in 2009 and 2010 (crop cycle 2) in CONBtC; in these two years Bt cotton was uprooted two months earlier than in the other three farming systems in order to grow an additional wheat crop (wheat 2) in the Rabi (winter) season (Figure 2). This was done to account for local practices; local conventional farmers noticed that Bt cotton matures earlier than non-Bt cotton, and produces the majority of the yield during the first three months of the harvesting period. Therefore, they started between 2007 and 2010 to grow another wheat crop before the start of the Zaid (summer) season, a practice which was also confirmed by Brookes & Barfoot [65]. Soybean was grown from July to October and followed by wheat from December to March. The land was prepared with bullock-drawn ploughs, harrows and levelers. Cotton was sown by hand at a rate of 0.91 plants m^{-2} (9'070 plants ha^{-1}). Soybean and wheat were sown with bullock-drawn seed drills. The inter row and intra row spacing were 30 cm and 4 cm, respectively for both soybean and wheat. In 2007, heavy monsoon rains led to severe waterlogging in the plots which stunted soybean growth and necessitated re-sowing the whole trial. Cultivars were selected according to local practice and availability. In cotton, these were Maruti 9632 (2007), Ankur 651 (2008), Ankur AKKA (2009) and JK Durga (2010) in all farming systems, except in CONBtC where isogenic Bt lines of the same hybrids were used. Non-GM soybean, variety JS-335, and non-GM wheat, variety LOK-1 were cultivated in all farming systems and years. The whole trial was irrigated and all plots received similar amounts of irrigation water; prior to sowing, flood irrigation was carried out on sunn hemp (green manure), cotton and wheat plots (Figure 1). After the monsoon, cotton received additional drip irrigation and two to three flood irrigations to ensure continuous water supply throughout the cropping season. Sunn hemp and wheat received three to four and four to five flood irrigations, respectively. Soybean was grown purely rainfed during the Kharif (monsoon) season. Weeding was done mechanically at 20 (cotton) and 45 (soybean, wheat) days after sowing, using bullock-drawn blade or tine harrows in all farming systems. In cotton, additional hand weeding was carried out. No hand weeding was carried out in soybean and wheat. No synthetic herbicides were applied in conventional farming systems except in soybean and wheat in 2009 and 2010, which reflects the situation of most smallholder cotton farmers in India [66]. Cotton was harvested by several manual hand pickings. Soybean and wheat were harvested manually with sickles, and bound to bundles which were removed from the field and subsequently threshed with a threshing machine.

In order to obtain data from each crop during each year, the layout was doubled with shifted crop rotation in two strips, resulting in a total of 32 plots, and 16 plots per strip (Figure S1). Each farming system was replicated four times in a randomized block design in each of the two strips. Plots are sized 16 m×16 m (= gross plot) and time measurements of activities were recorded for gross plots. The outermost 2 m of each plot served as a border, and yield data were only obtained in the inner sampling plot sized 12 m×12 m (= net plot) in order to avoid border effects. The distance between two plots within a strip and between the two strips is 6 m and 2 m, respectively. Data was obtained from 2007 to 2010. Data from 2007–2008 belongs to the complete crop rotation of the 1st crop cycle (cycle 1), and data from 2009–2010 to the 2nd crop cycle (cycle 2).

As Bt cotton was commercially released in India in 2002, no official approval of the study was required. The land needed for the farming systems comparison trial was purchased and belongs to bioRe Association. No protected species were sampled.

3 Data consolidation and economic calculations

Calculations of gross margins required consolidation of production costs. We only considered variable (operational) production costs in our study, excluding interest rates for credits. We included input costs, labor costs for field activities (including e.g. compost preparation), and costs associated with the purchase of inputs from the local market. Time measurements on gross plots and farmers' fields were complemented with data obtained in expert meetings with experienced farmers and local extension officers. Variable production costs for cotton (Table S2), soybean (Table S3), and wheat (Table S4) were cross-checked with the values reported by the Ministry of Agriculture, Government of

India [67]. Gross margins were obtained by subtracting the variable production costs from the gross return (= yield * price per unit). Prices (products, inputs, labor) corresponded to local market conditions and were adapted each year (Table 2, Table S2). A premium price for organic cotton was considered in 2010 only (after three years conversion period according to IFOAM standards).

4 Statistical analysis

Data exploration revealed four outliers which were removed from the dataset. The reason was heavy monsoon rains and subsequent water-logging in four plots in 2009 (Plots 11 and 27 (both BIOORG), and plots 12 and 28 (both BIODYN), Figure S1).

Yield and gross margin data of each crop, and of the complete crop rotation (cotton+wheat 2+soybean+wheat) were analyzed separately with linear mixed effect models using the function lme from the package nlme [68] of the statistical software R version 2.15.2 [69]. We checked our data for model assumptions graphically (normal Q-Q of fixed and random effects, Tukey-Anscombe and Jitter plots) and no violation was encountered. We used a model with *System*, *Cycle*, the interaction of *System×Cycle* and *Strip* as fixed effects, and *Year* (n = 4), *Block* (n = 4) and *Pair* (n = 16) as random intercepts.

The fixed effect *Cycle* was included in the model to account for repeated measures on the same plot (e.g. cotton on plot 1 in 2007 and in 2009) and allows a partial separation of *Cycle* and *Year* effects due to the shifted crop rotation in the two strips as proposed by Loughin [70] for long-term field trials. *Cycle* effects give an indication how the situation changes across the timeframe of the trial. However, as we only have two levels of *Cycle* (thus Df = 1 for *Cycle* in the ANOVA) at this stage of the trial, we have little statistical power to detect *Cycle* effects. The same applies to the fixed effect *Strip*. We nevertheless included *Cycle* and *Strip* into our model to separate the *System* effect from possible *Cycle* and *Strip* effects. To account for similar conditions of neighboring plots (e.g. Plots 1 and 17, 2 and 18, etc., Figure S1) we included the random intercept *Pair* with 16 levels.

For yield and gross margin data of the complete crop rotation, the random intercept *Year* was removed from the model, as data from two years were compiled. Significant *System×Cycle* interactions suggested that the main effects of *System* and *Cycle* have to be interpreted with caution; As the effects of the different systems were not consistent across cycles, we split the datasets and performed post-hoc multiple comparisons for the fixed effect *System* separately for each cycle (method: Tukey, superscript letters after cycle-wise values in Tables 3 and 4). In the case of gross margin data of soybean, no significant *System×Cycle* interaction was encountered. Therefore, we performed post-hoc multiple comparisons on the whole dataset of cycle 1 and cycle 2 together (superscript letters after average values in Table 4). We defined a difference to be significant if $P < 0.05$ ($\alpha = 0.05$).

Results and Discussion

1 Yield

Cotton yields (seed cotton, picked bolls containing seed and fiber) were, averaged across the four years, 14% lower in organic (BIOORG, BIODYN) compared to conventional farming systems (CON, CONBtC). This is in the same range as the findings of a study conducted in Kyrgyzstan [27]. The *System×Cycle* interaction had a significant effect ($P<0.001$) on cotton yields (Table 3). The difference in yield was very pronounced in cycle 1 (2007–2008, +42% yield increase in conventional farming systems), while yields were similar among all farming systems in cycle 2 (2009–2010)

Table 2. Domestic market prices of cotton, soybean and wheat, premium prices on organic cotton and prices per working hour 2007–2010 in Khargone district, Madhya Pradesh, India.

Year	Commodity				
	Cotton [INR kg^{-1}]	Cotton premium price [INR kg^{-1}]	Soybean [INR kg^{-1}]	Wheat [INR kg^{-1}]	Labor [INR h^{-1}]
2007	23.3	4.7 (n.c.)	15.5	10.4	7.5
2008	26.8	3.3 (n.c.)	20.0	11.0	9.0
2009	31.5	3.3 (n.c.)	22.5	12.0	11.3
2010	49.0	4.0 (c.)	22.5	12.0	12.5

n.c.: not considered in economic calculations (conversion = first three years, according to IFOAM standards), c.: considered in economic calculations; No premium exists for organic soybean and wheat due to local market structures; Exchange rate INR: USD = 50:1 (source: http://eands.dacnet.nic.in/AWIS.htm, stand October 2012).

(Figure 3). CONBtC consistently showed higher yields than the three other farming systems, except in 2010. This is in line with the findings of several international meta-studies, which also reported generally higher yields and increased profitability in Bt cotton compared to non-Bt cotton production [34,65,71]. However, cotton yield increases through the use of Bt seeds may vary greatly among regions (from zero in Australia, up to 30% in Argentina) due to e.g. varieties used in Bt and non-Bt production, and effectiveness of chemical plant protection in non-Bt production [65]. Glover [72] also points out that the performance and impacts of Bt crops have been highly variable, socio-economically differentiated and contingent on a range of agronomic, socio-economic and institutional factors, thus underlining that the contextual interpretation of results is of paramount importance. The cotton yields per hectare of CONBtC in cycle 1 in our study were in the same range reported by Konduru et al. [73]. The severe decline in yield observed for CONBtC in cycle 2 when compared to cycle 1 (Figure 3) can be partly explained by the fact that Bt cotton plants were uprooted two months earlier than plants in other farming systems in cycle 2 (see chapter 2.2). However, this does not explain the decline in yield observed from 2007 to 2010 in the CON farming system, in which cotton plants were not uprooted. In cycle 1 the cotton yields in CONBtC were 16% higher than in CON (Table 3) which could be due to both the effect of the Bt gene products on pests (as isogenic hybrids were used) as well as the higher input of fertilizer (166 kg N ha^{-1} vs. 146 kg N ha^{-1}) recommended for Bt cotton. The difference in yield between Bt and non-Bt cotton in our study was much smaller than the differences in yield reported by others for India [33,36,38–40,74], indicating that the chemical plant protection applied to the CON system in our experiment was relatively effective. In contrast to the conventional systems, both of the organic farming systems showed rather stable cotton yield throughout the entire experimental period (Figure 3). As cycle effects and the System×Cycle interaction are confounded by year effects, we have to consider that in 2009 and 2010 the cotton yield was generally lower than in 2007 and 2008, as was confirmed by statistical yield data of the state Madhya Pradesh [67]. Apparently, the conventional farming systems could not realize their yield potential due to the less advantageous growing conditions in cycle 2 (rainfall and water logging in the harvest period October – December, Figure 1). The organic farming systems however, were not affected by the disadvantageous conditions in cycle 2 (Figure 3). An additional nitrogen fixing green manure pre-crop, planted before cotton in the organic systems in cycle 2 (Figure 2), may have contributed to the observed stability in yield in these systems through the consistent provision of nitrogen to the plants. Cotton yields of

future crop cycles will thus determine whether productivity in conventional systems will reach their initial high level as well as determine whether yields of organic farming systems will start to increase. A yield depression is usually observed during the conversion to organic farming in India [75]. However, in our trial no such trend was oberserved between cycle 1 and cycle 2.

Non-GM soybean and wheat varieties were cultivated in both CON and CONBtC systems. In 2007, average soybean yields across all farming systems were 45% lower compared to the other three years (Figure 3), as the whole trial had to be re-sown due to severe water logging. Soybean yields were, averaged across the four years, 7% lower in organic compared to conventional farming systems. The System×Cycle interaction had a significant effect (P<0.05) on soybean yields (Table 3). No significant difference in yield could be identified between farming systems in cycle 1. However, in cycle 2 CON and CONBtC showed significantly higher yields than BIOORG (Table 3). This is likely due to higher pest incidences and thus lower yields in organic systems in 2009. BIODYN produced similar soybean yields as both conventional systems throughout the experimental period (P>0.05). The 1% and 11% lower yields in organic farming systems in cycle 1 and 2, respectively, are considerably lower than the 18% lower yields reported for organic soybean in the Karnataka region [76]. These results indicate similar productivity of conventional and organic soybean production systems under subtropical (semi-arid) conditions and suggest that in similar settings no further yield gains can be achieved through the provision of synthetic inputs compared to organic management practices. The smaller difference in yield between conventional and organic soybean - when compared to cotton and wheat (see below) - could be explained by considering the plant type. Soybean is the only legume crop in the crop rotation, possessing the ability to fix atmospheric N, thereby avoiding potential nitrogen shortage for optimal plant growth. These results confirm the findings of Seufert et al. [17] whose meta-analysis showed a lower yield gap between conventional and organic legume crops when compared to non-legume crops, and indicate that cotton and wheat yields in organic farming systems in our trial may be restricted by nitrogen limitation in the soil.

Wheat yields were, averaged across the four years, 15% lower in organic compared to conventional farming systems, which is similar to the 20% yield gap reported for Uttarakhand [76]. The System×Cycle interaction had a significant effect (P<0.001) on wheat grain yield. Similar to cotton, there was a significant yield gap between conventional and organic farming systems in cycle 1 (+37% yield increase in conventional farming systems), but not in cycle 2 due to both slightly lower yields in the conventional systems and slightly higher yields in the organic systems compared to cycle

Table 3. Mean yields [kg ha^{-1}] of cotton, soybean and wheat, and total productivity per cycle and across four years (2007–2010) in the farming systems compared in central India.

Farming system	Crop								Total productivity of crop rotation	
	Seed cotton	SEM	Wheat 2 grains	SEM	Soybean grains	SEM	Wheat grains	SEM	Seed cotton + Wheat 2 grains + Soybean grains + Wheat grains	SEM
Cycle 1 (2007–2008)										
BIODYN	2'047 [c]	68	-	-	1'399 [a]	158	2'997 [c]	153	6'443 [b]	104
BIOORG	2'072 [c]	49	-	-	1'536 [a]	192	2'831 [c]	121	6'440 [b]	187
CON	2'700 [b]	141	-	-	1'483 [a]	155	4'262 [a]	221	8'444 [a]	146
CONBtC	3'133 [a]	176	-	-	1'473 [a]	195	3'730 [b]	272	8'336 [a]	254
Cycle 2 (2009–2010)										
BIODYN	1'894 [a]	108	-	-	1'807 [ab]	87	3'338 [a]	207	7'039 [b]	268
BIOORG	1'942 [a]	103	-	-	1'739 [b]	117	3'303 [a]	191	6'984 [b]	239
CON	1'614 [a]	43	-	-	1'993 [a]	108	3'273 [a]	175	6'880 [b]	119
CONBtC*	1'834 [(a)]	179	1'573	169	1'997 [a]	161	3'481 [a]	182	8'885 [a]	390
Average (2007–2010)										
BIODYN	1'971	64	-	-	1'603	104	3'167	132	6741	270
BIOORG	2'007	56	-	-	1'638	114	3'067	125	6712	257
CON	2'157	157	-	-	1'738	113	3'767	187	7'662	455
CONBtC	2'484	207	(787)	-	1'735	140	3'605	161	8'610	376
ANOVAs of linear mixed effect models										
Source of variation	P value	Df			P value	Df	P value	Df	P value	Df
System (S)	<0.001	3	-	-	0.102	3	<0.001	3	<0.001	3
Cycle (C)	<0.001	1	-	-	0.066	1	0.686	1	0.912	1
Strip	0.141	1	-	-	0.472	1	0.960	1	0.002	1
S×C	<0.001	3	-	-	0.039	3	<0.001	3	<0.001	3

SEM: standard error of the mean, BIODYN: biodynamic, BIOORG: organic, CON: conventional, CONBtC: conventional with Bt cotton, different superscript letters indicate significant difference between farming systems within one Cycle (Tukey test, $P<0.05$), * in 2009 and 2010 Bt cotton was uprooted 2 months earlier to grow a second wheat crop (wheat 2) to reflect common practice of local Bt cotton farmers (for the sequence of crops in different farming systems see Figure 2), P value and degrees of freedom (Df) of fixed effects in linear mixed effect models, random factors in the model: Year (n = 4), Block (n = 4), Pair (n = 16), for total productivity random factor Year was excluded as data from two years were compiled.

Table 4. Mean gross margins [INR ha^{-1}] of cotton, soybean and wheat, and total gross margin per cycle and across four years (2007–2010) in the farming systems compared in central India.

Farming system	Crop								Total gross margin of crop rotation	
	Seed cotton	SEM	Wheat 2 grains	SEM	Soybean grains	SEM	Wheat grains	SEM	Seed cotton + Wheat 2 grains + Soybean grains + Wheat grains	SEM
Cycle 1 (2007–2008)										
BIODYN	38'243 [c]	2'226	-	-	19'211	4'210	26'044 [c]	1'584	83'498 [b]	8'021
BIOORG	38'676 [c]	1'203	-	-	21'830	4'858	24'420 [c]	1'291	84'926 [b]	8'268
CON	51'792 [b]	2'779	-	-	18'401	4'093	37'099 [a]	2'221	107'292 [a]	4'546
CONBtC	60'811 [a]	4'851	-	-	18'147	4'719	31'361 [b]	2'832	110'319 [a]	8'308
Cycle 2 (2009–2010)										
BIODYN	62'786 [a]	7'217	-	-	32'176	1'683	30'764 [a]	2'374	125'726 [a]	6'721
BIOORG	64'490 [a]	6'714	-	-	30'812	2'278	30'443 [a]	2'181	125'745 [a]	9'354
CON	42'962 [b]	4'653	-	-	28'949	3'724	24'773 [b]	2'028	96'683 [b]	6'701
CONBtC*	43'810 [(b)]	2'995	4'837	2'190	29'399	4'644	27'037 [b]	2'117	105'082 [b]	5'232
Average (2007–2010)										
BIODYN	50'514	4'918	-	-	25'694 [ab]	2'758	28'404	1'507	104'612	7'461
BIOORG	51'583	4'789	-	-	26'321 [a]	2'865	27'432	1'450	105'335	7'008
CON	47'377	2'852	-	-	23'675 [b]	2'780	30'936	2'155	101'988	3'089
CONBtC	52'310	3'165	(2'418)	-	23'773 [b]	3'311	29'199	1'797	107'701	3'849
ANOVAs of linear mixed effect models										
Source of variation	P value	Df			P value	Df	P value	Df	P value	Df
System (S)	0.115	3	-	-	0.006	3	0.022	3	0.298	3
Cycle (C)	0.046	1	-	-	0.158	1	0.606	1	<0.001	1
Strip	0.001	1	-	-	0.469	1	0.805	1	<0.001	1
S×C	<0.001	3	-	-	0.150	3	<0.001	3	<0.001	3

SEM: standard error of the mean, BIODYN: biodynamic, BIOORG: organic, CON: conventional, CONBtC: conventional with Bt cotton, different superscript letters indicate significant difference between farming systems within one Cycle (Tukey test, $P<0.05$), * in 2009 and 2010 Bt cotton was uprooted 2 months earlier to grow a second wheat crop (wheat 2) to reflect common practice of local Bt cotton farmers (for the sequence of crops in different farming systems see Figure 2), P value and degrees of freedom (Df) of fixed effects in linear mixed effect models, random factors in the model: Year (n = 4), Block (n = 4), Pair (n = 16), for total gross margin random factor Year was excluded as data from two years were compiled.

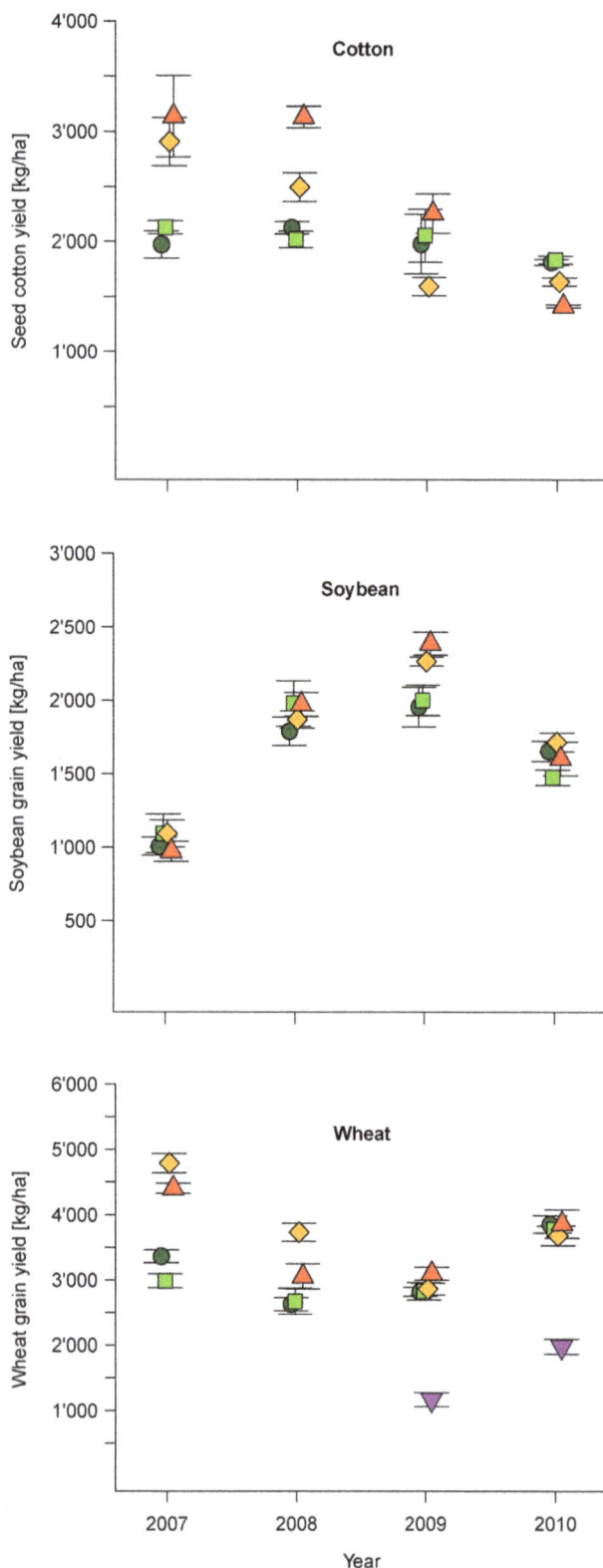

different panels of the graph.

1 (Table 3, Figure 3). For both soybean and wheat no yield differences were detected between CON and CONBtC farming systems, except for significantly higher wheat yields in CON in cycle 1 (Table 3).

Regarding the total productivity per crop rotation in terms of summed-up dry matter yields of cotton, soybean and wheat (including wheat 2 in CONBtC in cycle 2), a significant effect ($P<0.001$) of the *System×Cycle* interaction was found (Table 3). Both of the conventional farming systems were significantly more productive (+30%, Table 3) than the organic farming systems in cycle 1. However, in cycle 2 only CONBtC showed significantly higher productivity (+28%, Table 3) when compared to the other three farming systems, due to the additional wheat crop (wheat 2). Differences in yield between BIODYN and BIOORG were minor and not statistically significant for all crops and total productivity of the whole crop rotation (Table 3). Unexpectedly, there was a significant *Strip* effect for total productivity across the whole crop rotation. This may be explained by the fact that different crops were cultivated on the two strips in a given year (Figure 2). The compilation of whole crop rotations (compiling years) thus led to the combination of the observed variability for each crop across the four years, and subsequently to the *Strip* effect becoming significant (Table 3).

In general, the first four years of the farming systems comparison trial in India revealed that there was a significant yield gap in cycle 1 (2007–2008) for cotton (−29%) and wheat (−27%), in organic compared to conventional farming systems, whereas in cycle 2 yields of the three crops were similar in all farming systems due to low yields in the conventional systems (Table 3). Because there was no clear trend of yield development for cotton and wheat during the four year period in any of the systems, observed results rather reflect growth conditions in respective years than long-term yield trends of cotton and wheat. However, the marginal yield gap between the BIOORG system and the conventional systems, and the par soybean yields of BIODYN and the conventional systems show that leguminous crops are a promising option for conversion to organic systems under the given conditions. The yield development across the whole crop rotation needs to be verified during future crop cycles.

2 Economic analysis

The production costs (i.e. labor and input costs) in our trial (Tables S2, S3 and S4) were in a similar range as reported by the Ministry of Agriculture, Government of India [67]. The variable production costs of conventional (CON, CONBtC) compared to organic (BIOORG, BIODYN) farming systems were on average 38%, 66%, and 49% higher in cotton, soybean and wheat (Table S2, S3 and S4). This is in agreement with findings for cotton in Gujarat, but contradicts findings for wheat in Punjab and Uttar Pradesh [53]. The main reason for the differences observed in our study were the higher input costs (fertilizer, pesticides) in the conventional farming systems, which is in accordance with the findings of a study conducted in Kyrgyzstan [27]. Labor costs were similar among all farming systems, as organic and conventional farming systems did not differ greatly with regard to time requirements of activities (Table S2, S3 and S4). For instance, weeding was done manually in all systems and no herbicides were applied in the conventional farming systems except for soybean and wheat in cycle 2, reflecting the common practice of most smallholder cotton farmers in India [66]. This practice, however, might change in the near future, as labor costs in Indian

Figure 3. Yield (mean ± standard error) 2007-2010 in cotton, soybean and wheat. Farming systems: (●) biodynamic (BIODYN), (■) organic (BIOORG), (♦) conventional (CON), (▲) conventional with Bt cotton (CONBtC), (▼) wheat after Bt cotton (wheat 2); In 2009 and 2010 Bt cotton was uprooted 2 months earlier to grow a second wheat crop (wheat 2) to reflect common practice of local Bt cotton farmers. Non-GM soybean and wheat varieties were cultivated in the CON and CON-BtC plots throughout the trial. Note the different scales on y-axes in the

agriculture are on the rise [77]. The variable production costs of the two organic farming systems were similar for all crops. This was also true for the two conventional farming systems, except for cotton, where the variable production costs of CONBtC were 17% higher compared to CON due to both the higher seed price of Bt cotton (Table S2) [34,36,47] and production cost of the additional wheat crop in cycle 2. The prices we present here for Bt seed material are in the same range as reported by Singh *et al.* [78].

Cotton was the most important cash crop and accounted for 48% of the total gross return in the crop rotation, irrespective of the system. The *System×Cycle* interaction had a significant effect ($P<0.001$) on cotton gross margins (Table 4). Due to much higher yields, conventional cotton led to 32% higher gross margins compared to organic cotton in cycle 1 (2007–2008), which is in accordance with several international meta-studies [34,65,71]. However, the opposite was true in cycle 2 (2009–2010), where we observed 32% lower gross margins in conventional cotton, supporting the findings of Bachmann [27]. The significant *Strip* effect for cotton gross margin can be explained by the highly variable cotton prices across the four years (Table 2).

For soybean, the *System×Cycle* interaction was not significant (Table 4) which allowed for an analysis of gross margin data across both cycles (see 2.4). Considerably higher gross margins were obtained in organic systems (+10%) compared to conventional systems between 2007 and 2010. The difference was statically significant for BIOORG (+11%, $P<0.05$) and almost significant for BIODYN (+8%, $P<0.1$). These results indicate that the slightly lower productivity of organic soybean was balanced out by lower production costs rendering soybean production considerably more profitable in organic systems when compared to conventional farming systems.

For wheat gross margins, the *System×Cycle* interaction was found to be significant ($P<0.001$, Table 4). Under organic farming, wheat obtained significantly lower gross margins in cycle 1 (−26%), but significantly higher gross margins (+18%) in cycle 2 (Table 4). The earlier removal of Bt cotton from the field in order to grow another wheat crop in CONBtC, before the start of the Zaid (summer) season in cycle 2, only provided minor economic benefits compared to CON (Table 4). This was mainly due to low yields of wheat 2 (<50% compared to regular wheat crop, Figure 3) and lower market prices for wheat compared to cotton (Table 2). Thus, the additional wheat crop could not compensate for the missed cotton yield of the last picking period with respect to economic profitability, a result contradicting total yield performance across all crops.

A highly significant (P<0.001) *System×Cycle* interaction was found for the total gross margin per crop rotation. In cycle 1, favorable weather conditions allowed for the realization of the anticipated yield potential in conventional farming systems, and thus led to both higher cotton and wheat yields (Figure 3), and concomitantly significantly higher gross margins (+29%, Table 4). However, In cycle 2 the gross margins of the organic farming systems were significantly higher (+25%) (Table 4, Figure 4) due to par yields as measured in the conventional systems (Figure 3), but lower variable production costs (Tables S2 and S4). If the premium price in 2010 would not have been considered, the total gross margin per crop rotation in cycle 2 would not be substantially lower and still be significantly higher in organic farming systems (statistical analysis not shown). This could imply that, in favorable years (e.g. good yield, high price for commodities, etc.), premium prices are not required for achieving comparable economic returns in organic and conventional farming systems. However, the premium is needed in unfavorable

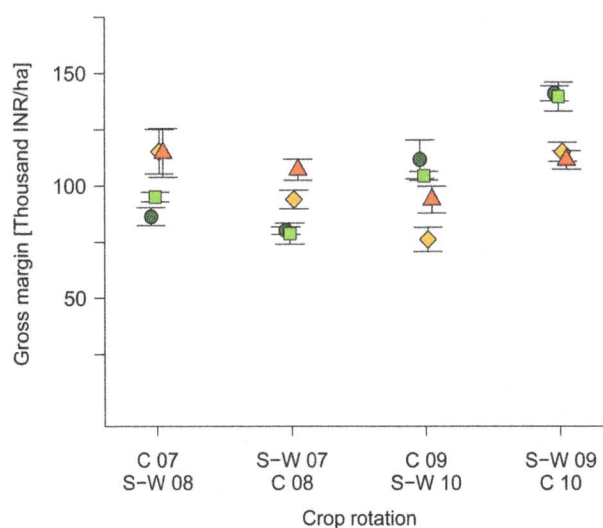

Figure 4. Gross margins (mean ± standard error) of four crop rotations. Farming systems: (●) biodynamic (BIODYN), (■) organic (BIOORG), (◆) conventional (CON), (▲) conventional with Bt cotton (CONBtC) (includes wheat cultivated after Bt cotton on the same plots in 2009 and 2010); C = cotton, S-W = soybean-wheat; Exchange rate Indian rupee (INR): US Dollar (USD) = 50:1 (stand October 2012), premium price on organic cotton only in 2010.

years in order to compensate for yield gaps, and to avoid that organic farmers sell their produce to the local conventional market. The significant *Strip* effect for total gross margin can be explained by compiling the individual gross margins of the different crops, thereby transferring the significant *Strip* effect of cotton to the total gross margin (Table 4).

The results of cycle 2 suggest that under certain conditions, organic farming can be an economically viable and less capital-intensive production system compared to conventional farming systems, which is in accordance with the findings by Ramesh *et al.* [76] and Panneerselvam *et al.* [79]. However, long-term studies are needed in order to substantiate these findings. Moreover, the viability of organic farming systems strongly depends on farmers having access to knowledge, purchased inputs such as organic fertilizers, pesticides and non-GM seeds, and assuming that there is a market demand and well developed certification system. These are vital components for the economic profitability of organic farming systems [27] especially against the backdrop of increasing labor costs in Indian agriculture [77]. The costs for organic certification are substantial in case individual farmers have to undergo this process, and premium prices may also have to cover these costs. Up to now, certification costs are usually covered by the cotton organization that is purchasing seed cotton from smallholders (here: bioRe India Ltd.). This includes extensive testing of seeds and seed cotton for GM contamination, as well as the implementation of Tracenet, an internet based electronic service offered by the Agricultural and Processed Food Products Export Development Authority (APEDA) for facilitating certification of organic export products from India which comply with the standards of the (National) Programme for Organic Production (NPOP/NOP). This is a big challenge of certified organic cotton compared to fair trade cotton [15], and further organic cotton initiatives rely on cost-efficient and trustful certification and education programs as well.

3 Transferability of field trial results

So far little is known about the comparative performance of cotton-based farming systems under organic and conventional management. To our knowledge, this is the first publication comparing the agronomic and economic performance of biodynamic, organic, conventional and conventional with Bt cotton-based farming systems. The few studies published to date compared either organic vs. conventional [9,28,80] or conventional vs. conventional with Bt cotton production systems [36].

By including two organic (BIOORG and BIODYN) and two conventional (CON and CONBtC) farming systems in our trial, we were able to cover a wide range of current cotton-based farming systems in central India (see 2.2). Forming close links to local partners and having practitioners in the steering committee of our systems comparison trial guaranteed that the various agronomic activities that were involved represented local farmers' practice. Due to the cooperative initiative of bioRe, cotton farmers are trained in compost preparations, and organic inputs are purchased collectively and distributed among the farmers. Farmers associated with bioRe may not face the various problems commonly observed during conversion to organic farming [75] to a similar extent as do farmers without affiliation to similar institutions. This is likely due to the experienced and well-functioning extension service of bioRe. Drip irrigation is strongly promoted and subsidized by the Indian government and is not specific to our experiment. However, a direct extrapolation of our results to the reality of smallholder farmers is not possible due to the fact that farmers are confronted with several obstacles not considered in our study; These are for example market access, access to inputs and know-how and in particular costs associated with the organic certification process (see 3.2). One also has to consider that yield estimates from optimally managed trial plots are usually higher than the average yield of smallholder farmers. This is due to the fact, that such optimal crop management, as it was applied in this farming systems comparison trial, might not always be possible under the real-world smallholder conditions. This is especially true as the trial was conducted on a fertile Vertisol soil. Based on a survey of more than 1'000 smallholders in Madhya Pradesh, the average yield levels in the respective time period (2007–2009) were 1'416 kg ha^{-1}, 1'285 kg ha^{-1}, and 2'426 kg ha^{-1} for seed cotton, soybean, and wheat [67], as compared to 2'585 kg ha^{-1}, 1'761 kg ha^{-1}, and 3'658 kg ha^{-1} found in our trial. In a survey performed between 2006 and 2008 among 700 smallholder cotton farmers in India, average yields of 1'743 and 2'048 kg ha^{-1} seed cotton were reported for conventional (without Bt) and Bt cotton, respectively [36], as compared to 2'700 kg ha^{-1} and 3'133 kg ha^{-1} in our trial in the same time period (2007–2008). Thus, our yields might be generally overestimated, but within the range of other field trials in India [78].

The following examples show that comparative findings on yield and economics between organic and conventional cotton are highly contextual. Eyhorn et al. [9] surveyed more than 50 conventional (without Bt cotton) and 30 organic cotton farmers in the Nimar Valley, Madhya Pradesh, India, during the period of 2003–2004. Their findings support our results of cycle 2 (years three and four, 2009–2010): yields of cotton and other cash crops were on par with conventional farmers, but with economic benefits for the organic farmers due to lower production costs. These findings also underline the practical relevance of our results for cotton production in the in the smallholder context in Madhya Pradesh. Likewise, Venugopalan et al. [81] reported similar or slightly higher cotton yields in an organic compared to a non-organic system under low input and semi-arid conditions in the Yavatmal district, Maharashtra, India (observation phase 2001–2005). In a long-term trial under rainfed conditions in Nagpur (Maharashtra, India), Menon [82] reported a yield gap of 25% of organic cotton compared to the modern method of cultivation (= conventional without Bt cotton) within the first six years after conversion (1994–2000). Thereafter (2002–2004), the organic farming systems outyielded the conventional systems by up to 227 kg seed cotton ha^{-1} [28].

However, our findings from India are in contrast to the results from a survey on cotton farms in Northern San Joaquin Valley, California [80]. There, averaged over a six-year observation period, yields of organic cotton were 19 and 34% lower ($P<0.05$) than those of cotton under conventional and integrated pest management (reduced insecticide input). It has to be taken into account that for two out of the six years assessed in their study, different varieties were compared under conventional and organic management. Production costs of organic cotton were 25 and 60% higher than those of cotton under conventional and integrated pest management, respectively. This was mainly due to the higher labor costs for manual weeding. In our trial, there was less difference in weed control, as manual weeding is still the common practice of most smallholder cotton farmers in India [66]. This example underlines the contextual nature of the findings concerning the agronomic and economic performance between organic and conventional cotton, which was also pointed out by Seufert et al. [17] and de Ponti et al. [16] for other crops than cotton.

Building on unique panel data on Indian cotton farming of smallholder farmers between 2002 and 2008, Kathage and Qaim [36] showed that the use of conventional Bt cotton led to a 24% yield increase and a 50% gain in cotton profit compared to smallholders growing conventional non-Bt cotton. In contrast to the systematic farm survey by Kathage and Qaim [36], our study represents a pairwise comparison of cotton-based farming systems under identical pedo-climatic conditions. While the study by Kathage and Qaim [36] can better depict the actual situation on real farms for a given region, our results can better represent the potential outcomes that are achievable under optimal conditions with respect to inputs and knowledge access. In our study, there were comparatively little differences between the CONBtC and the CON farming systems regarding cotton yield (+16 and +14% in cycle 1 and 2, respectively) and cotton gross margin (+17 and +2% in cycle 1 and 2, respectively). Furthermore, it needs to be taken into account that with the introduction of Bt cotton to India in 2002, the provincial governments began to subsidize Bt cotton considerably, especially between the years of 2002 and 2008. This led to the rapid spread of Bt cotton and the breakdown of the non-Bt cotton seed chain. The relatively weak performance of the non-Bt cotton in the farm survey by Kathage and Qaim [36] could partly be explained by the poor quality of non-Bt cotton seeds, as propagation of non-Bt cotton was abandoned and led to limited availability of non-Bt cotton seeds from old stocks of probably poor quality and mainly older cultivars [59,83].

In contrast to Kathage and Qaim [36], our study also includes other cash crops such as soybean and wheat as part of cotton-based farming systems. These are essential components for enabling long-term cotton cropping and for securing livelihoods of smallholders, as they enable the distribution of risks. According to our findings, the investigated organic farming systems also showed a significant yield gap compared to the conventional farming systems in wheat in cycle 1, as well as for total productivity per crop rotation (including additional wheat crop after Bt cotton (wheat 2)) in cycle 2. Furthermore, a smaller yet significant yield gap was observed for soybean in cycle 2 for the BIOORG system (but not for the BIODYN system) (Table 3). Nevertheless, as in our study organic farming systems were less capital-intensive than conventional ones for all crops, they may be of particular interest to smallholder farmers who often do not have the financial means to purchase inputs and would thus need to seek loans. If this can be verified in on-farm trials, organic farmers might be less exposed to financial risks associated with fluctuating market prices of synthetic fertilizers and crop protection products [27,79]. Additional on-farm investigations have been started in order to classify regional farms into several farm typologies with corresponding levels of available production factors. This should allow for the assessment of the perspectives of each farm type regarding conversion to organic farming systems. If organic farming is to be adopted more widely, more inter- and transdisciplinary research giving focus to the problems and benefits of organic management practices needs to be undertaken [84]. Furthermore, large efforts have to be made to gather and disseminate knowledge on production techniques. Intensifying research on organic farming systems to a similar extent as was the case for research on GM crops [85] may help to provide additional relevant information to policy makers, advisors and farmers about comparative advantages and limitations of different cotton production systems.

Conclusions

With this publication we respond to the urgent need for farming systems comparison trials in the tropics and subtropics [17,26]. Here we show results from the conversion period (first four years after inception of the trial) of cotton-based farming systems representative for Vertisol soils in Madhya Pradesh, central India. Due to the short-term nature of our results and the observed *System × Cycle* interactions (no clear trend of system performance over time) for yield and gross margin data of cotton and wheat, definitive conclusions about the comparative agronomic and economic performance of the investigated farming systems cannot be drawn. However, our results show that organic soybean productivity can be similarly high as in conventional systems at lower input levels, which can make organic soybean production - as part of cotton-based crop rotations - more profitable. Future research will bring further insights on agronomic and economic performance of the different farming systems after the conversion period, thus providing indications on the long-term sustainability across the whole crop rotation. Furthermore, the effects of the farming systems on biodiversity, soil fertility, other ecological co-benefits such as climate change mitigation by means of C sequestration, and product quality need to be elucidated.

Supporting Information

Figure S1 Experimental design of the farming systems comparison trial in Madhya Pradesh, India. Farming systems: biodynamic (BIODYN), organic (BIOORG), conventional (CON), conventional with Bt cotton (CONBtC), CONBtC includes wheat cultivated after Bt cotton on the same plots in 2009 and 2010, open squares belong to Strip 1, closed squares belong to Strip 2, distance between two plots within a strip = 6 m, distance between the two strips = 2 m.

Table S1 Fertilizer and plant protection practices in the farming systems compared in central India (2007–2010). BIODYN: biodynamic, BIOORG: organic, CON: conventional, CONBtC: conventional with Bt cotton, Ntotal: total nitrogen, OF: organic fertilizers (compost, FYM and castor cake), Ntotal includes only fertilizer derived N, nutrient inputs by green manures were not considered, DAP: Diammonium phosphate, MOP: muriate of potash, SSP: single super phosphate, 1Beavicide®: organic

pesticide containing Beauveria bassiana,2GOC: slurry made from rotten garlic, onion and chili with water, 3NeemAzal®: insecticide made from neem kernels, 4Top Ten: slurry made from leaves of ten wild plants and water, 5Verelac: organic pesticide containing Verticillium lecanii.

Table S2 Detailed list of variable production costs in cotton of the farming systems compared in central India (2007–2010). [1] in the text, BIODYN and BIOORG are referred to consistently as organic farming systems. [2] in the text, CON and CONBtC are referred to consistently as conventional farming systems. [3] figures include time for preparation of organic fertilizers to account for their market value. [4] figures represent subsidized prices for mineral fertilizers set by the Government of India. [5] longer time required for soil cultivation in CON and CONBtC due to soil compaction. [6] figure includes application of biodynamic preparations. [7] figures include uprooting cotton and removing the straw from the field. [8] figures include time required to purchase inputs (organic/synthetic) from the market and to produce organic (natural) pesticides and biodynamic preparations.

Table S3 Detailed list of variable production costs in soybean of the farming systems compared in central India (2007–2010). [1] in the text, BIODYN and BIOORG are referred to consistently as organic farming systems. [2] in the text, CON and CONBtC are referred to consistently as conventional farming systems. [3] figures include time for preparation of organic fertilizers to account for their market value. [4] figures represent subsidized prices for mineral fertilizers set by the Government of India. [5] longer time required for soil cultivation in CON and CONBtC due to soil compaction. [6] figure includes application of biodynamic preparations. [7] figures include removing soybean bundles from the field and threshing. [8] figures include time required to purchase inputs (organic/synthetic) from the market and to produce organic (natural) pesticides and biodynamic preparations.

Table S4 Detailed list of variable production costs in wheat of the farming systems compared in central India (2007–2010). [1] in the text, BIODYN and BIOORG are referred to consistently as organic farming systems. [2] in the text, CON and CONBtC are referred to consistently as conventional farming systems. [3] figures include time for preparation of organic fertilizers to account for their market value. [4] figures represent subsidized prices for mineral fertilizers set by the Government of India. [5] longer time required for soil cultivation in CON and CONBtC due to soil compaction. [6] figure includes application of biodynamic preparations. [7] figures include removing wheat bundles from the field and threshing. [8] figures include time required to purchase inputs (organic/synthetic) from the market and to produce organic (natural) pesticides and biodynamic preparations.

Acknowledgments

Special thanks go to Andreas Gattinger (Research Institute of Organic Agriculture, FiBL) for helping to prepare the manuscript and for his many valuable inputs. We thank Kulasekaran Ramesh (Indian Institute of Soil Science, IISS), Padruot Fried (Swiss Federal Agricultural Research Station Agroscope, ART), Monika Schneider, Franco Weibel, Andreas Fliessbach (Research Institute of Organic Agriculture, FiBL), Georg Cadisch (University of Hohenheim), and Philipp Weckenbrock (die Agronauten) for fruitful discussions. The field and desktop work of the whole bioRe Association team is also gratefully acknowledged. We thank Christopher Hay, Ursula Bausenwein and Tal Hertig for the language editing of the manuscript. We acknowledge the inputs by Fränzi Korner and Bettina Almasi regarding statistical analysis and data interpretation. Finally, we sincerely thank the anonymous reviewers for their very constructive and helpful comments and suggestions.

Author Contributions

Conceived and designed the experiments: DF CZ PM. Performed the experiments: DF RV CZ. Analyzed the data: CA MM. Wrote the paper: CA DF MM PM CZ RV.

References

1. Trewavas A (2002) Malthus foiled again and again. Nature 418: 668–670.
2. Badgley C, Moghtader J, Quintero E, Zakem E, Chappell MJ, et al. (2007) Organic agriculture and the global food supply. Renewable Agriculture and Food Systems 22: 86–108.
3. Singh RB (2000) Environmental consequences of agricultural development: a case study from the Green Revolution state of Haryana, India. Agriculture Ecosystems & Environment 82: 97–103.
4. Pimentel D (1996) Green revolution agriculture and chemical hazards. Science of the Total Environment 188: 86–98.
5. IAASTD (2009) International assessment of agricultural knowledge, science and technology for development (IAASTD): Executive summary of the synthesis report. Washington, DC: Island Press. 606 p.
6. Kilcher L (2007) How organic agriculture contributes to sustainable development. In: Willer H, Yussefi M, editors. The world of organic agriculture - Statistics and emerging trends 2007. Rheinbreitbach: Medienhaus Plump. pp. 82–91.
7. Altieri MA (1999) The ecological role of biodiversity in agroecosystems. Agriculture Ecosystems & Environment 74: 19–31.
8. Valkila J (2009) Fair Trade organic coffee production in Nicaragua - Sustainable development or a poverty trap? Ecological Economics 68: 3018–3025.
9. Eyhorn F, Ramakrishnan M, Mäder P (2007) The viability of cotton-based organic farming systems in India. International Journal of Agricultural Sustainability 5: 25–38.
10. Lyngbaek AE, Muschler RG, Sinclair FL (2001) Productivity and profitability of multistrata organic versus conventional coffee farms in Costa Rica. Agroforestry Systems 53: 205–213.
11. Mendez VE, Bacon CM, Olson M, Petchers S, Herrador D, et al. (2010) Effects of Fair Trade and organic certifications on small-scale coffee farmer households in Central America and Mexico. Renewable Agriculture and Food Systems 25: 236–251.
12. Borlaug NE (2000) Ending world hunger. The promise of biotechnology and the threat of antiscience zealotry. Plant Physiology 124: 487–490.
13. Trewavas AJ (2001) The population/biodiversity paradox. Agricultural efficiency to save wilderness. Plant Physiology 125: 174–179.
14. Nelson L, Giles J, MacIlwain C, Gewin V (2004) Organic FAQs. Nature 428: 796–798.
15. Makita R (2012) Fair Trade and organic initiatives confronted with Bt cotton in Andhra Pradesh, India: A paradox. Geoforum 43: 1232–1241.
16. de Ponti T, Rijk B, van Ittersum MK (2012) The crop yield gap between organic and conventional agriculture. Agricultural Systems 108: 1–9.
17. Seufert V, Ramankutty N, Foley JA (2012) Comparing the yields of organic and conventional agriculture. Nature 485: 229–U113.
18. Reganold JP (2012) Agriculture Comparing apples with oranges. Nature 485: 176–176.
19. Watson CA, Walker RL, Stockdale EA (2008) Research in organic production systems - past, present and future. Journal of Agricultural Science 146: 1–19.
20. Reganold JP, Glover JD, Andrews PK, Hinman HR (2001) Sustainability of three apple production systems. Nature 410: 926–930.
21. Mäder P, Fliessbach A, Dubois D, Gunst L, Fried P, et al. (2002) Soil fertility and biodiversity in organic farming. Science 296: 1694–1697.
22. Hepperly P, Douds Jr. D, Seidel R (2006) The Rodale farming systems trial 1981 to 2005: longterm analysis of organic and conventional maize and soybean cropping systems. In: Raupp J, Pekrun C, Oltmanns M, Köpke U, editors. Longterm field experiments in organic farming. Bonn: International Society of Organic Agriculture Resarch (ISOFAR). pp. 15–32.
23. Fliessbach A, Oberholzer H-R, Gunst L, Maeder P (2007) Soil organic matter and biological soil quality indicators after 21 years of organic and conventional farming. Agriculture Ecosystems & Environment 118: 273–284.
24. Teasdale JR, Coffman CB, Mangum RW (2007) Potential long-term benefits of no-tillage and organic cropping systems for grain production and soil improvement. Agronomy Journal 99: 1297–1305.
25. Birkhofer K, Bezemer TM, Bloem J, Bonkowski M, Christensen S, et al. (2008) Long-term organic farming fosters below and aboveground biota: Implications for soil quality, biological control and productivity. Soil Biology & Biochemistry 40: 2297–2308.
26. Gattinger A, Muller A, Haeni M, Skinner C, Fliessbach A, et al. (2012) Enhanced top soil carbon stocks under organic farming. Proceedings of the National Academy of Sciences of the United States of America 109: 18226–18231.
27. Bachmann F (2012) Potential and limitations of organic and fair trade cotton for improving livelihoods of smallholders: evidence from Central Asia. Renewable Agriculture and Food Systems 27: 138–147.
28. Blaise D (2006) Yield, boll distribution and fibre quality of hybrid cotton (Gossypium hirsutum L.) as influenced by organic and modern methods of cultivation. Journal of Agronomy and Crop Science 192: 248–256.
29. Rasul G, Thapa GB (2004) Sustainability of ecological and conventional agricultural systems in Bangladesh: an assessment based on environmental, economic and social perspectives. Agricultural Systems 79: 327–351.
30. FAO (2013) FAOSTAT database on agriculture. Available: http://faostatfaoorg. Accessed 11 October 2013.
31. Herring RJ (2008) Whose numbers count? Probing discrepant evidence on transgenic cotton in the Warangal district of India. International Journal of Multiple Research Approaches 2: 145–159.
32. James C (2012) Global status of commercialized biotech/GM crops: 2012. ISAAA Brief 44. ISAAA: Ithaca, NY. ISBN: 978-1-892456-53-2. Available: http://www.isaaa.org/resources/publications/briefs/44/executivesummary/pdf/Brief%2044%20-%20Executive%20Summary%20-%20English.pdf. Accessed 11 October 2013.
33. Qaim M KS (2013) Genetically Modified Crops and Food Security. PLoS ONE 8(6): e64879 doi:101371/journalpone0064879.
34. Finger R, El Benni N, Kaphengst T, Evans C, Herbert S, et al. (2011) A Meta Analysis on Farm-level Costs and Benefits of GM Crops. Sustainability 3: 743–762.
35. Krishna VV, Qaim M (2012) Bt cotton and sustainability of pesticide reductions in India. Agricultural Systems 107: 47–55.
36. Kathage J, Qaim M (2012) Economic impacts and impact dynamics of Bt (Bacillus thuringiensis) cotton in India. Proceedings of the National Academy of Sciences of the United States of America 109: 11652–11656.
37. Stone GD (2011) Field versus Farm in Warangal: Bt Cotton, Higher Yields, and Larger Questions. World Development 39: 387–398.
38. Crost B, Shankar B, Bennett R, Morse S (2007) Bias from farmer self-selection in genetically modified crop productivity estimates: Evidence from Indian data. Journal of Agricultural Economics 58: 24–36.
39. Bennett R, Kambhampati U, Morse S, Ismael Y (2006) Farm-level economic performance of genetically modified cotton in Maharashtra, India. Review of Agricultural Economics 28: 59–71.
40. Qaim M, Subramanian A, Sadashivappa P (2010) Socioeconomic impacts of Bt (Bacillus thuringiensis) cotton. In: Zehr UB, editor. Biotechnology in Agriculture and Forestry 65: Cotton. pp. 221–240.
41. ICAR (2002) All India coordinated cotton improvement project: Annual report 2001-02. Coimbatore: Indian Council for Agricultural Research. 97 p.
42. Singh NB, Barik A, Gautam HC (2009) Revolution in Indian Cotton. Directorate of Cotton Development, Ministry of Agriculture, Department of Agriculture & Cooperation, Government of India, Mumbai & National Center of Integrated Pest Management, ICAR, Pusa Campus, New Delhi. Available: www.ncipm.org.in/NCIPMPDFs/Revolution_in_Indian_Cotton.pdf. Accessed 11 October 2013.
43. Gruère GP, Sun Y (2012) Measuring the contribution of Bt cotton adoption to India's cotton yields leap. IFPRI Discussion Paper 01170. Available: http://www.ifpri.org/sites/default/files/publications/ifpridp01170.pdf. Accessed 11 October 2013.
44. Kranthi K (2011) 10 years of Bt in India - Three-part feature story on the history of Bt cotton in India.Cordova: Cotton Media Group. Available: http://www.cotton247.com/article/27520/part-ii-10-years-of-bt-in-india. Accessed 11 October 2013.
45. Aris A, Leblanc S (2011) Maternal and fetal exposure to pesticides associated to genetically modified foods in Eastern Townships of Quebec, Canada. Reproductive Toxicology 31: 528–533.
46. Marvier M, McCreedy C, Regetz J, Kareiva P (2007) A Meta-Analysis of Effects of Bt Cotton and Maize on Nontarget Invertebrates. Science 316: 1475–1477.
47. Azadi H, Ho P (2010) Genetically modified and organic crops in developing countries: A review of options for food security. Biotechnology Advances 28: 160–168.
48. Qaim M (2009) The economics of genetically modified crops. Annual Review of Resource Economics: 665–693.
49. Jacobsen S-E, Sørensen M, Pedersen S, Weiner J (2013) Feeding the world: genetically modified crops versus agricultural biodiversity. Agronomy for Sustainable Development: 1–12.
50. Radhakrishnan S (2012) 10 years of Bt cotton: False hype and failed promises. Coalition for a GM-Free India. Available: http://indiagminfo.org/wp-content/uploads/2012/03/Bt-Cotton-False-Hype-and-Failed-Promises-Final.pdf. Accessed 11 October 2013.
51. Herring RJ, Rao NC (2012) On the 'failure of Bt cotton' - Analysing a decade of experience. Economic & Political Weekly XLVII: 45–54.
52. Gruere G, Sengupta D (2011) Bt Cotton and Farmer Suicides in India: An Evidence-based Assessment. Journal of Development Studies 47: 316–337.
53. Charyulu K, Biswas S (2010) Economics and efficiency of organic farming vis-à-vis conventional farming in India. Ahmedabad: Indian Institute of Management. Available: http://www.iimahd.ernet.in/publications/data/2010-04-03Charyulu.pdf. Accessed 11 October 2013.

54. IFOAM (2012) The IFOAM norms for organic production and processing: Version 2012. Bonn: Die Deutsche Bibliothek. 134 p.

55. Koepf HH, Petersson BD, Schaumann W (1976) Biodynamic Agriculture: An Introduction. Anthroposophic Press, Hudson, New York. 430 p.

56. Kanwar JS (1988) Farming systems in swell-shrink soils under rainfed conditions in soils of semi-arid tropics. In: Hirekerur LR, Pal DK, Sehgal JL, Deshpande CSB, editors. Transactions of International Workshop on Swell-Shrink Soils. National Bureau of Soil Survey and Land Use Planning, Nagpur. pp.179–193.

57. Ministry of Agriculture GOI (2011a) State-wise estimates of area and production of cotton released. Bt cotton constitutes about 90% of total area under cotton cultivation. Press release of Department of Agriculture & Cooperation, Government of India, Mumbai. Release ID: 73448. Available: http://pib.nic.in/newsite/erelease.aspx?relid = 73448. Accessed 11 October 2013.

58. Choudhary B, Gaur K (2010) Bt Cotton in India: A Country Profile. ISAAA Series of Biotech Crop Profiles. ISAAA: Ithaca, NY. ISBN: 978-1-892456-46-X. Available: http://www.isaaa.org/resources/publications/biotech_crop_profiles/bt_cotton_in_india-a_country_profile/download/Bt_Cotton_in_India-A_Country_Profile.pdf. Accessed 11 October 2013.

59. Nagarajan P (2012) Fiber production report for India 2010-11. Textile exchange. Available: http://farmhub.textileexchange.org/upload/library/Farm%20and%20fiber%20report/Regional%20Reports%20-%20India%20-%20English%202010-11-FINAL.pdf. Accessed 11 October 2013.

60. IFOAM (2006) The IFOAM norms for organic production and processing: Version 2005. Bonn: Die Deutsche Bibliothek. 136 p.

61. Demeter International e.V. (2012) Production standards for the use of Demeter, biodynamic® and related trademarks. Demeter International production standards: The standards committee. 45 p. Available: http://www.demeter.net/certification/standards/production. Accessed 11 October 2013.

62. ICAR (2009) Handbook of Agriculture. New Delhi: Indian Council of Agricultural Research. 1617 p.

63. Beej SA, Hamara BA (2012) A decade of Bt cotton in Madhya Pradesh: A report. India Environment Portal, Centre for Science and Environment (CSE), National Knowledge Commission (NKC), Government of India. Available: http://www.indiaenvironmentportal.org.in/files/file/MP-DECADE-OF-BT-COTTON-2012.pdf. Accessed 11 October 2013.

64. Carpenter-Boggs L, Kennedy AC, Reganold JP (2000) Organic and biodynamic management: Effects on soil biology. Soil Science Society of America Journal 64: 1651–1659.

65. Brookes G, Barfoot P (2011) The income and production effects of biotech crops globally 1996-2009. International Journal of Biotechnology 12: 1–49.

66. Majumdar EG (2012) CICR technical bulletin - Mechanisation of cotton production in India. Nagpur: Central Institute for Cotton Research (CICR). Available: http://www.cicr.org.in/pdf/mechnaisation_cotton.pdf. Accessed 11 October 2013.

67. Ministry of Agriculture GOI (2011b) Cost of cultivation/production & related data. Directorate of Economics and Statistics, Department of Agriculture & Cooperation, Government of India, Mumbai. Available: http://eands.dacnet.nic.in/Cost_of_Cultivation.htm. Accessed 11 October 2013.

68. Pinheiro J, Bates D, DebRoy S, Sarkar D and the R Development Core Team (2013) nlme: Linear and Nonlinear Mixed Effects Models.R package version 3.1-110. Available: http://cran.r-project.org/web/packages/nlme/nlme.pdf. Accessed 11 October 2013.

69. R Core Team (2012) A language and environment for statistical computing. R Foundation for Statistical Computing, Vienna, Austria. ISBN 3-900051-07-0. Available: http://www.R-project.org/. Accessed 11 October 2013.

70. Loughin TM (2006) Improved experimental design and analysis for long-term experiments. Crop Science 46: 2492–2502.

71. Carpenter JE (2010) Peer-reviewed surveys indicate positive impact of commercialized GM crops. Nature Biotechnology 28: 319–321.

72. Glover D (2010) Is Bt Cotton a Pro-Poor Technology? A Review and Critique of the Empirical Record. Journal of Agrarian Change 10: 482–509.

73. Konduru S, Yamazaki F, Paggi M (2012) A Study of Indian Government Policy on Production and Processing of Cotton and Its Implications. Journal of Agricultural Science and Technology B 2: 1016–1028.

74. Sadashivappa P, Qaim M (2009) Bt cotton in India: Development of benefits and the role of government seed price interventions. AgBioForum 12: pp. 172–183. Available: http://agbioforum.org/v12n2/v12n2a03-sadashivappa.pdf. Accessed 11 October 2013.

75. Panneerselvam P, Halberg N, Vaarst M, Hermansen JE (2012) Indian farmers' experience with and perceptions of organic farming. Renewable Agriculture and Food Systems 27: 157–169.

76. Ramesh P, Panwar NR, Singh AB, Ramana S, Yadav SK, et al. (2010) Status of organic farming in India. Current Science 98: 1190–1194.

77. Ministry of Agriculture GOI (2011b) Agricultural wages in India. Directorate of Economics and Statistics, Department of Agriculture & Cooperation, Government of India, Mumbai. Available: http://eands.dacnet.nic.in/Cost_of_Cultivation.htm. Accessed 11 October 2013.

78. Singh RJ, Ahlawat IPS, Kumar K (2013) Productivity and profitability of the transgenic cotton-wheat production system through peanut intercropping and FYM addition. Experimental Agriculture 49: 321–335.

79. Panneerselvam P, Hermansen JE, Halberg N (2011) Food Security of Small Holding Farmers: Comparing Organic and Conventional Systems in India. Journal of Sustainable Agriculture 35: 48–68.

80. Swezey SL, Goldman P, Bryer J, Nieto D (2007) Six-year comparison between organic, IPM and conventional cotton production systems in the Northern San Joaquin Valley, California. Renewable Agriculture and Food Systems 22: 30–40.

81. Venugopalan MV, Rajendran TP, Chandran P, Goswami SN, Challa O, et al. (2010) Comparative evaluation of organic and non-organic cotton (Gossypium hirsutum) production systems. Indian Journal of Agricultural Sciences 80: 287–292.

82. Menon M (2003) Organic cotton re-inventing the wheel. Hyderabad: Booksline, SRAS Publications. Available: http://www.ddsindia.com/www/PDF/Organiccotton_Cover_Text.pdf. Accessed 11 October 2013.

83. Nemes N (2010) Seed security among organic cotton farmers in South India. University of Hohenheim: Department of Rural Communication and Extension. Available: http://www.organiccotton.org/oc/Library/library_detail.php?ID = 305. Accessed 11 October 2013.

84. Forster D, Adamtey N, Messmer MM, Pfiffner L, Baker B, et al. (2012) Organic Agriculture—Driving Innovations in Crop Research. In: Bhullar G, Bhullar N, editors. Agricultural Sustainability - Progress and Prospects in Crop Research: Elsevier. pp 21–46.

85. Vanloqueren G, Baret PV (2009) How agricultural research systems shape a technological regime that develops genetic engineering but locks out agroecological innovations. Res Policy 38: 971–983.

22

Increasing Seriousness of Plant Invasions in Croplands of Eastern China in Relation to Changing Farming Practices

Guo-Qi Chen[1], Yun-He He[1,2], Sheng Qiang[1]*

1 Weed Research Laboratory, Nanjing Agricultural University; Nanjing, China, **2** Department of Landscape Architecture, Zhejiang Agricultural and Forestry University, Hangzhou, China

Abstract

Arable areas are commonly susceptible to alien plant invasion because they experience dramatic environmental influences and intense anthropogenic activity. However, the limited reports on relevant factors in plant invasion of croplands have addressed single or a few invasive species and environmental factors. To elucidate key factors affecting plant invasions in croplands, we analyzed the relationship between 11 effective factors and changes in composition of alien plants, using field surveys of crop fields in Anhui Province conducted during 1987–1990 (historical dataset) and 2005–2010 (recent dataset), when rapid urbanization was occurring in China. We found that in the past few decades, the dominance and richness of alien plant populations approximately doubled, despite differences among the 4 regions of Anhui Province. Among the 38 alien invasive plant species observed in the sites, the dominance values of 11 species increased significantly, while the dominance of 4 species decreased significantly. The quantity of chemical fertilizer and herbicide applied, population density, agricultural machinery use, traffic frequency, and annual mean temperature were significantly related to increased richness and annual dominance values of alien plant species. Our findings suggest that the increase in alien plant invasions during the past few decades is primarily a result of increased application of chemical fertilizer and herbicides.

Editor: Fei-Hai Yu, Beijing Forestry University, China

Funding: This research was supported by National Basic Research and Development Program (2009CB1192), China National Natural Science Foundation (31070482), Graduate Student Research and Innovation Program of Jiangsu Province (CXZZ11.0647) and the 111 Project (B07030). The funders had no role in study design, data collection and analysis, decision to publish, or preparation of the manuscript.

Competing Interests: The authors have declared that no competing interests exist.

* E-mail: wrl@njau.edu.cn

Introduction

With their high levels of available resources and anthropogenic disturbance, agricultural areas are particularly susceptible to alien plant invasions [1,2].The continuous introduction and expansion of invasive plants into arable areas makes the management of alien crop weeds increasingly challenging. It is critically important to measure the seriousness and potential trends of alien weed invasions in arable lands, in order to form strategies for the management of these species [3].

Invasive plant species are generally capable of rapid adaptation to altered environments, climate, and human disturbance [1,2], and shifts in farming practices may facilitate plant invasions in arable areas [4]. For example, overuse of herbicides may lead to outbreaks of resistant or tolerant invasive weed populations; 346 herbicide-resistant biotypes of 194 weed species have been described [5], of which many (e.g., *Lolium multiflorum*, *Sorghum halepense*, and *Conyza canadensis*) are highly invasive. Long-term application of chemical fertilizer may promote alien plant invasions in agricultural areas, as habitats with higher levels of nutrient resources tend to be more susceptible to plant invasions [6,7]. Kovacs-Hostyanszki et al. [8] found that fertilizer had a negative impact on the richness of weed species with lower nitrogen preference, and on the coverage of native weeds. In addition, use of agricultural machinery may disperse seeds or other propagules (e.g., rhizomes) of invasive plant species over great distances [9,10]. Climate change also facilitates range expansion of many invasive plant species [3], such as *Sorghum halepense* [11] and *Carduus nutans* [12]. Comparative studies examining the relative influence of the wide variety of factors involved in plant invasion in arable areas are scarce.

In China, large numbers of invasive plant species have established and spread rapidly in arable areas since the 1980s [13]. Changes in the dynamics of plant invasions have occurred concomitantly with changes in agrochemical inputs, land use [13], farming methods [14], and climate [15]. Before 1980, weed control practices in China were limited to manual removal, tillage, and adjustments in crop rotation. Herbicide application began during the 1980s, and expanded rapidly to become the primary weed management strategy after the middle 1990s [14]. In addition, rapid urbanization was accompanied by abandonment of rural areas and farming by large numbers of people, leading to increased mechanization and fertilizer use. The effects of climate change have also become more apparent since the 1980s [15]. The combination of these diverse changes in environmental factors and farming practices may lead to profound changes in plant community dynamics. However, few studies have addressed the invasion of crops by alien weeds in China. In this study, we assessed weed species richness and dominance in croplands in Anhui Province, a typical agricultural province in eastern China.

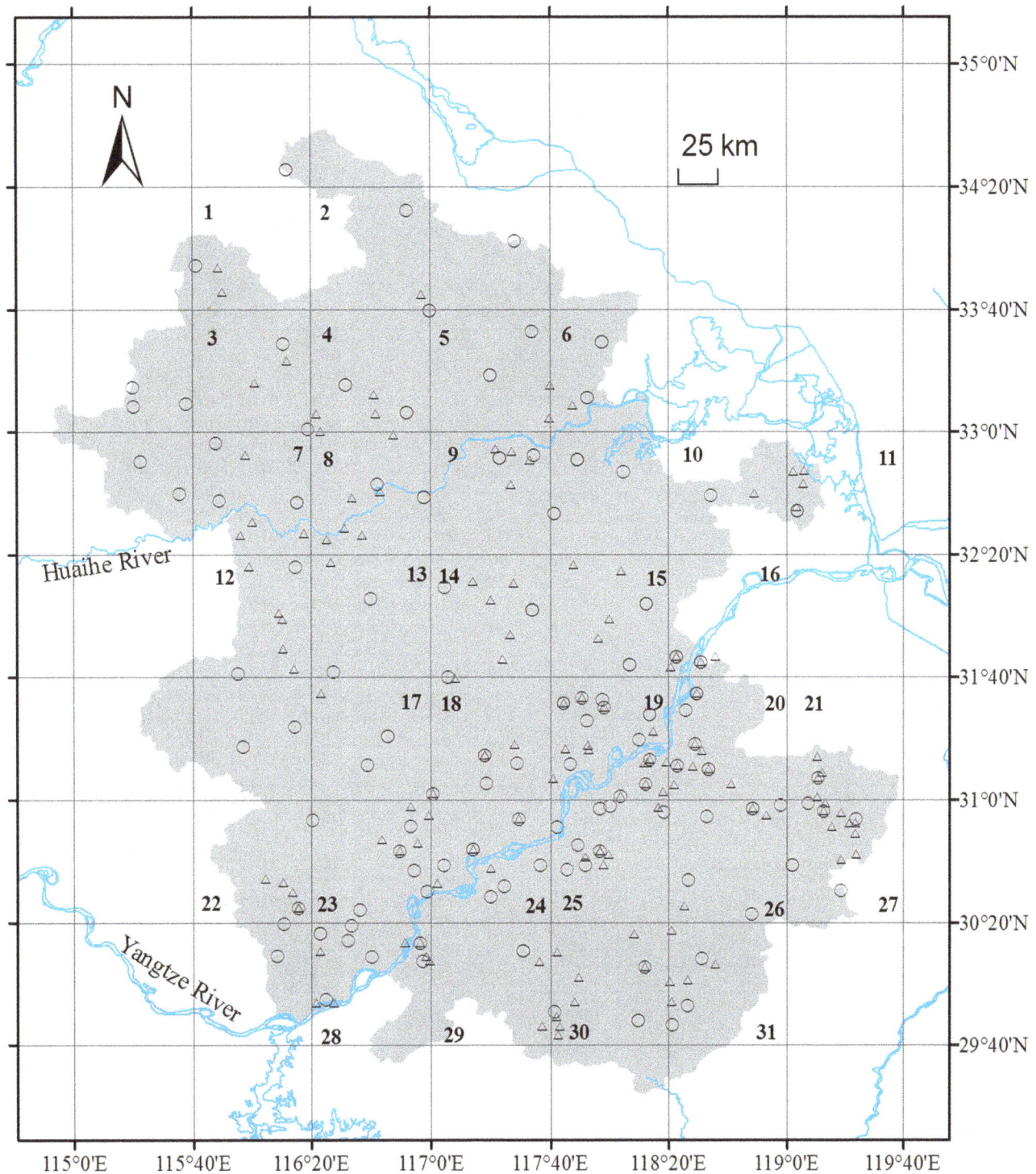

Figure 1. Sites surveyed during 1987–1990 (Δ) and 2005–2010 (○) in summer crop fields in Anhui Province, China.

We hypothesized that increasing applications of chemical herbicide and fertilizer, traffic frequency, and population density may promote alien plant invasions in croplands. Climate change may also have a strong influence on alien weed invasions in cropland. To test this hypothesis, we analyzed 2 datasets obtained from field surveys of summer crops (wheat and oilseed rape) conducted during 1987–1990and 2005–2010, and explored the key factors

responsible for changes in alien plant species richness and dominance.

Materials and Methods

Ethics Statement

Here, by conducting field surveys, we studied weed communities in croplands in Anhui Province, China. No specific

Table 1. Visual scoring method for weed dominance value in crop fields.

		Maximum height in field	
Code	>80 cm	20cm–80 cm	<20cm
0.1	1–3 stems or total coverage <0.1%	<10 stems or total coverage <1%	<15 stems or total coverage <2%
0.5	4–10 stems or total coverage 0.2%–0.9%	11–15 stems or total coverage 1%–2%	16–30 stems or total coverage 3%–5%
1	11–15 stems or total coverage 1%–2%	16–30 stems or total coverage 3%–5%	31–60 stems or total coverage 6%–10%
2	16–30 stems or total coverage 3%–5%	31–60 stems or total coverage 6%–10%	61–100 stems or total coverage 11%–25%
3	31–60 stems or total coverage 6%–10%	61–100 stems or total coverage 11%–25%	101–200 stems or total coverage 25%–50%
4	61–100 stems or total coverage 11%–25%	101–200 stems or total coverage 25%–50%	201–500 stems or total coverage 50%–90%
5	>100 stems or total coverage >25%	>200 stems or total coverage >50%	>500 stems or total coverage >90%

permissions were required and the field studies did not involve endangered or protected species.

Study Area

Anhui Province (29°24' to 34°57'N lat, 114°53' to 119°39'E long) is located in eastern China with a total area of approximately 1.4×10^5 km^2 (Fig. 1). The province differs in climate from north to south. Average annual temperature ranges from 14 to 17°C, average rainfall from 800 to 1800 mm y^{-1}, and average frost-free period from 200 to 250 d [16].

Data Collection

Two datasets from field surveys made approximately 15 years apart in Anhui province (Fig. 1) were used to estimate changes in weed species richness and dominance. Field surveys were conducted from 1987 to 1990 ("historical dataset," by Sheng Qiang [17]) and from 2005 to 2010 ("recent dataset," this study, by Guo-Qi Chen and Yun-He He) using the same methods. All surveys were conducted after crops (wheat, *Triticum aestvum* L. or oilseed rape, *Brassica napus* L.) flowered, but before harvest. Eighty-three sites cultivated with oilseed rape and 47 sites with wheat

Table 2. Mean values of environmental factors of 130 historical sites and 147 recent sites examined in this study.

Factor	Historic	Recent
Crop type (oilseed rape or wheat)	–	–
Crop rotation (wet-dry crop or just dry crops) [A]	–	–
Mean temperature of the coldest month (January, °C)	3.79	3.09*
Mean temperature of the hottest month (July, °C)	28.45	28.52 NS
Annual mean temperature (°C)	15.88	16.47*
Annual mean precipitation (mm)	1671.02	1239.29*
Population density (people/km^2)	329.54	485.30*
Traffic frequency (Freight turnover (10^4 ton km/km^2))	7.02	341.74*
Net cropland agricultural machinery power (kw/ha.)	4.01	12.11*
Net cropland chemical fertilizer applied (kg/ha.)	374.66	782.20*
Net cropland herbicide applied (kg/ha.)	0.53	9.77*

For each environmental factor, mean values of historical and recent datasets in 31 grids (see Figure 1) were compared with paired-sample t-tests. Note: [A]: the preceding crops in fields with wet-dry crop rotation were wet rice, while those in lands with dry crop rotation were dry crops such as soybean, corn, and cotton. "*": P<0.05 and "NS": not significant.

were surveyed from 1987 to 1990, and 78 sites cultivated with oilseed rape and 69 sites with wheat were surveyed between 2005 and 2010. In Anhui Province, the amount of arable land per capita is approximately 0.06 ha [16], and most croplands are divided into small units with clear margins. At each survey site, we established ten 0.06-ha quadrats (approximately 666 m^2) [17] in which we recorded all weed species and their dominance values. Dominance values were divided into 7 categories according to relative coverage, abundance, and height of each weed species (Table 1). This method is one of the most common protocols employed in arable weed field surveys in China [17].

In order to determine the key factors affecting the increasing dominance and richness of invasive plant species in the surveyed summer croplands, we analyzed 11 environmental factors relating to crop type, crop rotation, climate, and human disturbance and farming mode (Table 2). We obtained these data for each site and survey year from records in the Anhui Statistical Yearbook [16]for the cities in which field sites were located. We obtained data on herbicide application from the Yearbooks and from local Plant Protection Stations, which were government organizations, which were responsible for consisting farmers cultivating crops, as well as introducing pesticides to farmers. Two types of crop rotations were employed in the surveyed croplands: (1) wet-dry rotation consisting of rice followed by wheat (oilseed rape); and (2) dry crop rotation, which was just2 dry crops in succession, such as corn (soybean) – wheat (oilseed rape) rotation.

Data Analysis

Data matrices for the historical and recent datasets included alien weed richness and dominance value and environmental

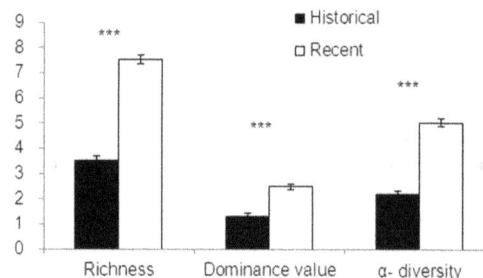

Figure 2. Comparisons between the historic and recent datasets in richness (number of species per site), dominance value, and α-diversity of overall alien weed species in summer crop fields in Anhui Province, China. Note: "***": P<0.001.

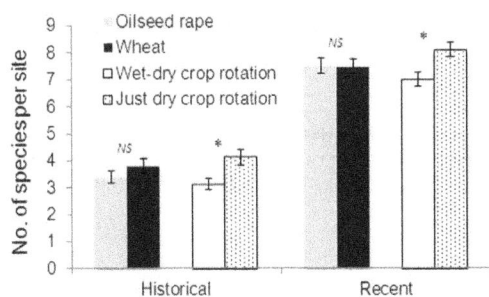

Figure 3. Overall dominance values of alien crop weeds in different groups of summer croplands surveyed in Anhui Province, China.

factors for each site. Mean dominance values by species were calculated for each site from the 10 quadrats to obtain a site-species data matrix. Shannon-Weiner α-diversity [18] was then calculated for both datasets. For each site, the number of alien weed species was estimated as the richness value for each site, and the sum of the dominance values of all alien weed species was calculated as overall dominance value.

To better test the relationship between changes in alien weed invasions and changes in environmental factors, we divided the surveyed area (Anhui Province) into small grids, each of which covered a geographic area of $40' \times 40'$. The surveyed area included 31 grids that contained both the historical and recent survey sites (see Fig. 1), which were used as the sample units for the analyses. For each index in a given grid, we used the average value of the sites located in that grid. We then calculated the change in each index for a given grid by the value in the recent dataset minus that for the related historical data. For each environmental factor, mean values of historical and recent datasets in the 31 grids were compared using paired-sample t-tests. On the basis of the data from these grids, we employed redundancy analysis (RDA) to test the relationship between changes in environmental factors and changes in dominance values of alien weed species, using the "vegan" add-on package in the R 2.12.1 Language and Environment for Statistical Computing [19].

Stepwise regression models [20,21] were used to explore relationships between changes in environmental factors and changes in total richness and dominance of alien weeds. The best-fit models were selected using Akaike's information criterion (AIC) [22,23]. Changes in total richness and dominance values, amount of herbicide and chemical fertilizer applied, and agricultural machinery power were calculated as ratios of recent to historical data. Changes in precipitation and temperature were calculated by subtracting historical values from recent values.

Results

Human disturbance factors and mean annual temperature increased significantly ($P<0.05$) from 1987 to 2010, while mean temperature of the coldest month and annual precipitation decreased significantly (Table 2). Compared with the historical dataset, the amount of chemical fertilizer applied doubled, the amount of chemical herbicide applied increased by 18 times, and traffic frequency was 49 times greater in the recent dataset. Moreover, both the overall richness and dominance values of alien weeds doubled.

In the historical dataset, 24 alien weed species from 17 genera and 10 families were recorded (Table 3). Among these, 9 species (37.5%) had frequencies higher than 10% and 2 (8.3%) had

frequencies higher than 50%. The average number of alien weed species per site was 3.65 (Fig. 2). In the recent dataset, 35 alien weed species from 26 genera and 13 families were recorded (Table 3). Among these, 17 (48.57%) had a frequency higher than 10% and 6 (17.14%) had a frequency higher than 60%. Overall richness, dominance value, and α-diversity of alien weeds in the recent dataset were significantly higher ($P<0.001$) than the relative values in the historical dataset (Fig. 2). Wheat fields had a significantly higher ($P<0.01$) occurrence of alien weeds than oilseed rape fields in the historical dataset, but not in the recent dataset (Fig. 3). Croplands with wet-dry crop rotation (rice as the preceding crop) showed significantly lower dominance of alien weeds than those with dry crop rotation (another dry crop as the preceding crop) for both datasets, while the difference in the recent dataset was significantly lower ($P<0.01$).

According to the RDA results, the 31 grids could be organized into 4 groups (Fig. 4), and increasing dominance of alien weed species was significantly related to changes in 8 environmental factors (Figs. 4 and 5). These factors included annual precipitation and mean temperature, mean temperature of the hottest (July) and coldest (January) month, amount of herbicide applied, amount of chemical fertilizer applied, population density, and traffic frequency. Among these 8 factors, increases in population density and traffic frequency showed the greatest influences on changes in alien species invasions in north Anhui (grids 1 to 5); the dominance values of several alien weed species, including *Avena fatua* and *Euphorbia helioscopia*, increased significantly ($P<0.05$) with increased population density and traffic frequency (Fig. 5). Increased annual precipitation and mean January temperature facilitated invasion of alien weeds, including *Veronica persica*, *Alternanthera philoxeroides*, and *Conyza* spp., in south Anhui (grids 29 to 31). The increase in mean July temperature showed a negative influence on *A. fatua*, *V. sativa*, and *E. helioscopia*. A few species (e.g., *Veronica persica* and *Conyza canadensis*) benefited from increases in annual mean temperature and herbicide application, while species with higher invasiveness (*Geranium caroliniamum*, *A. fatua*, *V. specie*, *A. philoxeroides*, *Erigeron annuus*, *C. bonariensis*, and *C. sumatrensis*) benefited from increased fertilizer application.

Stepwise regression models revealed relationships between changes in alien weed richness and dominance and environmental factors. Six of the 8 environmental factors were significantly related to an increased richness of alien weeds (Table 4); the factor with the greatest influence was the increase in the amount of chemical fertilizer applied. The model identified 4 environmental factors that affected change in overall weed dominance, among which 3 (increased herbicide application, traffic frequency, and mean temperature of the hottest month) showed significant effects (Table 4). The greatest (positive) effect was due to the amount of herbicide applied.

Discussion

Plant invasion is a major economic problem in agricultural fields, and a variety of factors may influence the expansion of alien plants [24]. We collected data from field surveys that measured changes in diversity and dominance of alien agricultural weeds in relation to a historical survey. The most important factors identified as being correlated with increased diversity and dominance of alien weeds were the amount of chemical fertilizer and herbicide applied in agricultural areas. This finding has implications for the ways in which further plant invasions might be managed.

Table 3. Alien weed species, and their frequencies and change in dominance value (DV), among all sites surveyed between 1987–1992 (historical, 130 sites) and 2005–2010 (recent, 147 sites).

Species	Frequency recent (%)	Frequency historical (%)	Change in DV
Geranium caroliniamum	90.79	22.96	0.554*
Conyza canadensis	85.53	59.26	0.152*
Alternanthera philoxeroides	69.74	8.89	0.210*
Vicia sativa	65.79	65.93	−0.041 NS
Veronica persica	61.84	13.33	0.022 NS
Erigeron annuus	61.18	26.67	0.129*
Avena fatua	58.55	47.41	−0.029*
Euphorbia helioscopia	39.47	18.52	0.152*
Sonchus asper	28.95	0.74	0.014*
Daucus carota	26.97	4.44	0.044*
Conyza bonariensis	26.97	1.48	0.086*
Bidens frondosa	25.00	0	0.045*
Conyza sumatrensis	20.39	0.74	0.060*
Aster subulatus	14.47	1.48	0.011*
Veronica polita	12.50	34.81	−0.208*
Coronopus didymus	11.18	0	0.007 NS
Cyperus rotundus	11.18	5.19	0.005 NS
Veronica arvensis	7.24	17.04	0.019 NS
Bidens pilosa	4.61	0.74	0.014 NS
Lepidium virginicum	3.29	0	0.005 NS
Aeschynomene indica	3.29	0	0.003 NS
Sonchus oleraceus	3.29	1.48	0.004 NS
Phytolacca americana	3.29	0	0.004 NS
Plantago virginica	2.63	0	0.006 NS
Bromus catharticus	2.63	0	0.012 NS
Ambrosia artemisiifolia	2.63	0	0.003 NS
Lolium temulentum	1.97	7.41	−0.006 NS
Thlaspi arvense	1.97	8.89	−0.040 NS
Amaranthus retroflexus	1.97	0	0.000 NS
Euphorbia maculata	1.32	0	0.000 NS
Crassocephalum crepidioides	1.32	0.74	0.003 NS
Lolium multiflorum	0.66	0	0.002 NS
Solidago canadensis	0.66	0	0.001 NS
Chenopodium ambrosioides	0.66	0	0.000 NS
Amaranthus tricolor	0.66	0	0.000 NS
Veronica hederaefolia	0	2.96	−0.007 NS
Coreopsis drummondii	0	0.74	0.000 NS
Veronica peregrina	0	9.63	−0.005*

Note: "*": P<0.05 and "NS": not significant.
Note: Change in DV for each species in each grid was calculated by the DV in the recent dataset minus that in the corresponding historic dataset.

Human Factors

Factors associated with human activities are frequently found to facilitate plant invasions [24]. Together, our results implied that increasing traffic frequency and population density, as well as changed farming practices, have significant influences on the occurrence of alien crop weeds.

Firstly, increased application of herbicides negatively affected the richness but positively affected the dominance of alien weeds in the surveyed area. Herbicide application can be an effective means of addressing invasion by the most serious weed species, and chemical weed control has greatly improved the efficiency of weed management in agricultural areas [14]. Many agricultural weed species are controlled simultaneously by herbicides, resulting in a decrease in species richness with increased herbicide application [25], as found in our study. However, many invasive plant species are capable of adapting to herbicides [1], such that chemical weed control may promote the dominance of invasive plants.

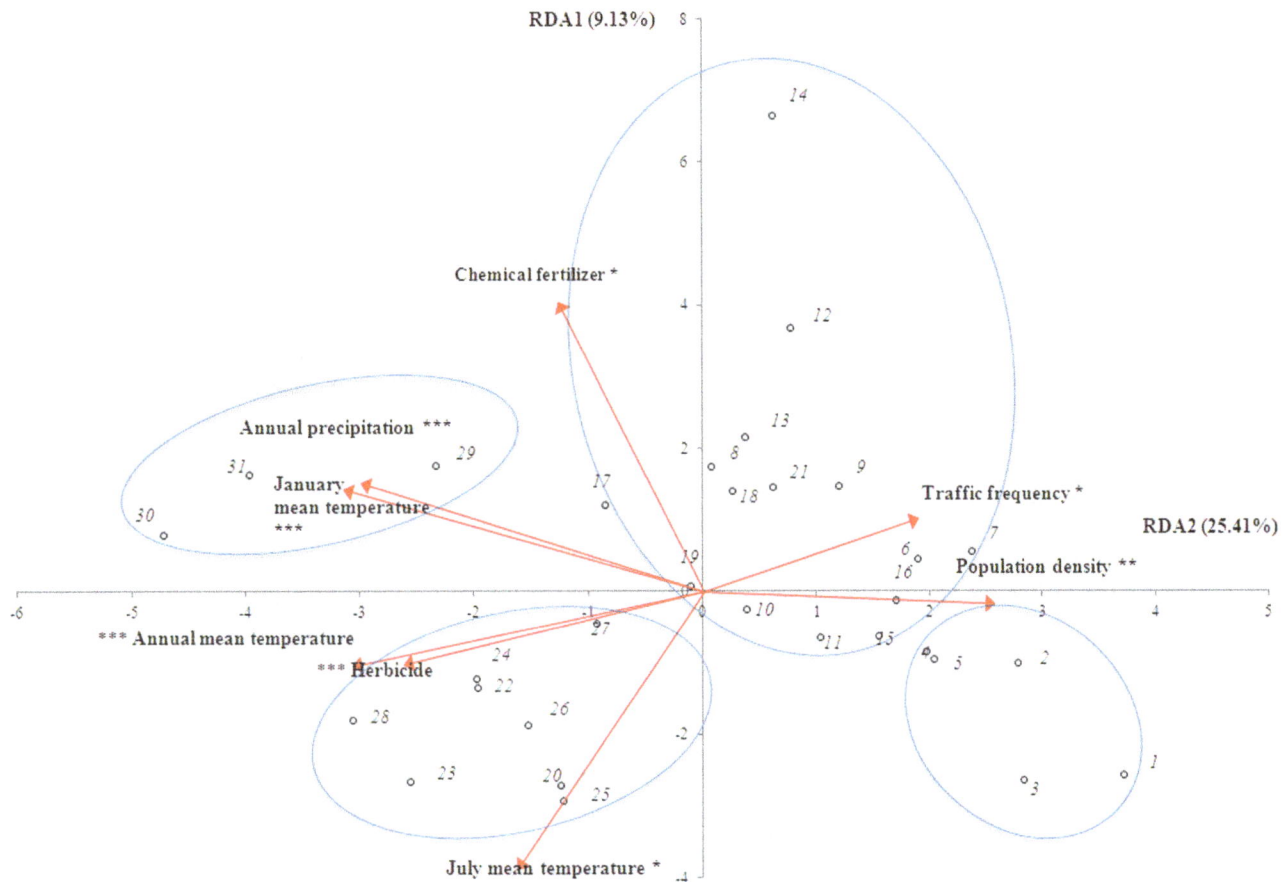

Figure 4. Redundancy analysis (RDA) showing the 8 significant environmental factors and the 31 geographic grids in Anhui Province, China (see Figure 1). RDA was conducted to analyze the relationship between changes in environmental factors and changes in the dominance values of alien weed species in croplands.

Many invasive plant species are also less sensitive to herbicides than are native plants or crop species. The active ingredients contained in the herbicides applied in our survey area primarily include 2, 4–D (2, 4-dicholrophenoxyacetic acid), MCPA (2-methyl-4-chlorophenoxy acetic acid), metsulfuron methyl, tribenuron-methyl, chlorsulfuron, isoproturon, fenoxaprop-P-ethyl, and quizalofop-p-ethyl [26–28]. *G. caroliniamum* showed higher tolerance than other species to most of the herbicides, [29,30]. In the surveyed areas, *G. caroliniamum* increased in frequency from 23% to 91%, and its dominance value per site increased by 6 times from the historical to the current survey period. Similarly, *E. helioscopia* is reported to be insensitive to 2, 4–D, MCPA, and tribenuron-methyl [31,32]; although increased herbicide application negatively impacted the dominance of this species, *E. helioscopia* maintained a high dominance value.

Long-term application of herbicides can result in selection and rapid distribution of herbicide-resistant weed species [33]. Our results suggested that the increased dominance of *C. canadensis* and *V. persica* was positively correlated with increased herbicide application (Fig. 5). Many invasive agricultural weeds show high genetic or phenotypic plasticity, and thus can become herbicide resistant in a short time, as has been reported for *C. canadensis* [34], *A. fatua* [35], *V. sativa* [36], and *Lolium multiflorum*. Ten years of intense glyphosate use (3.7 kg ha^{-1} y^{-1}) in fruit orchards in Chile led to selection of glyphosate-resistant *L. multiflorum* populations [37]. The most-serious herbicide-resistant weed species in South America, such as *S. halepense*, *C. canadensis*, and *L. multiflorum*, are all

invasive species [38]. Further, many serious herbicide-resistant weed species in North America, including *Ambrosia artemisiifolia*, *Ambrosia trifida*, *S. halepense*, *Amaranthus retroflexus*, and *C. canadensis* [33,39] are serious invaders in countries outside of their native ranges [40]. Therefore, developing ecological weed management that is less reliant on herbicide application is of high importance [4,41]; integrated agricultural practices such as rice-fish [42] and rice-duck [43] co-culture systems, and well-designed crop rotation and intercropping [44], could be part of the solution to this challenge.

The application of chemical fertilizer and use of agricultural machinery may promote alien plant invasions. Our results suggested that the increasing amount of chemical fertilizer applied was the most significant factor related to the increase of alien plant species richness (Table 4), and that increased quantities of chemical fertilizers applied to crops was positively related to increased dominance of several invasive plant species in the surveyed croplands. Combine harvesters can disperse weed seeds and other propagules over great distances, as shown for *A. fatua* [10] and *A. philoxeroides* (according to our field observation). Agricultural mechanization in Anhui Province is still at a low level, but is developing quickly [16]; hence, the potential influence of agricultural machinery on alien plant invasions deserves more attention.

In addition, our study suggested that alien plant invasions increased more rapidly in oilseed rape fields than in wheat fields. This could be the result of changes in the composition of alien

Table 4. Results of the stepwise regression models used to test the relationships between changes in alien weed species richness and dominance, and changes in environmental factors between the 2 datasets surveyed in different time periods.

Parameter	Estimate	SE	t-value	P
Alien weed species richness				
Net cropland chemical fertilizer applied	0.008	0.001	7.349	<0.001
Net cropland herbicide applied	−0.004	0.001	−5.660	<0.001
Population density	0.844	0.258	3.274	0.004
Net cropland agricultural machinery power	0.023	0.007	3.057	0.006
Annual mean temperature	−0.899	0.347	−2.592	0.017
Traffic frequency	0.950	0.448	2.119	0.046
Mean temperature of the coldest month	−0.180	0.089	−2.013	0.057
Mean temperature of the hottest month	0.048	0.030	1.606	0.123
ΔAIC with null model	−60.34			
Overall dominance value of alien weeds				
Net cropland herbicide applied	0.002	0.000	4.168	<0.001
Mean temperature of the hottest month	−0.137	0.054	−2.526	0.018
Traffic frequency	−0.446	0.188	−2.366	0.026
Mean temperature of the coldest month	0.328	0.169	1.948	0.063
ΔAIC with null model	−17.28			

The changes in Akaike's information criteria between final and null models are also shown.

plant species in the surveyed croplands. In the historical dataset, the most dominant alien species was *A. fatua*, followed by *V. sativa*, *Veronica polita*, and *V. persica*, all of which are archaeophytes (introduced before 1840). These 4 alien species account for 86% of total alien dominance value, and each of these species tended to occur in wheat fields. Nevertheless, several neophytes (introduced after 1840) in the surveyed areas spread quickly, particularly *G. caroliniamum*, *A. philoxeroides*, and *C. canadensis*, all of which tended to occur in oilseed rape fields. The continuous introduction and expansion of alien neophytes have resulted in plant invasions in oilseed rape areas becoming as serious as those in wheat fields. Other types of cropland that currently have lower rates of plant invasion may increasingly be faced with similar problems.

Alien plant invasions in lands with wet-dry crop rotation were less serious than those in lands with dry crop rotation, but increased faster. Alien invasions in areas with wet-dry crop rotation were very low in the historical dataset. One explanation may be that the annual shift in soil moisture between wet and dry planting periods caused a barrier against maintenance of weed seed banks [45], particularly for alien species that lack a long history of co-evolution with crop cultivation in China. Recently, some perennial alien plant species with rapid vegetative reproduction have invaded croplands and spread quickly, such as *A. philoxcroides* [46], which is highly adapted to different moisture conditions [47]. Seeds of some alien plant species, including *G. caroliniamum*, *V. sativa*, and *V. persica* [48] can remain viable in wet-

dry crop rotation systems, and frequently infest this type of cropland. Moreover, some invasive weeds, such as *C. canadensis* and *E. annuus* [49], produce large amounts of small seeds and maintain large seed banks from which they readily disperse into croplands. Well-designed crop rotation systems could help to manage weed invasions. If possible, wet-dry crop rotation system showed higher resistance to alien plant invasions. Moreover, invasive weed species that are well adapted to both moist and dry soil should be more carefully monitored and controlled, and regular field investigations on crop weed communities should be conducted every few decades, as well the distribution of serious invasive plants should be monitored.

In addition, increases in human population density and traffic frequency both showed positive influences on the richness of alien plant species, consistent with studies that have shown correlations between anthropogenic disturbance and plant invasion [50–52]. Our results also suggested that the increased dominance of alien plants in areas with greater increases in traffic frequency was less pronounced in north Anhui Province. In north Anhui Province, several archaeophytes have been major agricultural weeds over long time periods, and the occurrence of these species did not increase significantly. For example, the most dominant alien weed species in the historical dataset were *A. fatua*, *V. sativa*, *V. persica*, *V. polita*, and *Thlaspi arvense*. Among these species, the dominance of *A. fatua*, *V. polita*, and *T. arvense* decreased significantly and those of *V. sativa* and *V. persica* did not change significantly.

Physical Factors

Our results suggested that climatic change plays an important role in promoting the invasivity of alien plants in agricultural areas. Temperature and precipitation are widely known to be key factors in determining the distribution of plant species [3,53,54]. The increase in mean annual temperature and annual precipitation were positively correlated with increased dominance of several common invasive plants. However, the increase in mean temperature of the hottest month was negatively associated with dominance of most alien plant species. Located in the subtropical and temperate zones, Anhui Province has a climate with high precipitation, warm winters, and mild summers [16], which favor the establishment of a broad range of invasive plant species. Thus, the threat of plant invasions in this area may worsen in the future. The temperature in Anhui Province has shown a clear warming trend in recent decades [55], and annual precipitation has been highly variable with more frequent rainstorms [56]. A warming climate may enable the distribution of many invasive plants to expand [3]. Thus, cropping systems in many regions are likely to experience new vulnerabilities to exotic plant invasions in the future [11]. Additionally, more frequent rainstorms may result in increased flooding, which may further promote plant invasions by dispersing large numbers of seeds across a large area, as well as causing habitat fragmentation [57,58].

Conclusion

With increasing applications of herbicide and chemical fertilizer, higher population density, an upgraded traffic system, and the influence of climate change, plant invasions in crop areas have approximately doubled during the past few decades in Anhui Province, China. Differences in the seriousness of plant invasions among different types of cropping systems are fading. Much more attention should be focused on studying the current and potential distributions of invasive plants in agricultural areas, and to assess the risks of plant invasions in various croplands, particularly in areas that currently experience low rates of invasion. Considering that current patterns of biological invasion may better reflect

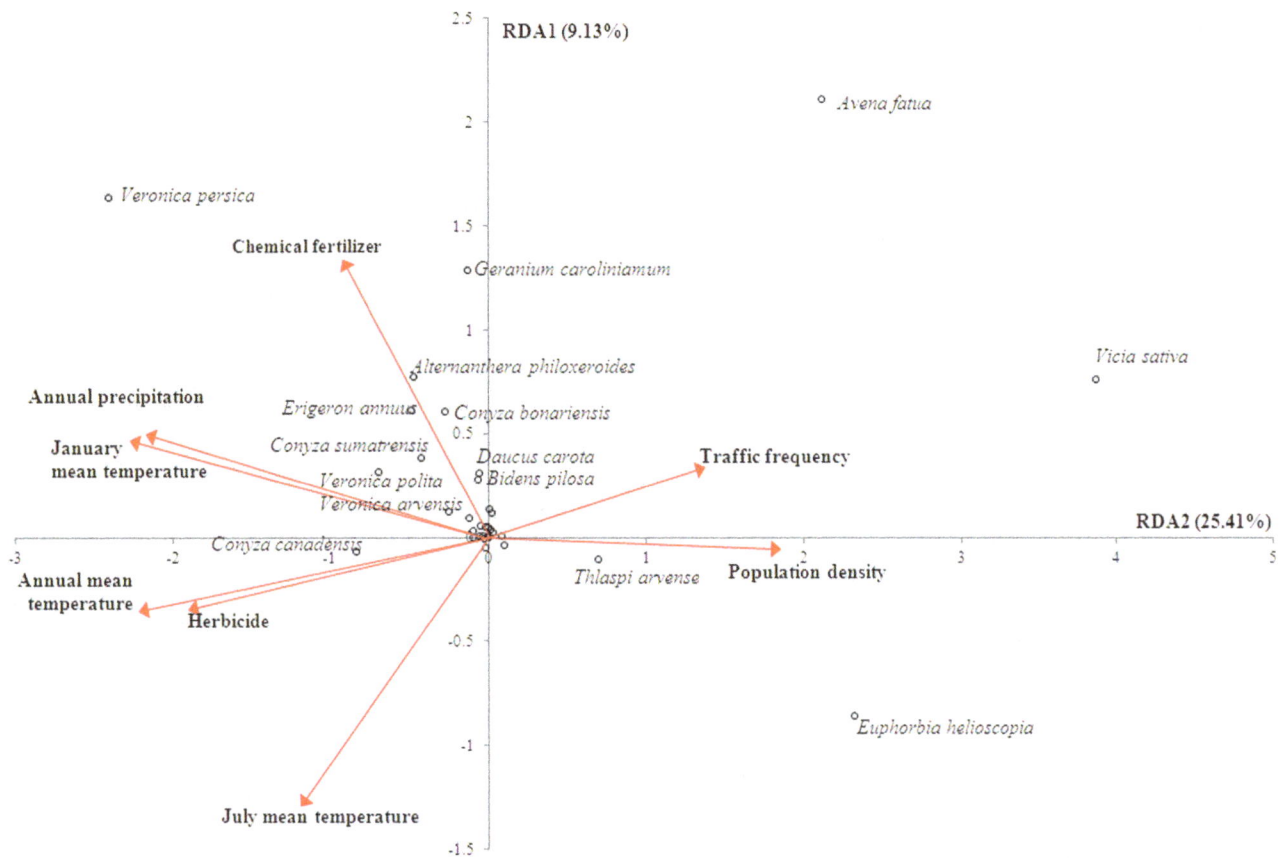

Figure 5. Redundancy analysis (RDA) showing the 8 significant environmental factors and alien weed species in croplands in Anhui Province, China. Note: species with lower correlations with RDA axes are not shown.

historical than recent human activities [59], the potential for crop weed invasions in this area could be a serious threat in the next several decades and beyond. Integrative weed management with reduced application of chemical fertilizers and herbicides deserves much more attention.

Acknowledgments

We thank Kerin Bentley, Joan West and Sandra Hoffberg from the University of Georgia (USA) for their help with preparing the manuscript.

Author Contributions

Conceived and designed the experiments: SQ Y-HH G-QC. Performed the experiments: SQ Y-HH G-QC. Analyzed the data: G-QC. Contributed reagents/materials/analysis tools: G-QC. Wrote the paper: G-QC Y-HH SQ.

References

1. Clements DR, DiTommaso A, Jordan N, Booth BD, Cardina J, et al. (2004) Adaptability of plants invading North American cropland. Agriculture, ecosystems & environment 104: 379–398.
2. Seastedt T (2007) Plant ecology - Resourceful invaders. Nature 446: 985–986.
3. Clements DR, Ditommaso A (2011) Climate change and weed adaptation: can evolution of invasive plants lead to greater range expansion than forecasted? Weed Research 51: 227–240.
4. Petit S, Boursault A, Le Guilloux M, Munier-Jolain N, Reboud X (2011) Weeds in agricultural landscapes. A review. Agronomy for sustainable development 31: 309–317.
5. HRAC (2012) International Survey of Herbicide Resistant Weeds.
6. DeGasperis BG, Motzkin G (2007) Windows of opportunity: Historical and ecological controls on Berberis thunbergii invasions. Ecology 88: 3115–3125.
7. Richardson D, Pyšek P (2006) Plant invasions: merging the concepts of species invasiveness and community invasibility. Progress in Physical Geography 30: 409.
8. Kovacs-Hostyanszki A, Batary P, Baldi A, Harnos A (2011) Interaction of local and landscape features in the conservation of Hungarian arable weed diversity. Applied Vegetation Science 14: 40–48.

9. Blanco-Moreno J, Chamorro L, Masalles R, Recasens J, Sans F (2004) Spatial distribution of Lolium rigidum seedlings following seed dispersal by combine harvesters. Weed Research 44: 375–387.
10. Shirtliffe SJ, Entz MH (2005) Chaff collection reduces seed dispersal of wild oat (Avena fatua) by a combine harvester. Weed Science 53: 465–470.
11. McDonald A, Riha S, DiTommaso A, DeGaetano A (2009) Climate change and the geography of weed damage: Analysis of U.S. maize systems suggests the potential for significant range transformations. Agriculture, Ecosystems & Environment 130: 131–140.
12. Zhang R, Jongejans E, Shea K (2011) Warming increases the spread of an invasive thistle. PLoS One 6.
13. Ding JQ, Mack RN, Lu P, Ren MX, Huang HW (2008) China's booming economy is sparking and accelerating biological invasions. BIoscience 58: 317–324.
14. Zhang ZP (2003) Development of chemical weed control and integrated weed management in China. Weed Biology and Management 3: 197–203.
15. Wang Z, Ding Y, He J, Yu J (2004) An updating analysis of the climate change in China in recent 50 years. Acta Meteorologica Sinica 62: 228–236 (in Chinese with English Abstract).
16. Statistics Bureau of Anhui Province (1991–2011) Anhui statistical yearbook. Beijing: China Statistics Press. (in Chinese)

17. Qiang S (2005) Multivariate Analysis, Description, and Ecological Interpretation of Weed Vegetation in the Summer Crop Fields of Anhui Province, China. Journal of Integrative Plant Biology 47: 1193–1210.

18. Ponce C, Bravo C, de León DG, Magaña M, Alonso JC (2011) Effects of organic farming on plant and arthropod communities: A case study in Mediterranean dryland cereal. Agriculture, Ecosystems & Environment 141: 193–201.

19. R Development Core Team (2011) R: A language and environment for statistical computing. Vienna, Austria: R Foundation for Statistical Computing.

20. Whittingham MJ, Stephens PA, Bradbury RB, Freckleton RP (2006) Why do we still use stepwise modelling in ecology and behaviour? Journal of Animal Ecology 75: 1182–1189.

21. Mundry R, Nunn CL (2009) Stepwise model fitting and statistical inference: turning noise into signal pollution. The American Naturalist 173: 119–123.

22. José-María L, Armengot L, Blanco-Moreno JM, Bassa M, Sans FX (2010) Effects of agricultural intensification on plant diversity in Mediterranean dryland cereal fields. Journal of Applied Ecology 47: 832–840.

23. Burnham KP, Anderson DR (2002) Model selection and multimodel inference: a practical information-theoretic approach. New York: Springer.

24. Wang R, Wang JF, Qiu ZJ, Meng B, Wan FH, Wang YZ (2011). Multiple mechanisms underlie rapid expansion of an invasive alien plant. New Phytologist 191: 828–839.

25. Dekker J (1997) Weed diversity and weed management. Weed Science: 357–363.

26. Yu EL (2010) Investigation and strategies of control of weed in wheat and rapeseed in Anhui Province. Anhui Agri. Sci. Bull. 16: 172–173.(in Chinese with English abstract)

27. Zhang BW, Fang WH (2002) Research and application of weed control strategies in oilseed rape fields in Anqing City. Pesticide 41: 12–14. (in Chinese)

28. Zhang QY, Xu J, Zhou QF, Hu M (2003) Research progress on chemical weed control in wheat fields in Anhui Province. Plant Protection Technology and Extension 23: 34–35. (in Chinese)

29. Hu YZ (2010) Experiments on controlling broadleaved weeds such as *Geranium caroliniamum* in wheat fields with herbicides. Anhui Agri. Sci. Bull. 16: 143–143. (in Chinese)

30. Wei YB (2012) A study on chemical control against *Geranium caroliniamum* - a serious weed species in wheat fields. Anhui Agri. Sci. Bull. 18: 68–69. (in Chinese)

31. He JR, Han BR, Zhen ZZ, Xu W, Du JH (2008) Compositions and integrative management of weed communities in winter wheat fields in Suqian City. Anhui Agri. Sci. Bull. 14: 41–43.(in Chinese)

32. Song AY (2011) A preliminary study on the succession of weed community of wheat fields in Xiaoxian county.Anhui Agri. Sci. Bull. 17: 33–35. (in Chinese)

33. Powles SB (2008) Evolved glyphosate-resistant weeds around the world: lessons to be learnt. Pest Management Science 64: 360–365.

34. Mueller TC, Massey JH, Hayes RM, Chris L, Stewart CN Jr (2003) Shikimate accumulates in both glyphosate-sensitive and glyphosate-resistant horseweed (*Conyza canadensis* L. Cronq.). Journal of Agricultural and Food Chemistry 51: 680–684.

35. Beckie HJ, Thomas AG, Légère A, Kelner DJ, Van Acker RC, et al. (1999) Nature, occurrence, and cost of herbicide-resistant wild oat (*Avena fatua*) in small-grain production areas. Weed Technology: 612–625.

36. Nandula VK, Foy CL, Orcutt DM (1999) Glyphosate for Orobanche aegyptiaca Control in Vicia sativa and Brassica napus Weed Science 47: 486–491.

37. Perez A, Kogan M (2003) Glyphosate-resistant *Lolium multiflorum* in Chilean orchards. Weed Research 43: 12–19.

38. Vila-Aiub MM, Vidal RA, Balbi MC, Gundel PE, Trucco F, et al. (2008) Glyphosate-resistant weeds of South American cropping systems: an overview. Pest Management Science 64: 366–371.

39. Powles SB, Preston C (2006) Evolved Glyphosate Resistance in Plants: Biochemical and Genetic Basis of Resistance 1. Weed Technology 20: 282–289.

40. Holm LG, Pancho JV, Herberger JP (1979) A geographical atlas of world weeds. New York: John Wiley and Sons.

41. Thomas AG, Légere A, Leeson JY, Stevenson FC, Holm FA, et al. (2011) Weed community response to contrasting integrated weed management systems for cool dryland annual crops. Weed Research 51: 41–50.

42. Xie J, Hu L, Tang J, Wu X, Li N, et al. (2011) Ecological mechanisms underlying the sustainability of the agricultural heritage rice–fish coculture system. Proceedings of the National Academy of Sciences 108: E1381–E1387.

43. Huang Y, Wang H, Huang H, Feng Z, Yang Z, et al. (2005) Characteristics of methane emission from wetland rice–duck complex ecosystem. Agriculture, ecosystems & environment 105: 181–193.

44. Zhu Y, Chen H, Fan J, Wang Y, Li Y, et al. (2000) Genetic diversity and disease control in rice. Nature 406: 718–722.

45. Wu J, Zhou H (2000) Seed bank of weeds in paddy fields. Chinese Journal of Rice Science 14: 37–42.

46. Wang N, Yu FH, Li PX, He WM, Liu J, et al. (2009) Clonal integration supports the expansion from terrestrial to aquatic environments of the amphibious stoloniferous herb *Alternanthera philoxeroides*. Plant Biology 11: 483–489.

47. Pan XY, Geng YP, Zhang WJ, Li B, Chen JK (2006) The influence of abiotic stress and phenotypic plasticity on the distribution of invasive *Alternanthera philoxeroides* along a riparian zone. Acta Oecologica 30: 333–341.

48. Zuo RL, Qiang S, Li RH (2007) Relationship between weed seeds dispersed by irrigation water and soil weed seedbank of paddy field in rice-growing region. Chinese Journal of Rice Science 21: 417–424 (in Chinese with English Abstract).

49. Li YH (1998) Weed flora in China. Beijing: China Agriculture Press. (in Chinese)

50. Huang QQ, Wang GX, Hou YP, Peng SL (2011) Distribution of invasive plants in China in relation to geographical origin and life cycle. Weed Research 51: 534–542.

51. Gavier-Pizarro GI, Radeloff VC, Stewart SI, Huebner CD, Keuler NS (2010) Housing is positively associated with invasive exotic plant species richness in New England, USA. Ecol Appl 20: 1913–1925.

52. Seipel T, Kueffer C, Rew LJ, Daehler CC, Pauchard A, et al. (2012) Processes at multiple scales affect richness and similarity of non-native plant species in mountains around the world. Global Ecology and Biogeography 21: 236–246.

53. Freckleton RP, Stephens PA (2009) Predictive models of weed population dynamics. Weed Research 49: 225–232.

54. Graziani A, Steinmaus SJ (2009) Hydrothermal and thermal time models for the invasive grass, *Arundo donax*. Aquatic Botany 90: 78–84.

55. Xu XW, Sun MY, Fang YY, He XQ, Xue F, et al. (2011) Impact of climatic change on rice production and response strategies in Anhui Province. Journal of Agro-Environment Science 30: 1755–1763 (in Chinese with English abstract).

56. Xie WS, Tian H (2011) Characteristics of rainstorm in Anhui in the past 50 years. Meteorological Science and Technology: 160–164 (in Chinese).

57. Price JN, Berney PJ, Ryder D, Whalley RDB, Gross CL (2011) Disturbance governs dominance of an invasive forb in a temporary wetland. Oecologia 167: 759–769.

58. Touchette B, Romanello G (2010) Growth and water relations in a central North Carolina population of *Microstegium vimineum* (Trin.) A. Camus. Biological Invasions 12: 893–903.

59. Essl F, Dullinger S, Rabitsch W, Hulme PE, Hulber K, et al. (2011) Socioeconomic legacy yields an invasion debt. Proc Natl Acad Sci U S A 108: 203–207.

PERMISSIONS

LIST OF CONTRIBUTORS

Matteo Campioli and Niki Leblans
Department of Biology, University of Antwerp, Wilrijk, Belgium

Anders Michelsen
Department of Biology, University of Copenhagen, Copenhagen, Denmark
Center for Permafrost (CENPERM), University of Copenhagen, Copenhagen, Denmark

Shenzhong Tian, Tangyuan Ning, Hongxiang Zhao, Bingwen Wang, Na Li, Huifang Han and Zengjia Li
State Key Laboratory of Crop Biology, Shandong Key Laboratory of Crop Biology, Shandong Agricultural University, Taian, Shandong PR, China

Shuyun Chi
College of Mechanical and Electronic Engineering, Shandong Agricultural University, Taian, Shandong PR, China

Quansheng Chen
State Key Laboratory of Vegetation and Environmental Change, Institute of Botany, Chinese Academy of Sciences, Beijing, China

Xingguo Han
State Key Laboratory of Vegetation and Environmental Change, Institute of Botany, Chinese Academy of Sciences, Beijing, China
State Key Laboratory of Forest and Soil Ecology, Institute of Applied Ecology, Chinese Academy of Sciences, Shenyang, China

Huifen Zheng
State Key Laboratory of Vegetation and Environmental Change, Institute of Botany, Chinese Academy of Sciences, Beijing, China
Graduate University of Chinese Academy of Sciences, Beijing, China

Qi Li, Xiaotao Lü, Qiang Yu, Haiyang Zhang and Wenju Liang
State Key Laboratory of Forest and Soil Ecology, Institute of Applied Ecology, Chinese Academy of Sciences, Shenyang, China

Cunzheng Wei
State Key Laboratory of Vegetation and Environmental Change, Institute of Botany, Chinese Academy of Sciences, Beijing, China

State Key Laboratory of Forest and Soil Ecology, Institute of Applied Ecology, Chinese Academy of Sciences, Shenyang, China
Graduate University of Chinese Academy of Sciences, Beijing, China

Nianpeng He
Institute of Geographic Sciences and Natural Resources Research, Chinese Academy of Sciences, Beijing, China

Paul Kardol
Department of Forest Ecology and Management, Swedish University of Agricultural Sciences, Umeå, Sweden

Etienne Laliberté
School of Plant Biology, The University of Western Australia, Crawley, Western Australia, Australia.

E. Carol Adair
National Center for Ecological Analysis and Synthesis, University of California Santa Barbara, Santa Barbara, California, United States of America

Sarah E. Hobbie
Department of Ecology, Evolution and Behavior, University of Minnesota, Saint Paul, Minnesota, United States of America

Tara Joy Massad
Department of Ecology and Evolutionary Biology, Tulane University, New Orleans, Louisiana, United States of America
Program on the Global Environment, University of Chicago, Chicago, Illinois, United States of America

Lee A. Dyer and Gerardo Vega C.
Department of Biology, University of Nevada, Reno, Nevada, United States of America

Yunfeng Peng, Xuexian Li and Chunjian Li
Key Laboratory of Plant-Soil Interactions, Ministry of Education, Department of Plant Nutrition, China Agricultural University, Beijing, China

Hai-Lin Zhang, Jian-Fu Xue, Zhong-Du Chen and Fu Chen
College of Agronomy and Biotechnology, China Agricultural University, Key Laboratory of Farming System, Ministry of Agriculture, Beijing, China

Xiao-Lin Bai
College of Agronomy and Biotechnology, China Agricultural University, Key Laboratory of Farming System, Ministry of Agriculture, Beijing, China
Patent Examination Cooperation Center of the Patent Office, SIPO, Beijing, China

Hai-Ming Tang
Soil and Fertilizer Institute of Hunan Province, Changsha, China

Adam S. Davis
United States Department of Agriculture/Agricultural Research Service, Global Change and Photosynthesis Research Unit, Urbana, Illinois, United States of America

Jason D. Hill
Department of Bioproducts and Biosystems Engineering, University of Minnesota, St. Paul, Minnesota, United States of America

Craig A. Chase
Leopold Center for Sustainable Agriculture, Iowa State University, Ames, Iowa, United States of America

Ann M. Johanns
Department of Economics, Iowa State University Extension and Outreach, Osage, Iowa, United States of America

Matt Liebman
Department of Agronomy, Iowa State University, Ames, Iowa, United States of America

Hongwei Nan, Qing Liu, Jinsong Chen, Xinying Cheng, Huajun Yin, Chunying Yin and Chunzhang Zhao
Key Laboratory of Mountain Ecological Restoration and Bioresource Utilization & Ecological Restoration Biodiversity Conservation Key Laboratory of Sichuan Province, Institute of Biology, Chinese Academy of Sciences, Chengdu, China

Shannon M. Murphy
Department of Biological Sciences, University of Denver, Denver, Colorado, United States of America

Gina M. Wimp and Danny Lewis
Biology Department, Georgetown University, Washington, DC, United States of America

Robert F. Denno
Department of Entomology, University of Maryland, College Park, Maryland, United States of America

Anja Vogel
Institute of Ecology, Friedrich Schiller University Jena, Jena, Germany

Michael Scherer-Lorenzen
Michael Scherer-Lorenzen, Faculty of Biology – Geobotany, University of Freiburg, Freiburg, Germany

Alexandra Weigelt
Alexandra Weigelt, Institute of Biology, University of Leipzig, Leipzig, Germany

Evan Siemann
Department of Ecology and Evolutionary Biology, Rice University, Houston, Texas, United States of America

Christopher A. Gabler
Department of Ecology and Evolutionary Biology, Rice University, Houston, Texas, United States of America
Department of Biology and Biochemistry, University of Houston, Houston, Texas, United States of America

Wenyi Dong, Xinyu Zhang, Huimin Wang, Xiaoqin Dai, Xiaomin Sun and Fengting Yang
Key Laboratory of Ecosystem Network Observation and Modeling, Institute of Geographic Sciences and Natural Resources Research, Chinese Academy of Sciences, Beijing, People's Republic of China

Weiwen Qiu
The New Zealand Institute for Plant and Food Research Limited, Christchurch, New Zealand

Joanna L. Nelson and Erika S. Zavaleta
Environmental Studies Department, University of California Santa Cruz, Santa Cruz, California, United States of America

Aurore Philibert, Chantal Loyce and David Makowski
INRA, UMR 211 Agronomie, Thiverval Grignon, France
AgroParisTech, UMR 211 Agronomie, Thiverval Grignon, France

Claudio de Sassi and Jason M. Tylianakis
School of Biological Sciences, University of Canterbury, Christchurch, New Zealand

Shenzhong Tian, Tangyuan Ning, Hongxiang Zhao, Bingwen Wang, Na Li and Zengjia Li
State Key Laboratory of Crop Biology, Shandong Key Laboratory of Crop Biology, National Engineering Laboratory for Efficient Utilization of Soil and Fertilizer Resources, Shandong Agricultural University, Taian, Shandong, China

Yu Wang
Shandong Rice Research Institute, Jinan, Shandong, China

Shuyun Chi
College of Mechanical and Electronic Engineering, Shandong Agricultural University, Taian, Shandong, China

Xiaoqin Dai, Zhu Ouyang, Yunsheng Li and Huimin Wang
Key Laboratory of Ecosystem Network Observation and Modeling, Institute of Geographic Sciences and Natural Resources Research, Chinese Academy of Sciences, Beijing, China

Wenming He
Key Laboratory of Aquatic Botany and Watershed Ecology, Wuhan Botanical Garden, Chinese Academy of Sciences, Moshan, Wuchang, Wuhan, Hubei Province, China
Graduate University of Chinese Academy of Sciences, Beijing, China

Fang Chen
Key Laboratory of Aquatic Botany and Watershed Ecology, Wuhan Botanical Garden, Chinese Academy of Sciences, Moshan, Wuchang, Wuhan, Hubei Province, China
International Plant Nutrition Institute, Wuhan, China

Valentin H. Klaus, Norbert Hölzel and Till Kleinebecker
University of Münster, Institute of Landscape Ecology, Münster, Germany

Daniel Prati, Barbara Schmitt and Markus Fischer
University of Bern, Institute of Plant Sciences, Bern, Switzerland

Ingo Schöning and Marion Schrumpf
Max-Planck-Institute for Biogeochemistry, Jena, Germany

Dionys Forster and Christian Andres
International Division, Research Institute of Organic Agriculture (FiBL), Frick, Switzerland

Rajeev Verma
Research Division, bioRe Association, Kasrawad, Madhya Pradesh, India

Christine Zundel
International Division, Research Institute of Organic Agriculture (FiBL), Frick, Switzerland
Ecology Group, Federal Office for Agriculture (FOAG), Bern, Switzerland

Monika M. Messmer and Paul Mäder
Soil Sciences Division, Research Institute of Organic Agriculture (FiBL), Frick, Switzerland

Guo-Qi Chen and Sheng Qiang
Weed Research Laboratory, Nanjing Agricultural University; Nanjing, China

Yun-He He
Weed Research Laboratory, Nanjing Agricultural University; Nanjing, China
Department of Landscape Architecture, Zhejiang Agricultural and Forestry University, Hangzhou, China

Index

www.ingramcontent.com/pod-product-compliance
Lightning Source LLC
Chambersburg PA
CBHW080524200326

41458CB00012B/4328